12.43

DC 5126-71

P9-AFZ-874

WITHDRAWN

JUN 1 0 2024

DAVID O. McKAY LIBRARY
BYU-IDAHO

GEAR DESIGN AND APPLICATION

Edited by
NICHOLAS P. CHIRONIS
SENIOR ASSOCIATE EDITOR, *Product Engineering*

McGRAW-HILL BOOK COMPANY

New York San Francisco Toronto London Sydney

GEAR DESIGN AND APPLICATION

Copyright © 1967 by McGraw-Hill, Inc. All Rights Reserved. Printed in the
United States of America. No part of this publication may be reproduced, stored
in a retrieval system, or transmitted, in any form or by any means, electronic,
mechanical, photocopying, recording, or otherwise, without the prior written per-
mission of the publisher. *Library of Congress Catalog Card Number* 67–20656

10787

1234567890HDBP75432106987

PREFACE

It is surprising to note that in this age of missiles, supersonic jets, and high-speed automation, the subject of gearing—which has always had an irresistible fascination for countless mechanical designers from Leonardo da Vinci onward—is continuing to develop new and unusual forms of gear systems. Today there are so many interesting gear arrangements, gear systems, and gear mechanisms, many of which are little known to designers, that there is compelling need for a compilation of such material.

The information in this book is drawn largely from *Product Engineering* material written by over ninety leading gear experts. It has been compiled and integrated to serve two functions: to describe and illustrate as many unusual, yet practical, gear systems and mechanisms as possible, and to provide short-cut techniques and formulas for simplifying the design of both standard and nonstandard gear systems. Much of the material cannot be found elsewhere. The book contains, for example:

· An exclusive yet very simple method for designing the smallest gear set to meet given horsepower, speed, and life requirements.

· The largest selection of planetary-gear arrangements—with their speed-ratio and efficiency equations and their stress-cycle, torque, and volume requirements.

· The largest assortment of gear mechanisms, many from foreign sources.

· Exclusive design formulas for the recently developed Wildhaber-Novikov gear system that employs circular-flank teeth, for a wide variety of noncircular and logarithmic gears, and for the little-known twin-eccentric and twin-worm gear systems.

· Design data on modified-tooth gear systems, including systems with enlarged-tooth pinions, and full-recess spur, helical, and worm gears.

· Details of all-steel friction drives, including multiroller planetary drives that can operate at speeds of 500,000 rpm, with minimum noise and vibration.

· Tabulated data on an exclusive helical-gear system that uses a special helix angle to simplify all the necessary calculations.

• Details of a new computer system that employs "memory disks" to spin out "instant" gear design in which the designer types his requirements into the computer's typewriter in simple abbreviated English, and in minutes gets back complete gear specifications.

• A series of comparison tables incorporating the characteristics of 16 right-angle gear systems, including the Revacycle-bevel, Beveloid, Coniflex-bevel, Spiroid, Planoid, Helicon, crossed-helical, and other lesser-known gear systems.

• Unusual ways to employ worm gears in precision drives, including the meshing of a cylindrical worm with a standard spur gear, and an hour-glass worm with a gear that has straight-sided triangular teeth.

• Exclusive computer-derived tolerance tables for over-pin measurements to meet AGMA requirements.

• Details of the new AGMA classifications system for gears.

• A new method for finding the synchronic index of gears—the number of turns that the input will make before a complex gear train returns to its original alignments.

• Details on European methods of cushioning gear drives.

• Formulas for splines, precision sprockets, precision knurls, and ratchets.

• A large assortment of gear-fastening and gear-shifting techniques.

Many of the contributors are well-known to gear designers. Here you will find material by Darle Dudley, George Michalec, Ernest Wildhaber, Elliot Buckingham, Sigmund Rappaport, Paul Dean, Jr., Wells Coleman, Gene Shipley, John Glover, Professor M. F. Spotts, and many others. The material should prove highly useful to both gear designers and to students of applied gearing.

Nicholas P. Chironis

CONTENTS

Preface   iii

Chapter 1.   Parallel Gear Systems   1

How Increased Hardness Reduces the Size of Gear Sets   2
How Change of Size Affects Other Gear Factors   13
New Equations and Charts Pick Off Lightest-weight Gear Trains   25
New Formulas Give Quick Estimate of Size of Spur Gear Drives   33
Simplified Helical Gear Design   35
Computer Memory Disk Speeds Gear Design   39

Chapter 2.   Right-angle Gear Systems   43

Which Right-angle Gear System?   44
Latest Curves and Formulas for Design of Bevel Gears   56
Guide to Worm Gear Types   69
Design Chart for Crossed-helical Gears   78
When Crossed-helical Gears Get You out of Tight Spots   79
Design of Tapered Gears   83
Design of Face Gears   87
Nomographs Speed Gear-force Analysis   92

Chapter 3.   Planetary Gear Systems   93

Efficiency and Speed-ratio Formulas for Planetary Gears   94
A Simple Method of Determining Ratios in Planetary Gear Trains   104
Planetary Gear Design Details   106
Volume Requirements of Epicyclic Gear Systems   110
Formula for Epicyclic Gear Train of Large Reduction   114
Novel Concept Gives Record Gear Ratios   115
Levers Balance Planetary-gear Loads   118
The Harmonic Drive—High-ratio Gearing   120

Chapter 4.   Modified-tooth Gear Systems   123

Design of Novikov Gears   124
Design of Full and Semi-recess Action Gears   136
Enlarged-tooth Pinions   144
Two-tooth Gear Systems   148
What You Can Do with Nonstandard Spur Gears   149
Charts for Designing Long-short Addendum Spur Gears   152
Avoiding Gear Tooth Interference   154

Chapter 5.   Noncircular Gear Systems   157

Design Guide for Noncircular Gears   158
Elliptical Gears for Cyclic Speed Variations   166
Design of Twin Eccentric Gears   169

Chapter 6.   Gear Specifications, Inspection, and Tests **175**

New AGMA Classification System for Gears   176
Obtaining Accurate Gear Center Coordinates   183
7 Rules Simplify Instrument Gear Specifications   186
A Workable Approach to Gear Tolerances   188
Tolerances for Over-pin Gear Measurements   190
Pin Measurements for 20-deg Spur Gears   196
Profile Diagrams—A Graphical Way to Specify Gear Errors   198
12 Ways to Load-test Gears   201
Calculating Design Data from Sample Gears   207
How to Substitute Helicals for Foreign Module Gears   208

Chapter 7.   Gear Ratios and Tooth Geometry **209**

How to Compute the Synchronic Index of Gear Trains   210
Five Ways to Find Gear Ratios   214
Whole Number Solutions for Finding Gear Ratios   218
Matrix Arithmetic for Finding Gear Ratios   220
How to Plot Involute Curve Tooth Profiles   222
Contact Ratio Charts for Spur Gears   224
Basic Gear Geometry and Tooth Proportions   226

Chapter 8.   Angular Errors and Backlash Control **227**

Equations for Angular Errors in Gears   228
Effect of Mounting Tolerances on Backlash   234
Lost Motion in Gear Trains   236
Ways to Control Backlash in Gearing   237
More Ways to Control Backlash   242

Chapter 9.   Gear Efficiency, Life, and Vibration **243**

How to Predict Efficiency of Gear Trains   244
Friction of Gear Teeth   246
New Formulas and Charts Help Predict Operating Life of Gears   249
Cushioned Gear Drives   259
Noise Standards for Gears   262
Critical Speeds of Geared Shafts   263
Bearing Loads on Geared Shafts   265

Chapter 10.   Gear Materials and Lubrication **267**

Compendium of Materials for Gears   268
When to Specify Gears Made by Special Processes   280
Instrument-gear Materials That Wear Least   289
Design of Nylon Resin Spur Gears   294
Powder Metals for Gears   298
Stronger Gears with Silicon Bronzes   299
Shot Peening and Nitriding of Gears   300
Comparison of Gear Heat-treatment Methods   301
Comparison of Surface Coatings for Gears   302
Typical Methods of Providing Lubrication for Gear Systems   304
Pressurized Oil Systems for Gear Drives   306
Powder Lubrication—New, with Wide-open Possibilities   308

Chapter 11.   Splines, Sprockets, and Friction Drives **309**

How to Design Involute Splines   310
Allowable Stresses and Load Ratings for Splines   316
Racks for Spindles and Sleeves   320
Sheet Metal Gears, Sprockets, Worms, and Ratchets   322
Design of Precision Sprockets   324
How to Specify Precision Knurls   326
Design of All-Steel Friction Drives   328
Multi-roller Planetary Friction Drives   330

Chapter 12.   Fastening and Shifting Techniques for Gears **333**

15 Ways to Fasten Gears to Shafts 334
Small Shaft Becomes Its Own Broach for Retaining Gears 339
Ways to Fasten Gears to Plates 340
Gear-shift Arrangements 342
Gears and Friction Disk Make a Fast-reversing Drive 344
Locomotive Gear-shift Arrangement 346
Multispeed Gear Arrangement 346

Chapter 13.   Gear Mechanisms and Special Drives **347**

Geared Machinery Mechanisms 348
Cycloid Gear Mechanisms 354
Cam-controlled Planetary Gear System 360
Intermittent Spur Gears 361
Special Gearing Devices 362
Special-motion Gear Mechanisms 364
One-way Output from Gear Reducers 366
Twinworm Gear Devices 368
Equations for Three-gear Drives 370
Gear Arrangements for Amplifying Motion 372

Index 373

1

PARALLEL GEAR SYSTEMS

How increased hardness reduces the
Size of gear sets

Here's a new design procedure—the simplest we've seen yet—with key tables, that leads to the smallest gear set for given horsepower requirement

DARLE W. DUDLEY, Manager, Mechanical Transmission, Mechanical Technology Inc, Latham, NY

THE most important criterion for the size of a gear set is the hardness of the pinion and gear teeth. We're talking, of course, of power-transmitting gears, not gears for instruments. Such gears are usually made of carbon or alloy steel with hardened teeth. But there's a bigger bonus in going to harder teeth than most engineers are aware of.

It is often possible to reduce by half, the length, width, and height of a gear box, by simply changing from steel gears with a low hardness value to full-hard gear teeth. This is an 8:1 reduction in gear weight, which means substantial savings in material, machinery, storage, and shipping costs of the gearing and the housing.

One result of this technique is the light, compact, single-reduction helical-gear set shown in Fig 1, which has been successfully designed to transmit 20,000 hp with long life. Fig 2 shows the change in size of a helical gear set, designed for a certain horsepower and speed ratio, and with 200 bhn (brinell hardness number) teeth, 350 bhn teeth or 600 bhn (full-hardened) teeth.

Lower derating factors

There is another reason for going to harder teeth. The resulting smaller gear set will have lower moments of inertia and pitch-line velocities. This means that lower "derating" factors can be applied to the design.

Generally, all gear sets are rated downward from the computed value of load capacity to take into consideration such detrimental factors as shock loads, shaft misalignment, tooth errors. But a derating factor is not a safety factor; it is the basis for transforming a theoretical design to a practical unit. A normal speed reducer, for example, may be derated by a factor of 4:1. With full-hardness gears, the derating factor may be reduced to 3 or 2½:1. Hence, the 8:1 reduction in weight may well wind up as a 10:1 reduction. Derating factors are given later in the design procedure.

Are industries following this concept?

Some are—but some can go to much harder gears:

• **Aircraft and automotive industries** have shifted almost entirely to full-hard teeth.

• **Marine gearing,** on the other hand, has generally stayed with low-brinell gears. But there is increased interest in harder gears for naval and marine use.

• **Truck, tractor and agricultural-equipment manufacturers** are in the medium-hard to full-hard range. In certain cases there's definitely room for improvement.

• **Speed-reducer manufacturers** on the whole tend to be in the low-hardness range. But some companies, such as General Electric (Fig 1) and Philadelphia Gear (Fig 3) have been active in building full-hard teeth and a general trend toward harder teeth is in the making.

What are the disadvantages?

There is the added cost in heat treatment, of course, and the difficulty in machining. The teeth frequently have to be ground. Also, with large gears

1—Helical gear set rated at 20,000 hp continual operation, and employing hardened gears to obtain lightweight design. Designed and built at General Electric Lynn, Mass., where the author headed the Advance Gear Engineering Department.

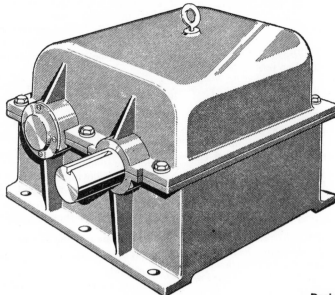

Design A
Weight 8000 lb
Hardness 200 bhn

Design B
Weight 2800 lb
Hardness 350 bhn

Design C
Weight 1000 lb
Hardness 600 bhn

you need large furnaces, unless you go to induction hardened gears (Fig 4).

One worry that a speed-reducer manufacturer expressed is that some purchasers tend to favor larger, bulkier units in the supposition that they're getting more for their money. This tendency is diminishing as engineers become aware of the value of high-hardened teeth.

Surface durability vs power capacity

Why do very hard teeth carry so much more load than you would expect? The answer is that the power rating of a gear set is primarily based on the *surface durability* of gear teeth, and not on their *beam strength,* which is the resistance of the tooth to bending. But it will be seen later that if you do make gears hard you must select a low diametral pitch so as to obtain sufficiently large teeth to assure adequate beam strength.

Now, what is surface durability? For the most part it is a measure of the resistance of the tooth profile to a phenomenon known as pitting. Of the four modes of gear-tooth failure (Fig 5) pitting is the main culprit. The factors which influence the mode of failure are given in Table I, next page. In general:

Pitting can be considered a *fatigue* type of failure. If the load is high enough the surface of the pinion will be eaten away with pits after some millions of cycles. Since pitting is a surface type of failure and tends to develop slowly, the ability to resist pitting is called the *surface endurance* of the teeth or simply, *surface durability.*

Scoring is generally not a problem in industrial gearing provided that ac-

2—Comparative reduction in the size of a helical gear set by going from normally hardened gears (200 bhn) to full-hardened gears (600 bhn). (Same horsepower.)

3—Both these gear reducers have the same catalog rating. The reduction in size of the unit on the left was obtained by going to full hardened and ground gearing. Philadelphia Gear Corp.

curacy and lubrication are good and that the gears are not run at too high a speed. Oils selected for aircraft gears are very thin and speeds and loads tend to be high. Even here the design is usually set by surface durability considerations. Scoring is avoided by using a very good surface finish (20 microin. rms or better), by profile modification, and sometimes by special break-in procedures.

Abrasive wear is generally not encountered except in a few cases of very slow-speed gears or gears operating with sand, dirt, ashes, or acids in the lubricant.

Surface durability vs compressive stress

Compressive strength is the key factor that influences surface durability. The compressive stress between two contacting convex-curved surfaces, such as gear teeth, is proportional to the square-root of the load. This relationship is based on the well-known Hertzian stress equation which for a steel spur pinion meshing with a steel gear with 20-deg pressure angle teeth is

$$s_c = 5715 \sqrt{\frac{W_t}{Fd}\left(\frac{m_G + 1}{m_G}\right)} \quad (1)$$

where s_c = compressive stress, psi
W_t = tangential driving force, lb
F = net contacting face width, in.

d = pinion pitch diameter, in.
m_G = speed ratio

= $\dfrac{\text{number gear teeth}}{\text{number pinion teeth}}$

= $\dfrac{\text{pinion rpm}}{\text{gear rpm}}$

Thus, an increase in the ability of the gear material to take compressive strength greatly increases the load capacity because the permissible load on the gear teeth is proportional to the *square* of the compressive strength.

Compressive stress vs hardness

The final step is to relate hardness to compressive strength. Each gear steel has its own physical characteristics, but in general these two relationships hold true for all steels, including carbon and alloy steel:

There is an almost straight line relation between brinell hardness and tensile strength (Fig 6). The compressive strength of the material although not shown on the chart, follows the tensile strength quite closely. It too will have this straight line relationship with brinell hardness.

There is a practical top limit of hardness for a steel, which depends on its carbon content. Fig 7 gives the approximate maximum hardness for different carbon contents of steel. This holds

true whether it is a carbon steel, such as 1040, or an alloy steel, such as 4140. Both have 0.4% carbon content, hence the maximum hardness that can be obtained will be in the vicinity of 530 bhn.

Keep in mind, however, that the hardness of the steel gear will also depend upon the skill with which the steel has been heat-treated. Improper quenching, or tempering at the wrong temperature, will completely defeat the concepts presented here. The larger gears require alloying elements such as nickel, chromium, and molybdenum to aid them in developing the potential hardness of their carbon content.

K-factors come into the picture

To take into consideration the type of application, the operating conditions, and other derating items, we make use of a K-factor, which represents the factors under the radical in Eq 1. Thus:

$$K = \frac{W_t}{Fd}\left(\frac{m_G + 1}{m_G}\right) \quad (2)$$

But Eq 1 assumes that the maximum loading is at the pitch-line of the teeth. Also it is restricted to 20-deg spur gears and does not include derating factors. A more up-to-date version of Eq 1 is

$$s_c = C_k \sqrt{KC_D} \quad (3)$$

5—Four modes of gear failure. Pitting is by far the most frequent and is influenced by the compressive strength of the material, which in turn primarily depends on the hardness value.

Table I	How sizing factors influence
Mode of failure	Prime variables
Pitting	Hardness
	Pitch diameter
	Face width
Tooth breakage	Diametral pitch
	Pitch diameter
	Face width
	Hardness
Scoring	Surface finish
	Kind of oil
	Oil temperature
	Pinion RPM
	Diametral pitch
	Profile modification
	Tooth proportions
	Pitch diameter
	Face width
Abrasive wear	Abrasives or acids in oil
	Very slow pitch line speed
	Surface finish
	Hardness

4—**Induction hardening** of 85 in. dia. gear with 26 in. face width. Made of 4340 steel casting hardened to 50-54 R_c. Rated at 15,000 lb tangential loading. Philadelphia Gear Corp

the type of gear failure
Secondary variables
   Diametral pitch
   Pressure angle
   Tooth proportions
   Pressure angle
   Tooth proportions

   Pressure angle
   Hardness

   Diametral pitch
   Pitch diameter
   Pressure angle

C_k = **Contact stress factor,** which takes into consideration the point of contact between mating teeth and the elasticity of the material. Values of C_k versus the number of teeth for pinion and gear have been recently obtained with the aid of a computer.

C_D = **Overall derating factor** for surface durability which, in turn, has the following terms in it:

$$C_D = C_o C_s C_m C_f / C_v$$

See Table IV for the author's list of C_D factors. You can also compute your own from the following data:

C_o = **Overall factor,** which makes allowances for roughness or smoothness of operation, and is based on service factors. It may go as high as 2.25 with heavy shock and is taken as 1.0 with no shock (Table V). For enclosed drives, use the service factors, Table XIV, in place of the C_o values in Table V.

C_s = **Size factor.** Taken as 1.0 when proper choice of material is made and case depth or hardness is adequate. Face widths over 10 in. should be derated 10 to 20%.

C_f = **Surface condition factor.** Taken as 1.0 if appropriate surface finish from cutting, grinding, etc, is obtained.

C_m = **Load-distribution factor.** Ranges from 1.0 with almost-perfect alignment and rigid mounting to over 2.0 with poor alignment (Table VI).

C_v = **Dynamic factor.** Considers tooth spacing, profile errors, inertia of rotating parts, pitchline velocity, transmitted load per inch of face width, lubricant properties. Ranges from 1.0 for high-accuracy gears operating at low speeds to 0.3 with heavy, inaccurate gears at high speeds (Fig 8).

For a quick evaluation of the overall derating factor, C_D, you can go directly to Table IV which gives typical values for a variety of applications.

How K factors reduce design time

One way of applying the K-factor is to obtain or compute values for s_c, C_k, and C_D, and solve for K in Eq 3. Then, by means of Eq 2, a value is computed for the gear-size factor, Fd. This product, in turn, is broken down into separate values for face width and pitch diameter with the aid of other tables. This, in effect, forms the basis for the design procedure which follows.

Values of K-factor are also available from tests, as in Fig 9, or from evaluations of successful designs, Table VII. These factors can form the basis for preliminary estimates or evaluations of comparative designs, or can be employed directly in Eq 2 to determine gear size.

K-factors vs hardness

In Fig 9, design curves are given which relate K to brinell hardness for helical and spur gears:

Curve A is the upper limit for well proportioned helical gears designed for 10 million (pinion) cycles of life. It is assumed that there is no appreciable dynamic load and that the gear set is made to a high order of precision so that it does not have to be derated for misalignment or other inaccuracies.

Curve B is the upper limit for well proportioned spur gears of high accuracy and precision.

Curve C shows the approximate average capacity of spur test gears that are actually run to destruction on test rigs. The data for test gears are based on tests where the gears were made to high accuracy and operated with a test set up having low masses and flexible shaft connections so that the dynamic overload was small. Other derating factors were also small.

Note that curve C is in the center of the scatter band of failures. The data for the band has been compiled from

Table II — Contact stress factors C_k for spur and helical gears					
		Spur		Helical	
		Pressure angle		Normal pressure angle	
Number of teeth		**20°**	**25°**	**18½°**	**16½°**
				Helix angle	
				23°	35°
pinion	gear				
N_P	N_G	C_K	C_K	C_K	C_K
18	25	6045	5540	4527	4677
	35	6175	5660	4475	4627
	80	6365	5820	4383	4540
19	25	5990	5490	4518	4670
	35	6115	5610	4466	4620
	80	6300	5770	4375	4537
20	25	5940	5450	4508	4660
	35	6060	5565	4457	4610
	80	6240	5725	4365	4530
22	25	5870	5385	4490	4643
	35	5975	5500	4440	4596
	80	6145	5650	4350	4515
25	25	5790	5315	4466	4622
	35	5885	5420	4419	4574
	80	6040	5565	4330	4495
30	30	5765	5290	4407	4565
	80	5940	5470	4300	4470
	275	6010	5570	4235	4410
35	35	5755	5280	4365	4527
	80	5875	5415	4280	4450
	275	5940	5500	4215	4392
50	50	5740	5265	4280	4445
	80	5785	5325	4240	4410
	275	5840	5395	4177	4355
80	80	5735	5260	4200	4373
	275	5775	5330	4138	4320

test experience at General Electric and from the reported test results of other experimenters. For reliability, work to design limits at or below the lower edge of the scatter band.

Some kinds of steel will show a good tendency to work harden. The short section of Curve C shown dashed indicates the potential for increasing gear capacity by a careful break-in procedure so that the very surface of the tooth is work hardened *before* the tooth is subjected to its full design loads. Alloy steel with nickel shows strong work-hardening characteristics.

Note also that helical gears tend to carry more K-factor than spur gears due to their more favorable geometry. This explains why design Curve A is so much higher than design Curve B.

Procedure of Design

Follow this simple design procedure . . . and come up with a compact gear set

Given—An aircraft gas turbine, with an output speed of 20,000 rpm, is to drive an electric generator at 8000 rpm. The generator has a maximum continuous rating of 60 hp. (It has the continuous rating of 40 hp and a momentary overload rating of 120 hp, but the determining load should be based at 60 hp.)

Find—The smallest spur gear set to meet the above operating conditions.

From the given data:

Pinion speed, ω_p = 20,000 rpm

Speed ratio, $m_G = \dfrac{20,000}{8000} = 2.5$

Step 1—Calculate pinion torque, T_p

$$T_P = \frac{63,025 \text{ hp}}{\omega_p}$$

continued, next page

Table III — Contact stress factors C_k for bevel gears.

Number of teeth		Straight and Zerol bevel gears Shaft angle 90° Pressure angle 20°	Spiral bevel gears Shaft angle 90° Pressure angle 20° Spiral angle 35°
pinion	gear	C_K	C_K
15	25	8520	7570
	35	8731	7264
	50	8967	6837
	80	8946	6404
18	25	8157	7502
	35	8309	7276
	50	8569	6818
	80	8480	6426
20	25	8003	7449
	35	8129	7305
	50	8315	6789
	80	8279	6634
22	25	7889	7408
	35	7985	7293
	50	8148	6764
	80	8088	6413
25	25	7766	7379
	35	7819	7230
	50	7959	6727
	80	7868	6396
30	30	7649	7278
	50	7754	6674
	100	7507	6248
35	35	7565	6957
	50	7616	6627
	100	7333	6622
50	50	7338	6475
	100	6989	6121

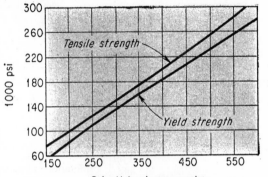

6—Brinell hardness vs tensile and yield strength. This almost straight-line relationship holds true for carbon and alloy steels.

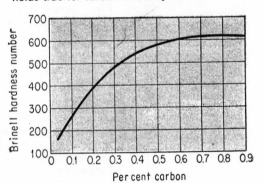

7—Maximum hardness vs carbon content of steel. This shows that you need about 0.06% carbon to get up to a full hardness of 600 bhn.

8—Dynamic factors vs speed for helical precision gears: Curve 1—no dynamic loads; Curve 4—high dynamic loads.

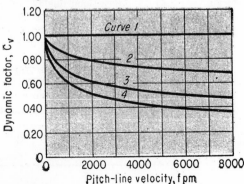

$$= \frac{63,025(60)}{20,000} = 189 \text{ in.-lb}$$

Step 2—Calculate the driving force, W_t

$$W_t = \frac{2T_P}{d} = \frac{2(189)}{d} = 378/d$$

Step 3—Solve for required K from Eq 2

$$K = \frac{378}{Fd^2} \left(\frac{2.5 + 1}{2.5} \right)$$

$$K = \frac{529}{Fd^2}$$

Step 4—Determine the minimum hardness and compressive strength for the steel used.

Assume case-carburized steel of 60 Rockwell C (approximately 587 bhn). Therefore from Table VIII (or Table IX for bevel gears) the minimum compressive stress will be

$$s_c = 200,000 \text{ psi}$$

Step 5—Determine the number of teeth for the pinion, N_p.

A high number of teeth means a small tooth size. Thus could result in tooth breakage when teeth are hardened. The approximate maximum number of teeth that you can select for a given hardness, and still have adequate tooth strength, is given in Table X. (This table assumes a good structure in the steel and no abrasive wear. If you feel that the actual conditions do not meet this level, use about two-thirds the tabulated values.) For $m_G = 2.5$, and 600 bhn (interpolating):

$$N_p = 26 \text{ teeth}$$
therefore $N_G = 65$ teeth

Step 6—Determine the contact-stress factor, C_k

From Table II (or Table III for bevel gearing), for $N_p = 26$ and $N_G = 65$, C_k will be approximately (interpolating from the values listed):

$$C_k = 5800$$

Step 7—Estimate the overall derating factor

From Table IV. Assume

$$C_D = 1.5$$

Step 8—Compute the allowable K

Table IV — Overall derating factors

Derating factors				Conditions of application and gears
Spur	Helical	Bevel Straight	Bevel Zerol and spiral	
1.3	1.3	1.3	1.1	Aircraft quality (AGMA Quality 12), smooth driving and driven apparatus, low masses and flexible shafts, face widths of spur of single helical not over 50% of d.
2.3	2.3	2.3	1.9	Same as above except moderate shock in driving and driven apparatus.
2.8*	2.2	2.8*	1.8	High quality gears finished by grinding, shaving or lapping, (AGMA Quality 10), smooth driving and driven apparatus, moderate masses and reasonably flexible shafts, face widths of spur or single helical up to 100% of d.
5.0*	3.9	5.0*	3.2	Same as above but moderate shock in driving and driven apparatus.
3.4†	2.8*	3.4†	2.8*	Good quality cut gears (AGMA Quality 8), smooth driving and driven apparatus, masses may be heavy and shafts stiff, face widths of spur or single helical up to 100% of d, face widths of double helical up to 200% of d.
6.0†	5.0*	6.0†	5.0*	Same as above but moderate shock in driving and driven apparatus

*Based on pitch line velocity of 6000 fpm or less. †Based on pitch line velocity of 4000 fpm or less.

Table V — Overload Factors, C_o

Power source	Character of load on driven machine		
	Uniform	Moderate shock	Heavy shock
Uniform	1.00	1.25	1.75
Light shock	1.25	1.50	2.00
Medium shock	1.50	1.75	2.25

For speed-decreasing drives only. For speed-increasing drives of spur and bevel gearing add $0.01 (N_G/N_P)^2$ to the factors. For enclosed drives use the service factors in Table XIV in place of C_0.

Table VI — Load-distribution Factor, C_m

Condition of support	Face width			
	2" and under	6"	9"	16" and over
Accurate mountings, low bearing clearances, minimum elastic deflection, precision gears	1.3	1.4	1.5	1.8
Less rigid mountings, less accurate gears, contact across full face	1.6	1.7	1.8	2.2
Accuracy and mounting such that less than full face contact exists	Over 2.2			

Solving for K in Eq 3:

$$K = \frac{s_c^2}{C_k^2 C_D} = \frac{200,000^2}{(5800)^2(1.5)}$$

$$K = 794$$

Step 9—Compute the width-diameter factor, Fd^2.

From Steps 3 and 8

$$Fd^2 = \frac{529}{794} = 0.666$$

Step 10—Obtain the recommended F/d ratios

From Table XI (or Table XII for bevels). Assume from the table that

$$F = d/2$$

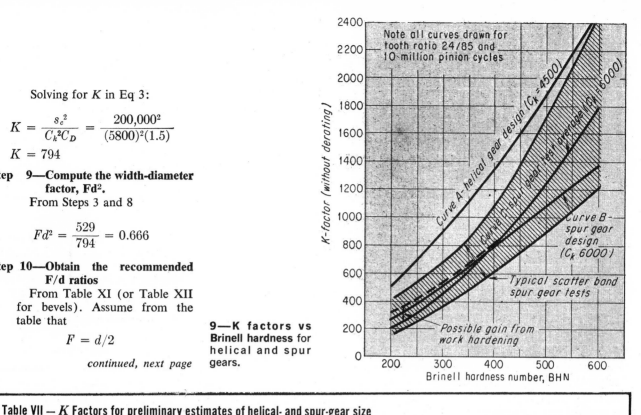

9—K factors vs Brinell hardness for helical and spur gears.

Table VII — K Factors for preliminary estimates of helical- and spur-gear size							
Application	**Service characteristics**		**BHN of material**		**Pitch-line speed, fpm**	**Accuracy**	**K factor**
	Driver	**Driven**	**Pinion**	**Gear**			
Turbine driving generator...		......	225	180	Over 4,000	High precision	80-110
			575	575			400-800
Engine driving compressor..		......	225	180	Over 4,000	High precision	45-70
			575	575			220-425
Motor driving compressor...		......	225	180	Over 4,000	High precision	55
Industrial drives	Uniform	Uniform	575	575	1,000	Commercial	500-1,000
			350	300	1,000	Commercial	350-450
			210	180	1,000	Commercial	170-250
	Uniform	Uniform	575	575	3,000	Commercial	475-750
			300	300	3,000	Commercial	275-375
			210	180	3,000	Commercial	125-200
Large industrial gears— hoists, kilns, mills......	Uniform	Moderate shock	225	180	1,000 max	Generated	80-100
	Uniform	Moderate shock	260	210	......	Generated	130-170
Aircraft (single pair)......	Engine	Propeller	60 R_C	60 R_C	10,000	High ground	1,000 (at take-off)
Aircraft planetary	Engine	Propeller	60 R_C	60 R_C	3,000-10,000	Ground	700 (at take-off)
Automotive transmission		In low gear	60 R_C	60 R_C	......	High	1,500
Small commercial	Uniform	Uniform	350 BHN	Phenolic-laminated	Under 1,000	Commercial	75
				Nylon	Under 1,000	Commercial	50
Small gadget	Uniform	Uniform	200	Zinc-alloy die casting	Under 1,000	Commercial	25
			200	Brass or aluminum	Under 500	Commercial	25
			Brass or aluminum	Brass or aluminum	Under 500	Commercial	15

Table VIII — Hardness vs compressive stress for spur and helical gears

Material	Minimum surface hardness	s_c, psi
Steel	Through hardened	
	180 bhn	85- 95,000
	240 bhn	105-115,000
	300 bhn	120-135,000
	360 bhn	145-160,000
	440 bhn	170-190,000
	Case carburized	
	55 R_C	180-200,000
	60 R_C	200-225,000
	Flame or induction hardened	
	50 R_C	170-190,000
Cast Iron		
AGMA Grade 20	—	50-60,000
AGMA Grade 30	175 Bhn	65-75,000
AGMA Grade 40	200 Bhn	75-85,000
Nodular Iron	165-300 Bhn	10% less than steel
Tin Bronze	40,000	30,000
Aluminum Bronze	90,000	65,000

Table IX — Hardness vs compressive stress for straight bevel, Zerol and spiral bevel gears

Material	Heat treatment	Minimum surface hardness Brinell	Minimum surface hardness Rockwell C	s_c, psi
Steel	Carburized (case-hardened)	625	60	250,000
	Carburized (case-hardened)	575	55	200,000
	Flame or induction hardened	500	50	190,000
	Hardened and tempered	440	—	190,000
	Hardened and tempered	300	—	135,000
	Hardened and tempered	180	—	95,000

Table X — Recommended maximum number of teeth to assure adequate tooth strength

Ratio m_G	High speed gearing brinell hardness				Industrial gearing brinell hardness			
	200	300	400	600	200	300	400	600
	N_P	N_P	N_P	N_P	N_P	N_P	N_P	N_P
1	80	50	39	35	50	37	29	26
1.5	67	45	32	30	45	30	24	22
2	60	42	28	27	42	27	21	20
3	53	37	25	25	37	24	18	18
4	49	34	24	24	34	23	17	17
5	47	32	23	23	32	22	17	17
7	45	31	22	22	31	21	16	16
10	43	30	21	21	30	20	16	16

Note — Hardness values are minimum for gear. Pinion should be as hard or harder than gear. Hardnesses of 200, 300 and 400 assume through hardened steel. The 600 bhn value assumes appropriate case carburizing.

Step 11—Determine the pitch diameters and face width

Combining the values from Steps 9 and 10: Pitch diameter of pinion

$$d^3 = 0.666(2) = 1.333$$
$$d = 1.100 \text{ in.}$$

Pitch diameter of mating gear

$$D = m_G d = (2.5)(1.100)$$
$$= 2.750 \text{ in.}$$

Face width

$$F = d/2 = 0.550 \text{ in.}$$

These are minimum values and would be enlarged slightly to permit use of a standard pitch. Also, when the design is finalized, check against all applicable AGMA standards as there are many design limitations and rules that are beyond the scope of this article.

Machinability vs hardness

Check Table XIII for the machinability of a progressive range of hardnesses used for steel and iron gears.

Cast iron gears are often employed with no more heat treatment than casting and annealing. The casting itself serves as a sort of quench. The best grades of cast iron usually have some alloy content and are given a quench and draw treatment after casting to establish better hardness and structure.

For low-capacity applications, steel gears are often made directly from hot-rolled or cold-rolled material as furnished by the steel mill. The hardness may be in the range of 200 to 250 bhn. The higher performance steel gears are given quench and draw treatment after rough machining the blank. Assuming proper alloy content for section thickness of the blank and proper carbon content steel gears may be through hardened up to 350 bhn with good results. The structure obtained is usually quite uniform and there is no great hazard of high residual stresses from heat treatment. The general rule of thumb in the metalworking trade is that the limit of machinability is at 350 bhn (38R_c).

Techniques for high hardness gearing

If you want through-hardened steel gears, then the limit is about 400 bhn. At this hardness, the gears would be quite hard to machine and it is necessary to slow down the feeds and cutting speeds, and to use special cutting oils and rugged machine tools to avoid vibration of the cutting tool. There is also some danger of detrimental residual stresses being left in the blank.

You can through-harden the richer alloy steels with carbon content of

0.4% or more and get around 500 bhn. It is almost impossible to finish cut this hardness but, of course, teeth can be finished by grinding. This hardness level is quite treacherous to work with because of serious problems of residual stresses in the blank and also because the material is rather brittle and notch sensitive. This tends to make the teeth weaker in beam strength than they should be, even though the load-carrying capacity of the tooth surface may be about as high as it should be for the hardness. Some success has been achieved at this hardness level by flame-hardening teeth.

In the range of 600 bhn or 60 R_c the most common practice is to case carburize the teeth. Pick a low-carbon material so that the cased tooth has a core in the range of 250 to 400 bhn. The carburized case is rich in carbon and develops a full martensitic hardness. The cased tooth is given a quench and draw treatment either as part of the carburizing cycle or as a subsequent treatment after carburizing. The teeth distort somewhat because of the heat-treating cycle and it is usually necessary to grind them to obtain suitable accuracy. In the case of rather small carburized gears for automobile transmissions, very close control of the process plus allowances for changes in the tooth shape before carburizing make it possible to omit the final grinding and run the gears either as-hardened or with a light honing or lapping.

Induction hardening of special steels can produce localized hardening of teeth in the range of 600 bhn. This is a special process requiring considerable technical skill and development work. Induction hardening is not widely employed in the gear trade—but a few companies have mastered the art and have produced large quantities of high capacity gears by this method.

The highest hardness of all can be obtained by nitriding. Special steel compositions properly nitrided can develop a case in the range of 65 to 70 R_c—but the nitride case tends to be shallower than the carburized case. If the pitch of the teeth is not too coarse and the nitrided case is backed up by a high-core hardness these teeth will demonstrate very high load-carrying capacity. In certain cases where abrasive wear or marginal lubrication problems exist the nitrided case has demonstrated a peculiar ability to stand up better than other methods of achieving high hardness.

Manufacturing limitations of hardened gears

Cast iron gears or steel gears that are finished by gear tooth cutting can

Table XI — Recommended ratios of face width to pinion pitch diameter for spur and helical gears	
F = ¼d	Maximum face width when alignment is a serious problem.
F = ½d	Alignment problems often make it impractical to use more face width.
F = d	Maximum face width for spur or single helix gearing; alignment must be very good.
F = 2d	Maximum face width for double helix gearing; alignment must be very good.

Table XII — Recommended ratios of maximum face width to bevel-pinion pitch diameter. (For 90 deg. shaft angle)

	Face width based on 0.3 cone	Face width based on limiting value of 10 in. divided by 1 diametral pitch Number of pinion teeth		
Ratio	distance	15 teeth	20 teeth	25 teeth
1	0.212d	0.667d	0.500d	0.400d
1.5	0.270d	0.667d	0.500d	0.400d
2	0.335d	0.667d	0.500d	0.400d
3	0.474d	0.667d	0.500d	0.400d
4	0.618d	0.667d	0.500d	0.400d
5	0.765d	0.667d	0.500d	0.400d
6	0.912d	0.667d	0.500d	0.400d
7	1.061d	0.667d	0.500d	0.400d

Note — Use whichever relation above gives the smallest face width.

Table XIII — Hardness vs machinability for steel or iron gears

Hardness		Machinability	Comments
Brinell	Rockwell C		
150-200	...	Very easy	Very low hardness. Minimum load carrying capacity.
200	...	Easy	Low hardness, moderate load capacity. Widely used for industrial gears
250	24		
250	24	Moderately hard to cut	Medium hardness. Good load capacity. Widely used in industrial work
300	32		
300	32	Hard to cut, often considered limit of machinability	High hardness. Excellent load capacity. Used in lightweight, high-performance jobs
350	38		
350	38	Very hard to cut. Many shops cannot handle	High hardness. Load capacity excellent provided heat-treatment develops proper structure
400	43		
500	51	Requires grinding to finish	Very high hardness. Wear capacity good. May lack beam strength
550	55		
587	58	Requires grinding	Full hardness. Usually obtained by case carburizing. Very high load capacity for aircraft, automobiles, tanks
...	63		
...	65	May be surface-hardened after final machining	Superhardness. Usually obtained by nitriding. Very high load capacity
...	70		

be produced in a wide variety of sizes. In fact, a few manufacturers can even produce gears as large as 200 in. dia with as much as 50 in. of face width and at a hardness as high as 350 bhn.

In the very large sizes the furnace facilities to heat and quench the blank are fairly limited; you often have to buy the blank already heat treated and quenched from one of the larger steel mills. In sizes of gears ranging from 1 to 48 in. diameter there are innumerable companies that have facilities to heat treat and cut the teeth. But if a high order of precision and control of metallurgical quality is required, there are only a very few companies that have the engineering and laboratory facilities and the technical know-how to achieve top-notch results.

As pointed out earlier, the case carburized gear must generally be finished by grinding. The availability of suitable gear grinding equipment is a limiting factor on the size of case-hardened gears. Talking in generalities only, there is a considerable amount of equipment available to grind spur, helical, and bevel gears up to 12 in. dia. The equipment becomes limited in quantity for gears up to 24 in. dia. Above 24 there is a great scarcity of grinding capacity. For instance, there is almost no bevel machinery to grind bevel gears above 30 in. dia., and there are perhaps a half dozen to a dozen machines in the whole country that will grind spur and helical gears to 72 in. dia.

There is also a serious problem in making large case carburized gears. If the gear blank distorts sufficiently to require grinding almost through the case to clean up the gear, then all the value of carburizing is lost. A gear, like a chain, is no stronger than its poorest tooth.

Speaking again in somewhat general terms, it is relatively easy to carburize gears to 12 in. dia. and control distortion so that they will clean up all right in finish grinding. At 24 in. dia. the problem becomes critical and considerable skill in heat treating is required, plus the proper use of quenching dies and other devices to control the quenching cycle. A few gears have been successfully carburized as large as 48 in. dia. and with as much as 10 in. face width. If the pitch on such a gear is around 1 or 2 the problem is not too bad. However, if the pitch is around 5 or 6, the case on the tooth cannot be made too deep (for reasons not discussed in this article) and it is extremely difficult to successfully grind the part and achieve the right amount of stock removal. Those who have done this kind of work have had to do

a considerable amount of expensive development-type experimentation and exercise great technical skill to achieve success.

There is some experimental work going on now which should make it possible in the near future to successfully carburize and grind gears in the range of 72 to 100 in. dia.

Table XIV — Service factors for electric-motor drives

[Use this table in place of Table V when drive is enclosed. For multi-cylinder engines, the service factor is 25% higher than shown. For intermittent duty (3 hr/day), the service factor is 25% lower than shown.]

Application	10-hr service	24-hr service
Agitators		
Pure liquids	1.00	1.25
Liquids and solids	1.25	1.50
Blowers		
Centrifugal and vane	1.00	1.25
Lobe	1.25	1.50
Compressors		
Centrifugal	1.00	1.25
Lobe, rotary	1.25	1.50
Reciprocating, single-cylinder	1.75	2.00
Reciprocating, multi-cylinder	1.25	1.00
Conveyors		
Uniformly, loaded screw	1.00	1.25
Reciprocating or shaker	1.75	2.00
Crushers		
Ore or stone	1.75	2.00
Fans		
Centrifugal	1.00	1.25
Induced draft	1.25	1.50
Large industrial	1.25	1.50
Light (small diameter)	1.00	1.25
Machine Tools		
Auxiliary drives	1.00	1.25
Main drives	1.25	1.50
Punch press (geared)	1.75	2.00
Mills		
Rotary ball, dryers, coolers, kilns, pebble, rod	1.25	1.50
Tumbling barrels	1.75	2.00
Pumps		
Centrifugal	1.00	1.25
Proportioning	1.25	1.50
Reciprocating	1.25	1.50
Rotary: Gear, lobe, vane	1.00	1.25

Effects of Size on Gear Design Calculations

How scale effects influence center distance, accuracy of angular transmission, and contact ratio in very fine pitch gearing. For each of these factors, methods of designing to minimize them are presented and discussed.

PAUL M. DEAN, Jr.
Mechanical Technology, Inc, Latham, NY

A MACHINE ELEMENT of given capacity may be enlarged or reduced proportionally in all of its dimensions to achieve a new component of different size and capacity. This approach, design by similitude, frequently is used in the design of gearing because most of the tooth proportions and all of the angles remain the same for all pitches of gearing of any given number of teeth. Since most of the formulas, tables, and curves pertaining to gear design apply to a gear of one diametral pitch, sometimes called the unit gear, the design can be carried out as a one diametral pitch unit, and after the tooth proportions are established, the appropriate values can be divided by the diametral pitch of the gearing required.

However, in the case of pinions with less than about 15 teeth or in diametral pitches finer than about 64, this approach involves considerable risk, particularly if dimensions are subsequently given tolerances for machining purposes. A machining tolerance may produce an effect on operation—called the "scale effect"—that is out of all proportion to the size of the tolerance.

Strict adherence to existing gear design standards does not alleviate the situation since the tolerances are usually applied after the basic design has been established.

Tolerances That Create Scale-Effects

There are three major groups of tolerances, Table I, that contribute to the scale-effect. These are tolerances on center distance, blank dimensions, and tooth dimensions. They may produce tooth interference, error in the transmission of angular motion, rough running gears having too low a contact ratio, or gearing of inadequate strength.

In the first grouping in Table I are not only the tolerances on distance between bearing centers, but also the tolerances introduced by the bearings, some of which support the pinion and others the gear. In some cases, these tolerances accumulate on assembly.

In the second grouping, the most important tolerances—at least in the case of spur and helical gears—are those on the outside diameter. When "topping" cutters are not used, the

Table I—Scale Effects Resulting From Manufacturing Tolerances

	Manufacturing Tolerance	Specific Effect	General Effect
Tolerances on Mounting Distance	On center of bores that locate the bearings supporting the shafts and gears.	(a) Tolerances permitting increase over nominal center distance introduce backlash and loss of contact ratio. (b) Tolerances permitting decrease under nominal center distance produce tooth interference or binding.	The net buildup of these tolerances effects the center distance. If the total is too large, there may be excessive backlash or the gearing may run rough or "ratchety"; if too small, the gearing may bind and run unevenly.
	Fit of shafts within bearings and bearings within bores.	(c) Clearance causes increase in center distance. See (a) above.	
	Within bearings.		
Tolerances on Gear Blank	On size of outside diameter (top lands) of gear blank.	(d) If tolerances permit outside diameters larger than nominal, tip interference may occur with the mating gear. (e) If tolerances permit outside diameter to become too small, a loss of contact ratio may occur.	May result in gearing that does not run smoothly.
	On runout of outside diameter of gear blank.	(f) May cause loss of contact ratio at one point and tip interference at another on one gear; a combination of (d) and (e) above.	Results in gearing that does not transmit angular motion smoothly and are noisy at high speeds.
	Of bore and shaft, to establish fit of blank on shaft.	(g) If too loose a fit, set screws or keys used to fasten gear may hold gear eccentric causing difficulties under (f) above.	
Tolerances on Gear Teeth	On concentricity of pitch circle with bore axis	(h) If excessive, will cause errors in the transmission of angular motion.	Results in gearing that transmits angular motion unevenly. This may cause noisy, rough operation at high speeds.
	On tooth thickness.	(i) If teeth are too thin, excessive backlash and errors in transmission of angular motion will occur. (j) Undersize teeth are produced by feeding cutter to too great a depth. This results in undercut teeth, loss of strength, or excessive backlash.	

tolerances on size as well as on concentricity are important. These have an effect on contact ratio if they result in a gear with too small an outside radius.

In the third group are tolerances that govern the running qualities of the gears: backlash, accuracy of angular transmission of motion, and noise. Tooth thickness and concentricity are the most important.

To properly analyze the scale-effects on a design, the designer should establish the machine tolerances that result from the machine practices used in his shop. Most manufacturing processes are subject to two types of errors: the first is brought about by imperfections within the gear cutting machine itself, which in gearing manifests itself as runout, tooth-to-tooth errors, profile errors, and variations in size from one part to the next in a given batch of parts. The second type of error is due to the inability of the machine operator to set a machine to cut to an exact size; this results in variance in the size of parts from one batch to another. To these may be added errors of measurement that occur during inspection.

These variations are inherent in all machining processes and are, in general, reasonably constant for any given machine tool. As a result, a small part produced on a particular machine will have errors of about the same magnitude as those found in a large part produced on the same machine. For example, the angular spacing error in a gear will be about the same for the small as the large gear, since the same index gear is used for both. The runout or concentricity will be about the same for both, since the same spindles and centers are used. The size variations will also be similar since it is about as easy for the operator to control depth on a small gear as on a large one. The net result is that the smaller gear will vary from the nominal in all of its dimensions a far greater amount in proportion to its size than will a larger gear made to the same tolerances.

The tolerances on tooth thickness in Table III are divided into two groups. The first is for use when an adjustment can be made at assembly to compensate for size variations in both the gears and their mountings. Frequently the gear is made with a nominal tooth thickness, and half the tolerance is taken on each side of nominal.

The second group of tolerances, for gears mounted on fixed centers, is for use when the gears are mounted on bored centers. In this case the nominal tooth thickness or the center distance should be adjusted to compensate for the errors inherent in the gear. Usually this tolerance is specified as minus except in the cases of pinions with small numbers of teeth. These tolerances lead to scale-effects particularly in the cases of the commercial classes of gears. In the case of a very fine pitch pinion, a reduction in tooth thickness of 0.004 in. for tooth thickness and an allowance of 0.006 in. for total composite error (0.004 + 0.006 × 2 tan 20 deg = 0.0085 in.) is a large percent of the total circular tooth thickness (0.0158 in.) of a 100 diametral pitch gear. It can be reasonably concluded that satisfactory gearing of 64 diametral pitch or finer cannot be made to commercial 1 or 2 classes because of these tolerances.

Influence at Center Distance on Contact Ratio and Backlash

Generally, the maximum and the minimum probable center distance at which the gears may be operated is established by using the root-mean-square of all of the tolerances on the machine surfaces that combine to make up the center distance. In the simplest cases, where the gears are held between two plates, the maximum and the minimum center distances are made up of tolerances on (1) the center distance between the bores that locate the bearings supporting the shafts, and (2) the eccentricity and radial looseness within the bearings.

In some designs, the center distance is determined by a large number of machined surfaces. Such a case is shown in Fig. 1, which is a cross section of an aircraft actuator. This assembly, although not necessarily typical, represents a good example for analysis of center distances in a complex assembly (see Table IV).

The lettered columns in this table show the following: (A) the item; (B) the tolerance, usually on a diameter; (C) the plus or minus amount the tolerance can contribute to center distance (D) the square of the tolerance contributing to the minimum center distance; and (E) the square of the tolerance contributing to the maximum center distance. The values in columns (D) and (E) are added algebraically and their sums shown below the columns. Below these are the square roots of the sums, which are the probable variations from nominal center distance (+0.0021,−0.0019).

The tolerances and the resulting center distances are not unreasonable for gears of 24 diametral pitch. A 100 diametral pitch pinion would cause serious trouble, however. If its tooth thickness and outside diameter were adjusted to accommodate the minimum center distance, it would be running at a low contact ratio at maximum center distance. Its working tooth height is only 0.020 inch. A 0.004 in.

Table II—Tolerances on Fine Pitch Gearing

Class of Gear	Total Composite Error, in.	Tooth-to-Tooth Composite Error, in.
Commercial 1	0.006	0.002
2	0.004	0.0015
3	0.002	0.001
4	0.0015	0.0007
Precision 1	0.001	0.0004
2	0.0005	0.0003
3	0.00025	0.0002

Data taken from ASA B6.11—1951

Table III—Suggested Tolerances on Tooth Thickness for Fine Pitch Gearing

Class of Gear	Tolerance on Tooth Thickness, in.		Gears Mounted on Fixed Centers*
	Gears Mounted on Adjustable Centers		
	20 through 48 diametral pitch	49 diametral pitch and finer	
Commercial 1	0.005	0.002	0.0044
2	0.004	0.002	0.0024
3	0.004	0.002	0.0018
4	0.003	0.002	0.0008
Precision 1	0.003	0.001	0.0004
2	0.002	0.001	0.0003
3	0.002	0.001	0.0002

* These values will result in variations in backlash as shown in Fig. 2.

Note: Quality designations for fine-pitch gears have recently been changed as follows, although the former is still in use. For more information, see p 176.

Former designation	Present quality no.
Commercial 1	5 or 6
Commercial 2	6 or 7
Commercial 3	8
Commercial 4	9
Precision 1	10 or 11
Precision 2	12
Precision 3	13 or 14

variation from one assembly to the next would represent a change equal to 20 percent of the tooth height, a serious matter with fine pitch pinions containing low numbers of teeth.

There are several things about this example that are worthy of note.

(a). Most bearing catalogs indicate bearing and shaft tolerances that produce an interference fit. This fit cannot contribute to the maximum or minimum center distance between gearing.

(b). These catalogs also indicate a tolerance on runout between the bore and the raceway of the inner ring, and this eccentricity contributes to the instantaneous center distance when the inner face is the moving portion of the bearing.

(c). Radial clearance between the inner and the outer bearing races usually acts to allow an increase in the center distance due to the reaction between the two meshing gears.

(d). Bearing catalogs indicate a tolerance on runout between the raceway and the outer diameter of the outer ring. This eccentricity contributes to the center distance at which the gears will operate.

(e). These catalogs also indicate a bore tolerance on the hole into which the bearing is fitted. Usually, the tolerance will provide clearance between the bearing and housing. When clearance exists, the gear reaction acts to increase the center distance to the limit allowed by the clearance.

(f). Many manufacturers of pilot-mounted motors specify a value of runout between the motor shaft and the pilot surface. It is customary to measure this by holding the motor shaft fixed (as between centers) and rotating the motor about its shaft. A dial indicator running on this surface will show the runout (twice eccentricity).

(g). A clearance fit is usually specified on the hole into which the pilot of the motor is to fit. This clearance allows the motor to be shifted somewhat and as a result has a direct effect on center distance. The value shown assumes a tolerance of $+0.0005$ and -0.0000 in. on the bore and of $+0.000$ to -0.002 in. on the diameter of the pilot on the motor.

(h). In an assembly such as Fig. 1, a tolerance must be placed on the location of the bore to receive the motor relative to the rabbet on the end-plate.

(i). Usual drafting practice is to specify a tolerance on the concentricity of the several cylindrical surfaces of a shaft. If these are all machined during one "set-up", they will have little or no eccentricity relative to each other, but this is not always possible.

If the designer wishes to reduce the spread between maximum and minimum backlash, three courses are open. First, the center distance may be made adjustable so that the gears can be moved until the desired mesh is obtained. This entails a method of moving the parts through small distances, and then being able to fix them securely when the desired distance has been reached. This method has an advantage in that it offers a means of eliminating the effects of tolerances both on the size of the teeth and on the distance between gear centers. The

scale-effects of these tolerances are, therefore, non-existent in this type of design.

The second alternative is to use fixed centers but to closely control part tolerances. This may be done by selective assembly or by highly refined machining practice, and is preferred for gearing where service under adverse conditions is expected and where simplicity of design is required, since long trains of gearing do not lend themselves well to adjustable center construction.

The third alternative, useful where only small torques are involved, is to use split (scissor) gearing. Such gearing eliminates the need for close control of tooth thickness or center distance, but cannot be used to transmit large torques.

Influence of Scale Effect on Accuracy of Angular Transmission

The success of a fine pitch gear train for instrument applications is usually determined on the basis of its ability to transmit angular motion accurately and to transmit power without undue vibration, or to a combination of both.

Accurate transmission of angular motion may be further classified into two typical situations. First, the gearing may be required to operate with an absolute minimum of lost motion, the actual angular position of the driving and the driven member at any one instant not being of importance. An example of this is the tuning drive in

a radio set. Here the actual angular relationship, at any one instant, between the hand knob and the dial is not important, but any backlash between the knob and dial is intolerable. Second, the gearing may be required to transmit angular motion with great accuracy. Examples of this are mechanisms that indicate the position of a gun, rudder or the distance between rolls of a steel mill. If uniform, a small amount of lost motion may be tolerable.

The problem of maintaining an absolute minimum of lost motion is

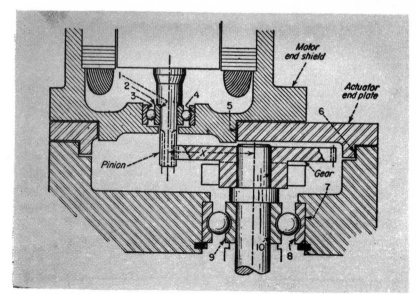

Fig. 1—First stage gearing in an aircraft type actuator. Analysis of the center distance, X, between the pinion on the motor shaft and the gear is shown in Table IV.

Table IV—Center Distance Analysis of Actuator Gearing

Item No.	Item	Surface Designation (See Fig. 1)	Tolerance (Clearance) in.	Equivalent Change in Center Distance	Minimum Center Distance Variation	Maximum Center Distance Variation
	Letters, *a*, *b*, *c*, etc, following items refer to the alphabetical listing given on pg 132.	(A)	(B)	(C)	(D) col. (C)^2 $\times 10^{-6}$	(E) col. (C)^2 $\times 10^{-6}$
I	Fit of motor shaft in inner race of ball bearing.........................(*a*)	(1)	0.0000	0.0000	0.00	0.00
II	Eccentricity of inner ring of bearing . . (*b*)	(1–2)	0.0002	0.0001	−0.01	+0.01
III	Radial clearance in bearing.........(*c*)	(2–3)	0.0004	0.0002	+0.04	+0.04
IV	Eccentricity of outer ring of bearing . . (*d*)	(3–4)	0.0004	0.0002	−0.04	+0.04
V	Fit of outer race of bearing into motor end shield.........................(*e*)	(4)	0.0008 to 0.0000	0.0004 to 0.0000	0.00	+0.16
VI	Concentricity of axis of bore with axis of rabbet on end shield...............(*f*)	(4–5)	0.0020	0.0010	−1.00	+1.00
VII	Clearance between rabbet of motor end shield and bore of actuator end plate . . (*g*)	(5)	0.0025	0.00125	−1.56	+1.56
VIII	Concentricity of motor bore in actuator end plate and rabbet...............(*h*)	(5–6)	0.0010	0.0005	−0.25	+0.25
IX	Distance between axis of gear shaft bore and rabbet in actuator housing.......(*i*)	(6–7)	0.0020	0.0010	−1.00	+1.00
X	Clearance between bore, for bearing and bearing outer race.................(*e*)	(7)	0.0008	0.0004	+0.16	+0.16
XI	Eccentricity, outer race of bearing. . . . (*d*)	(7–8)	0.0004	0.0002	−0.04	+0.04
XII	Radial clearance in bearing.........(*c*)	(8–9)	0.0004 0.0000	0.0002 0.0000	0.00	+0.04
XIII	Eccentricity, inner race of bearing. . . . (*b*)	(9–10)	0.0002	0.0001	−0.01	+0.01
XIV	Fit of gear shaft in bearing race......(*a*)	(10)	0.0000	0.0000	0.00	0.00
XV	Eccentricity, bearing journal and surface locating gear.........................	(10–11)	0.0002	0.0001	−0.01	+0.01
XVI	Fit of gear on shaft...................	(11)	0.0002	0.0001	−0.01	+0.01

Total ($\times 10^{-6}$) −3.73   +4.33

Probable value: $\dfrac{+0.0021 \text{ in.}}{-0.0019 \text{ in.}}$

Note: The maximum and minimum values are based on the assumption that the separating forces between pinion and gear are sufficient to hold the shafts at maximum separation within the limits of the clearance in bearings and their mountings.

usually solved by using "spring-loaded" gears, either split scissor gears or movable "spring-loaded" centers. This type of solution works well if the transmitted load is light and serves to eliminate most of the scale-effect. In many cases, gears of relatively poor accuracy may be used. However, such gearing will not transmit accurate angular motion unless gears of the very highest accuracy are used. It, also, has the further disadvantage of requiring a spring load on each gear tooth greater than the load to be transmitted. This, in turn, necessitates wider or harder gears if tooth wear is to be comparable to gears that are not spring loaded.

If the gears are not spring loaded, the problem of transmitting angular motion with great accuracy requires maximum accuracy on gear size and minimum tooth spacing error. The following discussion covers the major factors to be considered when designing gear trains to accurately transmit angular motion, and the method of analyzing the requirements and the resultant scale-effects:

1. The accumulated tooth spacing error must be held to a minimum. Accumulated tooth spacing error may be defined as "the maximum accumulation of pitch errors obtained by algebraic addition." In effect, it represents the greatest angular error that any one tooth can have relative to any other tooth on a gear. This error should be measured relative to the journals that locate the gear in its mountings.

2. The thickness of the teeth of the gear must be held to close tolerances if there is no provision for adjusting the center distance at assembly.

3. The center distance must be held to close tolerances if non-adjustable.

4. The bearings that support the gears must be of high quality and be correctly mounted.

The method usually used to calculate the lost motion or accuracy of angular transmission in a fine pitch gear train is based on the assumption that all errors will be manifested in the form of backlash. This simplifies the determination of the magnitude of each of the scale-effects that influences lost motion, and, since each stage of the gear train is evaluated in terms of the last stage, its contribution to the total backlash is readily apparent.

The backlash that can exist at the mesh of any particular pair of AGMA Class of Fine Pitch Gearing operating on various center distances and held to the tolerances shown in Table II and III is plotted in Fig. 2. The backlash is independent of the size of the gears. Scale-effects on the accuracy of angular transmission can make themselves felt as a given value of backlash; a small diameter gear will be able to turn through a larger angle than a large diameter gear when turning within the backlash.

The values of tooth thickness are assumed because no standards have been established for these as yet. The following equations may be used to

establish backlash for other combinations of tooth tolerance:

For gears on adjustable centers

$$\Delta B = 4 \tan \phi(E) \qquad (1)$$

For gears on fixed centers

$$\Delta B = 2 \tan \phi[2E + T] + 2t \qquad (2)$$

where E is the total composite error of each gear of a meshing pair, ΔB is the difference between maximum and minimum backlash, ϕ is the operating pressure angle, T is the tolerance on center distance, and t is the tolerance on tooth thickness of each gear of a meshing pair. Thus, the magnitude of backlash that exists between mating gears is a function of the relative tooth thickness and space widths of each of the gears at their operating pitch circles.

The tooth thickness or space width of each gear can be measured in a number of ways, the most popular of which for fine pitch gearing is the rolling check, Fig. 3. The type of chart obtained from this rolling check, Fig. 4, offers a convenient way of describ-

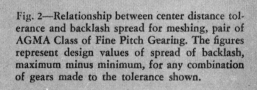

Fig. 2—Relationship between center distance tolerance and backlash spread for meshing, pair of AGMA Class of Fine Pitch Gearing. The figures represent design values of spread of backlash, maximum minus minimum, for any combination of gears made to the tolerance shown.

Fig. 3—Schematic diagram of rolling fixture for checking relative tooth thickness and space width. This device holds a master gear of known proportions and the gear to be measured in tight contact, and indicates or records changes in center distance as the work gear is rotated.

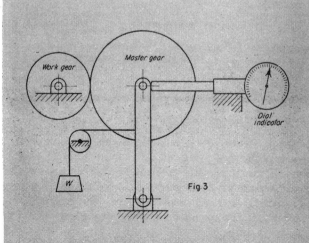

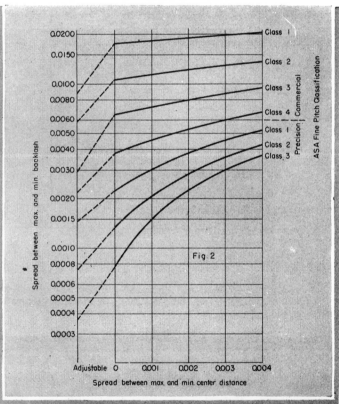

Fig. 4 (Right)—Charts obtained by rolling check of gears mounted on adjustable centers, (A); non-adjustable centers, (B). Chart (C) shows the mesh at maximum center distance tolerance for gears on non-adjustable centers.

ing the problem of evaluating the back-lash that results from the combination of tooth thickness tolerance, center distance, and pitch line runout. The charts in Fig. 4(A) represent a gear and pinion that has been brought into mesh and the center distance adjusted at one point so that there is no back-lash. The maximum backlash that can theoretically occur is represented by the sum of the total composite tooth errors of the gears (adjusted by 2 tan ϕ). This type of design eliminates the scale-effects of tooth thickness and center distance tolerances, and results in gearing having a minimum of lost motion.

In Fig. 4(B) the tolerance on center distance is assumed to be zero. The wavy lines show the tooth-to-tooth and

the total composite errors. The solid horizontal lines show the minimum permissible limits of tooth thickness. These are the limits that apply to any gear of a given design. They repre-sent the sum of the total composite error and the tolerance on tooth thick-ness adjusted by 2 tan ϕ. The maxi-mum possible backlash in this case is equal to the sum of the total composite errors and tooth thickness tolerances of each of the gears adjusted by 2 tan ϕ.

In Fig. 4(C) the effect of tolerance on center distance is included. The backlash in this case is the same as in Fig. 4(B), plus that resulting from the effect of the center distance toler-ance.

As previously mentioned, the first step in calculating the backlash that

results from the scale-effects of tooth thickness, concentricity, and center dis-tance is to obtain the backlash in each pair of gears (See Fig. 5). Assume the pitch and numbers of teeth as shown, and obtain the backlash for each mesh from Fig. 2. Then let

Y = shaft to be rotated
Z = shaft to be held fixed
S = speed of Y
G, G', G'' = either member of any pair of gears in the train
B, B', B'' = backlash (in inches) at the pitch line between G and its mating gear.
S, S', S'' = speed of the shaft on which G is mounted
D, D', D'' = pitch diameter of G in inches
θ = angle through which Y can be rotated as a result of the lost motion in the train
c = constant of proportionality

To get θ in radians, $c = 1.0$
To get θ in degrees, $c = 57.296$
To get θ in minutes, $c = 3438.$

$$\theta = cS \sum \frac{B}{\frac{1}{2}D \times S} + \frac{B'}{\frac{1}{2}D' \times S'} +$$

$$\frac{B''}{\frac{1}{2}D'' \times S''} + \cdots + \cdots \quad (3)$$

The term "speed" as used above is the velocity of one shaft relative to another. Usually the slowest shaft is assigned the value of 1, and the other shafts geared to this are assigned speeds proportional to their ratios.

Assuming the motor shaft (Z) in Fig. 5 is to be fixed, determine: (a) the amount of lost motion each mesh contributes to the total lost motion at the output shaft (Y); (b) the lost motion (in minutes) at the output shaft; (c) the lost motion at the motor if the output shaft is held fixed.

The equation for backlash is most useful when solved in tabular form as shown below since the amount of lost motion contributed by each mesh is made clearly evident.

(a)

Gear No.	Backlash at Mesh	Pitch Radius	Shaft Speed	$\frac{B}{\frac{1}{2}D \times S}$
1	0.0015	1.50	1	0.0010
3	0.0030	0.625	6	0.0008
5	0.0030	0.625	24	0.0002
Total				0.0020 radians

(b) $0.0020 \times 3438 \times 1 = 7.14$ minutes
(c) $0.0020 \times 3438 \times 192 =$
    1374.0 minutes or 22.85 deg.

Each of the values in the last column above is the amount of lost motion con-tributed to the single speed shaft by each of the individual gear meshes. Even though the unit speed (out-

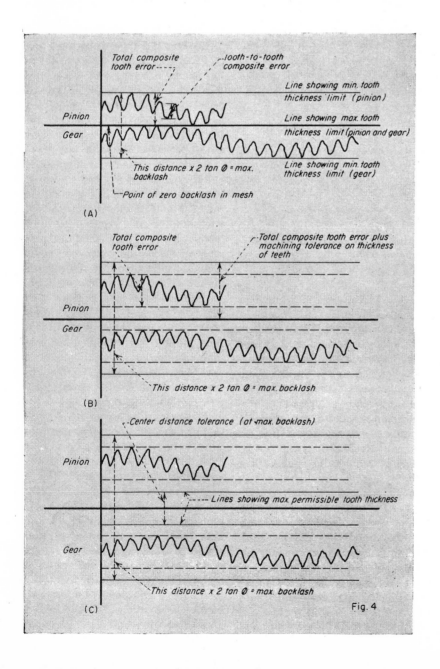

Total composite tooth error----,
tooth-to-tooth composite error

Line showing min. tooth thickness limit (pinion)

Pinion

Gear

Line showing max. tooth thickness limit (pinion and gear)

This distance x 2 tan Ø = max. backlash

Line showing min. tooth thickness limit (gear)

-Point of zero backlash in mesh

(A)

Total composite tooth error

Total composite tooth error plus machining tolerance on thickness of teeth

Pinion

Gear

This distance x 2 tan Ø = max. backlash

(B)

-Center distance tolerance (at max. backlash)

Pinion

---- Lines showing max. permissible tooth thickness

Gear

This distance x 2 tan Ø = max. backlash

(C)

Fig. 4

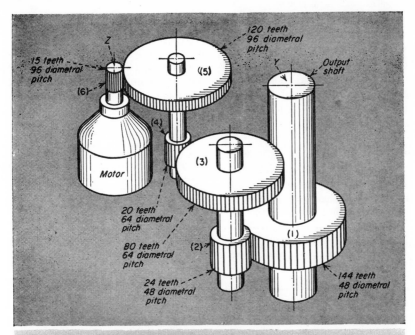

Fig. 5—Typical fine pitch gear train and proportions of various operating members. See text for method of calculating backlash in the train.

Gear Proportions

Gear	No. Teeth	Pitch	Pitch Radius	Backlash at Mesh	Gear Ratio	Center Distance Tolerance	Shaft Speed
1	144	48	1.50	0.0015	6:1	adj.	1S
2	24	48	0.25				6S
3	80	64	0.625	0.0030	4:1	±0.0005	6S
4	20	64	0.1563				24S
5	120	96	0.625	0.0030	8:1	±0.0005	24S
6	15	96	0.078				192S

put) mesh was made adjustable, it is still by far the largest contributor to the total backlash in the gear train. Thus, it may be concluded that the motor mesh might even be reduced in accuracy with little increase in the overall lost motion at the output. In this case, the analysis indicates that from the standpoint of backlash, gears of relatively low accuracy may be used.

Several other problems should be considered before the lower class of gearing is specified in the high speed meshes Nos. 3-4 and 5-6. First is the problem of gears running at high pitch line velocities. The AGMA suggests the following classes of Fine Pitch Gearing for different operating speeds in surface fpm.

Up to 80 fpm   Commercial Class 1 or 2
Up to 400 fpm   Commercial Class 3 or 4
Up to 2,000 fpm Precision Class 1
Over 2,000 fpm   Precision Class 2 or 3

Suppose that in the above example the single revolution shaft were to operate up to 50 rpm. This would require motor speeds up to 10,600 rpm on the motor shaft. The peripheral velocity of the motor pinion under these conditions would be 435 fpm. A Commercial No. 4 or Precision No. 1 gear would be indicated, depending on load and noise considerations. In some cases, this limitation might be the governing consideration rather than backlash.

Another problem is that of the scale-effects of the center distance tolerance and the larger tolerances on contact ratio that are allowed on the lower classes of fine pitch gearing. Suppose, for example, that the gearing had excessive lost motion for the required application. If members 1 and 2 did not carry much power, the gears could be spring loaded. The gear train would

then have a theoretical lost motion of 0.00085 + 0.00021 or 0.00106 radians, only 51 percent as much as the previous value.

Another approach would be to make the low speed gear larger, thus reducing the scale-effect. This could be done by using the following ratios: 24/288, 48 pitch; 20/75, 64 pitch; and 15/64, 96 pitch.

The lost motion calculation would appear as follows:

Gear No.	Backlash at Mesh	Pitch Radius	Shaft Speed	$\frac{B}{\frac{1}{2}D \times S}$
1	0.0015	3.00	1	0.000500
3	0.0030	0.586	12	0.000427
5	0.0030	0.3335	45	0.000200

Total = 0.001127 radians
0.001127 × 3438 = 3.87 minutes

Thus, this last design represents an improvement of approximately 50 percent over the original design as far as lost motion is concerned.

Suppose that the original design had been based fixed centers. It would then have contributed 0.002 radians and the total would be 0.003 radians. This is 150 percent of the initial design.

In addition to the backlash produced by the scale-effects of tooth thickness, there is an effect on the point width of the tooth and on its strength. In Fig. 6 is shown a 100 diametral pitch 10 tooth pinion. The tooth thickness represented by the outer outline is 0.002 in. more than that of the inner outline. The outer outline represents a tooth designed in accordance with the ASA Standard B6.7—1950. This figure shows the large effect on tooth point width that is produced by a not unusual tolerance on tooth thickness. Any further reduction in tooth thickness would reduce the outside diameter, thus affecting contact ratio.

The scale effects of tooth thickness on tooth strength are pronounced (see Fig. 7). In this example, a gear of 10 teeth 100 diametral pitch designed in accordance with ASA Standard B6.7 is assumed to have 0, 0.001 in., and 0.002 in. backlash. Only the trochoidal portions of the fillets are shown for the latter two cases. By a construction similar to that shown, the "Y" factors which are used in the Lewis equation to establish the beam strength of a tooth for each tooth thickness are found to have the following values:

1. 0.264 for a standard addendum,

resulting in a theoretical tooth strength of 100 percent.

2. 0.214 for an error of 0.001 in. on tooth thickness, giving a theoretical tooth strength of 81 percent.

3. 0.180 for an error of 0.002 on tooth thickness, resulting in a potential strength of 68 percent.

Thus, each 0.001 in. reduction in tooth thickness lowers pinion strength by approximately 20 percent. Conversely, an increase in thickness—especially in the case of very fine pitch gearing—may cause binding.

Influence of Scale Effect on Outside Diameter

The outside diameter of a gear need not bear a fixed relationship to its number of teeth and pitch. This fact is commonly used to advantage in the design of pinions with small numbers of teeth (to avoid undercut) and of gearing requiring given gear ratios on non-standard center distances. There are two conditions that limit the amount that the outside diameter can be made different than standard. The minimum outside diameter for a gear of given pitch and number of teeth is limited by the diameter of the undercut portion of the tooth, Fig. 8. The maximum outside diameter is limited by the diameter at which the teeth come to a point.

The outside diameters of gears having small numbers of teeth or fine pitches are very susceptible to scale-effects. The ASA Standard B5.11—1950 recommends a series of values for outside diameters for pinions with small numbers of teeth. The dash line in Fig. 9 shows the ASA values recommended to avoid dangerous undercut of pinion teeth. This curve is based on one diametral pitch.

The value ΔD is a term frequently used in the design of fine pitch gearing. It is the difference between the "standard" outside diameter D_o (defined below) and a modified outside diameter.

$$D_o = D + 2 \qquad (4)$$

$$D = \frac{N}{Pd} \qquad (5)$$

where N is the number of teeth in the pinion or gear, Pd is the diametral pitch of pinion or gear, D'_o is the outside diameter other than D_o, and ΔD equals D'_o minus D_o.

The Standard previously referred to, ASA B5.11—1950, recommends that the outside diameter of a 15 tooth pinion of one diametral pitch should be 17.2453 in. The standard outside diameter D_o would be 17.00 in. Therefore, ΔD is 17.2453 minus 17.0 or 0.2453. For the 15 tooth motor pinion, Fig. 5, the outside diameter would be 17.2453/96 or 0.1796 in. basic. If this were a precision Class I pinion, its minimum outside diameter at the point of lowest runout, Table V, would be:

$$0.1796 - \left[\frac{0.001}{2} + \frac{0.001}{2} \right] = 0.1786 \text{ inch}$$

The value ΔD is divided by the diametral pitch of the pinion or gear gives the amount in inches that the outside diameter is larger or smaller

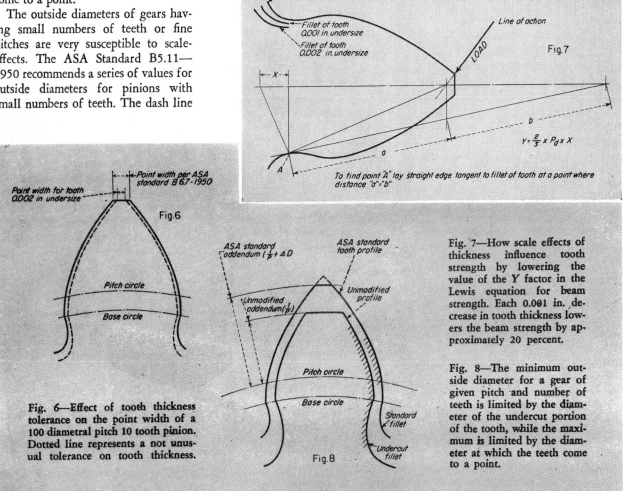

Fig. 6—Effect of tooth thickness tolerance on the point width of a 100 diametral pitch 10 tooth pinion. Dotted line represents a not unusual tolerance on tooth thickness.

Fig. 7—How scale effects of thickness influence tooth strength by lowering the value of the Y factor in the Lewis equation for beam strength. Each 0.001 in. decrease in tooth thickness lowers the beam strength by approximately 20 percent.

Fig. 8—The minimum outside diameter for a gear of given pitch and number of teeth is limited by the diameter of the undercut portion of the tooth, while the maximum is limited by the diameter at which the teeth come to a point.

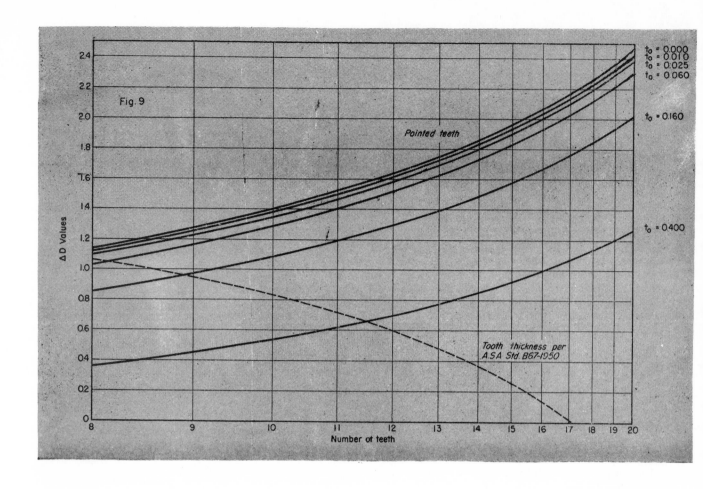

Fig. 9

Pointed teeth

$t_0 = 0.000$
$t_0 = 0.010$
$t_0 = 0.025$
$t_0 = 0.060$

$t_0 = 0.160$

$t_0 = 0.400$

Tooth thickness per A.S.A. Std. B67-1950

ΔD Values

Number of teeth

than standard. This value, and values of ΔD for all numbers of teeth for pinions having from 8 through 17 teeth, may be found on the curves presented in Fig. 9. As the whole depth of fine pitch gearing is increased or decreased the same amount as the outside diameter in most cases, the value ΔD represents twice the change in center distance at which the gear will operate correctly. Fig. 9 shows the amount of ΔD that may be added to any gear of a given number of teeth in order to achieve (a) pointed teeth, or (b) teeth with specific values of top land.

The outside diameter of a cut gear is produced either by the lathe or screw machine used to produce the gear blank, or by the hob used to cut the teeth if a topping hob is used. Each of these methods involves scale-effects.

Gears cut without topping are subject to the effects of diameter and runout tolerances on the outside diameter, Table V. In the case of fine pitch gearing particularly with small numbers of teeth, these tolerances make the difference between teeth with a definite top land and teeth which come to a point.

The sketch in Fig. 10 shows how the point width of a tooth varies when the outside diameter circle is 0.002 in. oversize and also when this diameter is 0.002 in. undersize. This layout was made for a 10 tooth 100 diametral pitch pinion.

Gears with pointed teeth actually have a lesser outside diameter than the blank from which they were cut. Thus, the gears are subject to detrimental scale-effects as far as the contact ratio is concerned as well as suffering an inherent decrease in tooth load carrying capacity.

Gears cut with topping hobs avoid the scale-effects of the runout and size tolerances on the blanks. These gears are subject to the size tolerance ground into the topping hob. In general, the hob manufacturer desires a tolerance

of 0.002 in. or 0.003 in. on the relationship between the part of the hob that produces the top land of the gear tooth and the part that controls the tooth thickness. Thus, two gears of a given tooth thickness cut by two different hobs will show rather serious scale-effects on contact ratio in the very fine pitch range. The amount that a gear tooth can be altered in tooth thickness at the pitch line to obtain a given width at the tip is shown in Fig. 9.

This represents the limiting outside diameter and is of importance on pinions with very small numbers of teeth (10 or less). Here a small change in tooth thickness will make a large change in outside diameter. Thus, the contact ratio, which is already critical, will be greatly affected.

How Scale Effects Influence Contact Ratio

The scale-effects on center distance and on outside diameter of pinion and gear may cause a dangerously low contact ratio. This will manifest itself as rough running gearing. For smooth operation the gear mesh should have a contact ratio of 1.2 or higher in ac-cordance with a recommendation of the American Gear Manufacturing Association. Contact ratio may be defined as the ratio of the base pitch to the effective length of the line of action, Fig. 11. As the center distance is increased or the outside diameters are

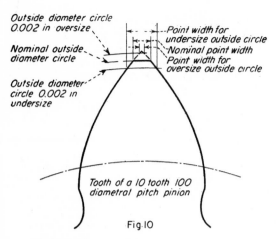

Fig. 9—Relationship between $\triangle D$ and thickness of tooth at the tip. The value $\triangle D$ represents twice the change in center distance at which the gear will operate correctly.

Fig. 10—Teeth on gears that are not hobbed at the tip are influenced by the diameter and runout tolerances.

Fig. 11—Base pitch and line of action define contact ratio. As the center distance is increased or the outside diameters are decreased, the effective length of the line of action become shorter, thus reducing the contact ratio.

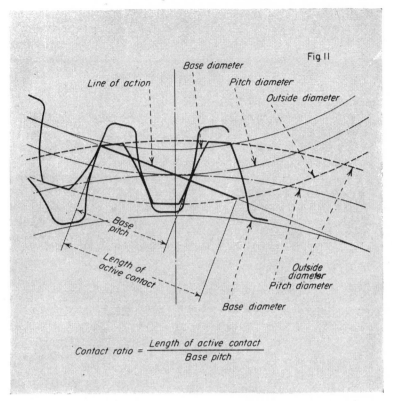

$$\text{Contact ratio} = \frac{\text{Length of active contact}}{\text{Base pitch}}$$

decreased, the effective length of the line of action becomes shorter, thus reducing the contact ratio. When calculating the minimum contact ratio that is possible for a given pair of gears, the maximum possible center distance and the minimum effective outside radius should be used. The equation is as follows:

$$m_n = 1/p_b \left[\sqrt{(D_o'/2)^2 - (D_b'/2)^2} + \sqrt{(D_o/2)^2 - (D_b/2)^2} - C' \sin \varphi_1 \right] \quad (6)$$

where m_n is the contact ratio, D_o is the outside diameter of the driving gear, D_o' is the outside diameter of the driven gear, D_b is the base circle diameter of the driving gear, D_b' is the base circle diameter of the driven gear, C_1 is the operating center distance, ϕ_1 is the operating pressure angle, and

$$p_b \text{ is } \pi D_b/N. \quad (7)$$

The operating pressure angle is obtained from

$$\cos \phi_1 = \frac{C \cos \phi}{C_1} \quad (8)$$

where

C = standard center distance

$$= \frac{N' + N}{2Pd} \quad (9)$$

C_1 = operating center distance
ϕ = standard pressure angle.

Recommended procedure for finding the contact ratio for a gear train can be illustrated by an analysis of the drive shown in Fig. 5.

The standard center distance from Eq (9) will be

$$C = \frac{15 + 120}{96 \times 2} = \frac{135}{192} = 0.703125$$

From the previous analysis made on this gear train the maximum tolerance on operating center distance was found to be 0.0021 in. The maximum operating center distance C_1 is

$$0.703125 + 0.0021 = 0.705225$$

From Eq (8) the operating pressure angle is

$$\cos \phi_1 = \frac{C \cos \phi}{C_1} = \frac{0.703125 \times 0.93969}{0.705225}$$

$$= 0.9368918$$
$$\phi_1 = 20.464° \quad \sin \phi_1 = 0.3496203$$

The base pitch from Table VI is 0.0308. The outside radius is 0.078 in.

for the pinion and 0.625 in. for the gear (see above). The contact ratio (m_n) from Eq (6) is

$$m_n = [\sqrt{(0.0893)^2 - (0.0734)^2} + \sqrt{(0.6336)^2 - (0.5873)^2} - 0.705225 \times 0.3496203] \div 0.0308$$

$$m_n = 1.31$$

If standard values of outside diameter (nominal) are used, the value of the contact ratio for various numbers of teeth in pinions and gears can be read from Fig. 12. As a convenience in designing fine pitch gearing, special charts for each diametral pitch and pressure angle can be prepared to show the relationship between contact ratio and the number of teeth in the gear and pinion and the center distance at which they operate, Fig. 13. This is very convenient in establishing the scale-effects of center distance on contact ratio. The chart shown is based on outside diameter tolerances on gear and pinion as shown in Table V.

Contact ratio is very sensitive to

scale-effects. It is reduced by an increase in center distance. It may also be reduced by a reduction in the outside diameter of the gear or pinion. This may occur as a result of runout or tolerance on size and may be aggravated by excessive brushing to remove burrs, thus rounding the tips of the teeth. If the undercut portion of the tooth extends too far up the flank of the tooth, the contact ratio will also be reduced.

If the actuator design which has been used to illustrate the examples in this article is examined by means of the curve in Fig. 13, it will be seen that as a result of the center-distance tolerance for the mesh 5-6, the contact ratio will be within reason based on the center distance 0.0021 in. as found previously. If, however, this design had been based on a pinion of 9 or 10 teeth, for example, the center distance tolerance would be found to be too large as shown by Fig. 13. The contact ratio for a 10 tooth pinion operating at a center distance 0.002 in. above nominal is seen to be 1.14 which is dangerously low. In such a case the scale-effects are of such a magnitude as to make the design undesirable.

The following general conclusions may be reached concerning the scale-effects of machining tolerances on very fine pitch gearing.

(A) Gearing of diametral pitches finer than 64 should not be designed to ASA Classification Commercial Classes 1 and 2 unless a study of scale-effects is made, or previous experience has definitely shown the design to be satisfactory.

In general, scale-effects are of concern mostly to the designer who is used to the design and application of gearing larger than 48 diametral pitch and who finds himself confronted with a design requiring extremely small gearing.

(B) Gearing of diametral pitches finer than 100 should not be used unless a study of scale-effects is made or previous experience has shown the design to be satisfactory.

(C) For gear trains that must transmit angular motion accurately, the low speed gearing should be of the greatest accuracy or else provisions should be made to minimize lost motion and angular error of the teeth at these meshes.

(D) Manufacturing and inspection procedures must be compatible with the size and design requirements of the very fine pitch gearing. Close attention must be given to center distances, outside diameter, and tooth size. Such minor details as rounding of tips of gear teeth during burr brushing become important contributions to scale-effects if not closely controlled.

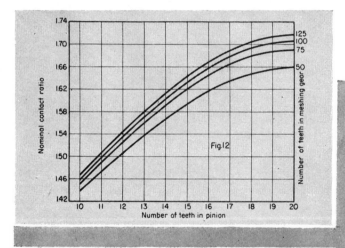

Fig. 12—Contact ratios for various numbers of teeth in pinions and gears, based on standard values of nominal outside diameter.

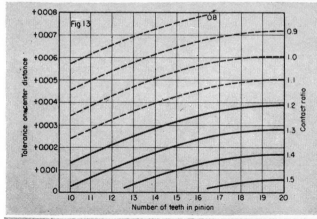

Fig. 13—Special charts, such as the one above, can be prepared to show the relationship between center distance tolerance and contact ratio for each diametral pitch and pressure angle.

Table V—Tolerances on Gear Blank Dimensions

Class	Outside Diameter	
	Size Tolerance in.	Runout Tolerance *
Commercial 1, 2, 3		
20–40 Dia Pitch	+0.000−0.004	0.003
41–80 Diametral Pitch	+0.000−0.003	0.002
81 and Finer Diametral Pitch	+0.000−0.002	0.001
Commercial Class 4 and Precision Classes 1, 2, 3		
20–40 Dia Pitch	+0.000−0.003	0.002
41–80 Dia Pitch	+0.000−0.002	0.0015
81–Finer Dia Pitch	+0.000−0.001	0.001

* (Total Indicator Reading, in.)

Table VI—Base Pitch for Diametral Pitch

Diametral Pitch	Base Pitch, in.	Diametral Pitch	Base Pitch, in.
20	0.1476	48	0.0615
24	0.1229	64	0.0491
32	0.0923	80	0.0369
40	0.0738	96	0.0308

New equations and charts pick off

Lightest-weight gears

Large gears or miniature, marine drives or aircraft instrumentation—this method will select the lightest gear train that meets requirements.

R. J. WILLIS JR, Advance Gearing, General Electric Co, West Lynn, Mass

NOT long ago, weight was a critical factor mainly in aircraft design. Today, weight reduction is the top design goal for a wide range of products, from 24-ft marine gears to compact washing machines. Weight reduction usually means volume reduction, which in turn lowers cost of materials, handling, and shipping.

The method presented here pinpoints the gear sizes and ratios that will permit the lightest possible design —while still meeting the horsepower and surface durability requirements of the application. It avoids the trial-and-error approach that frequently is employed at the beginning of a design. It also gives you a close estimate of the total weight of the drive, including weight of shafts, bearings, and housing—before your design is on the drawing board.

Eight gear systems are analyzed, including star and planetary systems. The method is applicable to spur, helical, and bevel gears—and to composite designs consisting of, say, a simple reduction system coupled to planetary gears. Three problems drawn from actual industrial applications show how the curves and equations are used.

Design equations

The design equations are based on the following assumptions:

1) The weight of a gear drive is proportional to the solid rotor volume (Fd^2) of the individual gears in the drive. This may seem an oversimplification—but a specific volume of material is necessary to carry a load at a particular reduction ratio. The fact that the gears usually are not solid disks, but may contain holes and webbed cross sections, is taken into consideration *after* the initial analysis which determines the optimum first-reduction gear ratio.

2) Surface durability factor, K, of each gear mesh is a constant. Typical K values are provided on p 9.

The relationship between the overall gear dimensions, the gear ratio and the horsepower capacity is

$$(C.D.)^2 F = \frac{31,500 P (m_g + 1)^3}{K n_p m_g} \quad (1)$$

Symbols are defined on p 26. Some typical values for K are listed in Table VII, p 9. These are for 10 hr per day operation. For continuous service divide the K values by 1.25; also, for heavy shock conditions, divide by 1.5.

The center distance is equal to

$$C.D. = \tfrac{1}{2}(d_p + d_g)$$
$$= \frac{d_p}{2}(m_g + 1) \quad (2)$$

Hence, Eq 1 becomes

$$Fd_p^2 = \frac{126,000 P}{K n_p}\left(\frac{m_g + 1}{m_g}\right) \quad (3)$$

Since $P = T n_p / 63,000$, Eq 3 becomes

$$Fd_p^2 = \frac{2T}{K}\left(\frac{m_g + 1}{m_g}\right) \quad (4)$$

The solid rotor volume of a gear is $V = \pi F d^2 / 4$. Hence Eq 4 is a function of the solid rotor volume of the input pinion. The weight of a gear system is approximately proportional to the sum of the solid rotor volumes of its gears; therefore a term similar to Eq 4 can be written for the mating gear. For the gear: $d_g = d_p m_g$. Therefore

$$Fd_g^2 = Fd_p^2 m_g^2 \quad (5)$$

If this pinion and gear are the only two gears in the gear train, as with the simple offset, Fig 1A, the total weight

1 .. Eight gear systems

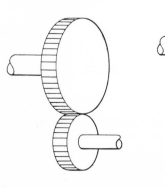

(A) OFFSET

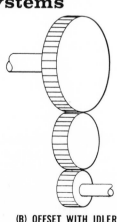

(B) OFFSET WITH IDLER

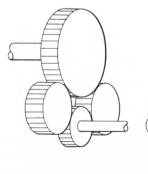

(C) OFFSET WITH TWO IDLERS

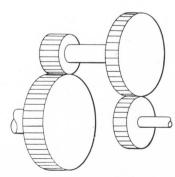

(D) DOUBLE REDUCTION

will be a function of their sum:

$$\Sigma F d^2 = F d_p{}^2 + F d_g{}^2 \quad (6)$$

or, by combining Eq 4, 5, and 6:

$$\Sigma F d^2 = \frac{2T}{K}\left(\frac{m_g + 1}{m_g}\right)$$
$$+ \frac{2T}{K}\left(\frac{m_g + 1}{m_g}\right)m_g{}^2 \quad (7)$$

For simplicity, let

$$C = 2T/K \quad (8)$$

Then, by substitution and simplification, we obtain

$$\sum \frac{F d^2}{C} = 1 + \frac{1}{m_g} + m_g + m_g{}^2 \quad (9)$$

where $\Sigma F d^2/C$ is the weight factor.

This general approach for obtaining the weight factor is now applied to a gear system with three gears: pinion, idler, and gear, Fig 1B. Here m_{g1} = ratio of pinion to idler, m_{g2} = ratio of idler to gear, and M_o = system over-all ratio = $m_{g1} \times m_{g2}$. Therefore $d_i = d_p\, m_{g1}$.

Hence instead of Eq 4 we have

$$F d_i{}^2 = \frac{2T}{K}\left(\frac{m_{g1} + 1}{m_{g1}}\right)m_{g1}{}^2 \quad (10)$$

Since $d_g = M_o\, d_p$, then

$$F d_g{}^2 = \frac{2T}{K}\left(\frac{m_{g1} + 1}{m_{g1}}\right)M_o{}^2 \quad (11)$$

Hence

$$\Sigma F d^2 = C\left(\frac{m_{g1} + 1}{m_{g1}}\right)$$
$$+ C\left(\frac{m_{g1} + 1}{m_{g1}}\right)m_{g1}{}^2$$
$$+ C\left(\frac{m_{g1} + 1}{m_{g1}}\right)M_o{}^2 \quad (12)$$

The weight factor for the pinion-idler-gear arrangement becomes

$$\sum \frac{F d^2}{C} = 1 + \frac{1}{m_{g1}} + m_{g1}$$

$$+ m_{g1}{}^2 + M_o + \frac{M_o{}^2}{m_{g1}} \quad (13)$$

However, there is only one value for m_{g1} for any value for M_o that will provide the least weight. To determine this, the first derivative of Eq 13 is taken with respect to the input ratio, m_{g1}, and the resulting equation is set equal to zero. Thus

$$\frac{d(\Sigma F d^2/C)}{d(m_{g1})} = 1 + 2m_{g1}$$

$$- \frac{1}{m_{g1}{}^2} - \frac{M_o{}^2}{m_{g1}{}^2} = 0 \quad (14)$$

Solving results in

$$2m_{g1}{}^3 + m_{g1}{}^2 = M_o{}^2 + 1 \quad (15)$$

Thus, by knowing the over-all ratio, we obtain from Eq 15 the input ratio for which weight will be minimum.

The design curves

As an example of how to apply Eq 15, suppose the over-all ratio is 5; then from the equation, m_{g1} = 2.18. But Eq 15 cannot be solved directly, so a plot of this equation is given in Fig 2. These minimum-weight curves are applicable to spur, helical, or bevel gears with arrangements shown in Fig 1A to 1C.

It will be shown later that the weight factor, $\Sigma F d^2/C$, can be employed to predict with a fair degree of accuracy the total weight of the gear assembly. The weight factor is obtained by first solving for m_{g1} in Eq 15 (for a given value of M_o) and then employing both m_{g1} and M_o in Eq 13. To reduce computation time, a total-weight curves have been established, Fig. 3, which directly give the weight factor for values of the over-all ratio (it is assumed that m_{g1} has been optimized).

The same techniques are applied to the other gear systems illustrated in Fig 1, and the resulting minimum-

SYMBOLS

b = number of branches (or planets) in an epicyclic gear

$C = 2T/K$, in.³

C.D. = center distance, in.

d_g = gear pitch diameter, in.

d_i = idler pitch diameter, in.

d_p = pinion (or planet) pitch diameter, in.

d_r = ring gear pitch diameter, in.

d_s = sun gear pitch diameter, in.

F = face width, in.

P = input horsepower

K = surface durability factor, lb/in.²

m_g = ratio of input pinion to its mating gear

m_{g1} = ratio of input to idler

m_{g2} = ratio of idler to gear

m_s = ratio of planet gear to sun pinion

M_o = over-all ratio

$M_o{}'$ = system composite over-all ratio

n = arbitrary number of ratio splits

n_p = pinion speed, rpm

N_i = number of teeth in idler

N_g = number of teeth in gear

N_p = number of teeth in pinion

P_d = diametral pitch

T = input torque, lb-in.

U = unit load, lb/in. of face

V = solid rotor volume, in.³

W_t = tangential driving force, lb

W_t/F = tooth load, lb/in.

θ = bevel gear pitch-cone angle, deg

Fig. 1 (continued)

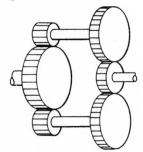
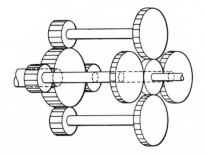
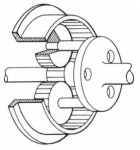
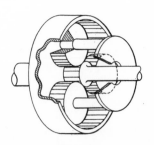

(E) DOUBLE-REDUCTION DOUBLE-BRANCH   (F) DOUBLE-REDUCTION FOUR-BRANCH   (G) PLANETARY   (H) STAR

For offset gears

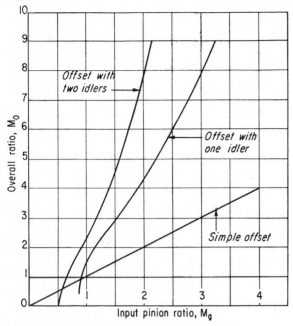

2 . . MINIMUM WEIGHT CURVES

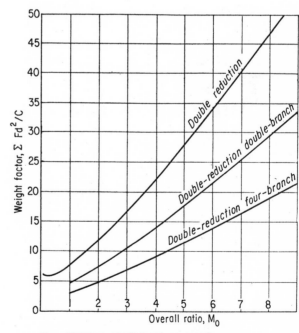

3 . . TOTAL WEIGHT CURVES

For double-reduction gears

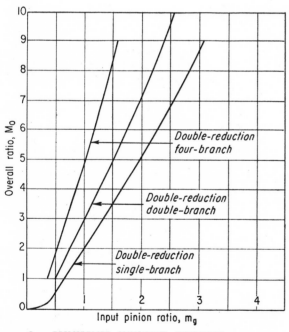

4 . . MINIMUM WEIGHT CURVES

5 . . TOTAL WEIGHT CURVES

weight equations are summarized in Table I.

In the case of the simple offset system the geometry of the arrangement governs the situation. The input pinion ratio and the over-all ratio are identical ($m_g = M_o$) because only one gear mesh is involved and

it is impossible to obtain a "minimum weight" based on the ratio split. The curve of the weight factor vs the over-all ratio (Fig 3), however, does apply.

Multi-branch systems

The minimum-weight curves for the double-reduction systems are shown in

Fig 4; Fig 5 gives the total-weight curves which can also be employed to compare one system to another before a design even reaches the drawing board. The multi-branch double-reduction systems, Fig 1E and F, offer multiple-torque paths. For example, each shaft in the four-branch system,

TABLE I .. MINIMUM-WEIGHT EQUATIONS

SIMPLE OFFSET (THEORETICAL)	$2 m_g^3 + m_g^2 = 1$
OFFSET WITH IDLER	$2 m_g^3 + m_g^2 = M_o^2 + 1$
OFFSET WITH TWO IDLERS	$2 m_g^3 + m_g^2 = \dfrac{M_o^2 + 1}{2}$
OFFSET WITH n IDLERS	$2 m_g^3 + m_g^2 = \dfrac{M_o^2 + 1}{n}$
DOUBLE REDUCTION	$2 m_g^3 + \dfrac{2 m_g^2}{\left(\dfrac{M_o + 1}{M_o}\right)} = \dfrac{M_o^2 + 1}{\left(\dfrac{M_o + 1}{M_o}\right)}$
DOUBLE-REDUCTION DOUBLE-BRANCH	$2 m_g^3 + \dfrac{2 m_g^2}{\left(\dfrac{M_o + 1}{M_o}\right)} = \dfrac{M_o^2 + 1}{2\left(\dfrac{M_o + 1}{M_o}\right)}$
DOUBLE-REDUCTION FOUR-BRANCH	$2 m_g^3 + \dfrac{2 m_g^2}{\left(\dfrac{M_o + 1}{M_o}\right)} = \dfrac{M_o^2 + 1}{4\left(\dfrac{M_o + 1}{M_o}\right)}$
DOUBLE-REDUCTION n BRANCHES	$2 m_g^3 + \dfrac{2 m_g^2}{\left(\dfrac{M_o + 1}{M_o}\right)} = \dfrac{M_o^2 + 1}{n\left(\dfrac{M_o + 1}{M_o}\right)}$
PLANETARY (THEORETICAL)	$2 m_s^3 + m_s^2 = \dfrac{0.4(M_o-1)^2 + 1}{b}$
STAR (THEORETICAL)	$2 m_s^3 + m_s^2 = \dfrac{0.4 M_o^2 + 1}{b}$

TABLE II .. TOTAL-WEIGHT EQUATIONS

OFFSET	$\sum Fd^2/C = 1 + \dfrac{1}{m_g} + m_g + m_g^2$
OFFSET WITH IDLER	$\sum Fd^2/C = 1 + \dfrac{1}{m_g} + m_g + m_g^2 + \dfrac{M_o^2}{m_g} + M_o^2$
OFFSET WITH TWO IDLERS	$\sum Fd^2/C = \dfrac{1}{2} + \dfrac{1}{2m_g} + m_g + m_g^2 + \dfrac{M_o^2}{2 m_g} + \dfrac{M_o^2}{2}$
DOUBLE REDUCTION	$\sum Fd^2/C = 1 + \dfrac{1}{m_g} + 2m_g + m_g^2 + \dfrac{m_g^2}{M_o} + \dfrac{M_o^2}{m_g} + M_o$
DOUBLE-REDUCTION DOUBLE-BRANCH	$\sum Fd^2/C = \dfrac{1}{2} + \dfrac{1}{2m_g} + 2m_g + m_g^2 + \dfrac{m_g^2}{M_o} + \dfrac{M_o^2}{2m_g} + \dfrac{M_o}{2}$
DOUBLE-REDUCTION, FOUR-BRANCH	$\sum Fd^2/C = \dfrac{1}{4} + \dfrac{1}{4 m_g} + 2m_g + m_g^2 + \dfrac{m_g^2}{M_o} + \dfrac{M_o^2}{4m_g} + \dfrac{M_o}{4}$
PLANETARY	$\sum Fd^2/C = \dfrac{1}{b} + \dfrac{1}{b m_s} + m_s + m_s^2 + \dfrac{0.4(M_o-1)^2}{b m_s} + \dfrac{0.4(M_o-1)^2}{b}$
STAR	$\sum Fd^2/C = \dfrac{1}{b} + \dfrac{1}{b m_s} + m_s + m_s^2 + \dfrac{0.4 M_o^2}{b m_s} + \dfrac{0.4 M_o^2}{b}$

Fig IF, carries one quarter of the torque carried in the single shaft, Fig 1D. Multi-branch systems are common as marine gears; in aircraft engines, epicyclic systems with 36 branches have been employed successfully. It is interesting to note, therefore, that *the weight factor is reduced as the number of branches is increased.* Although the stress cycles to the pinion will increase in proportion to the number of branches, there is usually only a slight reduction in the permissible stress, as determined from an S-N fatigue curve.

Epicyclic systems

Two epicyclic gear systems are covered, Fig 1: 1) In the *planetary* system, G, the ring gear is stationary, input is to the sun gear, and output is taken from the cage which retains the planet gears. 2) In the *star* system, H, the cage is held stationary, input is again to the sun gear, and output is taken from the ring gear, which is now permittted to rotate.

In the star system the over-all ratio, M_o, is equal to the ratio of ring-gear to sun-gear diameters, whereas in the planetary system M_o is equal to this ratio plus one. Hence the minimum-weight equations for these two systems, Table I, are only theoretical in that the geometry of the systems is fixed by the over-all ratio. Fig 6 shows plots of these equations. The weight factor for these systems, however, has physical significance, and the method for determining the total-weight equations is as follows:

The ratio, m_s, between planet and sun gear is equal to

$$m_s = \frac{d_p}{d_s} = \frac{N_p}{N_s} = \frac{M_o}{2} - 1 \quad (16)$$

and for the star gear

$$m_s = \frac{d_p}{d_s} = \frac{N_p}{N_s} = \frac{M_0 - 1}{2} \quad (17)$$

where d_p = planet gear pitch diameter, d_s = sun gear pitch diameter, N = number of teeth.

If b is the number of branches (or planet gears), then Eq 4 becomes

$$Fd_s^2 = \frac{2T}{bK}\left(\frac{m_{g1} + 1}{m_{g1}}\right) \quad (18)$$

Also, since $m_s = d_p/d_s$, then

$$bFd_p^2 = bFd_s^2 m_s^2 \quad (19)$$

and the relationship between ring gear and sun becomes

$$Fd_r^2 = Fd_s^2 \left(\frac{d_r}{d_s}\right)^2 0.4 \quad (20)$$

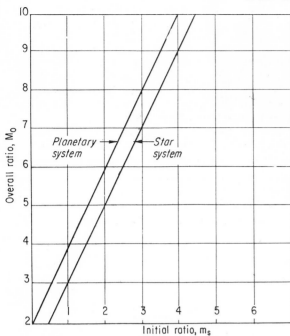

6 . . MINIMUM WEIGHT CURVES

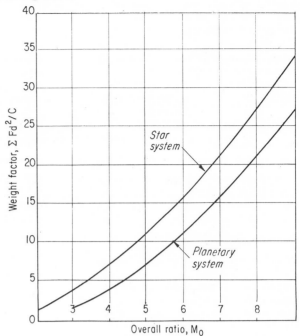

7 . . TOTAL WEIGHT CURVES

where d_r = pitch diameter of ring gear.

Factor 0.4 in Eq 20 is an arbitrary one based on our experience. It takes into consideration the weight of the cage structure and of the housing structure needed to give support and flexibility to the ring gear in a planetary system. It also accounts for the additional ring-gear structure and bearing support posts required in a star system.

The weight-factor equations for the planetary and star systems can now be obtained by employing a development similar to the one for Eq 5 through 9:

Planetary

$$\Sigma Fd^2/C = \frac{1}{b} + \frac{1}{bm_s} + m_s +$$
$$m_s{}^2 + \frac{0.4(M_o-1)^2}{b} +$$
$$\frac{0.4(M_o-1)^2}{bm_s} \qquad (21)$$

Star

$$\Sigma Fd^2/C = \frac{1}{b} + \frac{1}{bm_s} + m_s$$
$$m_s{}^2 + \frac{0.4M_o{}^2}{b} +$$
$$\frac{0.4M_o{}^2}{bm_s} \qquad (22)$$

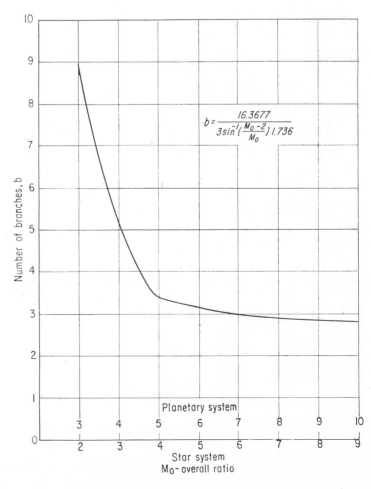

$$b = \frac{16.3677}{3\sin^{-1}\left(\frac{M_o-2}{M_o}\right)1.736}$$

8 . . OPTIMUM BRANCHES FOR EPICYCLIC GEARS

These total-weight equations are summarized with those of the other gear systems in Table II. Total-weight curves for the planetary and star systems are shown in Fig 7. Data for the curves are derived from Eq 21 and 22 by assuming that the systems contain the maximum number of branches that will fit into a planetary or star annulus. This can be determined as a function of over-all ratio, Fig 8.

Weight estimating

The value of $\Sigma Fd^2/C$ for any gear arrangement obtained from the appropriate curve is multiplied by the value for C to obtain a numerical value for ΣFd^2. In the normal gearing problem either the input horsepower and rpm or the torque is known. Thus, to solve for C in Eq 8 one must assume a value for K. Choice of the K factor depends on the materials, severity of service, life desired, Hertz stress levels, heat-treating techniques, and application. Hence the values for K on p 9 are only a guide.

A means of estimating the actual weight of complete gear systems is now available because our original premise was that the weight of a gear system is a function of total solid rotor volume. From our experience we have found that weight estimates are possible by multiplying the value of ΣFd^2 by the application factors listed in Table III. These factors take into consideration the gearing, shaft, bearings and immediate support structure. They do not include accessories.

Thus a planetary-gear assembly designed for aircraft may be less than one half the weight of a similar assembly for automotive use. (Multiply the total weight by, say, 0.25 for aircraft, and 0.60 for "commercial").

Composite gear assemblies

Data obtained from the curves can be applied to composite gear systems composed of two or more of the systems in Fig 1. For example, a transmission system frequently consists of an offset unit driving a planetary or star gear. A composite curve of ΣFd^2 for the over-all system vs system initial input pinion ratio can be simply constructed. This will indicate the ideal offset ratio to use for an over-all minimum-weight design. Also, different values of K factors can be utilized in each of the components. Hence for the offset gears

$$\Sigma Fd^2{}_{off} = \Sigma Fd^2/C_{off} \times \frac{2T}{K_{off}} \quad (23)$$

and for the epicyclic gear

$$\Sigma Fd^2{}_{ep} = \Sigma Fd^2/C_{ep} \times \frac{2Tm_g}{K_{ep}} \quad (24)$$

To construct a composite system curve, Fig 9, let the abscissa represent the initial system input ratio, in this instance the ratio of the offset unit, m_{g1}. The ordinate will then represent values of ΣFd^2. From the minimum-weight curve for the simple off-

set, Fig 2, select whole number values of m_g (here m_g is taking the place of M_o) and read off the corresponding values of $\Sigma Fd^2/C$. The values of m_g are transferred to the composite curve, while corresponding values of $\Sigma Fd^2/C$ are multiplied by the calculated value of C for the offset unit. The plotted points are joined by a smooth line and labeled offset gear.

The epicyclic gear is handled in much the same manner except that for

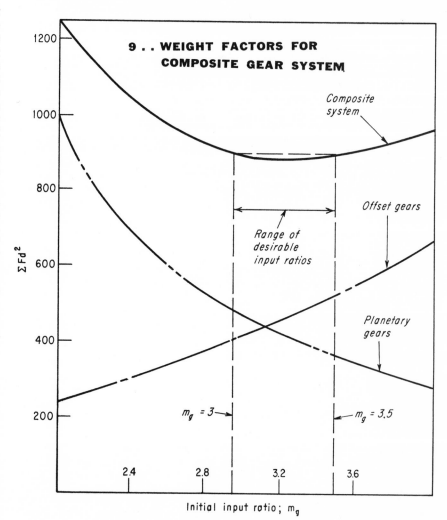

TABLE III .. **APPLICATION FACTORS FOR WEIGHT ESTIMATIONS**

Application	Factor	Typical conditions
Aircraft	0.25 to 0.30	Magnesium or aluminum casings; limited-life design; high stress levels; rigid weight control.
Hydrofoil	0.30 to 0.35	Lightweight steel casings; relatively high stress levels; limited life design; rigidity desired.
Commercial	0.60 to 0.625	Cast or fabricated steel casings; relatively low stress levels; unlimited-life design; solid rotors and shafts.

each chosen value of m_g for the offset gear, a value of M_o exists for the epicyclic gear so that the product of m_g and M_o will be equal to the composite system over-all ratio. In other words, $m_g \times M_o = M_o'$, where M_o' equals the system composite over-all ratio. The values of ΣFd^2 for the epicyclic gear are calculated as indicated by Eq 21; however, these values are plotted on the chart for corresponding values of m_g. The points for the epicyclic gear unit are joined by a smooth line and labeled accordingly. The values are either added graphically or tabulated and added mathematically. The resultant curve now represents the composite gear system.

Problem 1—Double reduction, in-line shafts

The double-reduction reverted gear system, page 72, for a general-purpose industrial application has the following torque and ratio requirements:

Input torque, $T = 1000$ lb-in.
Over-all ratio, $M_o = 7$

It is also required that the input and output shaft centerlines be in line. (A center-distance requirement such as this usually increases the

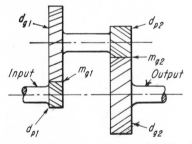

system weight—but the method is still practical). Thus

$$\tfrac{1}{2}(d_{p1} + d_{g1}) = \tfrac{1}{2}(d_{p2} + d_{g2})$$

To design for minimum weight, the initial ratio, m_{g1}, is obtained from the single-branch curve in Fig 4. For an over-all ratio of 7, a value of 2.52 for the initial input ratio is required. Therefore $M_{g1} = 2.52$, and

$$m_{g2} = \frac{M_o}{m_{g1}} = \frac{7}{2.52} = 2.777$$

The weight factor of the gear system is found from Fig 5, where $\Sigma Fd^2/C = 40$ when $M_o = 7$. In checking with page 9, assume for the present example a value of $K = 500$ for both meshes. Then, from Eq 8: $C = (2)(1000)/500 = 4$, and $\Sigma Fd^2 = (40)(4) = 160$ in.[3]

The total weight of the gearset may now be estimated by referring to Table III. If the gearset is for an industrial application we may estimate its weight as

Weight $= \Sigma Fd^2(0.60)$
$= 160(0.60) = 96$ lb

The second mesh is dealt with first because its gears will have greater volume requirements:

$$Fd^2_{p2} = \frac{2T}{K}\left(\frac{m_{g2}+1}{m_{g2}}\right)m_{g1}$$

$$= \frac{(2)(1000)}{500}\left(\frac{3.777}{2.777}\right)2.52$$

$$= 13.709$$

The tangential driving force is

$$W_{t2} = \frac{2T}{d_{p2}}m_{g1}$$

$$= \frac{(2)(1000)}{d_{p2}}(2.52) = \frac{5040}{d_{p2}}$$

A unit load of $U = 12,000$ lb/in. is assumed. Usually U is chosen as approximately one-third the bending stress, as determined from an S-N curve. Hence

$$U = 12,000 \text{ lb/in.} =$$

$$\frac{W_{t2}P_{d2}}{F_2} = \frac{5040P_{d2}}{F_2 d_{p2}}$$

Hence

$$F_2 d_{p2} = \frac{5040P_{d2}}{12,000} = 0.42P_{d2}$$

Since $Fd^2_{p2} = 13.709$, then

$$d_{p2} = 32.64/P_{d2}.$$

For a spur or helical mesh, we usually employ a limiting value of $F/d_p = 0.70$ in. This results in:

$d_{p2} = 2.70$ in. $P_{d2} = 12.222$
$d_{g2} = 7.50$ in.   C.D. $= 5.10$ in.
$F = 1.890$ in.

Because an integral number of teeth are required, the ratio above for d_{p2} should be changed to $d_{p2} = 33/P_{d2}$. A table is now constructed to analyze potential gears with various face-width/diametral-pitch ratios (Table

IV) — up to the limiting value of $P_{d2} = 12$.

All the gears in the table below will operate at the same root stress level (in this case at approximately 36,000 psi). Choice will depend upon other gear design parameters such as fatigue endurance limits. Assume that the 12-pitch set satisfies all of the conditions. Now, referring to the initial gear mesh:

$$Fd_{p1}^2 = \frac{2T}{K}\left(\frac{m_{g1}+1}{m_{g1}}\right) = 5.5873$$

$$W_{t1} = \frac{2T}{d_{p1}} = \frac{(2)(1000)}{d_{p1}} = \frac{2000}{d_{p1}}$$

$$U = 12,000 \text{ lb./in.}$$

$$= \frac{W_{t1}P_{d1}}{F_1} = \frac{2000P_{d1}}{Fd_{p1}}$$

$$F_1 d_{p1} = \frac{2000P_{d1}}{12,000} = 0.1666P_{d1}$$

Therefore $d_{p1} = 33.5239/P_{d1}$.

For design purposes use $d_{p1} = 33/P_{d1}$. As long as the value of Fd^2_{p1} is held at 5.5873, as indicated by the equation above, a minimum weight design will be obtained.

Regarding the center distance, from Table IV, C.D., $=$ C.D.$_2 = 5.194 = \tfrac{1}{2}(d_{p1} + d_{g1})$. Hence $d_{p1}(m_{g1} + 1) = 10.388$ in.
Thus, for the first gear pair:

$d_{p1} = 2.951$ in. $P_{d1} = 11.1826$
$d_{g1} = 7.436$ $F_1/d_{p1} = 0.2139$
$F_1 = 0.6315$ in.

A check of the ΣFd^2 weight values

$F_1 d_{p1}^2 =$
$\qquad (0.6135)(2.951)^2 = \quad 5.499$
$F_1 d_{g1}^2 =$
$\qquad (5.4993)(2.52)^2 = \quad 34.92$
$F_2 d_{p2}^2 =$
$\qquad 1.83(2.75)^2 = \quad 13.839$
$F_2 d_{g2}^2 =$
$\qquad 13.8393(2.777)^2 = \underline{106.785}$

$\qquad\qquad \Sigma Fd^2 \quad 161.044$ in.[3]
or $\qquad 161.044(0.60) = 96.5$ lb

Problem 2—Bevel gears

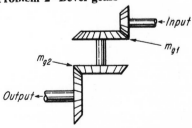

Assume that $T = 25,000$ lb-in. and $M_o = 4$.

Although bevel gearing is employed, the arrangement is basically a double-offset system. Therefore, from Fig 4, $m_{g1} = 1.66$, $m_{g2} = 2.41$. Then $T_2 =$

TABLE IV .. MINIMUM-WEIGHT

P_{d2}	9	10	12
d_{p2}	3.666	3.300	2.750
$F_2 d_{p2}$	3.78	4.20	5.04
F_2	1.031	1.273	1.83
F_2/d_{p2}	0.281	0.385	0.665
d_{g2}	10.185	9.166	7.638
C.D.	6.925	6.233	5.194

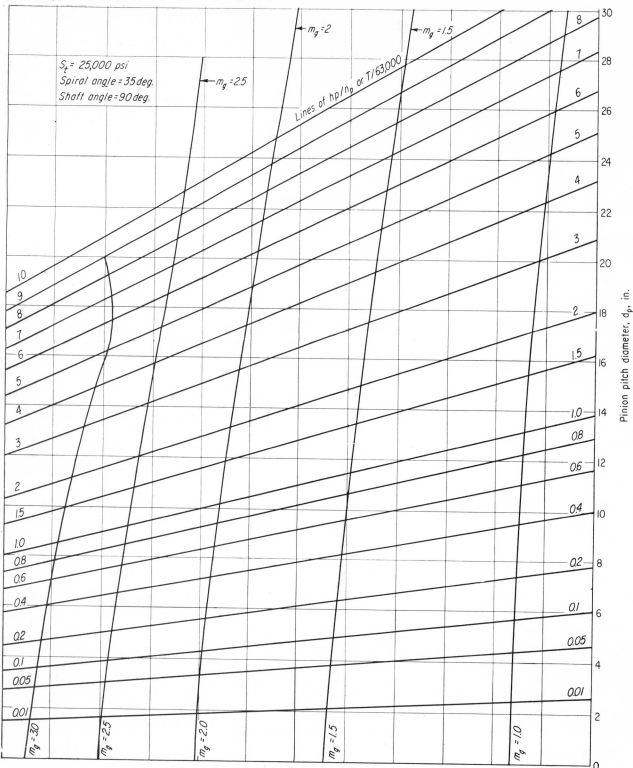

$S_t = 25,000$ psi
Spiral angle = 35 deg.
Shaft angle = 90 deg.

Lines of hp/n_p or T/63,000

Pinion pitch diameter, d_p, in.

$Tm_{g1} = 41,500$ lb-in.

The graph in Fig 10 will help estimate bevel-gear sizes. The value for $T/63,000$ for the first reduction is 0.397. Enter the curve from the bottom with the value for $m_{g1} = 1.66$ and move parallel to the ratio lines going upward until the value of 0.397 is reached. Move horizontally to the right and read the pinion pitch diameter. For the bevel gears the pitch diameters are given at the right.

	First Reduction	Second Reduction
d_p	7.80	8.20
d_g	12.80	19.75
F	2.26	3.20

The face widths were estimated from

$$F = 0.15 d_p / \sin \theta$$

where θ is the bevel gear pitch cone angle equal to

$$\theta = \tan^{-1} (1/m_g)$$
$$F_1 = 2.26 \text{ in.} \qquad F_2 = 3.20 \text{ in.}$$

	First Reduction	Second Reduction
Fd_p^2	137.5	215
Fd_g^2	387.3	1249
ΣFd^2	515.8	1464

System $\Sigma Fd^2 = 1979.8$ in.³

Use an appropriate multiplier from Table III to obtain the estimated weight for the bevel-gear system.

New formula quickly picks size of

Spur gear drives

The author successfully brings together nine key factors which influence gear size to obtain this useful formula for day-to-day design.

T. A. PARSSINEN, gear engineer, Joy Manufacturing Co., Claremont, NH

HERE'S a quick way to arrive at the diametral pitch —and hence the gear size—suitable for the horsepower, speed, and speed-ratio requirements of the application. The method is applicable to spur gear trains with 20-deg pressure angle teeth. Often the only recourse is to guess at the diametral pitch and then check the resulting number of teeth, the strength capabilities (to resist tooth breakage) and the durability capabilities (to resist pitting and wear, ie, gear life). Such a procedure must be repeated again and again until all factors blend satisfactorily. But, by making a series of well-grounded assumptions, and by combining several key equations, a design formula is obtained here which produces the diametral pitch directly.

Because of the assumptions, the formula is not foolproof—but it will produce a good first approximation for almost all cases, and it can be easily modified if you prefer other values for the assumptions.

Main assumption

Generally, the pinion should have as small a number of teeth as possible. There is little debate on this point. But can a number be selected which will hold true for the overwhelming majority of applications?

Analysis—and experience—has shown that for gears with either the 20-deg full-depth tooth or the 20-deg stub tooth, the minimum will range from about 14 to 16 teeth before undercutting becomes a factor. Therefore, select 16 as the number of pinion teeth. The pinion can be either the driver or the driven, depending upon whether the drive is speed reducing or speed increasing.

Derivation of formula

The two equations which are popularly employed for dynamic load capacity and for durability are (from Buckingham's *Manual of Gear Design,* Industrial Press):
Dynamic load capacity

$$W_d = W_t + \frac{0.05\, V\, (F\, C + W_t)}{0.05\, V + (F\, C + W_t)^{1/2}} \qquad (1)$$

Durability

$$W_w = d\, Q\, K\, F \qquad (2)$$

See list of symbols at the right for explanation of notations. The ratio factor is equal to

$$Q = \frac{2\, N_o}{N_o + N_i}$$

Also, by definition

$$m_g = N_o / N_i$$

where m_g is greater than unity in speed reducing drives, and less than unity in speed increasing drives. Hence

$$Q = \frac{2\, m_g}{m_g + 1} \qquad (3)$$

Based on the previous assumption of a 16-tooth pinion, the pinion pitch diameter becomes

$$d = 16/P \qquad (4)$$

The commonly accepted limit for face width ranges from $10/P$ to $12/P$; therefore, select the smaller face width:

$$F = 10/P \qquad (5)$$

By substituting Eq 3, 4, and 5 into Eq 2, the durability equation becomes

$$W_w = \frac{320\, m_g\, K}{P^2\, (m_g + 1)} \qquad (6)$$

The pitch-line velocity, ft/min, is

$$V = \frac{\pi\, d\, n}{12} = 0.262\, d\, n \qquad (7)$$

Symbols

d = pitch diameter of small gear, in.
C = deformation factor
F = face width, in.
H = desired horsepower capacity
K = durability factor for a mesh, see
n = output rpm for speed increasing drives, speed reducing drives or input rpm for
N_i = number of teeth in driving gear
N_o = number of teeth in driven gear
P = diametral pitch
Q = ratio factor, $2\, m_g / (m_g + 1)$
m_g = speed ratio, N_o / N_i
V = pitch line velocity, ft/min.
W_d = dynamic load on tooth, lb
W_t = tangential load, lb
W_w = durability load, lb

Combining Eq 7 with Eq 4 gives

$$V = \frac{4.19\,n}{P} \qquad (8)$$

The tangential load is

$$W_t = \frac{33,000\,H}{V} \qquad (9)$$

Also, for most commercially cut gears, the value for the deformation factor, C, can be safely taken as 1500. Hence, the tooth-strength equation, Eq 1, can now be written as

$$W_d = \frac{7880\,P\,H}{n}$$
$$+ \frac{\left(0.21\,\dfrac{n}{P}\right)\left(\dfrac{15,000}{P} + \dfrac{7880\,P\,H}{n}\right)}{\left(0.21\,\dfrac{n}{P}\right) + \left(\dfrac{15,000}{P} + \dfrac{7880\,P\,H}{n}\right)^{1/2}} \qquad (10)$$

It is desirable that dynamic load capabilities be equal to those of durability; hence

$$W_w = W_d \qquad (11)$$

Combining Eq 6, 10, and 11, and simplifying, results in the following design formula:

$$47\,n + 49.3\,P^2\,H - \frac{m_g\,K\,n}{P\,(1 + m_g)}$$
$$+ 423\left(\frac{24.6\,P^3\,H}{n} - \frac{m_g\,K}{1 + m_g}\right)\left(\frac{1.9}{P} + \frac{P\,H}{n}\right)^{1/2} = 0 \qquad (12)$$

Typical values for K are shown in Table VII, p 9. You can now solve for P when m_g, H, and n are known.

Numerical example

Given: a 50-hp, 3.53:1 speed-reduction gear drive. Input speed is at 80 rpm. The K factor is selected as 860. Find the diametral pitch.

Substituting these values into Eq 12 and simplifying gives

$$3760 + 2465\,P^2 - \frac{53700}{P}$$
$$+ (6510\,P^3 - 283,500)\left(\frac{1.9}{P} + 0.625\,P\right)^{1/2} = 0$$

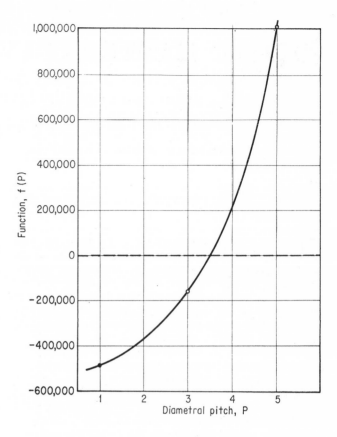

This equation cannot be solved directly. Therefore, select values for P, calculate values for the left side of the equation, then plot on a chart to find P which will satisfy the equation. Hence

For $P = 1$,

$$3760 + 2465 - 53700$$
$$+ (6510 - 283500)\,(1.9 + 0.625)^{1/2} = -488,400$$

Similarly for $P = 3$, the value for $f(P) = -162,940$; and for $P = 5$, the value for $f(P) = 1,057,600$.

These three points are plotted in the chart above. By drawing a smooth curve through the points, a value is found of approximately $P \approx 3.5$.

The actual design employs a gear set with 3.5 diametral pitch and full-depth 20-deg teeth. The unit has a 17-tooth pinion meshing with a 60-tooth gear.

Helical gear design

A helix angle whose cosine is a simple fraction permits rapid calculation of center distances and pitch diameters.

W. U. MATSON, Mechanical Engineering Consultant, Naval Research Laboratory

HELICAL gears are used when both high speed and high horsepower are required. Although the 45-deg helix angle is most popular for stock gears, as the gear can be used for either parallel or crossed shafts, the large helix angles—30 to 45 deg—impose high thrust loads on bearings when single helicals are used and unless precisely cut and installed increase gear backlash. These helix angles also increase gear weight without proportionally increasing either the strength of the gear or the power and load the helical gear can transmit.

As Table I indicates, based on the use of disks, weights would increase as the square of the diameter. Hence a disk for a 45-deg helix angle of the same normal diametral pitch would be $(1.4142)^2$, or twice as heavy for the same face width as a spur gear (0-deg helix angle). The table also points out the percentage of rise in thrust and bearing loads imposed by the higher-angle gears.

Smaller helix angles—below 30 deg—may increase gear wear slightly but improve backlash tolerance and give lower bearing loads. One helix angle —20° 21' 50.887"—adds a major advantage, ease of design.

Why a small helix angle

Thrust on the bearings caused by helix angles above 20 deg can be mitigated by double-helical or herringbone teeth. However, face width increases and manufacturing is complicated. Most ball or tapered roller bearings capable of being preloaded can be used with gears of about 20-deg helix angle, as the thrust is less than 50% of the tangential load.

Backlash in the plane of rotation

TABLE I - Effect of helix angle on various parameters

HELIX ANGLE ψ DEG	INCREASE IN PITCH DIA OVER STANDARD SPUR GEAR %	AXIAL PRESSURE ANGLE AT PITCHLINE, φ	TANGENTIAL PITCHLINE LOAD	
			THRUST %	BEARING %
*0	0	20	0	106.4
10	1.54	20°17'	17.6	106.6
20° 21' 50.877"	6.66	21°13'	37.1	107.3
30	15.47	22°48'	57.7	108.4
40	30.54	25°25'	83.9	110.7
45	41.42	27°14'	100.0	112.5
	$\%_{inc} = 100\,(1/\cos\psi - 1/\cos 0°\,)$	$\tan\varphi = \dfrac{\tan\varphi_n}{\cos\psi}$	$\%_T = 100\tan\psi$	$\%_B = 100\,\dfrac{1}{\cos\varphi}$

*Spur gear

NUMBER OF TEETH IN GEAR, N	PITCH DIAMETER FOR $P_N=1$ D_1	OUTSIDE DIAMETER FOR $P_N=1$ D_{o1}	MEASUREMENT OVER 1.728/P_N WIRES FOR $P_N=1$ M_1*	NUMBER OF TEETH IN GEAR, N	PITCH DIAMETER FOR $P_N=1$ D_1	OUTSIDE DIAMETER FOR $P_N=1$ D_{o1}	MEASUREMENT OVER 1.728/P_N WIRES FOR $P_N=1$ M_1*
17*	18.1333	20.1333	20.5249	65	69.3333	71.3333	71.7743
18	19.2000	21.2000	21.5948	66	70.4000	72.4000	72.8413
19	20.2666	22.2666	22.6641	67	71.4666	73.4666	73.9081
				68	72.5333	74.5333	74.9752
20	21.3333	23.3333	23.7334	69	73.6000	75.6000	76.0422
21	22.4000	24.4000	24.8026				
22	23.4666	25.4666	25.8714	70	74.6666	76.6666	77.1091
23	24.5333	26.5333	26.9402	71	75.7333	77.7333	78.1761
24	25.6000	27.6000	28.0089	72	76.8000	78.8000	79.2432
				73	77.8666	79.8666	80.3100
25	26.6666	28.6666	29.0773	74	78.9333	80.9333	81.3770
26	27.7333	29.7333	30.1457	75	80.0000	82.0000	82.4439
27	28.8000	30.8000	31.2139	76	81.0666	83.0666	83.5107
28	29.8666	31.8666	32.2821	77	82.1333	84.1333	84.5777
29	30.9333	32.9333	33.3502	78	83.2000	85.2000	85.6446
				79	84.2666	86.2666	86.7115
30	32.0000	34.0000	34.4182				
31	33.0666	35.0666	35.4860	80	85.3333	87.3333	87.7784
32	34.1333	36.1333	36.5539	81	86.4000	88.4000	88.8454
33	35.2000	37.2000	37.6217	82	87.4666	89.4666	89.9121
34	36.2666	38.2666	38.6894	83	88.5333	90.5333	90.9790
35	37.3333	39.3333	39.7571	84	89.6000	91.6000	92.0460
36	38.4000	40.4000	40.8247	85	90.6666	92.6666	93.1128
37	39.4666	41.4666	41.8923	86	91.7333	93.7333	94.1797
38	40.5333	42.5333	42.9599	87	92.8000	94.8000	95.2465
39	41.6000	43.6000	44.0274	88	93.8666	95.8666	96.3133
				89	94.9333	96.9333	97.3802
40	42.6666	44.6666	45.0949				
41	43.7333	45.7333	46.1623	90	96.0000	98.0000	98.4471
42	44.8000	46.8000	47.2297	91	97.0666	99.0666	99.5139
43	45.8666	47.8666	48.2971	92	98.1333	100.1333	100.5807
44	46.9333	48.9333	49.3645	93	99.2000	101.2000	101.6476
45	48.0000	50.0000	50.4318	94	100.2666	102.2666	102.7144
46	49.0666	51.0666	51.4990	95	101.3333	103.3333	103.7813
47	50.1333	52.1333	52.5663	96	102.4000	104.4000	104.8482
48	51.2000	53.2000	53.6336	97	103.4666	105.4666	105.9150
49	52.2666	54.2666	54.7008	98	104.5333	106.5333	106.9818
				99	105.6000	107.6000	108.0486
50	53.3333	55.3333	55.7680				
51	54.4000	56.4000	56.8352	100	106.6666	108.6666	109.1154
52	55.4666	57.4666	57.9022	101	107.7333	109.7333	110.1823
53	56.5333	58.5333	58.9695	102	108.8000	110.8000	111.2491
54	57.6000	59.6000	60.0367	103	109.8666	111.8666	112.3158
55	58.6666	60.6666	61.1037	104	110.9333	112.9333	113.3826
56	59.7333	61.7333	62.1709	105	112.0000	114.0000	114.4495
57	60.8000	62.8000	63.2380	106	113.0666	115.0666	115.5163
58	61.8666	63.8666	64.3050	107	114.1333	116.1333	116.5831
59	62.9333	64.9333	65.3722	108	115.2000	117.2000	117.6499
				109	116.2666	118.2666	118.7166
60	64.0000	66.0000	66.4392				
61	65.0666	67.0666	67.5062	110	117.3333	119.3333	119.7834
62	66.1333	68.1333	68.5733	111	118.4000	120.4000	120.8503
63	67.2000	69.2000	69.6404	112	119.4666	121.4666	121.9170
64	68.2666	70.2666	70.7073	113	120.5333	122.5333	122.9838
				114	121.6000	123.6000	124.0506

*For odd tooth gears—divide measurement over wires by 2 and make radial measurement.

increases with helix angle. For larger helix angles, greater precision of gear cutting is required where the backlash desired is small in the direction of rotation.

Manufacture by hobbing of gears with high helix angles is sometimes limited by the extent to which the hobs can be rotated or swiveled. Gears of high helix angle sometimes require special setups and equipment.

Face width is usually planned to give complete pitch-line contact overlap. For very small helix angles the face required to do this is large, as sin ψ becomes small in the denominator of the equation. The small faces afforded by large helix angles can not be used because of forces during cutting. But, face widths for gears of about 20-deg helix angle are of reasonable size for the various normal diametral pitches.

Ease of design can not be ignored.

pitch for full depth helical gears

Number of Teeth, Divide Values by Normal Diametral Pitch (Hob Cutter Pitch).

NUMBER OF TEETH IN GEAR, N	PITCH DIAMETER FOR $P_N=1$ D_1	OUTSIDE DIAMETER FOR $P_N=1$ D_{o1}	MEASUREMENT OVER 1.728/P_N WIRES FOR $P_N=1$ M_1*	NUMBER OF TEETH IN GEAR, N	PITCH DIAMETER FOR $P_N=1$ D_1	OUTSIDE DIAMETER FOR $P_N=1$ D_{o1}	MEASUREMENT OVER 1.728/P_N WIRES FOR $P_N=1$ M_1*
115	122.6666	124.6666	125.1174	162	172.8000	174.8000	175.2546
116	123.7333	125.7333	126.1842	163	173.8666	175.8666	176.3213
117	124.8000	126.8000	127.2510	164	174.9333	176.9333	177.3881
118	125.8666	127.8666	128.3177				
119	126.9333	128.9333	129.3845	165	176.0000	178.0000	178.4548
				166	177.0666	179.0666	179.5214
120	128.0000	130.0000	130.4513	167	178.1333	180.1333	180.5882
121	129.0666	131.0666	131.5180	168	179.2000	181.2000	181.6549
122	130.1333	132.1333	132.5848	169	180.2666	182.2666	182.7216
123	131.2000	133.2000	133.6516				
124	132.2666	134.2666	134.7182	170	181.3333	183.3333	183.7883
				171	182.4000	184.4000	184.8550
125	133.3333	135.3333	135.7850	172	183.4666	185.4666	185.9217
126	134.4000	136.4000	136.8519	173	184.5333	186.5333	186.9884
127	135.4666	137.4666	137.9186	174	185.6000	187.6000	188.0551
128	136.5333	138.5333	138.9854				
129	137.6000	139.6000	140.0522	175	186.6666	188.6666	189.1218
				176	187.7333	189.7333	190.1885
130	138.6666	140.6666	141.1188	177	188.8000	190.8000	191.2553
131	139.7333	141.7333	142.1856	178	189.8666	191.8666	192.3219
132	140.8000	142.8000	143.2524	179	190.9333	192.9333	193.3887
133	141.8666	143.8666	144.3191				
134	142.9333	144.9333	145.3859	180	192.0000	194.0000	194.4554
				181	193.0666	195.0666	195.5220
135	144.0000	146.0000	146.4527	182	194.1333	196.1333	196.5887
136	145.0666	147.0666	147.5193	183	195.2000	197.2000	197.6555
137	146.1333	148.1333	148.5861	184	196.2666	198.2666	198.7221
138	147.2000	149.2000	149.6529				
139	148.2666	150.2666	150.7196	185	197.3333	199.3333	199.7889
				186	198.4000	200.4000	200.8556
140	149.3333	151.3333	151.7864	187	199.4666	201.4666	201.9222
141	150.4000	152.4000	152.8532	188	200.5333	202.5333	202.9890
142	151.4666	153.4666	153.9198	189	201.6000	203.6000	204.0557
143	152.5333	154.5333	154.9866				
144	153.6000	155.6000	156.0534	190	202.6666	204.6666	205.1223
				191	203.7333	205.7333	206.1891
145	154.6666	156.6666	157.1201	192	204.8000	206.8000	207.2558
146	155.7333	157.7333	158.1869	193	205.8666	207.8666	208.3225
147	156.8000	158.8000	159.2536	194	206.9333	208.9333	209.3892
148	157.8666	159.8666	160.3203				
149	158.9333	160.9333	161.3870	195	208.0000	210.0000	210.4560
				196	209.0666	211.0666	211.5226
150	160.0000	162.0000	162.4538	197	210.1333	212.1333	212.5894
151	161.0666	163.0666	163.5205	198	211.2000	213.2000	213.6561
152	162.1333	164.1333	164.5872	199	212.2666	214.2666	214.7227
153	163.2000	165.2000	165.6540				
154	164.2666	166.2666	166.7206	200	213.3333	215.3333	215.7894
				300	320.0000	322.0000	322.4590
155	165.3333	167.3333	167.7874	400	426.6666	428.6666	429.1270
156	166.4000	168.4000	168.8542	500	533.3333	535.3333	535.7946
157	167.4666	169.4666	169.9208	600	640.0000	642.0000	642.4618
158	168.5333	170.5333	170.9876	700	746.6666	748.6666	749.1288
159	169.6000	171.6000	172.0543	800	853.3333	855.3333	855.7958
				900	960.0000	962.0000	962.4627
160	170.6666	172.6666	173.1210	1000	1066.6666	1068.6666	1069.1295
161	171.7333	173.7333	174.1878				

And the cosine of 20° 21' 50.887" is equal to 0.9375 = $1\tfrac{5}{16}$, a simple fraction. Thus, many design calculations can be done in longhand or with slide rule, eliminating some tedious calculations. For example, if a 20-deg angle is used the cosine value equals 0.93969262, much harder to manipu-late than $1\tfrac{5}{16}$, although the difference in helix angle is slight.

There are many other angles whose cosines are simple fractions such as $\tfrac{4}{5}$, $\tfrac{6}{7}$, $\tfrac{7}{8}$, etc. These range from helix angles of 36° 52' 11" to 16° 15' 36" and many have simple sine values. Many of these might be good choices.

Why standardize on an angle

Manufacturing and engineering can be simplied, as stocks of helical gears could be made and cataloged for the trade as is done with spur gears. In addition, expensive cam guides for gear shapers could be purchased with certainty of full use by gear manu-

EXAMPLE—DESIGN OF HELICAL GEARS — $\psi = 20°\ 21'\ 50.887''$
$P_N = 48,\quad PA = 20°,\quad N_P = 38,\quad N_G = 64$

STEP	FORMULA	PINION	GEAR
1	$P = (P_N)(15/16)$	45	45
2	PA = pressure angle (given)	20°	20°
3	P_N = normal diametral pitch (given)	48	48
4	N = number of teeth (given)	38	64
5	ψ = helix angle (given)	20° 21' 50.887"	20° 21' 50.887"
6	$D = D_1(\text{table})/P_N$	0.8444	1.4222
7	$a = 1/P_N$	0.02083	0.02083
8	Hand	Right	Left
9	Whole depth	Not required	Not required
10	$D_o = D_{01}(\text{table})/P_N$	0.8861	1.4638
11	$G = 1.728/P_N$	0.036	0.036
12	$M = M_1(\text{table})/P_N$	0.8949	1.4730
13	$F = 9.03/P_N = 0.188$	Use 0.250	Use 0.218
14	$C = (D_G + D_o)/2$	1.1333	1.1333

facturers. And, it is now possible to achieve enough precision to manufacture fully interchangeable helical gears rather than furnish them in pairs of special design. Mathematical tables such as those that follow can be used to simplify design and manufacturing calculations.

Tables reduce mathematics

Helical gears can be manufactured by hobbing with standard spur gear cutters as easily as spur gears, once the essential calculations have been made. For this reason, the following set of helical gears with a helix angle of 20° 21' 50.887" is suggested for use where smoothness, lack of vibration and quietness of drive are desired.

Tables have been prepared in terms of one normal pitch, with the effects on the dimensions caused by the helix angle already incorporated. To obtain the values to be tabulated for a gear, the one-normal-pitch dimensions are divided by the normal (or cutter) diametral pitch. Steps in using the table are:

1) Complete pitch diameters by either of these two methods:
   a) By formula:

$$D = N/P = N/P_N \cos\psi$$
$$= N/P_N(15/16)$$
$$= 16N/15P_N$$

   b) By table:
Choose value D_1 and divide by the cutter or normal diametral pitch, P_N.

2) Compute outside diameter of gears by either of these two methods:
   a) By formula

Add $2/P_N$ to pitch diameter

   b) By table:
Choose value D_{01} and divide by normal diametral pitch, for full-depth involute gears only.

3) Compute minimum face width, F. Very wide faces should be avoided.

$$F = \pi/P_N \sin\psi = 9.03/P_N$$

4) Compute measurement over wires. This has been done for one normal diametral pitch; hence divide table M_1 values by normal diametral pitch, P_N. Use radial measurements for odd-toothed gears.

5) Compute thrust. Any ball or tapered roller bearing of the same shaft bore that has provisions for axial preloading or thrust will usually be adequate to handle thrust of these helical gears. Where doubt exists, thrust load W_T equals $W \tan\psi$, which is 0.371 W. W is tangential pitch-line load. Bearings should be free of end play and kept close to gears for rigid support.

How to use tables

A design example is shown above. A summary of the formulas used for the example in order of gear tabulation for drawings would be:

1) Diametral pitch in plane of rotation, P

$$P = P_N \cos\psi$$

2) Normal pressure angle, ϕ_N
3) Normal diametral pitch, P_N
4) Number of teeth, N
5) Helix angle, ψ
6) Pitch diameter, D

$$D = D_1(\text{table})/P_N$$

7) Addendum, a

$$a = 1/P_N$$

8) Hand, one gear must be left, the other right
9) Whole depth, not required
10) Outside diameter, D_o

$$D_o = D_{01}(\text{table})/P_N$$

11) Wire diameter, G

$$G = 1.728/P_N$$

12) Measurement over wires, M

$$M = M_1(\text{table})/P_N$$

13) Minimum face width, F

$$F = 9.03/P_N$$

14) Center distance, C

$$C = (D_G + D_P)/2$$

All other values such as index gears, feed gears, machine constant, feed, material, etc, must be determined on the basis of gear application and machine type and cutter type to be used. This information must come from the gear consultant and is based on manufacturing equipment available for generating the gear teeth.

Engineer boils down. gear needs . . .   . . . types them into computer and . . .   . . . gets back complete specifications.

New computer memory disk spins out instant design

Bypassing the programming stage, it provides in minutes
all data needed for a gear train or other complex mechanism.
The engineer simply imposes his own design requirements

A new epoch of computerized mechanical design is opening up with the proving of a prepackaged IBM program for spur and helical gears. In this new era, the engineer can take a magnetic disk from the shelf and put it in the computer, then can type his special design requirements into the computer and get back, in minutes, all the data he and the shop need—in easily readable, abbreviated English.

Previous computerized gear programs were too cumbersome for the average engineer. They required translation into Fortran computer language, forcing the engineer to turn to a middleman (the programmer) to understand his needs and work out the mystic language to achieve the desired result. At best, the engineer faced a delay while the program was worked out; at worst, the program might miss the mark.

The prepared basic program for gear design short-cuts this procedure, eliminating the middleman. It is in the form of a set of punched cards, but these can be fed through the computer once to put the data on a magnetic disk.

More to come. The program was developed by IBM's Manufacturing Industry Development group at Endicott, N. Y., with David Frayne as project leader for the mechanical design work. According to Frayne, the program is available without charge to users of the IBM 1130, and similar programs are being developed for other mechanisms.

The 1130 computer may be purchased for about $65,000 or rented for $900 to $1100 a month. This may sound expensive for, say, solving one gear problem, but it could prove quite inexpensive if it enables (as it does) a staff of engineers to turn out instant designs of mechanisms by merely taking a mechanism-design record from the shelf. For example, packaged programs for statistical analysis are already available, and a spring-design program is on the way.

In field use so far, it seems to be the practice to wipe each disk clean, magnetically, after the program has been used, so the disk is ready for a new problem. This practice may be based on the cost of a disk— about $100—and the inconvenience of storing it. However, design engineers may prefer to have sets of instant-design disks on their bookshelves alongside their handbooks.

Program libraries. Computer-aided design is not new. In the field of mechanical design, companies were using the computer to design cams a dozen years ago, and the aerospace industry would be lost without computers for design aid. What is new, though, is the turn away from conventional program libraries, and the freeing of the new gear program from the special language of computers.

Parodoxically, the growth of conventional programs in recent years has tended to lash computerized design into a straitjacket. Companies were reluctant to throw away elaborate sets of program cards after a project was completed. They tended to try to adapt future projects to programs that were on file.

Typical gear design problem

The engineer types his requirements into the computer:

```
MECHANISM, 'PN, GN'
    GEAR, FINE;
    AGMA QUAL SIX;
    GEAR GEOMETRY, PRA 20., CX 1.875;
    GEAR SIZE TABLE ONE;
    GEAR SPEEDS, RMP 1500., RMG 300.;
    GEAR LOADS, TOR .75;
    DESIGN SPUR GEARS;
```

Now the computer does the rest:

INPUT DATA . . . SPUR GEAR PROGRAM

QUALITY CLASS . . . . =	6
PRESSURE ANGLE—DEGREES . . . . =	20.00000
APPROXIMATE CENTER DISTANCE—INCHES . . . . =	1.87500
INPUT REVOLUTION PER MINUTE . . . . =	1500.00000
OUTPUT REVOLUTION PER MINUTE . . . . =	300.00000
INPUT TORQUE—FT—RB . . . . =	0.75000
MAX. ALLOW DESIGN STRESS FOR PINION—PSI . . . . . =	30000.00000
MAX. ALLOW DESIGN STRESS FOR GEAR—PSI . . . . . . =	30000.00000
MAX. ALLOWABLE COMPRESSIVE STRESS—PSI . . . . . . =	100000.00000
MODULUS OF ELASTICITY—PINION—1000 PSI . . . . . =	29500.00000
MODULUS OF ELASTICITY—GEAR—1000 PSI . . . . =	29500.00000

COMPUTED DESIGN DATA—PINION

(Similar data for mating gear not shown)

MATERIAL TYPE IS NOT SPEC
AGMA QUALITY NUMBER IS 6
THE NORMAL RANGE FOR COARSE-PITCH GEARS IS (3-15)
THE NORMAL RANGE FOR FINE-PITCH GEARS IS (5-16)
THE HIGHER THE NUMBER THE MORE PRECISE THE GEARING

NUMBER OF TEETH . . . . =	30
DIAMETRAL PITCH . . . . =	48.00000
PRESSURE ANGLE—DEGREES . . . . =	20.00000
STANDARD PITCH DIAMETERS—INCHES . . . . =	0.62500
TOOTH FORM=AM. STD. FULL DEPTH INVOLUTE	
MAX. CIR. THICK. ON STD. PITCH CIRCLE—INCHES . . . =	0.03122
TESTING RADIUS—INCHES . . . . =	0.31050
TOLERANCE ON TESTING RADIUS—INCHES . . . . =+.000,−0.00400
MAX. COMPOSITE RADIAL VARIATION—INCHES . . . . . . =	0.00400
MAX. TOOTH TO TOOTH VARIATION—INCHES . . . . . . =	0.00150
OUTSIDE DIAMETER OF GEAR—INCHES . . . . =	0.66266
TOLERANCE ON OUTSIDE DIAMETER—INCHES . . . . . . =+.000,−0.00416

ADDITIONAL CALCULATED DATA

CLEARANCE—INCHES . . . . =	0.00616
FACE WIDTH OF TEETH INCHES . . . . =	0.18750
CIRCULAR PITCH . . . . =	0.06544
STANDARD CENTER DISTANCE—INCHES . . . . =	1.87500
TOLERANCE ON CENTER DISTANCE—INCHES . . . . =+.000,−0.00200
MAXIMUM BACKLASH . . . . =	0.01020
MINIMUM BACKLASH . . . . =	0.00150
OPERATING PRESSURE ANGLE—DEGREES . . . . =	20.00000
OPERATING CENTER DISTANCE—INCHES . . . . =	1.87500
NUMBER OF TEETH IN CONTACT . . . . =	1.61562
GEAR RATIO . . . . =	5.00000
CALCULATED REVOLUTION PER MINUTE—OUTPUT . . . =	300.00000
PITCH LINE VELOCITY—FPM . . . . =	245.43685
TORQUE TRANSMITTED BY PINION—FT-LB . . . . =	0.75000
TORQUE TRANSMITTED BY GEAR—FT-LB . . . . =	3.75000
EFFICIENCY OF GEAR DRIVE . . . . =	0.99873
WORKING COMPRESSIVE STRESS—PSI . . . . =	92196.89854
ACTUAL WORKING STRESS IN PINION TEETH—PSI . . . . =	24165.47270
ACTUAL WORKING STRESS IN GEAR TEETH—PSI . . . . . =	18750.08691
TRANSMITTED LOAD—LBS . . . . =	5.76000
CORRESPONDING DYNAMIC LOAD—LBS . . . . =	5.76000
MAX. ALLOWABLE DYNAMIC LOAD IN PINION—LBS . . . =	35.75349
MAX. ALLOWABLE DYNAMIC LOAD IN GEAR—LBS . . . . =	49.59693
MAX. ALLOWABLE LOAD FOR WEAR—LBS . . . . =	32.34904
OVERALL DERATING FACTOR . . . . =	1.00000

END OF COMPUTATION.

To minimize duplication of effort, companies a few years ago overcame their reluctance to provide data that might help their competitors and began to contribute their programs to IBM's program libraries. In such libraries, the design engineer can look over available programs in search of one that will meet his problem requirements.

However, there are drawbacks to using a borrowed program. In some instances, limitations of application were not readily evident. Perhaps a given design may not be valid for speeds over 10,000 rpm, but the program doesn't say so. When the borrower runs into trouble, it may take a long exchange of correspondence before he learns of this limitation. Moreover, the original purpose and structure of a program may have been half-forgotten by those concerned by the time a borrower tries to use the program.

New approach. As IBM sees it, the time is ripe for someone to work out packaged programs that are basic to design of a variety of mechanical components as well as to electronics, graphics, statistics, and optics. IBM is also seeing that the programs are doublechecked for any "bugs." The gear program has been verified at Farmingdale Agricultural & Technical College, Farmingdale, N. Y., which has a complete battery of up-to-date gear-cutting equipment.

In solving a gear problem, the engineer starts with IBM's set of pre-punched program cards (flow chart, opposite) or with the magnetic memory disk made by feeding the cards through a card reader. Or he could, of course, start directly with a prerecorded disk. Either way, he puts the packaged program in the computer, preparing the machine for his own special input data.

The input data, such as the distance between input and output shafts in the gear set, the desired shaft speeds, and the input torque or horsepower, can also be fed into the computer on punched cards or can be typed directly on the keyboard of the computer. This kind of typing doesn't demand the services of a specialist. The engineer can do it himself, using gear terms with which he is already familiar and

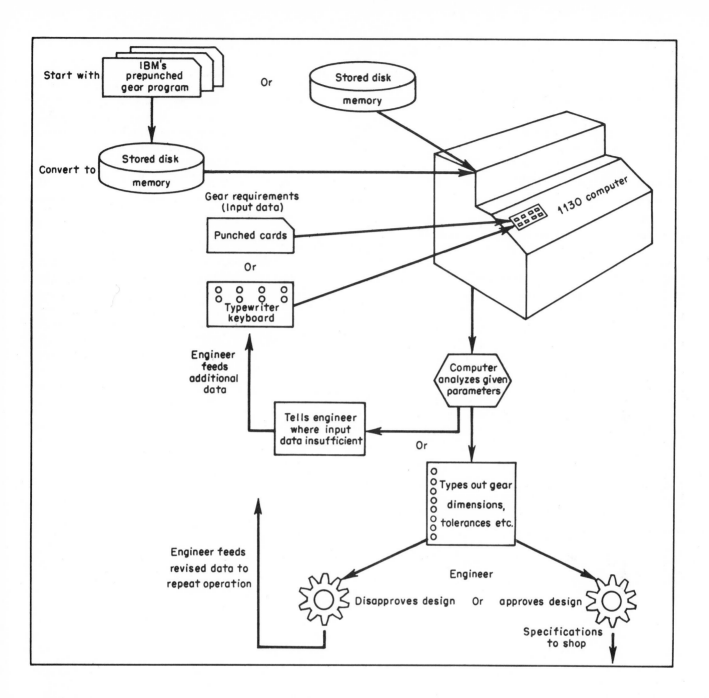

feeding them into the computer in any desired order.

The language is English, abbreviated to the first three letters of the longer words. For example, AGMA QUALITY SIX is typed AGMA QUA SIX (see specimen at left).

The computer analyzes the input data in the light of the programmed memory. If gear requirements can be met, the machine in seconds starts printing out the gear specs on the same keyboard and in the same simple language that was used for the input data. If requirements can't be satisfied, the computer similarly tells the engineer why the design failed or what additional information is needed.

What's in the program. The prepared program in the computer is essentially a minimum-weight design, somewhat like that developed by Darle Dudley of Mechanical Technology, Inc. (see the article, p 2). It considers both the gear structure (the tooth strength) and gear life (the tooth durability). It includes all the AGMA strength de-rating factors and tables, based on AGMA standards 220.02 for spur gears and 221.02 for helical gears.

Input data. The language for typing the input data into the computer is simple (note example at left):

A. MECHANISM, 'PN, GN',

These symbols, assigned by the user, represent "sample pinion" and "sample gear," respectively.

B. GEAR, FINE;

This entry specifies material's physical constants and whether fine- or coarse-pitch gears are to be used. Allowable stresses and physical constants of material are specified if they differ from standard values.

C. AGMA QUAL SIX;

This abbreviation represents AGMA's manufacturing tolerances.

D. GEAR GEOMETRY, PRA 20, CX 1.875;

On this line, pressure angle and approximate center distance are specified. The geometry entry can also include face width and number of teeth in pinion and gear. For helical gears, the helix angle and the angle between axes can be given.

E. GEAR SIZE TABLE ONE;

The program also contains an empirical relationship between standard diametral pitches and corresponding face widths, based on what an average gear-cutting shop can produce. In other words, selecting the number of teeth automatically determines the face width. The designer has the option of using his own table if he prefers. The table used is:

DP	120	96	80	64
FW	.0625	.0938	.1094	.125

DP	48	40	32	24	20
FW	.1875	.2188	.25	.3125	.375

F. GEAR SPEEDS, RMP 1500, RMG 300;

This entry lists the rpm's of the two gears.

G. GEAR LOADS, TOR .75;

This is the input torque. Optionally, the input horsepower may be given.

H. DESIGN SPUR GEARS

This instruction triggers the spur-gear design program.

Output data. The computed data comes from the keyboard machine in three main sections: (1) the typed input data, augmented by data based on the computer's selection of steel as the material for the gear set; (2) the computed geometry data for the pinion and gear, and (3) the additional calculated data necessary to evaluate the design.

With this printout in hand, the engineer has the option of accepting this design, modifying certain parameters to generate a new design, or re-analyze the design against changed values for speed or allowable stress.

He can also feed into this analysis one or more factors to allow for such considerations as dynamic load, enlarged-tooth systems, efficiency of planetary gear trains, etc.

IBM hopes to broaden the gear program to let the computer choose the gear type, not only between spur and helical gears but also among all the right-angle gear systems.

There is, in fact, no reason why the computer cannot some day be linked to direct control of a gear cutter.

2
RIGHT-ANGLE GEAR SYSTEMS

WHICH RIGHT-ANGLE GEAR SYSTEM?

Your guide to the 16 types that are now available. The authors discuss basic differences, then compare on the basis of power, efficiency, speed, ratio range and mounting sensitivity—to help you choose the best gear system for your application.

RICHARD C BRYANT
DARLE W DUDLEY,

ABOUT THE AUTHORS

Darle Dudley, well known for his book *Practical Gear Design*, published by McGraw-Hill, is editor-in-chief of the forthcoming *McGraw-Hill Gear Handbook* which brings together contributions from numerous sources in the gear industry. He is chairman of the AGMA Aerospace Gearing Committee, and recently received the Edward P Connell award "in recognition for his many and valuable contributions to the work of AGMA, of his service to the gear industry, and as an authoritative author of gearing literature."

Dick Bryant, a cum laude graduate from Tufts University, has done design appraisals on projects such as VTOL, STOL, helicopter, rocket and special amphibious-vehicle transmission systems. Several of these projects involve the selection of right-angle gearing. Before joining GE, he worked with General Motor's research staff on lubrication and wear problems.

There are 16 separate and distinct types of right-angle gear drives, each with its own advantages and limitations. This means that choice of the best type for a given job involves many considerations. What is optimum for one application most likely will not be optimum for another. For example, hypoid gearing is the almost universal choice for automotive rear-end drives because offset of the pinion centerline allows a low floor in the car. Also these gears are quiet, efficient, and can carry large amounts of power. But hypoid would not necessarily be the wisest choice for a hand-operated drill where it would be difficult to control the locating dimensions of the gears, and where capacity and efficiency are not so important.

This article compares the 16 systems on the basis of power, efficiency, speed, ratio range, and mounting

1 STRAIGHT BEVEL

2 CONIFLEX BEVEL

3 SPIRAL BEVEL

4 ZEROL BEVEL

sensitivity, and gives charts and tables on efficiency, horsepower ratings, manufacturing methods.

A LOOK AT THE RIGHT-ANGLE SYSTEMS

Two general classifications cover the right-angle gears:

Coplanar types, which have intersecting axes. In this category are the bevel-gear systems—straight bevel, Coniflex bevel, Zerol bevel, spiral-bevel and Revacycle bevel. Two other systems with intersecting axes are face and Beveloid gears (names capitalized indicate proprietary gear systems).

Offset types, which have nonintersecting or skew axes (their axes do not lie in a common plane). In this category are the crossed-helical, cylindrical-worm, Cone-Drive, hypoid, Spiroid, Planoid and Helicon. With proper design, face and Beveloid gears can also be employed as offset gears.

Each gear system is distinctive in the kind of manufacturing equipment required to make it and selection will depend on efficiency characteristics, mounting requirements, load-carrying capacity and availability of

manufacturing facilities. The tooth form employed is not important in itself from the standpoint of mathematical theory. Here's a brief description of each right-angle gear system:

Straight-bevel gears

The blanks are conical for both members. Tooth form approximates an involute in the normal section. The teeth taper. The cutting tool represents a crown rack having straight teeth.

These gears are the simplest and most commonly used form of bevel gearing. The machines to generate them, and more recently to grind them, were developed by the Gleason Works, Rochester, NY. More recent Gleason introductions are the Coniflex and the Revacycle bevel gears. Both are straight-bevel type with localized tooth contact produced by slightly relieving the ends of the teeth. They have the advantage of better ability to withstand small gear-mounting errors, or deflections under load, without unduly overloading the ends of the teeth (which may result in breakage). All straight-bevel gears produced on recent Gleason machines are Coniflex. Revacycle gears are made on

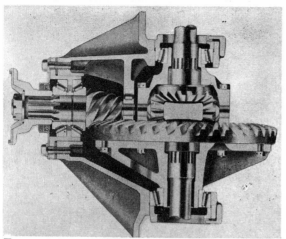

5 REVACYCLE BEVEL

6 FACE GEARS

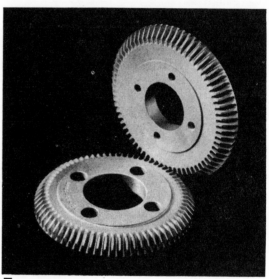

7 BEVELOID

special machines and are mostly for large-quantity applications, such as for automotive differentials.

Spiral bevels, Zerol bevels, hypoids

These gears are cut on conical blanks. Teeth of both members are tapered. The cutting tool represents a crown rack having a curved tooth. Spiral-bevel teeth have a slope like helical gears. Zerol teeth are curved with a zero slope at midpoint of the face width. Hypoid gears have the axes offset—otherwise they are similar to spiral-bevel gears.

Generally, the gear member of the pair is made first and the pinion member is then cut to match. The tooth form in the normal section approaches that of an involute.

Spiral-bevel gears have smoother, quieter action than straight bevels because the teeth have a lengthwise curvature and are cut at an angle to the gear axis.

Zerol-bevel gears avoid the overlapping tooth action of spiral-bevel gears because of the zero spiral angle. This also eliminates some of the thrust reaction on the mounting bearing. Bearing reactions do not depend on direction of rotation and the gears can be mounted interchangeably in the same mounting structure used to support straight-bevel gears.

Hypoid gears grew out of the automobile industry's need for a silent rear-end drive which would allow the driveshaft to be below the centerline of the rear axle. Special high-ratio hypoid gears are available for ratios above 10:1. These offer the advantage that both members can be ground, providing maximum accuracy.

Face gears

Here the pinion is a normal involute-toothed part. The gear has teeth cut on the end face of the blank by a cutting tool that is an exact duplicate of the pinion.

Face gears can be designed to mesh either "on center" (axis of the pinion lies in the same plane as the axis of the gear) or "off center" (pinion axis is offset, much as a hypoid pinion is offset). Because the pinion of the face-gear set is simply a spur gear, it can be produced by any of the standard spur-gear methods such as hobbing, shaping, or grinding. Also, the axial location of the pinion is not critical, which simplifies mounting. Face gears were developed by the Fellows Gear Shaper Co, Springfield, Vt.

Beveloid gears

Beveloid gears are basically involute gears with tapered tooth thickness, root and outside diameters. The appearance is much like a straight-bevel gear, but every transverse section represents a spur gear. This means that a Beveloid gear will mesh properly with any other involute gear derived from the same basic rack.

In right-angle gear systems, the Beveloid gear offers four major advantages:

1—Mounting dimensions are not critical.

2—The two gears do not require a common vertex.

3—Beveloid gears are individually interchangeable, without lapping or matching.

4—Backlash can be controlled or eliminated if desired.

8 CROSSED HELICAL

9 WORM

10 CONE DRIVE

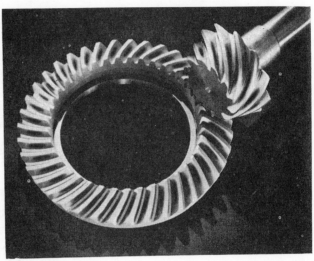

11 HYPOID

12 OFFSET FACE GEAR

Crossed-helical gears

Both members are involute helicoids made on cylindrical blanks. The teeth do not taper. One main advantage is mounting ease. Crossed-helical gears are not sensitive to axial movement of either member or to small changes in shaft angle or center distance. Low cost is another advantage—either member can be made by any of the methods for ordinary helical gears.

The teeth mesh is a point of contact, which limits the load that can be transmitted. However, this changes to line contact as the gears wear in.

Cylindrical-worm gears

This system employs a single-enveloping worm that resembles a threaded screw; the thread is either straight-sided or slightly curved by action of the straight-sided cutter that makes the worm (a worm employing an involute tooth form is made by De Laval Holroyd Co). The gear is hobbed on a "throated" cylindrical blank, with a hob that is an exact duplicate of the worm. The gear teeth are curved and each end of the tooth is thicker than the middle.

Low power capacity of a pair of Beveloid gears is the major limitation. This is because point contact is produced on the meshing surfaces. Another limitation is undercutting when small numbers of teeth or large cone angles are used.

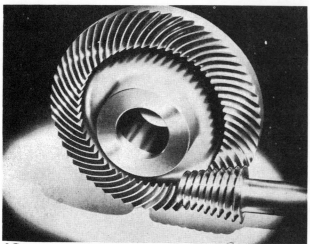

13 SPIROID

14 PLANOID

15 HELICON

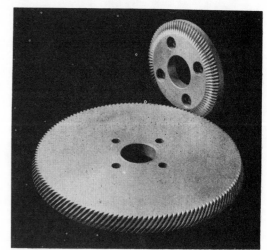

16 OFFSET BEVELOID

The cylindrical worm has greater load-carrying capacity than the crossed-helical type because line contact is produced instead of point contact. However, mounting dimensions are more critical because the throated member must be positioned to wrap properly around the pinion.

Cone-drive or double-enveloping worm gears

The worm has an hour-glass shape and is cut by a straight-sided cutter representing the central plane of the gear. The gear, in turn, is hobbed by a hob that is an exact duplicate of the worm.

Because both worm and worm wheel are throated to wrap around each other, more tooth surface is in contact at any instant. This increases the load capacity but also increases the rate at which heat is generated. All mounting dimensions of a double-enveloping worm drive must be accurately held because the axial locations of both members are critical along with center distances and shaft angles.

Spiroid gears

The pinion of this system has a screw thread on a tapered blank. The thread is essentially a straight-sided "V". The gear is cut by a tapered hob that is an exact duplicate of the pinion. Gear teeth are cut on a sloping end-face of the blank.

Spiroid gears are relatively new in the trade, but have found wide usage in high- and medium-ratio applications, and where backlash control is important.

Planoid gears

The Planoid gear has straight-sided teeth cut on a sloping end-face of the gear blank. The teeth taper. The pinion is cut by a cutter that duplicates the gear. Planoid gears are employed for low-ratio applications up to 8:1. They are relatively new in the field.

Helicon gears

The pinion has a screw thread on a cylindrical blank. The gear is cut by a hob that duplicates the pinion.

The gear teeth are on a slightly sloping endface of the blank. The Helicon and Spiroid are similar except for the taper in the pinion. Also Helicon gears can be manufactured for lower ratios.

COMPARISON—OPERATING CHARACTERISTICS

Table I, next page, compares gear systems with respect to power-to-weight ratio, efficiency range, ratio range, maximum horsepower, maximum pitch velocity,

Table 1—Characteristics of the 16 Right-angle Gear Systems

Class	Type	Power-to-Weight Ratio	Efficiency Range	Ratio Range	Maximum Hp (per mesh)	Maximum Pitch Line Velocity, ft/min.		Type of Contact	Mounting Sensitivity
						Commercial	Aircraft		
Coplanar	1—Straight bevel 2—Coniflex¹ bevel	Medium High	97–99.5	1:1 to 8:1	500	1000	10,000	Line	Critical in 3 directions
	3—Spiral bevel	Very High	97–99.5	1:1 to 8:1	5000	8000	25,000	Line	Critical in 3 directions
	4—Zerol¹ bevel	High	97–99.5	1:1 to 8:1	1000	1000	10,000	Line	Critical in 3 directions
	5—Revacycle¹ bevel	High	97–99.5	1:1 to 8:1	100	1000	—	Line	Critical in 3 directions
	6—Face gear	Medium High	97–99.5	3:1 to 8:1	200	4000	5000	Line	Critical in 1 direction
	7—Beveloid²	Low	97–99	1:1 to 8:1	150	4000	5000	Point	Not critical
Offset	8—Crossed helical	Low	50–95	1:1 to 100:1	100	4000	10,000	Point	Not critical
	9—Cylindrical worm	Medium High	50–98	3:1 to 100:1	750	5000	10,000	Line	Sensitive in 2 directions
	10—Cone-Drive³	Medium High	50–98	3:1 to 100:1	1000	4000	10,000	Line or area	Critical in 3 directions
	11—Hypoid, Low ratio	High	85–98	1:1 to 10:1	1000	6000	10,000	Line	Critical in 3 directions
	High ratio	Medium High	50–90	10:1 to 50:1	500	Same	Same	Same	Same
	12—Face Gear	Medium High	70–98	3:1 to 8:1	200	4000	5000	Line	Critical in 2 directions
	13—Spiroid⁴	Medium High	50–98	9:1 to 100:1	500	6000	10,000	Line	Critical in 2 directions
	14—Planoid⁴	High	70–98	1.5:1 to 8:1	1000	4000	10,000	Line	Critical in 3 directions
	15—Helicon⁴	Medium High	50–98	3:1 to 100:1	100	6000	10,000	Line	Critical in 2 directions
	16—Beveloid²	Low	70–98	1:1 to 8:1	150	4000	5000	Point	Not Critical

Registered Tradenames:
¹*Zerol, Coniflex* and *Revacycle*—Gleason Works, Rochester, NY; ²*Beveloid*—Vinco Corp, Detroit; ³*Cone-Drive*—Cone-Drive Div, Michigan Tool Co, Detroit; ⁴*Spiroid, Helicon* and *Planoid*—Spiroid Div, Illinois Tool Works, Chicago.

type of contact and mounting sensitivity. All are nominal values which can be exceeded at times by special design. These values, as in later tables, have been calculated by the use of AGMA standards where available, gear-manufacturers' data, and test data and engineering analyses compiled at the General Electric Co. Thus the values are guideposts representing the judgment of the authors rather than fixed values in the gear trade.

For example, a few worm gears for aircraft have been built with pitch-line speed above 20,000 ft/min. But these jobs have been critical on accuracy, surface finish and lubrication. Hence, 10,000 ft/min. is a good nominal upper limit.

Sensitivity in mounting

When gears are mounted they may have to be set precisely in one or more of the following directions:

- In direction of pinion axis
- In direction of gear axis
- In direction perpendicular to both axes

In addition it may be critical to have the axes exactly at 90° to each other.

When a gear is "critical" in a mounting direction the part generally has to be within 0.001 in. or less of true location. When a gear is not critical, errors in mounting of the order of 0.010 in. or more have usually only a negligible effect on gear tooth action. If errors in mounting between 0.001 and 0.010 can be tolerated, the gear may be called "sensitive."

The crossed-helical gear is often picked for low-horsepower jobs because it is quite insensitive to mounting errors. Beveloid gears are another type with low sensitivity. The face gear is critical only in one mounting direction and this is often an important consideration in choosing it over another type.

Power characteristics

The power a gear set can transmit depends on its ratio, speed, size, hardness of material, and quality. Even when the designer wants the most power and will consider any size, ratio or kind of material there is a fairly well-defined upper limit of horsepower for each gear type. The upper limit is set by limitations in available manufacturing equipment, pitch-line velocity and lubrication problems.

Table I can be used as a guide on power capabilities. If a gear system must handle 2000 hp, the table shows that a crossed-helical gear would be out of the question—and that a spiral bevel could undoubtedly be designed to do the job. This much horsepower is also too high for a worm gear or Cone-Drive. Spiral bevels are being considered on some special jobs up to 20,000 horsepower, although the table shows a nominal upper limit of 5000 hp for spiral bevels. It is also possible to build worm gears to handle substantially more horsepower than shown in the table.

Nominal power capabilities of right-angle gears, based on specific ranges of ratios, sizes and speeds, are given in **Tables 2 to 8**. The range shown is within the capability of the gear type, but generally the gears can

Table 2—Power Capacity of Straight- or Zerol-bevel Gears

(For case-carburized pinion and gear, 55 RC min, pitch-line speed not over 1000 fpm; 20° pressure angle, full-depth tooth; theoretical face width not to exceed 0.28 cone distance)

Ratio, m_G	Pinion pitch dia, in.	Gear pitch dia, in.	Number pinion teeth	Face width, in.	Rpm of pinion			
					100	720	1750	3600
					Hp at different pinion speeds			
1	1.0	1.0	22	3/16	0.094	0.611	1.34	2.38
2	1.0	2.0	20	5/16	0.156	1.01	2.22	3.95
4	1.0	4.0	17	9/16	0.277	1.80	3.96	7.04
8	1.0	8.0	14	11/16	0.334	2.18	4.78	8.49
1	2.0	2.0	23	3/8	0.523	3.14	6.73	....
2	2.0	4.0	20	5/8	0.864	5.18	11.1	....
4	2.0	8.0	17	1 1/8	1.54	9.24	19.8	....
8	2.0	16.0	14	1 3/8	1.86	11.1	23.9	....
1	4.0	4.0	24	3/4	2.93	16.1	....	....
2	4.0	8.0	21	1 1/4	4.85	26.7	....	....
4	4.0	16.0	18	2 3/16	8.41	46.4	....	....
8	4.0	32.0	15	2 5/8	10.0	55.1	....	....
1	8.0	8.0	30	1 9/16	16.7	....	....	....
2	8.0	16.0	26	2 1/2	26.5	....	....	....
4	8.0	32.0	23	3 7/16	36.2	....	....	....

Table 3—Power Capacity of Spiral-bevel Gears

(For case-carburized pinion and gear, 55 RC min; pitch-line speed not over 4000 fpm; 20° pressure angle, full-depth tooth; 35° spiral angle; theoretical face-width not to exceed 0.30 cone distance; face-width rounded to nearest lower 1/32 in.)

Ratio, m_G	Pinion pitch dia, in.	Gear pitch dia, in.	Number pinion teeth	Face width, in.	Rpm of pinion				
					100	720	1750	3600	10,000
					Hp at different pinion speeds				
1	1.0	1.0	22	3/16	0.139	0.906	1.99	3.53	7.73
2	1.0	2.0	20	5/16	0.239	1.55	3.41	6.06	13.3
4	1.0	4.0	17	9/16	0.501	3.26	7.15	12.7	27.8
8	1.0	8.0	14	11/16	0.598	3.89	8.54	15.2	30.4
1	2.0	2.0	23	13/32	0.725	4.35	9.32	15.7	....
2	2.0	4.0	20	21/32	1.21	7.28	15.6	26.4	....
4	2.0	8.0	17	1 5/32	2.50	15.0	32.2	54.3	....
8	2.0	16.0	14	1 13/32	13.3	18.8	40.2	67.9	....
1	4.0	4.0	24	27/32	4.11	22.6	41.9	73.8	....
2	4.0	8.0	21	1 5/16	7.23	39.8	73.8	130.	....
4	4.0	16.0	18	2 7/32	13.4	73.8	137.	240.	....
8	4.0	32.0	15	2 21/32	16.0	88.1	163.	287.	....
1	8.0	8.0	30	1 11/16	24.4	120.	225.	....	....
2	8.0	16.0	26	2 21/32	42.5	209.	393.	....	....
4	8.0	32.0	23	3 15/32	56.8	280.	525.	....	....

be used over broader ranges of ratio, size and load. The tables give the general trend of rating practice for a gear type rather than a complete listing.

Values are for normal industrial practice with gears made to a good level of accuracy. If the same-size gears are made of the highest quality possible for aircraft practice, it is possible in many cases to go to horsepower ratings two or three times greater. This is particularly true at the higher speeds. Industrial gears have their ratings reduced at higher speeds because of dynamic tooth-loading effects. Aircraft gears are extremely accurate, and very lightweight for their size. They can carry almost the same tooth stresses at high speed as at low. Aircraft gears have very little derating for speed while industrial gears usually are derated substantially.

The gears were calculated for AGMA Class 1 service, which varies somewhat for each type of gear. Generally speaking, Class 1 is for about 10 hours' continuous service per day without recurrent shock loading ($K_r = 1$). Starting torque is generally held to less than 300% of rated torque at full speed and full rated horsepower. A steam turbine driving a generator would normally be a good example of a shock-free drive. A piston engine geared to a rock crusher would represent a drive with a serious shock problem—it would probably be necessary to rate a gear drive at less than half of its Class 1 rating.

Tables 2 and 3 are for case-hardened bevel gears. Bevel gears of good alloy steel at a hardness level of 33 to 38 Rockwell C (300 to 350 BHN) have ratings only 50% of those shown.

Table 4 gives ratings for face-gear sets of medium hardness. The pinion of a face-gear set has about the same capacity per inch of face as the pinion of a spur-gear set. Face gears can be case-hardened but there is no equipment to grind them. In small sizes it is possible to hold the gear distortion to a low value.

If the face-gear set was case-carburized and had good accuracy, capacity would be 200% of that shown.

Table 5 gives the nominal capacity of crossed-helical gears. Because these have a point contact they are rather sensitive to the method of breaking-in. The ratings assume that load is applied gradually and the units are polished-in without any metal-pickup occurring. The gears broaden their contact area as they wear and thereby reduce the stress in the contact area. Table 5 is based on a case-hardened steel driver and a phosphor bronze-driven gear. No particular gain in capacity can be obtained by case-hardening both members. At slow speeds a gain in capacity (100% or more) can be had by making both members out of a good cast iron. Field experience with cast iron crossed helicals has not been very good. There is a danger of sand inclusions in cast iron and this can lead to an abrupt, premature failure.

Tables 6 and 7 show the capacity of worm gears—Table 6 is based on a hardened-steel worm and centrifugally cast phosphor bronze wheel, and Table 7 on a hardened steel worm and chill-cast phosphor bronze wheel. Cast iron is sometimes used at slow speed.

Table 4—Straight-face Gears
(For pinion of 300 BHN min and gear 255 BHN min)

Ratio m_g	Pinion pitch dia, in.	Gear pitch dia, in.	Number pinion teeth	Face width, in.	Rpm of pinion			
					100	720	1750	3600
					Hp at different pinion speeds			
1.5	1.000	1.500	20	.085	0.018	0.116	0.251	0.440
	1.400	2.100	22	1/8	0.040	0.249	0.523	0.875
	2.000	3.000	24	11/64	0.105	0.637	1.29	2.00
	3.000	4.500	30	1/4	0.365	2.10	3.98	5.38
	4.000	6.000	36	23/64	0.899	4.96	8.79	9.92
	6.000	9.000	36	35/64	2.66	13.5	20.6	.....
2.0	1.000	2.000	20	9/64	0.024	0.154	0.334	0.586
	1.400	2.800	22	13/64	0.0478	0.488	1.03	1.72
	2.000	4.000	24	19/64	0.194	1.18	2.38	3.70
	3.000	6.000	30	7/16	0.599	3.45	6.54	8.84
	4.000	8.000	36	37/64	1.38	7.62	13.5	15.2
	6.000	12.000	36	55/64	4.24	21.5	32.8	.....
4.0	1.000	4.000	20	3/8	0.073	0.466	1.01	1.77
	1.400	5.600	22	33/64	0.174	1.09	2.30	3.84
	2.000	8.000	24	23/32	0.491	2.97	6.01	9.34
	3.000	12.000	30	1 1/32	1.59	9.17	17.4	23.5

Table 6—Cylindrical Worm Gearing
(Sliding velocity not over 6000 fpm)

Ratio, m_g	Center distance, in.	Worm pitch dia, in.	Effective face width, in.	Rpm of worm				
				100	720	1750	3600	10,000
				Hp at different worm speeds				
5	2	0.825	1 15/32	0.30	1.6	2.8	3.5	5.0
8	2		1/2	0.20	0.9	2.1	2.6	3.9
15	2		15/32	0.15	0.7	1.4	1.9	3.4
25	2		15/32	0.10	0.5	1.0	1.4	2.9
50	2		15/32	0.03	0.3	0.5	0.7	1.0
5	4	1.525	7/8	2.1	9.7	14.5	18.5	23.0
8	4		15/16	1.8	6.8	10.6	14.2	19.0
15	4		15/16	0.9	4.6	7.0	8.8	11.3
25	4		15/16	0.6	3.0	4.8	6.7	8.6
50	4		15/16	0.4	1.6	2.6	4.2	5.2
5	8	2.800	1 11/16	12.0	44.5	66.6	86.0	....
8	8		1 13/16	8.0	32.5	48.0	65.0	....
15	8		1 13/16	5.5	23.0	33.5	47.0	....
25	8		1 13/16	4.0	15.2	22.0	34.0	....
50	8		1 13/16	2.1	8.0	11.3	20.0	....
5	16	5.100	3 1/8	65	190	265	350	....
8	16		3 3/8	50	140	210	265	....
15	16		3 3/8	36	100	145	185	....
25	16		3 3/8	23	66	100	130	....
50	16		3 3/8	12.5	30	50	75	....

Table 8 shows the nominal capacity of Spiroid gears, based on both members being case-hardened. The pinion can be ground, if need be, to achieve accuracy. The gear may be lapped after hardening. Spiroid sets can be made with a case-hardened pinion and a phosphor bronze gear. Ratings of these sets would be somewhat similar to cylindrical worm gears of the same size.

A hypoid gear has about the same rating as a spiral bevel gear of the same size. The hypoid pinion to go with a given hypoid gear is about 10% larger than a corresponding spiral bevel pinion. With this in mind

Table 5—Crossed-helical Gears

(For driver of 60 RC, and driven phos. bronze; pitch-line speed not over 4000 fpm; units given a short "break-in" run to polish contact area)

Ratio, m_G	Center distance in.	Helix angle of pinion, deg.	Pinion pitch dia. in.	Gear pitch dia. in.	Rpm of pinion 100	720	1750	3600	10,000
					Hp at different pinion speeds				
1	1.000	45	1.000	1.000	0.003	0.015	0.030	0.051	0.085
3	1.366	60	1.000	1.732	.006	.039	.082	.137	.228
5	1.9435	60	1.000	2.887	.010	.066	.139	.231	.385
8	1.572	75	1.000	2.144	0.19	.123	.260	.433	.720
1	2.000	45	2.000	2.000	.018	.117	2.45	.408	.....
3	2.732	60	2.000	3.464	.047	.312	.658	1.10	
5	3.887	60	2.000	5.774	.081	.528	1.11	1.85	
8	3.1435	75	2.000	4.287	.152	.988	2.08	3.46	
1	4.000	45	4.000	4.000	.143	.932	1.96	3.26	
3	5.464	60	4.000	6.928	.382	2.50	5.27	8.75	
5	7.7735	60	4.000	11.547	.646	4.22	8.90	14.8	
8	6.287	75	4.000	8.574	1.21	7.92	16.7	.....	
1	8.000	45	8.000	8.000	1.14	7.46	15.7	.....	
3	10.928	60	8.000	13.856	3.05	20.0	42.2	.....	
5	15.547	60	8.000	23.094	5.16	33.8	71.2	.....	

Table 7—Cone-Drive Worm Gearing

(Sliding velocity not over 6000 fpm)

Ratio, m_G	Center distance, in.	Worm pitch dia. in.	Rpm of worm 100	720	1750	3600	10,000
			Hp at different worm speeds				
5	2	0.830	0.35	1.72	2.96	4.20	6.20
15	2	0.830	0.15	0.77	1.40	1.86	2.65
50	2	0.850	0.05	0.25	0.45	0.66	0.98
5	4	1.730	3.0	12.9	19.8	25.6	34.5
15	4	1.550	1.27	5.95	9.64	13.0	17.8
50	4	1.660	0.39	1.82	3.07	4.22	5.60
5	8	3.450	24.4	84.4	119.2	148	....
15	8	2.940	10.4	40.9	61.2	74.5	....
50	8	2.900	3.28	12.8	19.5	24.5	....
5	16	5.200	172	480	620	.....	....
15	16	5.100	78	256	350	435	....
50	16	5.100	25	85	119	139	....
5	24	7.350	550	1420	1840	.....	....
15	24	7.350	252	715	934	.....	....
50	24	7.350	77.5	226	292	.....	....

Table 8—Spiroid Gears

(Pinion and gear case-hardened, 60 RC min; pitch-line speed not over 1700 fpm; based on tooth proportions recommended by Spiroid Div., Illinois Tool Works)

Ratio, m_G	Center distance, in.	Pinion outside dia. in.	Gear outside dia. in.	Rpm of pinion 100	720	1750	3600	10,000
				Hp at different pinion speeds				
10,250	0.500	0.437	1.500	0.0142	0.0697	0.129	0.1948	0.3354
14.667	0.500	0.421	1.500	0.0120	0.0589	0.109	0.1645	0.2834
25.500	0.500	0.423	1.500	0.0088	0.0432	0.080	0.1208	0.2080
47.000	0.500	0.427	1.500	0.0064	0.0313	0.058	0.0876	0.1508
10.250	1.000	0.853	3.000	0.0937	0.4601	0.852	1.287	....
14.667	1.000	0.821	3.000	0.0759	0.3726	0.690	1.042	....
25.500	1.000	0.827	3.000	0.0550	0.2700	0.500	0.7550	....
47.000	1.000	0.837	3.000	0.0377	0.1852	0.343	0.5197	....
71.000	1.000	0.761	3.000	0.0299	0.1469	0.272	0.4107	....
10.250	1.875	1.507	5.625	0.4961	2.435	4.51	6.810	
14.667	1.875	1.463	5.625	0.4125	2.025	3.75	5.663	
25.500	1.875	1.435	5.625	0.2915	1.431	2.65	4.002	
47.000	1.875	1.448	5.625	0.1804	0.8856	1.64	2.476	
71.000	1.875	1.308	5.625	0.1441	0.7074	1.31	1.978	
106.000	1.875	1.215	5.625	0.1188	0.5832	1.08	1.631	
10.200	3.250	2.395	9.750	2.101	10.31	19.1	.....	
14.667	3.250	2.465	9.750	1.760	8.64	16.0	.....	
25.500	3.250	2.319	9.750	1.21	5.94	11.0	.....	
47.000	3.250	2.344	9.750	0.7205	3.537	6.55		
71.000	3.250	2.090	9.750	0.5544	2.722	5.04		
106.000	3.250	1.933	9.750	0.4455	2.187	4.05		
10.167	5.125	3.342	15.375	7.007	34.40	63.7	.....	
14.250	5.125	3.342	15.375	5.918	29.05	53.8	.....	
25.333	5.125	3.092	15.375	3.949	19.39	35.9	.....	
47.500	5.125	2.910	15.375	2.255	11.07	20.5	.....	
71.000	5.125	3.097	15.375	1.727	8.478	15.7	.....	
106.000	5.125	2.841	15.375	1.386	6.804	12.6	.....	

Table 3 can be used to estimate hypoid capacity.

Planoid sets are somewhat new on the market. Early field experience indicates that they have about the same capacity as hypoid gears of the same size.

Helicon gears span the ratio ranges of both Planoids and Spiroids. If made of the same material, Helicons have about 50% of the capacity of Planoids and about 80% capacity of Spiroids (for same size and ratio).

Beveloid gears can fit many applications—hence they can vary all the way from a close approach to the line-contact conditions of a bevel set to the point contact of crossed helicals. They are generally made with both members case-hardened and ground. Their capacity ranges from about 20% to 70% of that of equivalent-size straight-bevel gears.

Efficiency ranges

Efficiency of a gear is a variable. Factors such as surface finish, kind of lubricant, temperature of lubricant, and accuracy of teeth all affect the nominal values shown in the design curves, next page. The curves are based on the assumptions that the accuracy and finish of the gears represent good gear-industry practice for the type involved, that the oil used is petroleum oil of normal viscosity, and that the temperature range is normal (say, in the 80 to 120-F range).

Intensity of tooth load is very important. Suppose a gear set of 100-hp input capacity had a loss of 10 hp. This would mean 90% efficiency. At essentially no load

Efficiency Curves for different ratios and angles

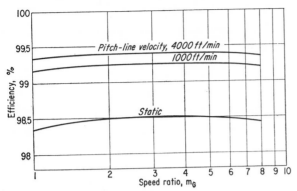

A—Spiral-bevel Gears

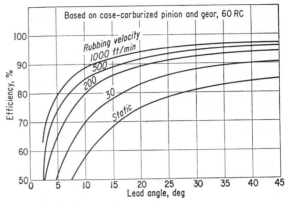

D—Spiroid Gears

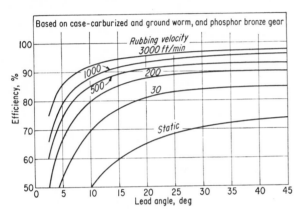

B—Cylindrical-worm Gears

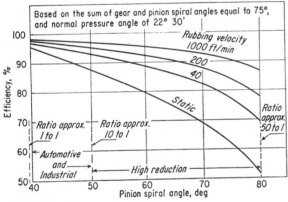

C—Hypoid Gears

such a gear set might still have a mesh loss of as much as 3 hp. If only 1 hp was transmitted through the gear set, its efficiency would have dropped all the way from 90% to 25%. Efficiency curves for gear sets generally assume that full rated power is being transmitted through the gear set. On these two pages:

Curves A are for ground spiral-bevel gears operating at full load capacity for industrial applications. Their efficiency would be somewhat higher if operating as high-load aircraft gears with a thin synthetic oil.

Efficiency of straight-bevel and Zerol-bevel gears is quite close to this data. Face gears also follow the same trend but are somewhat less efficient because they are not usually loaded as heavily for their size as ground

bevels. All these gear types are coplanar and have somewhat similar efficiency characteristics because of relatively low sliding velocity for a given pitch-line velocity.

Curves B show the efficiency of cylindrical worm gears as a function of worm lead angle. The efficiency drops off rapidly when the lead angle goes below 15°. However, in designs requiring low lead angles, horsepower is generally quite low and efficiency therefore is not too important a factor. In low-ratio units it is easy to design for a high worm lead angle but this becomes difficult at high ratios. **Curves E** give recommendations as to worm angle, center distance and ratio.

Crossed-helical gears (non-enveloping worm gears) follow a trend similar to that of cylindrical worm gears (Curves B), but tend to be less efficient because of their reduced load-carrying capacity. Double-enveloping worm gears such as Cone-Drive gears follow a similar trend to cylindrical worm gears.

Worm gears as a class all have a high sliding (or "rubbing") velocity with respect to the pitch-line velocity of the worm. The rubbing speed can be determined from:

$$v_r = \frac{0.262 n_w d}{\cos \lambda} \text{ fpm} \qquad (1)$$

where n_W = rpm of worm
d = pitch dia of worm, in.
D = pitch dia of gear

λ = lead angle; $\tan \lambda = \dfrac{D}{d m_G}$

m_G = speed ratio = $\dfrac{\text{No. gear teeth}}{\text{No. worm threads}}$

Curves C show the nominal efficiency of hypoid gears. Their efficiency pattern is somewhat similar to that of worm gears. The spiral angle of a hypoid pinion can be compared to the lead angle of worm by subtracting it from 90°. A hypoid pinion of 70° spiral is analogous to a worm with a 20° lead angle.

Hypoid gears have a lower rubbing speed in comparison to their pitch-line speed than do worm gears. This allows designing hypoid gears with better efficiency than worm gears to do the same job.

Curves D show some nominal efficiency data for Spiroid gears—the trend here is somewhat similar to that of worm gears. Their rubbing velocity is not quite as high with respect to their pitch-line velocity as that of worm gears. Lead-angle recommendations are given in **Curves F.**

Planoid gears have rubbing characteristics very similar to hypoids, and their efficiency could be expected to follow a similar pattern.

Helicon gears are for lighter loads. Their efficiency is somewhat poorer than that of a Spiroid or a Planoid, but they follow a similar trend of best efficiency at a low ratio and poorest efficiency at a high ratio.

Beveloid gears may approach coplanar gears quite closely. This tends to give them relatively high efficiency. However, their relatively low load-carrying capacity tends to lower their efficiency. Fully loaded Beveloids generally have efficiencies in the 95% to 99% range.

All of the efficiency curves mentioned above are based on mesh loss of the gear teeth. In a complete unit there are bearing losses; there may also be significant windage losses and oil-churning losses. Though these other losses must be considered by the designer in figuring over-all efficiency, oil-cooler capacity, etc., they are generally not important enough to enter into the choice of kind of gear for a given job. Bearing losses, for instance, are usually less than 0.5% in a high horsepower application, while in some low horsepower applications, bearing losses can be as high as 10%.

SELF-LOCKING CHARACTERISTICS

Most gear sets are employed for speed reduction—their characteristics as speed increasers are usually of little concern. In a few cases, however, right-angle gear sets serve mainly to increase speed. Superchargers, for example, run at high speed and generally require a speed-increasing drive.

The efficiency of gear drives for this purpose may be much poorer than when running as speed decreasers. In general, a helical tooth form and a large number of teeth on the pinion help make speed increasers run smoothly and efficiently. High-ratio, straight-bevel gears; high-ratio, straight-face gears; and worm gears with a low lead angle are the poorest to use for stepping up speed. In fact, worm gears have the ability to self-lock when driven backwards—although high-lead-angle worms are very successfully used as speed increasers. One of the early supercharged automobiles had a Cone-Drive worm gear set driving the supercharger.

In some cases the ability to self-lock when driven backward may be a prime design requirement. A winch, for instance, can be designed so that stopping the drive self-locks the gears and holds the load. This may permit the use of a design without a brake or a ratchet. The gears alone may be adequate to hold the load.

Best for self-locking are the single- or double-enveloping worm gears. Crossed helicals might be used but there is often some shock in self-locking. The stronger single- or still stronger double-enveloping worm gears are generally needed.

The exact lead angle at which self-locking occurs is somewhat variable, depending on kind of lubricant, tooth-surface finish and tooth proportions. The following rules will usually give satisfactory results.

Lead Angles for different ratios

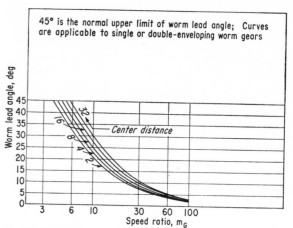

45° is the normal upper limit of worm lead angle; Curves are applicable to single or double-enveloping worm gears

E—Worm Gears

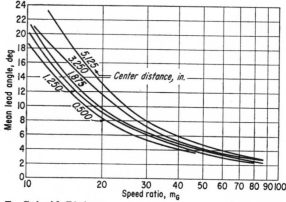

F—Spiroid Pinions

• For self-locking, the lead angle should not exceed 5°. In other words, the drive cannot be reversed with lead angles of 5° or less.

• If self-locking is desired, and the unit is already running when power is stopped and must then decelerate from speed, the helix angle should not exceed 3°.

• If appreciable vibration is present and a unit must stop when already running, the lead angle should not exceed 2° for self-locking.

• When a worm gear is to be driven backwards as a speed increaser, the lead angle should exceed 12°.

MANUFACTURING CONSIDERATIONS

Method of manufacturing can often determine which of several possible right-angle drives is best choice for a particular application. For example, if the quantity is low and cost is important, one might prefer a system manufactured with available or low-cost tooling.

A summary of common methods for manufacturing in the various right-angle gear systems is given in Table 9. It shows methods of producing both the pinion and gear member of each type, along with the processes currently being used to finish the teeth. Certain methods, such as casting or lapping, could apply to most types but are shown only where commonly used.

Machines designed and built by the Gleason Works

Table 9—Manufacturing and Finishing Methods

Gear Type	Member	Manufacturing Method	Refining Method
Straight bevel	Both	Forming rounded tools Generating planing tools Generating disk-mill cutters	. . . . Grind
Zerol bevel	Gear Both Both	Forming face mill cutters Generating planing tools Generating face-mill cutters	Grind . . . . Grind
Spiral bevel	Gear Both Both	Forming face-mill cutters Generating planing tools Generating face-mill cutters	Grind, lap Lap Grind, lap
Hypoid	Gear Both Both	Forming face-mill cutters Generating planing tools Generating face-mill cutters	Lap or grind Lap Grind or lap
High-reduction	Gear Pinion	Forming face-mill cutters Generating face-mill cutters	Grind or lap Grind or lap
Face gear	Pinion Gear	Milling Broaching Hobbing Shaping Shaping Casting	Shave Grind Lap Hone Lap
Crossed helical	Both	Hobbing Shaping Milling	Shave Grind Lap Hone
Beveloid	Both	Generating Hobbing Casting	Grind
Cylindrical worm	Worm Wheel	Milling Chasing Hobbing Hobbing	Grind Lap
Double-enveloping worm	Worm Wheel	Hobbing Hobbing	Grind Lap Grind Lap
Spiroid	Pinion Gear	Milling Chase Rolling Hobbing Casting Sintered Molded	Grind Lap Shave Lap
Planoid	Pinion Gear	Hobbing Milling Broaching	Lap Shave Grind Lap
Helicon	Pinion Gear	Hobbing Rolling Milling Chasing Hobbing Casting Sintered Molded	Grind Lap Shave Lap

are commonly used in the generation of either straight-bevel, Zerol-bevel, hypoid, or spiral-bevel gears and pinions. The fact that Zerol-bevel gears can be cut and ground on the same machines as spiral-bevel gears gives them one of their chief advantages over straight bevels.

The pinion of the face-gear set is merely a spur gear—hence, it can be made and finished by any of the methods used for parallel-axis spur gearing. The gear, however, must be shaper-cut or molded.

Both members of the crossed-helical pair can be manufactured with methods and equipment used for parallel-axis helical gearing. This is often very important if minimum tooling expense is desired.

Beveloid gears are generated on special machines, similar to hobbing machines, and are ground on machines similar to the hob-type grinders used to grind spur and helical gears.

With either cylindrical or double-enveloping worm gears, the worm wheel is generally made by hobbing, using either tangential- or radial-hob feed. The cylindrical worm is usually milled on a machine like a thread-milling machine or ground on a thread grinder. The double-enveloping worm is generally hobbed or shaped with a special attachment to obtain the throated form.

The Spiroid pinion can be manufactured on any machine set up for cutting a thread on a taper, such as a thread-milling machine with an angle block of some of the automatic thread-chasing machines. For high production, rolling the threads is very economical. The gear member must be machined in a hobbing machine, with a hob that duplicates the pinion.

The Planoid pinion is hobbed on a standard hobbing machine but with a special face-cutter mounted in the work spindle. The pinion to be cut is mounted in what is normally the hob spindle. The gear can be milled or broached one tooth at a time on standard machines.

Helicon pinions are made by milling, chasing, or hobbing in low-production quantities—or rolling where production is large. Their gears are hobbed with a hob which simulates the pinion. On ratios of less than 10:1, hobbing does not produce a satisfactory finish unless tangential-feed hobbing machines are used, so that below 10:1 the gear member is usually cast, molded or sintered.

EDITOR'S NOTE—The authors wish to thank the many individuals and organizations who helped in the preparation of this article, in particular: Wells Coleman of the Gleason Works; Fred Birtch, Cone-Drive Div of Michigan Tool; Fred Bohle and W. D. Nelson, Spiroid Div, Illinois Tool Works; Albert Beam, Vinco; J. E. Gutzwiller, Standard Products Div, DeLaval Steam Turbine Co. Photos of face and cross-helical gears are from Fellows Gear Shaper Co.

Latest curves and formulas for
Design of bevel gears

WELLS COLEMAN, Chief Gear Analyst, Gleason Works, Rochester, NY

BEVEL gears are the most efficient means of transmitting rotation between angularly disposed shafts. Power requirements may be in the thousands of horsepower, and in aircraft they have been successfully operated at pitchline speeds above 25,000 fpm.

Fortunately, though design goals are higher, the approach can be simpler. With the charts given here you can go directly to the proper range of gear sizes; with the rating formulas you can pinpoint the best gear rapidly. The data are based on two key factors, surface durability (pitting resistance) and strength (resistance to tooth breakage). Included are recommendations for diametral pitches, number of teeth, face widths, spiral angles, tooth proportions, mounting design, and gear lubrication—and a completely worked-out design problem.

There are three basic types of bevel gears—straight, spiral, and Zerol.

Power capacity of bevel gears

For case-carburized gears, 55 RC min, 20-deg pressure angle;

Ratio, m_G	Pinion pitch diameter, in.	Gear pitch diameter, in.	Number of pinion teeth	Face width, in.	STRAIGHT OR ZEROL BEVEL GEARS			
					Pinion speed, rpm			
					100	720	1750	3600
					Power capacity at different pinion speeds, hp			
1	1.0	1.0	22	3/16	0.094	0.611	1.34	2.38
2	1.0	2.0	20	5/16	0.156	1.01	2.22	3.95
4	1.0	4.0	17	9/16	0.277	1.80	3.96	7.04
8	1.0	8.0	14	11/16	0.334	2.18	4.78	8.49
1	2.0	2.0	23	3/8	0.523	3.14	6.73	.....
2	2.0	4.0	20	5/8	0.864	5.18	11.1	.....
4	2.0	8.0	17	1 1/8	1.54	9.24	19.8	.....
8	2.0	16.0	14	1 3/8	1.86	11.1	23.9	.....
1	4.0	4.0	24	3/4	2.93	16.1	.....	.....
2	4.0	8.0	21	1 1/4	4.85	26.7	.....	.....
4	4.0	16.0	18	2 3/16	8.41	46.4	.....	.....
6	4.0	32.0	15	2 5/8	10.0	55.1	.....	.....
1	8.0	8.0	30	1 9/16	16.7	.....	.....	.....
2	8.0	16.0	26	2 1/2	26.5	.....	.....	.....
4	8.0	32.0	23	3 7/16	36.2	.....	.....	.....

Yesterday's rule of thumb isn't good enough today.
With this systematic approach you can quickly
predict gear life for a given load capacity.

The three types

Straight bevels are the oldest, the simplest, and still the most widely used. Teeth are straight and tapered and, if extended inward, would intersect the gear axis. In recent years straight bevel-gear-cutting machines have been designed to crown the sides of the teeth in their lengthwise direction. They are known as Coniflex gears. The localized-tooth design tolerates small amounts of misalignment in the assembly of the gears and some displacement of the gears under load without concentrating the tooth contact at the ends of the teeth. As a result these gears are capable of transmitting heavier loads than the old-style straight bevel gears under the same conditions.

Spiral bevels have curved oblique teeth which contact each other gradually and smoothly from one end to the other, p 45. Cut a straight bevel into an infinite number of short face-width sections, angularly displace them relative to one another, and you have a spiral bevel gear. Well-designed spiral bevels have two or more teeth in contact at all times. The overlapping tooth action transmits motion much more smoothly and quietly than with straight bevel gears, which is why spiral bevels are superseding straight bevels in many applications.

Zerol bevels, p 45 have curved teeth similar to those of the spiral bevels but with zero spiral angle at the middle of the face width and little end thrust. Both spiral and Zerol gears can be cut on the same machines with the same circular face-mill cutters or ground on the same grinding machines. Both are produced with localized tooth contact which can be controlled for length, width, and shape.

Functionally, however, Zerol bevels are similar to the straight bevels and thus carry the same ratings (see Table below). In fact, Zerols can be used in the place of straight bevels without mounting changes.

Zerol bevels are widely employed in the aircraft industry, where ground-tooth precision gears are generally required. Because grinding equipment for straight bevel gears was not generally available, Zerol gears became widely accepted. Most hypoid cutting machines can cut spiral bevel, Zerol, or hypoid gears.

Use spiral bevel gears . . .

1) For high speeds—in excess of 1000 fpm or 1000 rpm. The only exception should be in differential or planetary gear sets where space does not permit antifriction bearings on the pinion. Grind teeth of gears which operate above 8000 ft/min to eliminate runout and other gear-tooth inaccuracies resulting from heat treatment.

2) For high loads — Spiral bevels have lower tooth loading than the straight or Zerol bevels because of larger radius of curvature on the tooth profile. Also, spiral bevels have two or more teeth in contact; hence tooth loads are more evenly distributed. The result is that the spiral bevels can carry more load without surface fatigue. Use ground-tooth gears where the gear blanks are subject to distortion by heat treat or where tooth loads are high.

3) For minimum gear size—But keep in mind that high-capacity spiral bevel gears require more rigid mountings and rolling-element bearings to absorb the higher thrust loads, which can add to the over-all weight.

4) For quiet operation—High precision of ground-tooth spiral bevels results in less noise and vibration.

5) For complete interchangeability—Because ground-tooth spiral bevels make possible duplication within very close limits, they are frequently used for slow speeds and light loads where interchangeability must be maintained.

Use straight- and Zerol-bevel gears . . .

1) When speeds are less than 1000 fpm—At higher speeds straight and Zerol bevel gears may be noisy. But ground Zerol bevel gears have been used successfully in aircraft planetary gear-sets (where noise was not a consideration) at speeds to 15,000 fpm.

2) When loads are light—Straight and Zerol bevel gears are also satisfactory for high static loads when surface wear is not a critical factor.

3) When space, gear weight, and mountings are premium—This includes planetary gear sets, where space does not permit inclusion of rolling-element bearings. In this case ground gears are a necessity.

theoretical face width not to exceed 0.28 cone distance.

SPIRAL BEVEL GEARS WITH 35-DEG SPIRAL ANGLE				
Pinion speed, rpm				
100	720	1750	3600	10,000
Power capacity at different pinion speeds, hp				
0.113	0.753	1.50	2.96	6.27
0.196	1.31	2.60	5.13	10.9
0.375	2.50	4.97	9.83	20.8
0.431	2.87	5.72	11.3	24.0
0.633	3.87	7.97	14.1	.....
1.11	6.80	13.9	24.7	.....
2.15	13.2	27.0	48.0	.....
2.40	14.7	30.2	53.5	.....
3.62	20.0	37.7	66.5	.....
6.44	35.5	67.0	118.	.....
12.1	66.8	126.	222.	.....
13.8	76.2	143.	254.	.....
21.6	109.	199.	.....	.....
37.4	190.	345.	.....	.....
51.0	259.	471.	.....	.....

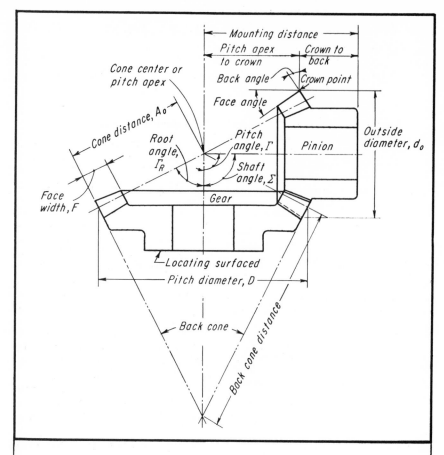

Symbols

a_0 = addendum, in.

A_0 = outer cone distance, in.

b_0 = dedendum, in.

c = clearance, in.

C_M = material factor

d = pinion pitch dia, in.

d_0 = outside dia, in.

D = gear pitch dia, in.

F = face width, in.

h_k = working depth, in.

h_t = whole depth, in.

K = circular thickness factor

I = durability geometry factor

J = strength geometry factor

m = speed ratio

m_F = face contact ratio

n = pinion speed, rpm

N_C = number of teeth in crown gear

N_G = number of gear teeth

N_P = number of pinion teeth

P = maximum operating horsepower, hp

P_d = diametral pitch

t = circular tooth thickness, pinion, in.

T = design pinion torque, lb-in.; also circular tooth thickness, pinion, in.

T' = maximum operating torque, or one-half peak pinion torque, or full peak pinion torque, lb-in.

V = pitch line velocity, ft/min

X_0 = pitch apex to crown, in.

δ = dedendum angle

γ = pinion pitch angle, deg

Γ = gear pitch angle, deg

γ_0 = pinion face angle

Γ_0 = gear face angle

γ_R = pinion root angle

Γ_R = gear root angle

ϕ = pressure angle, deg

Ψ = spiral angle, deg

Σ = shaft angle, deg

GEAR SIZE

Peak loads

First determine what fraction of the peak load to employ for estimating the gear size. This has been our experience:

If the total duration of the peak load exceeds ten million cycles during the total expected life of the gears, use the peak load for estimating the gear size.

If, however, the total duration of the peak load is less than ten million cycles, use one half the peak load, or the value of the highest sustained load, whichever is greater.

The pinion torque requirement (torque rating) can now be obtained as follows:

$$T = T' \qquad (1)$$

or

$$T = \frac{63,000\,P}{n} \qquad (2)$$

where T = design pinion torque, lb-in.

T' = maximum operating pinion torque, or one half peak pinion torque, or full peak pinion torque, as outlined above.

P = maximum operating horsepower

n = pinion speed, rpm

For general industrial gearing the preliminary gear size is based on surface durability (long gear life in preference to minimum weight). The design chart, Fig 1, is from durability tests conducted with right-angle spiral-bevel gears of case-hardened steel. Given pinion torque and the desired gear ratio, the chart gives pinion pitch diameter.

For other materials, multiply the pinion diameter given in Fig 1 by the material factor given in Table II.

Straight bevels and Zerol bevels will be somewhat larger. Multiply the values of pinion pitch diameter from Fig 1 by 1.3 for Zerol bevels and by 1.2 for Coniflex straight bevels. (Zerol and Coniflex are registered trademarks of the Gleason Works.)

For high-capacity spiral bevels (case-hardened, with ground teeth), the preliminary gear size is based on both surface capacity and bending strength. Based on surface capacity, the pinion diameter from Fig 1 should be multiplied by 0.80. Based on bending strength, the pinion diameter is given by Fig 2. Choose the larger pinion diameter of these two.

Statically loaded gears should be

designed for bending strength rather than surface durability. For statically loaded gears which are subject to vibration, multiply the pinion diameter from Fig 2 by 0.70. For statically loaded gears not subject to vibration, multiply the pinion diameter from Fig 2 by 0.60.

Tooth numbers

Although tooth numbers are frequently selected in an arbitrary manner, it has been our experience that for most applications the tooth numbers for the pinion from the charts, Fig 3 and 4, will give good results. Fig 3 is for spiral bevels and Fig 4 for straight and Zerol bevels. The number of teeth in the mating gear is of course governed by the gear ratio.

For lapped gears: Avoid a common factor in the numbers of teeth in the gear and mating pinion. This permits better and more uniform wear in the lapping process on hardened gears.

For precision gears: Accuracy of motion is of prime importance; hence the teeth of both pinion and gear should be hardened and ground. Also, use even ratios. Gears made for even ratios are easier to test, inspect, and assemble accurately.

Automotive gears: These are generally designed with fewer pinion

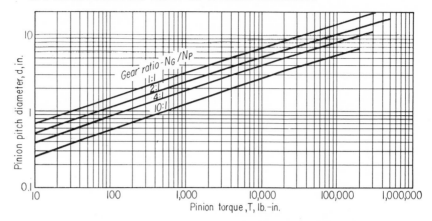

1 .. PITCH DIAMETERS BASED ON SURFACE DURABILITY

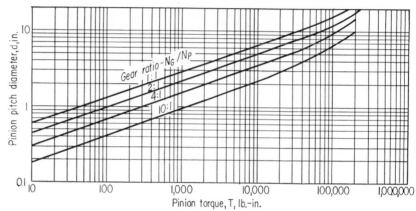

2 .. PITCH DIAMETERS BASED ON TOOTH BENDING STRENGTH

Table I .. Overload factors

Values in this table are for speed decreasing drives; for speed increasing drives add $0.01(N_G/N_P)^2$ to these factors.	POWER SOURCE	CHARACTER OF LOAD ON DRIVEN MACHINE		
		Uniform	Medium shock	Heavy shock
	Uniform	1.00	1.25	1.75
	Light shock	1.25	1.50	2.00
	Medium shock	1.50	1.75	2.25

Table II .. Material factors for gear mesh

GEAR		PINION		Material Factor C_M
	Minimum hardness		Minimum hardness	
Case-hardened steel	58 R_C	Case-hardened steel	60 R_C	0.85
Case-hardened steel	55 R_C	Case-hardened steel	55 R_C	1.00
Flame-hardened steel	50 R_C	Case-hardened steel	55 R_C	1.05
Flame-hardened steel	50 R_C	Flame-hardened steel	50 R_C	1.05
Oil-hardened steel	375-425 Brinell	Oil-hardened steel	375-425 Brinell	1.20
Heat-treated steel	250-300 Brinell	Case-hardened steel	55 R_C	1.45
Heat-treated steel	210-245 Brinell	Heat-treated steel	245-280 Brinell	1.65
Cast iron	....	Case-hardened steel	55 R_C	1.95
Cast iron	....	Flame-hardened steel	50 R_C	2.00
Cast iron	....	Annealed steel	160-200 Brinell	2.10
Cast iron	....	Cast iron	....	3.10

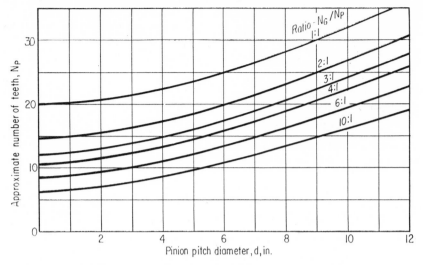

3 . . NUMBER OF TEETH FOR SPIRAL BEVEL GEARS

Ratio - N_G/N_P

Approximate number of teeth, N_P (vertical axis)

Pinion pitch diameter, d, in. (horizontal axis)

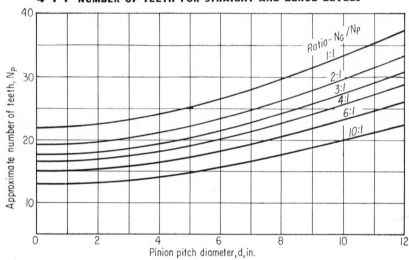

4 . . NUMBER OF TEETH FOR STRAIGHT AND ZEROL BEVELS

Ratio - N_G/N_P

Approximate number of teeth, N_P (vertical axis)

Pinion pitch diameter, d, in. (horizontal axis)

Table III .. Pinion teeth for automotive applications

Approximate ratio, N_G/N_P	Preferred number of pinion teeth, N_P	Allowable range, N_P
2.0	17	15-19
2.5	15	12-16
3.0	11	10-14
3.5	10	9-12
4.0	9	8-10
4.5	8	7-9
5.0	7	6-9
6.0	6	5-8
7.0	6	5-7
8.0	5	5-6

teeth. Table III gives suggested tooth numbers for automotive spiral bevel drives. The numbers of teeth in the gear and mating pinion should not contain a common factor.

Face widths

The face width should not exceed 30% of the cone distance for straight-bevel and spiral-bevel gears and should not exceed 25% of the cone distance for Zerol bevel gears. In addition, it is recommended that the face width, F, be limited to

$$F \leqq 10/P_d$$

where P_d is the diametral pitch. Practical values of diametral pitches range from 1 to 64.

The design chart in Fig 5 will give the approximate face width for straight-bevel and spiral-bevel gears. For Zerol bevels the face width given by this chart should be multiplied by 0.83. The chart is based on face width equal to 30% of cone distance.

Diametral pitch

The diametral pitch can now be determined by dividing the number of teeth in the pinion by the pinion pitch diameter. Thus

$$P_d = N_P/d$$

Because tooling for bevel gears is not standardized according to pitch, it is not necessary that the diametral pitch be an integer.

Spiral angle

The spiral angle of spiral-bevel gears should be so selected as to give a face-contact ratio, m_F, of at least 1.25. We have found that for smoothness and quietness, a face-contact ratio of 2.00 or higher will give best results.

The design chart, Fig 6, gives the spiral angle for various face-contact ratios. It is assumed that you have already determined the diametral pitch and face width to obtain the product, P_dF. The curves are based on the equation

$$m_F = P_dF[K_1 \tan \psi - K_2 \tan^3 \psi]$$

where
$\Psi =$ spiral angle
$K_1 = 0.2865$
$K_2 = 0.0171$.

The values for K_1 and K_2 are dependent upon the ratio of face width to outer cone distance of $F/A_0 = 0.3$.

Whenever possible, select the hand of spiral to give an axial thrust that tends to move both the gear and pinion out of mesh. As a second choice, select the hand of spiral to give an axial thrust that tends to move the pinion out of mesh.

Standard bevel systems

There are three standardized AGMA systems of tooth proportions for bevel gears: 20-deg straight bevel, spiral bevel, and Zerol bevel. There are also several special bevel-gear tooth forms which result in minor modifications to the above proportions. These special forms are used for manufacturing economy or to accommodate special mounting considerations. Because they are very closely tied to the method used in producing the gears, the means of achieving them and the effects they have on standard tooth proportions are beyond the scope of this article.

20-deg straight bevels

General proportions for this system are given in Table IV. The tooth form is based on a symmetrical rack, except where the ratio of tooth top lands on pinion and gear would exceed a 1.5 to 1 ratio. A different value of addendum is employed for each ratio to avoid undercut and to achieve approximately equal strength. If these gears are cut on modern bevel-gear generators they will have a localized tooth bearing. Coniflex gears have this tooth form. To provide uniform clearance, the face cone elements of the gear and pinion blanks are made parallel to the root cone elements of the mating member. This permits the use of larger edge radii on the generating tools, with consequent greater fatigue strength.

Note that the data in Table IV apply only to straight bevel gears that meet the following requirements:

1) The standard pressure angle is 20 deg. See Table V for ratios which may be cut with 14½, 22½, and 25-deg pressure angles.

2) The teeth are full depth. Stub teeth are avoided because of resulting reduction in contact ratio, which can increase both wear and noise.

3) Teeth with long and short addenda are used throughout the system (except on 1:1 ratios) to avoid undercut, increase strength, and reduce wear.

4) The face width is limited to one third the cone distance. The use of a greater face width results in an excessively small tooth size at the inner end of the teeth and, therefore, impractical cutting tools.

The American Gear Manufacturers Assn standard for this system is AGMA 208.02.

Spiral bevels

Tooth thicknesses (see Table IV) are proportioned so that the stresses in the gear and pinion will be approximately equal with a left-hand pinion driving clockwise or a right-hand pinion driving counterclockwise. These proportions will apply to all gears operating below their fatigue endurance limit. For gears operating above the endurance limit, special thickness proportions will be required. The standard for this system is AGMA 209.02.

The tooth proportions shown are based on the 35-deg spiral angle. A smaller spiral angle may result in undercut and a reduction in contact ratio. The data in this system do not apply to the following:

1) Automotive rear-axle drive gears, which normally are designed with

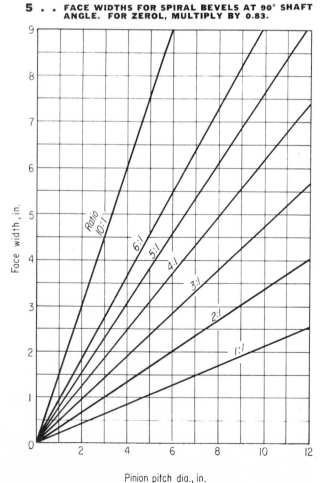

5 . . FACE WIDTHS FOR SPIRAL BEVELS AT 90° SHAFT ANGLE. FOR ZEROL, MULTIPLY BY 0.83.

Face width, in.

Ratio 10:1

6:1

5:1

4:1

3:1

2:1

1:1

Pinion pitch dia., in.

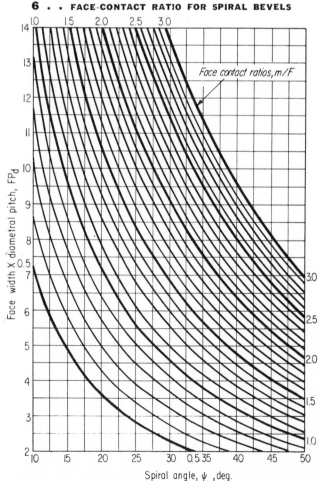

6 . . FACE-CONTACT RATIO FOR SPIRAL BEVELS

Face contact ratios, m/F

Face width X diametral pitch, FP_d

Spiral angle, ψ, deg.

Table IV..Tooth Proportions for Standard Bevel Gears

Item	STANDARD BEVELS (*see Table V for other cases*)		
	Straight	**Spiral**	**Zerol**
Pressure angle, ϕ, deg	20	20	20, 22½, 25
Working depth, h_k, in.	$2.000/P_d$	$1.700/P_d$	$2.000/P_d$
Whole depth, h_t, in.	$(2.188/P_d)+0.002$	$1.888/P_d$	$(2.188/P_d)+0.002$
Clearance, c, in.	$(0.188/P_d)+0.002$	$0.188/P_d$	$(0.188/P_d)+0.002$
Gear addendum, a_o, in.	$\dfrac{0.540}{P_d}+\dfrac{0.460}{P_d(N_G/N_P)^2}$	$\dfrac{0.460}{P_d}+\dfrac{0.390}{P_d(N_G/N_P)^2}$	$\dfrac{0.540}{P_d}+\dfrac{0.460}{P_d(N_G/N_P)^2}$
Face Width, F, in. (Use the smaller value from the two formulas)	$F \leqq \dfrac{A_o}{3}$ or $F \leqq \dfrac{10}{P_d}$	$F \leqq \dfrac{3A_o}{10}$ or $F \leqq \dfrac{10}{P_d}$	$F \leqq \dfrac{A_o}{4}$ or $F \leqq \dfrac{10}{P_d}$
Spiral angle, ψ, deg	None	25 to 35 (See note 1)	0
Minimum number of teeth (Note 2)	13	12	13
Diametral pitch range	no restriction	12 & coarser	3 & finer
AGMA reference number	208.02	209.02	202.02

Notes

1. 35 deg is the standard spiral angle. If smaller spiral angles are used, undercut may occur and the contact ratio may be less.

2. This is the minimum number of teeth in the basic system. See Table V for equivalent number of teeth in the gear member.

fewer pinion teeth than listed in this system.

2) Helixform and Formate (registered Gleason trademarks) pairs, which are cut with a nongenerated tooth form on the gear.

3) Gears and pinions of 12 diametral pitch and finer. Such gears are usually cut with one of the duplex cutting methods and therefore require special proportions.

4) Ratios with fewer teeth than those listed in Table V.

5) Gears and pinion with less than 25-deg spiral angle.

Zerol bevels

Considerations of tooth proportions to avoid undercut and loss of contact ratio as well as to achieve optimum balance of strength are similar to those for the straight-bevel gear system.

The Zerol system is based on tooth proportions (Table IV) in which the root cone elements do not pass through the pitch cone apex. The face cone element of the mating member is made parallel to the root cone element to produce uniform clearance.

The basic pressure angle is 20 deg. Where needed to avoid undercut, 22½-deg or 25-deg pressure angles are also used (see Table V). The face

Table V.. Minimum number of teeth

Pressure angle, deg	STRAIGHT		SPIRAL		ZEROL	
	Pinion	Gear	Pinion	Gear	Pinion	Gear
20 (standard)	16	16	17	17	17	17
	15	17	16	18	16	20
	14	20	15	19	15	25
	13	30	14	20		
			13	22		
			12	26		
14½	29	29	28	28	Not Used	
	28	29	27	29		
	27	31	26	30		
	26	35	25	32		
	25	40	24	33		
	24	57	23	36		
			22	40		
			21	42		
			20	50		
			19	70		
16	Not Used		24	24	Not Used	
			23	25		
			22	26		
			21	27		
			20	29		
			19	31		
			18	36		
			17	45		
			16	59		
22½	13	13	14	14	14	14
					13	15
25	12	12	12	12	13	13

width is limited to one quarter of the cone distance because, owing to the duplex taper, the small-end tooth depth decreases rapidly as the face width increases.

The standard for this system is AGMA 202.02.

Gear-dimension formulas

Table VI gives the formulas for bevel-gear blank dimensions. Tooth proportions are based on data from Table IV. A sample design problem later in the article illustrates the use of these formulas.

Rating formulas

Once the initial gear size has been determined from the above charts, the gears are checked for surface durability and strength, using the following two equations. Surface durability

is the resistance to pitting and involves the stress at the point of contact, using Hertzian theory. Strength is the resistance to tooth breakage and refers to the calculation of bending stress in the root of the tooth.

Surface durability:

$$ T = \frac{FIK_v}{2K_m} \left(\frac{S_c d}{C_p} \right)^2 \qquad (3) $$

Strength:

$$ T = \frac{FJK_v}{2K_s K_m} \left(\frac{S_t d}{P_d} \right) \qquad (4) $$

where

T = maximum allowable torque, lb-in. Use the smaller of the two values.

S_c = allowable contact stress. For recommended values see Table VIII.

S_t = allowable bending stress (also from Table VIII)

d = pinion pitch diameter at larger end of tooth, in.

P_d = diametral pitch at large end of tooth

F = face width, in.

C_p = elastic coefficient (see Table IX)

I = geometry factor (durability) from the design curves in Fig 9 and 10. Fig 9 is for spiral bevel gears with 20-deg pressure angle and 35-deg spiral angle, Fig 10 for straight-bevel and Zerol-bevel gears with 20-deg pressure angle.

J = geometry factor (strength) from the design curves in Fig 11 and 12. Fig 11 is for spiral-bevel gears with 20-deg pressure angle and 35-deg spiral angle. Fig 12 is for straight-bevel and Zerol-bevel gears with 20-deg pressure angle.

K_m = load distribution factor. Use 1.0 when both gear and pinion are straddle-mounted; use 1.1 when only one member is straddle-mounted. Somewhat higher values may be required if the mountings deflect excessively.

K_v = dynamic factor from the design curves in Fig. 13. Use curve 1 for high-precision ground-tooth gears, curve 2 for industrial spiral bevels, curve 3 for industrial straight-bevel and Zerol-bevel gears.

K_s = size factor from Fig 14.

If the computed value of T from either of the above torque equations is less than the design pinion torque,

7 . . CIRCULAR THICKNESS FACTORS, STRAIGHT AND ZEROL BEVELS

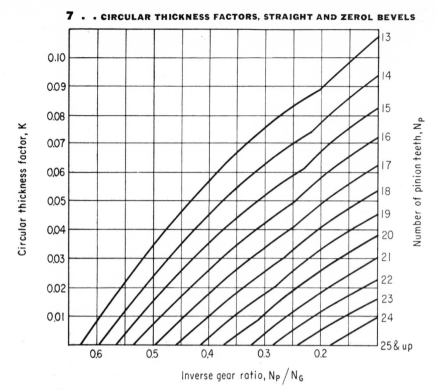

Circular thickness factor, K (vertical axis)

Inverse gear ratio, N_P/N_G (horizontal axis)

Number of pinion teeth, N_P (right axis): 13, 14, 15, 16, 17, 18, 19, 20, 21, 22, 23, 24, 25 & up

8 . . CIRCULAR THICKNESS FACTORS FOR SPIRAL BEVELS

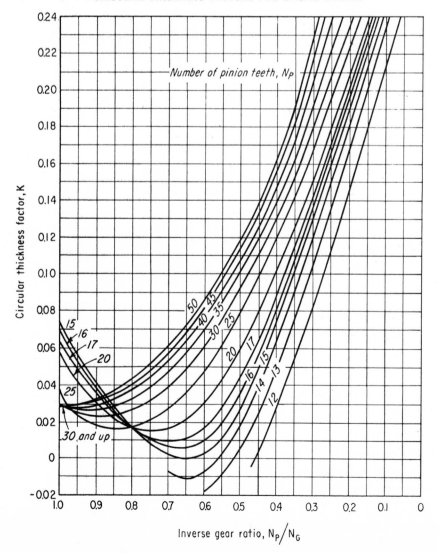

Circular thickness factor, K (vertical axis)

Inverse gear ratio, N_P/N_G (horizontal axis)

Number of pinion teeth, N_P

the gear sizes should be increased and another check should be made.

Design example

Select a bevel gear set to connect a small steam turbine to a centrifugal pump with the following specifications: The turbine is to deliver 29 hp at 1800 rpm to a centrifugal pump. The pump is to operate at 575 rpm.

Gear ratio:

$$m = 1800/575 = 3.13$$

Normal operating torque:

$$T' = 63,000(29)/1800$$
$$= 1015 \text{ lb-in.}$$

For a centrifugal pump driven by a steam turbine, only light shock with uniform load is anticipated. Therefore an overload factor of 1.25 is selected from Table I.

Design torque:

$$T = 1.25(1,015) = 1270 \text{ lb-in.}$$

Because the speed is above 1000 rpm, spiral-bevel gears are used.

Pinion pitch diameter: From Fig 1, for $T = 1,270$ lb.-in. and $N_G/N_P = 3.13$, $d = 2.2$ in. Because this is an industrial design, Fig 2 need not be consulted.

Number of teeth: From Fig 3, the pinion will have $N_P = 13$ teeth. Thus, for the gear, $N_G = 13(3.13) = 41$ teeth.

Face width: From Fig 5, the face width of both gears will be approximately $F = 1.1$ in.

Pitch line velocity:

$$V = \frac{3.14(2.2)(1800)}{12}$$

$$= 1030 \text{ ft/min}$$

The approximate size of the gear set has quickly been determined. Now check it for durability and strength using these factors:

$S_c = 200,000$ psi, from Table VIII, assuming that both pinion and gear are to be made from case-hardened steel
$C_p = 2800$, from Table IX
$I = 0.116$, from Fig 9
$K_m = 1.1$ for overhung pinion mounting
$K_v = 0.84$ from curve 2, Fig 13

Durability evaluation:

$$T = \frac{(1.1)(0.116)(0.84)(200,000)^2(2.2)^2}{(2)(1.1)(2800)^2}$$

$T = 1210$ lb-in.

Since the gears must be designed to carry 1270 lb-in. torque, the gear size should be increased slightly. To approximate the new size, multiply the

Table VI..Bevel-gear dimensions

Dimension *All dimensions in in. and deg*	Formulas, chart or table
1. Number of pinion teeth, N_P	Figs 3, 4; Tables III, V
2. Number of gear teeth, N_G	$N_G = m\,N_p$
3. Diametral pitch, P_d	Practical range, 1 to 64 P_d
4. Face width, F	Fig 5; Table IV
5. Working depth, h_k	Table IV
6. Whole depth, h_t	Table IV
7. Pressure angle, ϕ	Table IV
8. Shaft angle, Σ	Practical range, 10 to 180 deg

Dimension	Pinion	Gear
9. Pitch diameter; d, D	$d = \dfrac{N_P}{P_d}$ (Figs 1, 2)	$D = \dfrac{N_G}{P_d}$
10. Pitch angle; γ, Γ	$\gamma = \tan^{-1}\dfrac{\sin \Sigma}{\dfrac{N_G}{N_P} + \cos \Sigma}$	$\Gamma = \Sigma - \gamma$
11. Outer cone distance, A_o	$A_o = \dfrac{D}{2 \sin \Gamma}$	
12. Circular pitch, p	$p = \dfrac{3.1416}{P_d}$	
13. Addendum, a_{oP}	$a_{oP} = h_k - a_{oG}$	$a_{oG} =$ (Table IV)
14. Dedendum, b_{oP}	$b_{oP} = h_t - a_{oP}$	$b_{oG} = h_t - a_{oG}$
15. Clearance, c	$c =$ (Table IV)	
16. Dedendum angle, δ (See Note 1)	$\delta_P = \tan^{-1}\dfrac{b_{oP}}{A_o} + \Delta\delta$	$\delta_G = \tan^{-1}\dfrac{b_{oG}}{A_o} + \Delta\delta$
17. Face angle of blank; γ_o, Γ_o	$\gamma_o = \gamma + \delta_G$	$\Gamma_o = \Gamma + \delta_P$
18. Root angle; γ_R, Γ_R	$\gamma_R = \gamma - \delta_P$	$\Gamma_R = \Gamma - \delta_G$
19. Outside diameter; d_o, D_o	$d_o = d + 2\,a_{oP} \cos \gamma$	$D_o = D + 2\,a_{oG} \cos \Gamma$
20. Pitch apex to crown; x_o, X_o	$x_o = A_o \cos \gamma - a_{oP} \sin \gamma$	$X_o = A_o \cos \Gamma - a_{oG} \sin \Gamma$
21. Circular thickness; t, T (See Note 2)	$t = p - T$	$T = \dfrac{p}{2} - (a_{oP} - a_{oG})\dfrac{\tan \phi}{\cos \Psi} - \dfrac{K}{P_d}$

Notes:

1. The change in dedendum angle, $\Delta\delta$, is zero for straight bevel and spiral bevel gears; $\Delta\delta$ is given by Table VII for Zerol bevel gears.

2. Factor K is given by Fig 7 for straight bevel and Zerol bevel gears with 20 deg pressure angle, and by Fig 8 for spiral bevel gears with 20 deg pressure angle and 35 deg spiral angle. For other cases K can be determined by the method outlined in "Strength of Bevel and Hypoid Gears" published by the Gleason Works.

Table VII..Formulas for $\Delta\delta$ — Zerol bevel gears

Pressure Angle, deg	Change in Dedendum Angle, $\Delta\delta$, min	Note:
20	$\Delta\delta = \dfrac{6668}{N_C} - \dfrac{300}{F}\sqrt{\dfrac{1}{N_C\,P_d\,(\tan\gamma + \tan\Gamma)}} - \dfrac{14\,P_d}{N_C}$	$N_C = 2\,P_d\,A_o =$ number of teeth in crown gear.
22½	$\Delta\delta = \dfrac{4868}{N_C} - \dfrac{300}{F}\sqrt{\dfrac{1}{N_C\,P_d\,(\tan\gamma + \tan\Gamma)}} - \dfrac{14\,P_d}{N_C}$	
25	$\Delta\delta = \dfrac{3412}{N_C} - \dfrac{300}{F}\sqrt{\dfrac{1}{N_C\,P_d\,(\tan\gamma + \tan\Gamma)}} - \dfrac{14\,P_d}{N_C}$	

Table VIII..Allowable stresses

Material	Heat treatment	Minimum surface hardness	Contact Stress S_C, psi	Bending Stress S_t, psi
Steel	Carburized (*Case Hardened*)	55 R_C	200,000	30,000
Steel	Flame or Induction hardened (*Unhardened root fillet*)	50 R_C	190,000	13,500
Steel	Hardened and Tempered	300 Brinell	135,000	19,000
Steel	Hardened and Tempered	180 Brinell	95,000	13,500
Steel	Normalized	140 Brinell	65,000	11,000
Cast Iron	As Cast	200 Brinell	65,000	7,000
Cast Iron	As Cast	175 Brinell	50,000	4,600
Cast Iron	As Cast	. . . .	30,000	2,700

Table IX...Elastic coefficients, C_P

Pinion	GEAR MATERIAL				
	Steel	Cast iron	Aluminum	Aluminum bronze	Tin bronze
Steel, $E = 30 \times 10^6$	2800	2450	2000	2400	2350
Cast iron, $E = 19 \times 10^6$	2450	2250	1900	2200	2150
Aluminum, $E = 10.5 \times 10^6$	2000	1900	1650	1850	1800
Aluminum bronze, $E = 17.5 \times 10^6$	2400	2200	1850	2150	2100
Tin bronze, $E = 16.0 \times 10^6$	2350	2150	1800	2100	2050

trial pinion pitch diameter by the square root of the design torque divided by the allowable torque from the first trial.

New pinion pitch diameter:

$$d = 2.2 \sqrt{\frac{1270}{1210}} = 2.25 \text{ in.}$$

New face width, from Fig 5:

$$F = 1.125$$

All other values in Eq 3 remain the same. Use Eq 3 to again check the allowable torque:

$$T = \frac{(1.125)(0.116)(0.84)(200,000)^2}{(2)(1.1)(2800)^2}$$
$$\times (2.25)^2$$

$T = 1290$ lb-in.

This exceeds the required 1270 lb-in.

Strength evaluation

Now make a check of tooth strength, using these factors in Eq 4:

$J = 0.228$, from Fig 11

$K_s = 0.645$, from Fig 14

$S_t = 30,000$ psi, from Table VIII

$$P_d = \frac{N_P}{d} = \frac{13}{2.25} = 5.78$$

All other factors remain the same as for the durability evaluation, hence

$$T = \frac{(1.125)(0.228)(0.84)(30,000)}{(2)(0.645)(1.1)(5.78)}$$
$$\times (2.25)$$

$T = 1770$ lb-in.
(allowable for strength)

Safety factors:

$$\frac{1290}{1270} = 1.01 \text{ for surface durability}$$

$$\frac{1770}{1270} = 1.39 \text{ for strength}$$

The selected gears, therefore, are

correct for surface durability but are conservative for strength. In, say, aerospace applications, strength would dominate over durability and a smaller (and lighter) pinion and gear set would be selected. However, on heavily loaded gears where special surface treatment is given to increase the surface resistance to wear, actual test experience has shown that fatigue breakage in the root fillet rather than a breakdown of the tooth surface does occur. Thus, in applications such as aircraft and automotive, adequate fatigue strength must be assured.

The detail gear dimensions are now obtained.

Gear diameter:

$$D = \frac{N_G}{P_d} = \frac{41}{5.78} = 7.093 \text{ in.}$$

Spiral angle:

$$F P_d = (1.125)(5.78) = 6.5$$

The spiral angle is now selected with reference to Fig 6. From the curves the face contact ratio, m_F, will be 1.72 with a 35-deg spiral angle or 2.03 with a 40-deg spiral angle. If maximum smoothness and quietness is required, the 40-deg spiral angle is recommended. However, in this case the 35-deg spiral angle should give adequate smoothness. The lower spiral angle reduces the bearing loads and thereby reduces the cost of the unit.

Working depth, Table IV:

$$h_k = \frac{1.700}{5.78} = 0.294$$

Whole depth, Table IV:

$$h_t = \frac{1.888}{5.78} = 0.327$$

MOUNTING DESIGN

With regard to the mountings, the designer should keep three points in mind:

1) Designing the gear blanks, the shafts, bearings, and gear housings to provide the good rigidity as well as accuracy.

2) Designing the entire unit for ease of assembly.

3) Designing the blanks in a simple geometrical form for ease of manufacture.

The entire success of the bevel-gear drive depends not only on the design but also on care in manufacturing the unit. The gears must be assembled accurately.

Recommended methods for mounting bevel gears are shown in Fig 15 and 16, and poor vs good design points

text continued, page 68

9 . . DURABILITY FACTORS FOR SPIRAL BEVELS

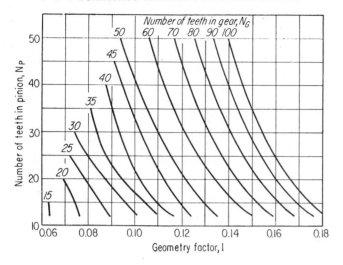

10 . . DURABILITY FACTORS, STRAIGHT AND ZEROL BEVELS

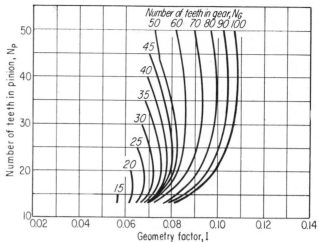

11 . . STRENGTH FACTORS FOR SPIRAL BEVELS

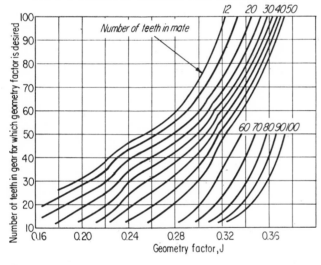

12 . . STRENGTH FACTORS, STRAIGHT AND ZEROL BEVELS

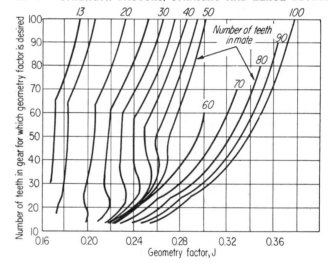

13 . . DYNAMIC FACTORS FOR ALL BEVELS

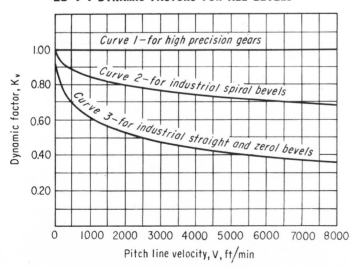

14 . . SIZE FACTORS FOR ALL BEVELS

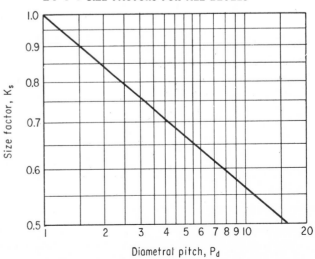

below. As a general rule rolling-friction bearings are superior to plain bearings for bevel gear mountings. This is especially true for spiral-bevel and hypoid gears because these types must be held within recommended limits of deflection and locked against thrust in both directions.

Gear lubrication

There are two methods recommended for lubricating bevel gears—the splash method and the pressure or jet method. The splash method, in which the gear dips in an oil sump in the bottom of the gear box, is satisfactory for gears operating at peripheral speeds up to 2000 fpm. At higher speeds churning of the oil is likely to cause overheating. For speeds above 2000 fpm a jet of oil should be directed on the leaving side of the mesh point to cover the full length of the teeth on both members. If the drive is reversible, jets should be directed at both the entering and leaving mesh.

Some present-day gear lubricants will operate continuously at temperatures of 200 F and above. However, 160 F is the recommended maximum for normal gear applications. Special oils are not normally required for bevel gears; the lubricants for spur and helical gears are also used for straight, Zerol and spiral bevels.

Typical mounting details

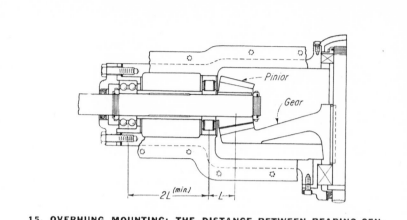

15. OVERHUNG MOUNTING: THE DISTANCE BETWEEN BEARING CENTERS SHOULD BE GREATER THAN TWICE THE OVERHUNG DISTANCE, L.

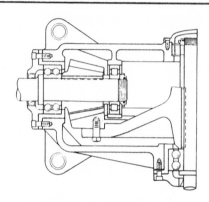

16. STRADDLE-MOUNTING FOR BOTH MEMBERS OF A SPIRAL BEVEL PAIR. FOR HIGH-RIGIDITY APPLICATIONS.

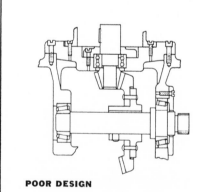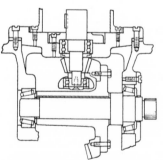

POOR DESIGN

☐ Face width too great, more than one third of cone distance.
☐ Metal at small end of pinion between teeth and bore too thin.
☐ Webbed steel ring gear adds to cost of material and machining.
☐ Use of rivets to hold gear to hub introduces danger of runout.
☐ A setscrew is inadequate to hold gear in correct axial position and tends to "cock" the gear.
☐ Pinion held on shaft only by fit.
☐ An overhung pinion cannot be held in line by one double-row bearing.
☐ No means of adjustment for gears.

GOOD DESIGN

☐ Face width reasonable, less than one third of cone distance.
☐ Sufficient metal at small end of pinion to provide strength.
☐ Section of ring gear simpler and more direct in design.
☐ Screws hold gear to hub.
☐ Gear positively held in position.
☐ Pinion locked in position by washer and screw.
☐ Pinion rigidly supported by addition of inboard bearing.
☐ Adjusting washers provided, to be ground to thickness required to correctly position gears.
☐ Pinion and bearings can be assembled as a complete unit.

3 are familiar, 3 little known in this

Guide to worm gear types

Though some look as if they wouldn't work, they do, and accurately too. In index drives you'll find them superior to power-transmission types

ELIOT K. BUCKINGHAM, *Buckingham Associates Inc, Springfield, Vt*

THERE are at least five distinct types of worm gear drives, each with its own advantages and limitations—a fact of which few engineers seem to be aware. A sixth type discussed here, actually a modification of one of the others, is also relatively little known. There is, too, a notable lack of standards on these types; only two of the six are covered by gear standards.

Yet all these worm drives can be manufactured on existing equipment, though some special fixtures or cutters may be required. And, although this article is slanted toward applications that transmit motion rather than power (where cumulative tooth-to-tooth error is of primary importance, eg, measuring equipment, data transmission, large-telescope drives, precision machine tools), the information is also of value to designers of speed reducers and other power-transmission units.

For simplicity it is assumed that all the worm drives are operating with the input and output shafts at right angles to each other. However, it is a simple matter to design for angles other than 90 deg.

Worms and gears

The action between a worm and its meshing gear is a combination of sliding and rolling, with sliding predominating in the higher reduction ratios. The teeth may or may not be of involute form. Also, the pitch diameters of the worm and gear are not proportional to the speed reduction ratio; a compact set can provide a large reduction ratio and speed.

To obtain a low coefficient of friction in sliding, dissimilar metals are normally selected. Because the enveloping member—worm or gear—is

1. Standard cylindrical worm meshing with a standard spur gear has only point contact and therefore is not suitable when transmitting heavy loads, but it is finding new applications in instruments and data transmission drives. This is the first of the six worm types compared in Table I, next page.

never completely formed during the generating process, it is usually made of the more plastic material so that it can be cold-worked to the correct form in use.

Worms are often made with more than one wrap-around thread, commonly referred to as a *start*. With some sets (Type I—Table I) only point contact results between the worm and gear (more on that later). To obtain line contact, the worm sets can be made "single enveloping" in which either the gear wraps around the worm (this is the conventional way—Type II), or the worm around the gear (Types IV and V). When it is the gear that wraps around the worm, its teeth take on a throated appearance, but the gear itself may or may not have topping stock. For example, a gear made without topping is commonly referred to as a *throated* gear, while some gears are made with topping.

When it is the worm that wraps around the gear, a less common type, it too takes on a throated configuration and is frequently called an *hourglass worm*.

Table I — Types of worm gear drives

Type	Name	Type of contact		Fig. No.	Type of worm		Type of mating gear		Characteristics
					Description	Configuration	Description	Configuration	Advantages
I	Cylindrical-worm and spur gear	Point		1	Cylindrical		Cylindrical—can be standard involute spur or helical gear		• Simple and economical • Center distance not critical • Alignment not critical • Worm can be designed to operate with existing spur or helical gear
II	Standard cylindrical-worm drive		Line	2	Cylindrical		Throated, with or without topping stock		• Most common type. Stocked by many manufacturers • Excellent power transmission when made with all-recess teeth • Axial position of worm not critical
III	Deep-tooth cylindrical-worm drive	Single enveloping	Line	3	Cylindrical		Throated, with or without topping stock		• Up to 11 teeth in contact — averages tooth position and profile errors • Can usually be designed to replace existing drive • May use multiple start worms for low ratios
IV	Enveloping-worm and spur drive		Line	4	Hourglass (throated)		Cylindrical—can be standard involute spur or helical gear		• Many teeth in contact—for indexing or timing (1/10 of teeth in gear) • Can use standard spur gear • Gear can be split and spring loaded to control backlash without affecting contact • Axial movement of gear does not affect accuracy • Worm can be generated on standard hobbing machine
V	Wildhaber worm drive		Line	7	Hourglass		Cylindrical, but with straight-sided teeth		• Same as Type IV except gear has a straight-sided tooth form • Can also be designed as a power drive by having only three teeth in contact • Worm can be generated with simple fly tool • Gear can be milled or ground on standard milling machine or ground from accurate index table or plate
VI	Hindly or double-enveloping worm drive	Double enveloping	Area	8	Hourglass		Throated		• Compact — can transmit high loads in small space and with close center distances

70

Worm gear sets can also be made *double-enveloping,* so that both members wrap around each other (Type VI). In such a case, the worm takes on the hourglass shape, and the gear the throated shape. This latter system is sometimes referred to as an *hourglass worm drive,* but as can be seen from Table I there are other systems that use hourglass worms.

TYPES OF WORM DRIVES

The characteristic of each drive is first discussed, then related equations and design procedures are given in a separate section that follows.

Cylindrical worm and spur—Type I

It is not too widely known that a standard single-start worm can mesh properly with a standard spur gear *with the shafts at right angles to each other,* Fig 1. This simple type of worm drive is finding applications where a high load capacity is not required. The straight worm has threads of constant form and lead along its length. It is actually a helical gear, and as the thread form is an involute helicoid, there will be true conjugate action in operations with a spur gear, or as is more frequently the case, a *helical* gear.

The contact is a point-contact between two curved surfaces, and so the angular relationship between the two shafts is not critical. Misalignment between the shafts merely shifts

Limitations
• Point contact — less than two teeth • Low power capacity • Not very accurate • Ratio must be 40/1 or higher for spur • Sum of number of teeth of gear and starts (threads) of worm must be 40 or more • Worm must be involute helicoid
• Axial alignment of worm gear critical • Commonly designed with pitch plane at center of thread depth. System then has 75% approach action. (Recess action preferred) • Any axial motion of worm or gear will spoil accuracy for indexing or timing • Only two to three teeth in contact. Very little averaging of errors
• Three teeth in contact gives maximum effectiveness for carrying load. Hence, added width does not add to load capacity • Axial alignment of gear critical • Axial motion of worm or gear affects accuracy • Special hob required
• Single start worms only for spur. Can use multiple starts with helical • Axial alignment of worm critical • Fellows type cutter required to make worm • Not recommended for high-power applications
• Single start worms only • Axial alignment of worm critical
• Axial alignment of gear and worm critical • Special tooling and equipment required to manufacture

Throated with topping

THROATED GEAR

2. Single enveloping worm drive. This is the "standard" type of worm drive that you often see employed in speed reducers and other power transmission applications. The worm is cylindrical, the meshing gear made is throated and machined either with topping (upper photo) or without topping (lower photo). Diagrams show the difference in configurations.

3. **Deep-tooth cylindrical worm drive** is similar to the systems shown in Fig 2 but with all dedendum teeth. Worm is a single start; pressure angle is low (10 deg). Approximately nine teeth are in contact as against three teeth for the gears in Fig 2. Manufactured by DeLaval-Holroyd Inc for precision indexing units.

4. **Single enveloping hourglass-worm drive.** The worm envelopes the gear and thus is the inverse of the types in Fig 2. The gear can be a standard involute spur or helical gear. The worm is shown being hobbed by a hob which is almost identical to the spur gear that will mesh with the worm. This particular set was designed for an instrument unit in the Apollo space program.

the contact point without affecting the nature of the contact, or the conjugate action. This characteristic can be used with advantage in instruments and data transmission applications. For example, an axial shaft in the gear does not cause a responding rotation in the worm, as it does in most the other types.

Another advantage of the cylindrical-worm and spur-gear drive (as it is sometimes called) is that both members can be manufactured on standard generating equipment, and inspected independently of each other. When a helical gear is employed, it can be generated with practically any value of helix angle within the complementary range of lead angles for worms. Also, both the worm and the gear can be ground after hardening, which is not easily possible with the standard worm drive, Type II.

Standard-form worm drive—Type II

This is the most common type of worm gearing in use today. It has the same, straight-cylindrical worm as in Type I, but the worm in this case meshes with a throated gear, Fig 2. Thus, the worm from a standard worm drive can mesh properly with a spur or helical gear—a fact frequently debated.

The contact between the threads of the worm and the teeth of the worm is a line contact.

Deep-tooth worm drive—Type III

This is similar to the conventional worm drive, Type II, but with extra deep teeth and low pressure angle (10-deg minimum), Fig 3. These modifications result in multiple contact, usually 8 to 12 teeth, and high recess action.

This eliminates the main disadvantage of the standard worm drive, Type II, which is that it is designed to operate with the pitch line at the center of the thread depth. This provides only a few teeth in contact (two to three teeth) and results in a conjugate action that is mostly approach—approximately 75% approach action and 25% recess action—which contributes to wear and noise. A full recess action is preferable, as is explained below. For a fuller treatment, see the author's previous article mentioned in the Editor's note.

The all-recess action of the Type III worm drive, is obtained by having the pitch line of the worm at the outside diameter of the gear, Fig 5. This is accomplished by increasing the lead of the worm or reducing the diameter of the gear. The pitch diameter, number of teeth, pressure angle, and center distance have not been changed, the worm is made larger and the gear smaller, so that the lead of the worm is equal to the circular pitch of the gear at the OD. The number of teeth in contact has not been changed, and the position and shape of the contact lines are still essentially the same, except that all contact is now recess.

But while an all-recess design might be satisfactory for some indexing applications, better results are obtained by also reducing the pressure angle from the standard 20-deg and 14½-deg angles normally employed. Fig 6 shows the contact diagram and basic rack of a 10-deg pressure angle worm with a double depth thread. This design gives nine teeth in simultaneous contact for the values used (256 teeth, 64 DP, 2.250-in CD).

It was this type of worm drive—with all-recess teeth and a 10-deg pressure angle—that was selected to precisely rotate the heavy 200-in. Mt Palomar telescope. It has also been employed in large range finders. De-Laval-Holroyd is presently making some Type III sets for Eastman Kodak.

Another advantage of this design is that it can be produced on standard equipment with a special hob. The large number of teeth in contact produce an averaging of tooth-to-tooth errors.

Enveloping-worm and spur-gear drive—Type IV

This drive consists of an hourglass-type of enveloping worm which drives a standard involute spur or helical gear, Fig 4. Properly designed, this geometry makes an excellent indexing drive, but it does not seem to be very widely known—although recent correspondence indicates increasing interest in Europe concerning this type. One reason may be that there are, as in the case of Types I, III, and V, no standard tables or procedure available for its design.

With an enveloping worm drive, there is no conjugate gear tooth action—although there is conjugate cam action—and hence no approach or recess. The worm is essentially a barrel cam, which pushes ahead the gear teeth as the worm rotates. The action may be compared to that of a screw and nut, except that the screw and nut have area contact, and the enveloping worm and spur gear have line contact. All relative motion between the worm and gear is pure sliding, always in the same direction for one direction of rotation of the

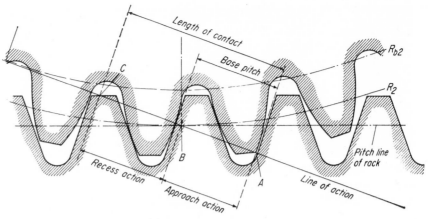

(A) Standard worm gear system

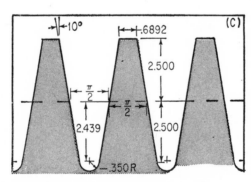

(B) All-recess worm gear system

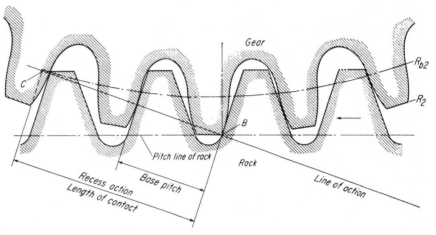

5. **Comparison of:** (A) standard involute and (B) full-recess action teeth in a worm gear system. The conjugate action of a worm drive is the same as a rack and gear. The contact will be the same whether the worm is rotated or moved axially. In the standard, the pitch plane is at the middle of the thread depth, in the all-recess it is at the OD of the worm gear. Recess action has a smoothing and polishing action. (C) Basic rack of a 10-deg PA worm with a double depth thread.

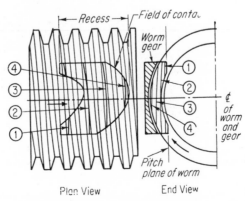

6. **Contact on a single start, Type II, all-recess worm drive.** With the worm rotating in the direction shown by the arrow in the end view, the thread advances in the direction marked by the arrow in the plan view. At the particular moment shown, there are four teeth in contact. The contact lines are marked 1,2,3, and 4. As the worm makes one revolution, contact line 1 moves to position 2; line 2 moves to position 3, etc. Thus there will be three teeth in contact all the time, and four teeth in contact for about three-fourths of the time. With the pitch plane at the outside diameter of the gear, there will be all recess action, and hence no reversal of sliding at the pitch plane, which would tend to break down any oil film.

worm, which makes for excellent resistance to wear and simplifies lubrication.

Because the contact line is to one side of the centerline of the worm, this is one of the few types of gear drives where a split, spring-loaded spur gear may be used to control backlash, without affecting the contact pattern or the performance of the drive. Also, because the gear is a straight spur gear, the drive can replace a splined connection. A wide-faced spur gear can travel across a worm, without affecting the accuracy of the indexing. The contact will be at a larger radius than with a spline, hence any errors due to the translation would have a lesser effect on the accuracy of the output. This feature could be well put to use in a hobbing machine; for instance, to replace the present standard-form worm that is now commonly used.

Other possible applications are large astronomical telescopes and radar screens where the worm gear is to be mounted on a large mass, often weighing several tons. As the telescope is rotated, changes in the center of gravity would tend to make the worm gear move axially, throwing the contact to one side of the worm gear, and introducing indexing errors. With an enveloping worm and spur gear, the spur gear could be allowed to shift axially at will, without affecting the performance. In this particular type of mechanism, it is much simpler to hold the worm rigid, and in position, than the gear.

Wildhaber worm drive—Type V

This drive was invented by Ernest Wildhaber back in 1922, about the same time that he invented his circular-arc helical gear drive, also referred to as the Novikov drive (see the article beginning on p 124).

The Wildhaber worm drive is similar to the enveloping drive of Type IV, with the exception that the tooth flanks of the gear are straight instead of involute, Fig 7. This gives a contact pattern that is very similar to that of the enveloping worm drive, Type IV, which shows, incidentally, that the form of the tooth—involute, straight-sided, or circular arc—is of little importance in a single enveloping drive.

The main advantage of a Wildhaber worm is that one can obtain extremely high accuracy in tooth spacing because the gear can be cut from a dividing head or indexing table—to within 0.25 sec of arc accuracy. Also, the system permits more flexibility in

design because no special hobs are required. A worm set with 63.9 DP is just as easy to make as one with 64 DP.

Double enveloping worm drive— Type VI

This system is also known as the *Hindley-worm drive* and with certain modifications, as the *Cone-Drive*. The latter is the proprietary name of the Cone Drive Gears Division of Michigan Tool Co.

As the name *double enveloping* implies, the worm and gear "wrap around" each other to produce a drive with an hourglass worm and a throated gear—the only one of the six with this combination, Fig 8. Also all teeth are straight sided, with flanks tangent to a common circle in any axial plane.

Although the Wildhaber and the double-enveloping drives have the same axial section and look similar, the double-enveloping worm cannot be properly meshed with a Wildhaber gear.

Proponents of the drive claim that the double enveloping design produces an *area contact,* rather than the *line contact* of Types II through V. This area appears on all teeth and engagement and results in high tooth contact and more compact designs. The analysis of the contact conditions of most of the worm drives—and certainly that of the double enveloping drive—is complex, and space precludes a fuller discussion in this article.

DESIGN AND MANUFACTURE
Cylindrical worm and spur—Type I

The form of the worm must be an involute helicoid. Any other form of helicoid will introduce momentary errors during operation. There are two easy methods to produce this form:

The first method is by hobbing, but as this may require a special hob, it is only justified when there is substantial production involved.

The second method is by grinding with a cone shaped cutter or wheel, with the axis of the cutter *parallel* to the axis of the worm—otherwise a true involute will not be produced. The line of action while milling or grinding lies inside the line of contact (throated line of operation with the spur gear). This means that the thread form on the worm must be ground deeper than the normal working depth of the thread to present the fillet interferrence with the spur gear.

To obtain all recess action on a spur and worm gear drive, the lead of the worm must approximately equal the circular pitch of the gear at the outside diameter, and the axial pressure angle of the worm must approximately equal the normal pressure angle of the gear at the operating pitch diameter.

Standard cylindrical worm drive, Type II

The throated gear is generated by a hob which is essentially a duplicate of the mating worm. The hob, however, should be slightly larger in diameter than the worm, to provide

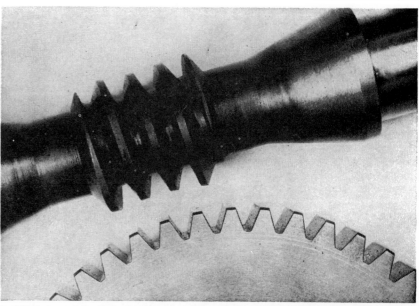

7. **Wildhaber worm drive** is similar to the system in Fig 4 but with straight-sided teeth for the gear. The teeth are sometimes designed to come to a point.

clearance both at the root of the teeth and at the outer edges. In other words, the center distance for generating the worm must always be larger than the center distance of operation. This rule applies to all worm drives.

During rotation, there is true conjugate gear tooth action, together with a camming action as the thread advances. The pitch surface of the worm is a plane, and the threads of the worm act as the teeth of a rack as they advance. Such rack teeth have a varying form across the face of the worm gear, and the larger the helix angle of the worm, the greater this change in rack form will be.

The pitch surface of the worm gear is a cylinder, which is tangent to the pitch plane of the worm. Neither of these pitch surfaces exists until the mating members are actually mounted in their operating positions. The axial pitch of the worm is constant for all diameters. The circular pitch of the worm, however, varies with its diameter and is equal to the circumference of its pitch circle, or cylinder, divided by the number of teeth in the gear. The pitch diameter, therefore, occurs where the circular pitch of the worm is identical to the axial pitch of the mating gear.

As with any other worm drive, the enveloping member should be made of the more plastic material. Thus, if the materials were to be steel and bronze, the worm would be made of bronze. When the drive is assembled, generation of the enveloping member, in this case the worm, will not be

complete. With proper design and manufacture, there will be contact on three or four teeth near the center of the region of mesh. As the gears are run-in, the more plastic material will cold-work, and the resultant plastic flow will gradually extend the contact towards the ends of the worm. In some cases, a carefully controlled lapping operation can accelerate this process. For some applications, particularly where the output shaft is driving an optical element, the loads may be so light that an artificial load must be employed to work in the gears.

Deep-tooth worm drive

Changing to all-recess action designs does not involve changes in the existing hobs or other generating tools for the gears. Instead, the outside diameter of throated worm gear blanks is made to the pitch diameter. For existing gear housings, the same gear blank is used but it is made with two or three more teeth than before. If the number of teeth in the gear must remain unchanged, the center distance is reduced from the conventional one to a value that brings the pitch plane of the worm substantially tangent to the outside diameter of the worm gear.

Enveloping worm and spur—Type IV

The design of the spur gear is unimportant. The worm is generated with a Fellows-type cutter, slightly larger but of the same design, and with the same circular pitch as the spur gear. One method is to have one more tooth in the cutter than in the spur

gear. Thus if a 144 tooth gear were required, the cutter could have 145 teeth. If adding one tooth to the cutter results in a high prime number, such as 269, then it would be better to use a 268 tooth cutter, slightly enlarged.

To generate the worm, the worm blank is mounted on the hob spindle of a hobbing machine, and the Fellows-type cutter on the work spindle. The cutter is fed the full depth of cut and allowed to pass through the worm blank, with both members rotating at the proper ratio. That is, the cutter moves across the worm blank in the direction of the axis of the cutter. The axis of the cutter and the worm blank are at right angles to each other. As long as the cutter and the gear are similar, there is little difference whether the gear is a standard form, or recess action.

The selection of the pressure angle is important. For good design, the momentary pressure angle should fall between 10 and 45 deg. Because the momentary pressure angle increases on one side of the centerline of the gear, and decreases on the other, maximum contact will be obtained when the nominal pressure angle is midway between the limits, or 27½ deg. Because this is not a standard, a 25-deg pressure angle gear system is preferred, with all other design features the same as a 20-deg pressure angle design. In a 27½-deg case, there are 25 teeth in contact, as opposed to 18 teeth making contact with the 20-deg pressure angle gear.

To drive a spur gear, the lead angle at the center of the throat should not exceed 4 deg, which essentially limits the design to single start worms, and also controls the minimum size of worm, as with the straight cylindrical worm and spur gear. The gear should have at least 40 teeth, and preferably more.

If a ratio smaller than 40/1 is required, then a worm with two or more starts may be used with a helical gear. But this restricts the axial movement of the helical gear for indexing accuracy, and any errors in the helix angle will affect the final performance. Hence, an hourglass worm and helical gear is a compromise, and should not be used for accuracy if a spur gear could be used instead.

Wildhaber drive

The main advantage of the Wildhaber drive is the ease by which it can be manufactured with accuracy. Compare it, for example, to the accuracy of fine pitch gears. Holding a 10 sec of arc on tooth positional

8. **Double enveloping worm drive** employs both an hourglass worm and a throated gear. Manufactured by Cone Drive Gears, Div of Michigan Tool Co.

errors is about the limit that can be obtained with the most accurate hobbing or grinding machines. For the Wildhaber drive, on the other hand, the gear can be cut from an indexing table (with a standard milling cutter or grinding wheel) to accuracies of 0.25 sec of arc or better.

The worm is generated with a straight-sided Fellows-type cutter, or with a single fly cutter as described earlier. Special hobs, or expensive tooling are not required; hence, the diametral pitch selected is unimportant. It is just as easy to manufacture a 59.98 DP drive as it is a 64 DP, which allows for more flexibility in design.

For maximum number of teeth in contact, the included angle of the space should be about 45 deg. This is equivalent to a 27½-deg pressure angle. This means that the gear may be cut with a standard 45-deg in-

DESIGN PROCEDURE FOR WORM DRIVES

Basic given data—to be used for all types of worm drives that follow:

m = Ratio = 256 to 1
C = Center distance = 2.250 in.
P_d = Diametral pitch = 64
N_G = number of teeth in gear = 256
N_W = number of threads in worm = 1
D = pitch diameter: D_1 = of worm = 0.500 in., D_2 = of gear = 4.000 in. D_{01} = of outside diameter of worm; D_{02} = of outside diameter of gear
F = face width
L = lead of worm
p = circular pitch of gear
h_k = working depth of thread and gear teeth = $2/P_d$ = $2/64$ = 0.03125
ϕ = pressure angle, deg
ϕ_c = pressure angle of cutter

Cylindrical worm and gear, Type I
Gear specifications

If standard form spur gear is used, then

$$\phi = 20 \text{ deg.}$$

$$D_{02} = \frac{N+2}{P} = \frac{256+2}{64} = 4.0312 \text{ in.}$$

$$F = 3/16 \text{ (this is arbitrary)}$$

Worm specifications

Make the lead approximately equal to the circular pitch of the gear at the OD.

$$L = \frac{\pi D_{02}}{64} = 0.0494703 \text{ at O.D.}$$

Because lead should be exact ratio, use

$$5/101 = 0.4950.$$

$$\cos \phi_c = \frac{p \cos \phi}{L} \qquad p = \frac{\pi}{64}$$

$$\cos \phi_c = \frac{\pi (0.93969262)(101)}{5} = 0.93176638$$

$$\phi_c = 21.29°$$
$$h_T = 0.036 \text{ (AGMA tooth depth for } 64P$$

Tooth thickness of cutter

$$= \frac{\pi - \text{arc thickness of gear at } D_2}{P_d}$$

$$= \frac{\pi - \frac{1}{2}\pi}{64} = 0.02454 \text{ in.}$$

This cutter will produce a "thin" thread; that is, the space will be thicker than the thread. All backlash should be cut into the gear.

Standard form worm and gear—Type II

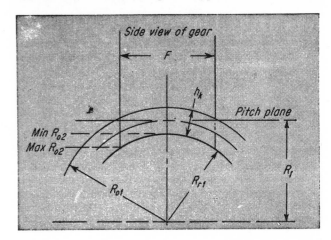

Assume $\phi = 14\frac{1}{2}$ deg
For exact ratio, let L = 3/61 = 0.04918

$$D_{01} = D_1 + h_k = 0.500 + 0.03125 = 0.53125$$

F should not be more than half this value. Use $F = 3/16$
From layout (above) of gear OD:

D_{02} (at throat) = $D_2 + h_k$ = 4.03125. This is the minimum OD.

If R_r = radius on worm to bottom of working depth of thread

$$R_r = C - \frac{4.03125}{2} = 0.234375$$

Max D_{02} of gear

$$= 4.03125 + 2[R_r - \sqrt{R_r^2 - (F/2)^2}]$$
$$= 4.070 \text{ in.}$$

Thus D_{02} can be anywhere from 4.032 to 4.070 in.

Full recess worm—Type III

$$\phi = 10 \text{ deg}$$

$$h_k = 2 \times \text{std depth} = \frac{4}{64} = 0.0625$$

$$D_{02} = D_2 = 4.000 \text{ in.}$$

From previous example, L = 3/61, F = 3/16
From layout of worm OD (that follows):

cluded angle milling cutter, and is essentially an index plate. This type of drive is being used by Dover Instruments Inc as an indexing drive on its gear checker, Fig 1, for checking gears to better than 5 sec of arc accuracy.

For manufacturing double-enveloping worm sets; the same cutter that is used to make the Wildhaber worm is used, except now the cutting operation is a pure in-feeding with no axial movement of the cutter. (As is the case in the other two hour-glass designs).

The gear is hobbed, with an hour-glass hob similar to the worm, but with thinner teeth. It is also in-fed to depth; then the gear rotated back and forth so that the side-cutting of the hob will obtain the proper tooth thickness. Extreme care must be taken in manufacture; the pair is usually lapped and sold as matched sets.

$$\text{Min } D_{01} = D_1 + 2\,h_k = 0.500 + (2)\,(0.0625) = 0.6250 \text{ in.}$$

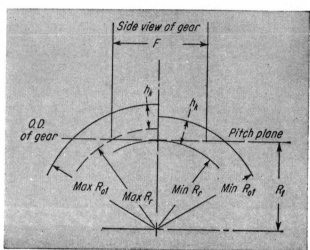

$$\text{Min } R_{r1} = R_1 = .250$$

$$\text{Max } R_{r1} = \sqrt{R_1{}^2 + \left(\frac{F}{2}\right)^2} = 0.2667$$

$$\text{Max } D_{01} = 2\,(0.2667 + 0.06250) = 0.6580$$

OD of worm could be 0.6250 to 0.6580 in.

Wildhaber worm drive, Type V

Let θ = tooth spacing angle = $360/N_G$
  (all angles are in degrees)
 α = one-half included angle of tooth
 β = one-half included angle of space
 R_2 = radius to equal tooth and space thickness
 ϕ_0 = nominal pressure angle
 R_b = radius of circle to which all tooth faces are tangent
 R_s = radius to sharp root of tooth
 R_p = radius to sharp point of tooth
 c = radial clearance between meshing teeth

Then the equations giving the necessary geometry factors are (see figure at right):

$$R_b = R_2 \sin \phi_0 \qquad R_{02} = R_p - c$$
$$R_p = R_b/\sin \alpha \qquad R_{r2} = R_s + c$$
$$R_s = R_b/\sin \beta \qquad h_s = R_p - R_s$$
$$\theta = 360/N_G \qquad h_t = R_{02} - R_{r2}$$

$$\beta = \alpha + \frac{\theta}{2} = \alpha + \frac{180}{N_G}$$

$$\phi_0 = \alpha + \frac{\theta}{4}$$

For conjugate action, design so that the minimum momentary pressure angle is between 10 to 45 deg. Contact must be within region of mesh. These are the same conditions imposed on the enveloping worm and spur:

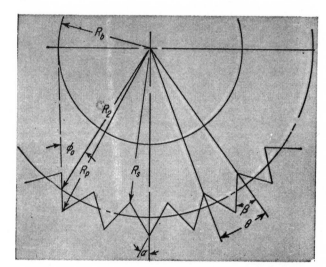

Ratio = 256/1 $CD = 2.250$ $N_G = 256$

Let $R_2 = 2.0$; $\beta = 30$ deg so that teeth on gear can be cut with a 60 deg angle cutter. Hence, from the above equations:

$$\theta = 360/256 = 1.40625$$

$$\alpha = 30 - \frac{1.40625}{2} = 29.266875$$

$$\phi_0 = 29.266875 - \frac{1.40625}{4} = 29.64843$$

$$R_b = 2\,(\sin \phi_0) = 0.98936$$
$$R_p = 0.98936/\sin \alpha = 2.0219$$
$$R_s = 0.98936/\sin \beta = 1.9787$$

Total tooth depth is $2.0219 - 1.9787 = 0.043$. Use 0.040; then

$$R_0 = 2.020 \qquad R_r = 1.980$$

Because $R_o < R_p$, and $R_r > R_s$, this design is satisfactory and the design of the gear is completed.

DESIGN CHART FOR CROSSED-HELICAL GEARS

For 90° shaft angle

WAYNE A. RING
Barber-Colman Company

IN PRELIMINARY LAYOUT WORK of gear drives, it is important to find whether the diametral pitch, gear ratio, and center distance, are possible. With this settled, the designer can proceed to calculate the specific helix angle.

This chart serves both purposes. For, if the Factor $2CP_n/N$ does not intersect the proper ratio curve R, this combination is not satisfactory. Also, in choosing the helix angle, a large amount of trial and error work is eliminated by this graph which gives a good approximation. In addition to helical gears, worm gear designs can be simplified with this chart.

This graph is based on the equation:

$$\frac{2CP_n}{N} = \frac{R}{\sin \alpha} + \frac{1}{\cos \alpha}$$

where: C = center distance, in.
P_n = normal diametral pitch
N = number of teeth in gear or worm wheel
n = number of teeth in pinion or threads in worm
R = ratio, n/N
α = helix angle of gear or worm wheel, deg.

EXAMPLE: Helical gears are to be designed for shafts 4 in. apart and at right angles. The details are: $P_n = 20$; $N = 64$; $n = 32$; thus $R = 0.50$ and Factor, $2CP_n/N = 2.5$. From the chart, two angles can be chosen. 20°-30′ or 58°-30′. The chosen angle, α, is substituted in the equation; it is altered to satisfy the equation. In this case, the exact angle is calculated to be either 20°-25′ or 58°-30′.

For greater efficiency and less wear, the gear and pinion helix angles should be nearly equal. Thus the gear angle chosen would be 58°-30′ and the pinion angle would become 31°-30′

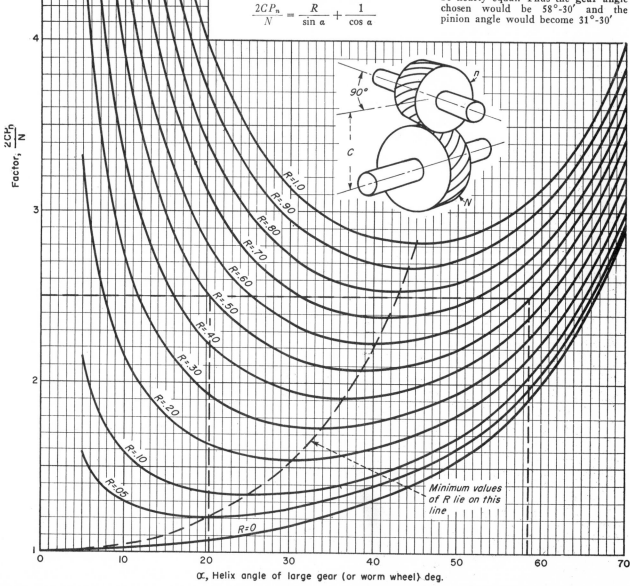

Factor, $\frac{2CP_n}{N}$

α, Helix angle of large gear (or worm wheel) deg.

R=1.0
R=.90
R=.80
R=.70
R=.60
R=.50
R=.40
R=.30
R=.20
R=.10
R=.05
R=0

Minimum values of R lie on this line

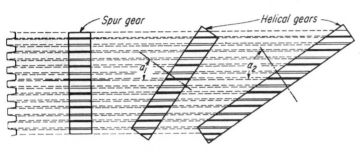

Developed Surfaces

BY VARYING HELIX ANGLE, spur gears can be replaced with helical gears to provide any number of workable pitch diameters, and add flexibility to design.

WHEN HELICAL GEARS GET YOU OUT OF TIGHT SPOTS

Their adaptability can overcome many difficulties in gear-train design that normally call for complicated, expensive measures.

SIGMUND RAPPAPORT, *kinematician,*
Ford Instrument Co, Div of Sperry Rand Corp
Adjunct professor, Polytechnic Institute of Brooklyn

Suppose that two shaft centers in a gear train are erroneously machined too far apart—or worst still, too close together. Or, suppose you wish to change the gear-speed ratio between two parallel shafts without changing the center distance. In each of the above problems, helical gears with the proper helix angle might do the trick.

There are even more design tricks possible: With non-parallel shafts, you can mesh gears of different size to operate at one-to-one ratio, or you can vary the relative sizes of the meshing gears to avoid, say, interference with another machine member.

And also, pairing helical with spur gears in an otherwise standard planetary system can boost the reduction ratio from, say 1:21 to 1:1600. I described this innovation in another article which has been given on p 115, but key points will be reviewed here.

Helix Angle Is the Key

The pitch diameter of a helical gear of a given pitch is not related in a simple linear proportion to the number of teeth—as is the case with spur gears—but is also dependent on the helix angle. In fact, for a given number of teeth and a given normal pitch, an infinite number of workable pitch diameters is possible. The smallest, based on a helix angle of zero, is identical with the pitch diameter of a spur gear of the same number of teeth and diametral pitch (see illustration above).

Although the tooth form and its proportions are based on the normal diametral pitch, the size of the helical gear—that is, its pitch diameter—is determined by the circular pitch, which changes with the helix angle.

Pitch dia D_s of a spur gear is found from

$$D_s = n/P$$

where n is the number of teeth and P is the diametral pitch.

Pitch dia D of a helical gear is

$$D = \frac{n}{P \cos \alpha}$$

where α is the helix angle (angle between the tooth direction and gear axis).

As can be seen from this equation, the minimum D

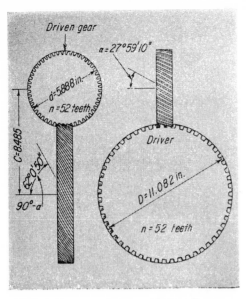

GEARS OF DIFFERENT PITCH diameters can give ratio of 1:1 by varying the helix angle.

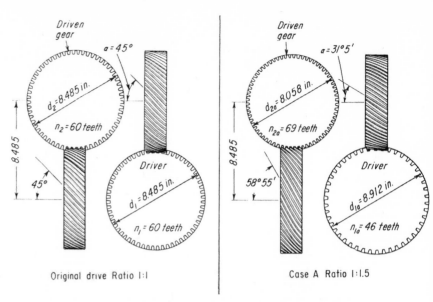

VERSATILITY of helical gears makes it possible to keep the same gearbox while varying the input-to-output ratios to suit requirements. Modifications are shown in Cases A, B and C.

occurs at $\alpha = 0$, which is a spur gear. Center distance between a pair of helical gears is a variable; and from these equations it can be seen that it is never smaller than the center distance between a pair of spur gears of the same pitch and number of teeth.

Compensating for Center-distance Errors

If two shaft centers are erroneously machined too far apart, the error can be compensated for by replacing spur gears with helical gears. The required helix angle α is then found from

$$\cos \alpha = C/C_e$$

where C is the center distance which should have been applied and C_e is the one actually put in.

An incorrect center distance sometimes can be compensated for when it is smaller than the theoretical one. Suppose two parallel shafts are to be connected by a pair of gears having 60 and 90 teeth, respectively, and a diametral pitch of 24. The pitch diameters of the gears are $60/24 = 2.5$ in., and $90/24 = 3.75$ in. Correct center distance is $\frac{1}{2}(2.5 + 3.75) = 3.125$ in. However, this distance is erroneously machined to 3.012 in. Spur gears can't be used because the next smaller spur gear pair of same ratio will have 56 and 84 teeth, respectively, with a required center distance of 2.917 in. But helical gears, with 56 and 84 teeth can operate at 3.012 in., providing that their helix angle equals:

$$\cos \alpha = 2.917/3.012$$
$$\alpha = 14° 25'$$

One of the pair of helical gears will be left-handed, the other right-handed. This will create a thrust that must be absorbed by the bearings.

How to Change Gear Ratios

Here's how helical gears permit you to change the gear ratio between two parallel shafts—yet still keep the

center distance unchanged. This problem sometimes occurs when a machine is being redesigned for a change in the size of the product.

Assume that the two shafts are connected by a pair of spur gears with diametral pitch of 12 and with 30 and 75 teeth respectively. The correct center distance for the shafts is 4.375 in.

It is now desired to change the ratio from 1:2.5 to 1:3.25. This necessitates a change in number of teeth. Dividing $(30 + 75)$ by $(1 + 3.25)$, gives $105/4.25 = 24.705$. Therefore, the pinion is selected to have 24 teeth. Number of teeth for the gear is $(24)(3.25) = 78$.

But spur gears of 24 and 78 teeth call for a center distance of 4.250 in. This problem is solved by employing a helical gear set with 24 and 78 teeth, and with a helix angle equal to:

$$\cos \alpha = 4.250/4.375$$
$$\alpha = 13° 44'$$

With Nonparallel Shafts

The versatile helical gear can also solve design problems when crossing shafts are involved. Consider the case where a gear train in an automatic machine contains two shafts crossed at 90°, connected in a 1:1 ratio by a pair of identical helical gears (45° helix angle). The gears have a diametral pitch of 10, and 60 teeth each. (Pitch dia = 8.485 in.).

At first, the drive had merely an oil pan; later, it was decided to enclose the drive totally in a cast iron housing, but there was too little clearance between the upper (driven) gear and the frame to accommodate the housing.

Thus, the following problem arises: Maintain the center distance, the normal diametral pitch and the ratio, but reduce the pitch diameter of the driven gear from 8.485 in. to about, but not more than 6 in. and increase the pitch diameter of the driver accordingly. The following relations hold (see illustrations above):

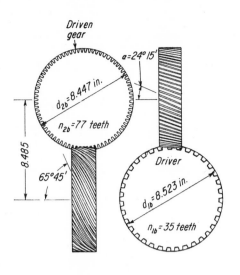

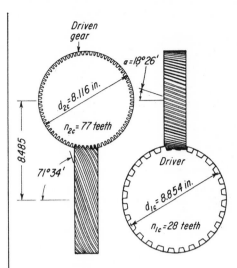

Case B Ratio 1:2.2 | Case C Ratio 1:2.75

$$d = \frac{n}{P \cos \alpha} \qquad (1)$$

$$D = \frac{n}{P \sin \alpha} \qquad (2)$$

$$C = \frac{d + D}{2} \qquad (3)$$

where

d = pitch dia of driven (upper) gear $\approx$ 6 in.
P = diametral pitch = 10
C = center distance = 8.485 in.
n = number of teeth
α = helix angle of driven gear
$(90 - \alpha)$ = helix angle of driver

Combining these three equations:

$$C = \frac{n}{2P} \left[\frac{(\sin \alpha + \cos \alpha)}{\sin \alpha \cos \alpha} \right] \qquad (4)$$

Substituting the value for n in Eq (1) into Eq (4) gives a solution for α:

$$\cotan \alpha = \frac{2C}{d} - 1 \qquad (5)$$

Using this value in Eq (1) seldom yields an integral value for n. But, because d was not rigidly established (it was only postulated that d be not larger than 6 inches) it is sufficient to take the nearest whole number of n. Here $n = 52$.

Eq (4) can be rearranged to give:

$$\frac{2CP}{n} = \frac{\sin \alpha + \cos \alpha}{\sin \alpha \cos \alpha}$$

which can be easily solved for α by squaring both sides and making use of the identity $2 \sin \alpha \cos \alpha = \sin 2\alpha$. Thus

$$\frac{C^2 P^2}{n^2} = \frac{1 + \sin 2\alpha}{\sin^2 2\alpha}$$

from which is found

$$\sin 2\alpha = \frac{n^2}{2C^2 P^2} - \sqrt{\left(\frac{n^2}{2C^2 P^2} \right)^2 + \frac{n^2}{C^2 P^2}}$$

In this equation substituting values for n, C and P, gives:

$\alpha = 27°59'10''$ from which we obtain

$$D = 11.082$$
$$d = 5.888$$

which yields the same center distance (8.485) as in the original pair.

Thus, a seeming paradox has been achieved of having two meshing gears of greatly differing size operating at a ratio of 1:1. This solution presents an additional bonus: axial thrust is reduced because the helix angle of the driven gear is less than 45°.

This solution would not have been possible if it were required to make the driver smaller and the driven gear larger, because beyond a certain point the helix angle exceeds the friction angle and the drive locks up.

Changing the Gear Ratio

This method can also be applied to use a given gearbox to accommodate a different gear ratio, while keeping the center distance constant. Here is an example:

Suppose you have a gearbox, built to accommodate two cross-shafts whose input-to-output ratio, R, is 1:1; and you want to use the same gearbox to produce the ratios listed below:

Case A — R_A = 1:1.5
Case B — R_B = 1:2.2
Case C — R_B = 1:2.75

Assume the original shafts have identical 60-tooth helical gears, with helix angle of 45°, diametral pitch $P = 10$ and pitch diameter of 8.485 in.

Find, for each of the three cases, the helix angle, α; the number of teeth in the driver, n_1; the number of teeth in the driven gear, n_2; the pitch diameter of the driver, d_1; and the pitch diameter of the driven gear, d_2.

These conditions should be met: center distance must remain 8.485 in., and gears should remain approximately the same size. continued, next page

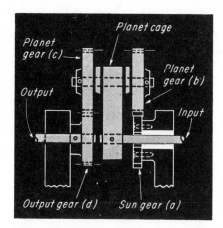

HELICAL GEARS in place of spur for one of the two sets boosts reduction ratio from 1:21 to 1:1600.

Since d_1 should be approximately equal to d_2 for each case, we find, with the use of Eq (1) and (2):

$$\frac{n_2}{P \cos \alpha} \approx \frac{n_1}{P \sin \alpha}$$

from which

$$R = \frac{n_1}{n_2} \approx \tan \alpha$$

This gives the following helix angles:

$$\alpha_A = 33° 41' \qquad \alpha_B = 24° 27' \qquad \alpha_C = 19° 59'$$

Next step is to determine n_1 and n_2 for Cases A, B and C.

Assume that $n_1 + n_2$ remains the same as in the original drive (120). Find whole-number values that will produce the desired ratio.

In Cases A and C the answers are:

Case A: $n_{1A} = 48$, $n_{2A} = 72$ **Case C:** $n_{1C} = 32$, $n_{2C} = 88$

But in Case B, there are no two whole numbers whose ratio is 1:2.2 and whose sum is 120. Since the ratio cannot be changed, the only alternative is to find integral solutions with a different total number of teeth:

$$n_{1B} = 35, \qquad n_{2B} = 77$$

But these data for numbers of teeth and corresponding helix angles would result in gear pairs with meshing gears of the same size:

$$d_{1A} = d_{2A} = 8.653 \text{ in.}$$
$$d_{1B} = d_{2B} = 8.458 \text{ in.}$$
$$d_{1C} = d_{2C} = 8.944 \text{ in.}$$

And if these gears were used with these diameters, the center distances for Cases A and C would be too large, and for Case B, too small.

This center distance for Case B can be enlarged to the required 8.485 in. by suitably modifying the helix angle, which is done most quickly by trial and error. Thus α_B is changed from 24°27' to 24°15', resulting in the pitch diameters

$$d_1 = 8.523 \text{ in.}$$
$$d_2 = 8.447 \text{ in.}$$

Half their sum yields the correct center distance of 8.485 in.

However, in Cases A and C, juggling of the helix angle cannot solve the problem because no solution can be found which simultaneously satisfies all requirements. Therefore, the sum of the teeth must be reduced, keeping the desired ratio.

Thus, for Case A:

$$n_{1A} = 46, \qquad n_{2A} = 69, \qquad \alpha_A = 31° 5'$$

resulting in

$$d_{1A} = 8.912 \text{ in.}, \qquad d_{2A} = 8.058 \text{ in.}$$

For case C:

$$n_{1C} = 28, \qquad n_{2C} = 77, \qquad \alpha_C = 18° 26'$$

with

$$d_{1C} = 8.854 \text{ in.}, \qquad d_{2C} = 8.116 \text{ in.}$$

The center distance of both these drives equals 8.485 in., as required by the problem. In all these cases the driving gear has the large helix angle, approaching the appearance of a multiple-start worm. This is not strange when one recalls that a single-start worm is nothing but a helical gear with one tooth.

Spur-helical Planetary

For the planetary system shown on this page the speed-reduction ratio is

$$R = 1 - \frac{ac}{bd}$$

where a, b, c and d are number of teeth or pitch dia of the respective gears. Here's where substituting a pair of helical gears for one of the spur-gear sets results in an extremely high ratio. For example, if $a = 39$, $b = 40$, $c = 41$, $d = 40$, the ratio becomes:

$$R = 1 - \frac{(39)(41)}{(40)(40)} = \frac{1}{1600}$$

To do this the helical pair (gears a and b) must have a helix angle equal to

$$\cos \alpha = \frac{a + b}{c + d} \qquad \alpha = 12° 45'$$

Limitation of an all-spur gear train is that the center distance between a and b must equal that between c and d. Hence for a similar train of $a = 40$, $b = 41$, $c = 40$, $d = 41$, the all-spur will produce a ratio of only

$$R = 1 - \frac{(40)(40)}{(41)(41)} = \frac{8}{1681} \approx \frac{1}{21}$$

Although a 1:1600 reduction is possible with a planetary system, the gain in torque is limited to about 40:1. This limitation also holds true for an all-spur planetary.

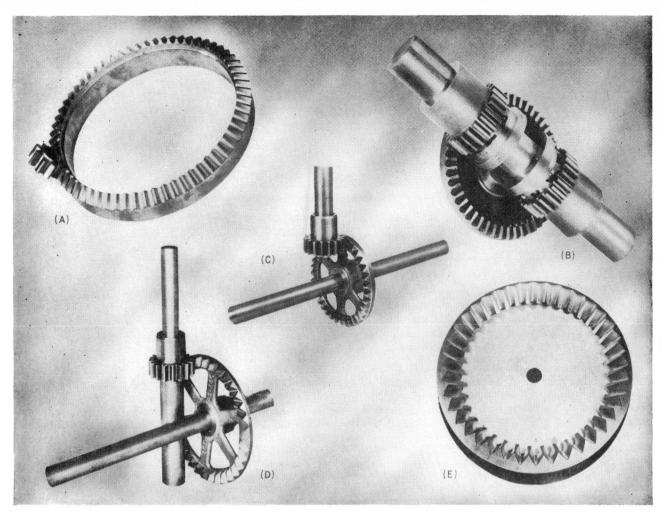

SOME BASIC types of tapered gears: (A), tapered gear and spur gear that mesh at a low shaft angle; (B), and (C), tapered (face) gears having 90 deg shaft angles; (D), offset face gear and pinion; and (E) internal tapered gear shown without pinion.

Effect of Axis Angle On Tapered Gear Design

GEORGE SANBORN
Chief Field Engineer

BEN BLOOMFIELD
Engineering Department
Fellows Gear Shaper Company

Relationships between the axis angle of tapered gear combinations and the amount of involute action, the face width of the tapered member, and the amount of offset of those members having axis that are non-intersecting.

A TAPERED GEAR differs from a conventional bevel gear in that it may mesh either with a tapered or cylindrical pinion. The axis of the gear and mating pinion may or may not intersect, and the taper angle may be anything up to 90 deg for an external gear, or up to 180 deg for a tapered internal gear. The tooth form may be helical or spur, depending on whether the teeth on the mating pinion are helical or spur. This article will be confined to a discussion of spur tapered gears.

Tapered and face gears having spur teeth, not on a helix, have many advantages. One is that the contact between teeth of mating members does not start across the entire face width of the gear as in conventional spur gears, but rather starts at one edge of the face and gradually creeps across. This is due to the angular relationship of the axes. On designs having a low axis angle, this advantage is not very great, and it is usually desirable to employ helical teeth if noise and smooth running are of prime importance. On high taper angle gears and face gears, the advantage in employing helical teeth becomes less unless quiet running action is important.

When running two tapered gears together, full contact bearings are not theoretically or actually obtained. The bearings are either restricted to the

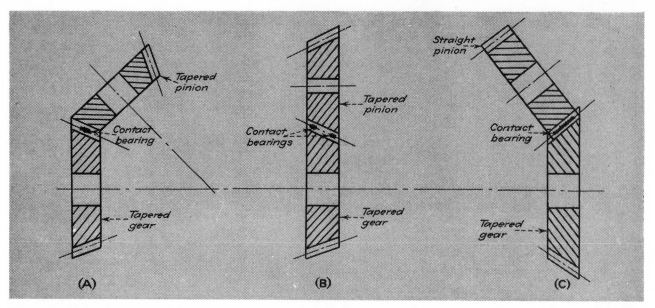

Fig. 1—Contact bearings on tapered gears meshing (A) with a tapered pinion at a low contact angle; (B) end-on-end with a tapered pinion; and (C) with a straight spur gear at a low contact angle. In the first case, the bearing surfaces are confined to center of the teeth; in the second, to the outer faces; and in the last, are across the face.

middle of the face or to the outer edges. For example, if two tapered gears of five inch diameter and ten degree taper angle are cut with a standard three inch gear shaper and run together on properly spaced centers at a shaft angle of approximately twenty degrees, the bearings would be confined to the center of the face. In this case, the large end of taper on one gear would mesh with the large end of the other gear, Fig. 1.

If these same two gears were meshed large end to small end on parallel shafts as shown at B, the bearings would be confined to the outer edges of the face. The nature of these bearings can be varied by changing the diameter of the gear shaper cutter employed for cutting. It might be possible to approach a full bearing in this manner, but this undoubtedly would result in an impractical diameter of cutter to be handled.

The most advantageous way of obtaining a full bearing on an angular gear set is to make one member with straight cylindrical elements and put the taper all in the other member, as indicated at C. In this case, the straight member is cut like any conventional gear. The tapered member is cut with a gear shaper cutter having the same number of teeth as that in the mating straight member. This gives theoretical full contact between the teeth of the two members.

If it is desired to relieve the bearing slightly at the outer edges of the face, this can be easily accomplished by using a cutter for the tapered member that is made oversize for its number of teeth, or a cutter with a few additional teeth. The amount of departure

is governed by the amount of relief desired. There is no formula for determining the amount of thinning of the teeth at the edges of the face when using cutters which have more teeth than the mate, or are oversized, or both. This must be determined by layout, and can be referred to the cutter manufacturer.

Another advantage of tapered gears is that cutting the conventional member requires no special operations or equipment. The tapered member requires some special tooling on the gear shaper, but the cost of the cutting operation itself is comparable to that of cylindrical gears of the same general size and pitch.

Also, the gears are quite easy to assemble. Since one member does not have any taper, it can be positioned anywhere along its axis without changing the bearing between the mating teeth except, if carried to extreme, the two gears would not be in mesh. When mating with external tapered gears, the straight member can be assembled from either side as far as the gear is concerned. The tapered member must be positioned along its axis quite accurately but this applies to one member only. Errors in location of the tapered member will affect the bearings between the teeth; however, this is not particularly sensitive, especially with lower angles of taper.

Finally, tapered gear sets are simple to design, especially those having one straight and one tapered member. The straight member is usually standard in every detail. The diameters of the tapered gear can be calculated in the same way as the diameter of a conventional gear having the same num-

ber of teeth and pitch, with this exception: The pitch diameter at the large end can be, as a working maximum, 10 percent larger than the corresponding pitch diameter of a conventional gear, and the pitch diameter at the small end can be, as a working minimum, 5 percent less than the pitch diameter of a conventional gear. A greater difference in the diameters of the two ends usually is found feasible after making a careful check, but the above rule is close enough for a preliminary design of the gears and surrounding parts, such as housings.

This general rule can be expressed in formula form as follows:

$$\text{Large diameter} \atop \text{(Approximate)} = \frac{1.1\,N \pm 2\,\text{Cos.}\,\phi}{P} \quad (1)$$

$$\text{Small diameter} \atop \text{(Approximate)} = \frac{0.95\,N \pm 2\,\text{Cos.}\,\phi}{P} \quad (2)$$

where N is the number of teeth in the tapered gear, ϕ is the shaft or axis angle between the axes, and P is the theoretical diametral pitch of the pinion. In Eqs (1) and (2), the signs between the terms in the numerator are plus for external gears and minus for internal gears.

The approximate face width of the gear along the angle of taper then is found by dividing the difference in the diameters by twice the sign of the angle, or

$$\text{Face width (Approximate)} = \atop \frac{\text{Large Diameter} - \text{Small Diameter}}{2\,\text{Sin}\,\phi} \quad (3)$$

Limitation on Face Width

The tapered gear has a definite limitation on face width. The teeth of the tapered gear do not have a uniform

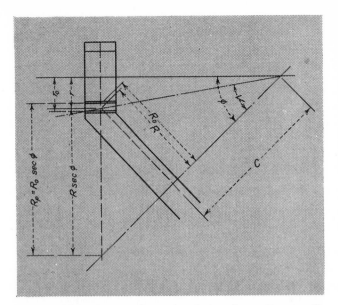

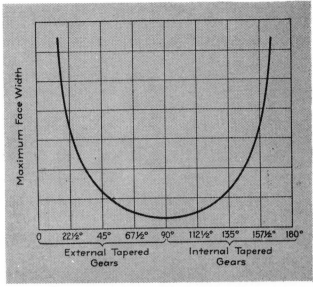

Fig. 2—Dimensional and angular designations: C, cone distance; O, shaft angle; R_o, outside radius of the gear; and r and r_o are the active pitch radius and outside radius of the pinion.

Fig. 3—The greater becomes the angle of taper, the less becomes the maximum face width on the tapered member, as indicated by this graphical presentation of the relationship.

cross-section across the face width, due to the fact that the pressure angle of the teeth vary; they have a low pressure angle at the small end of the taper and a high pressure angle at the large end of the taper. Consequently, an excessively wide face, resulting in a great difference between the small and large diameters, will cause the teeth to (1) come to a point near the large end; (2) be undercut severely at the small end; (3) or both. Although a good rule for a preliminary determination of these diameters has been given, it is always advisable to have designs of this kind checked by the tool vendor. The following formulas can be used to determine the pressure angle of the tapered gear and the involute action between the tapered gear and mating cylindrical gear, which will be the pinion, at a given cone distance.

Let ϕ = Shaft angle, or angle between the axes.

V = Pitch cone angle of tapered gear.

K = Gear ratio = $\dfrac{\text{No. pinion teeth}}{\text{No. gear teeth}}$

r = Active pitch radius of pinion.

R = Active pitch radius of tapered gear.

a = Base radius of pinion.

a_t = Nominal base radius of tapered gear.

R_c = Nominal active pitch radius of tapered gear.

C = Cone distance of tapered gear at active pitch point.

C_s = Cone distance of tapered gear at small end.

C_1 = Cone distance of tapered gear at large end.

PA = Pressure angle.

Then

$$\tan V = \frac{\sin \phi}{K + \cos \phi} \qquad (4)$$

$$R = \frac{r}{K} = C \tan V \qquad (5)$$

$$R_c = R \sec \phi \qquad (6)$$

$$a = \frac{\text{No. teeth in pinion}}{\text{Theoretical diametral pitch of pinion}} \times \text{Cos (Theoretical } PA \text{ of pinion).} \qquad (7)$$

$$a_t = \frac{a}{K} \sec \phi \qquad (8)$$

Cos PA at small end =

$$\frac{a \left(1 + \dfrac{\text{Sec } \phi}{K}\right)}{C_s \tan V \, (K + \text{Sec } \phi)} = \frac{a + a_t}{r + R_c} \qquad (9)$$

Cos PA at large end =

$$\frac{a \left(1 + \dfrac{\text{Sec } \phi}{K}\right)}{C_1 \tan V \, (K + \text{Sec } \phi)} \qquad (10)$$

To find the amount of involute action for any given cone distance C, indicated in Fig. 2:

Let r_o = Outside radius of the pinion.

R_o = Outside radius of the tapered gear.

R_p = Outside radius of the tapered gear in transverse plane of pinion.

N = Number of teeth on pinion.

Then $R_p = R_o \sec \phi$

Involute action =

$$\frac{\sqrt{r_o^2 - a^2} - \sqrt{R_p^2 - a_t^2} - (r + R_c) \sin PA}{\dfrac{2 \pi a}{n}} \qquad (11)$$

or, expressed differently,

Involute action =

$$\frac{\sqrt{r_o^2 - a^2} + \sec \phi \sqrt{\dfrac{R_o^2 - (a)^2}{K}}}{2 \pi a/n} - \frac{C_s \tan V \, (K + \sec \phi) \sin PA}{2 \pi a/n} \qquad (12)$$

It is always advisable to keep the involute action at more than 1.00 for at least half of the face width. For low angles of taper it is quite often possible to have more than an involute action of 1.00 for the entire face width, and even with face gears (90 deg taper angle), the good designer will get an involute action of more than 1.00 for more than half the face width.

Obviously, the greater the angle of

Fig. 4—Tapered gears having a low shaft angle and intersecting shafts.

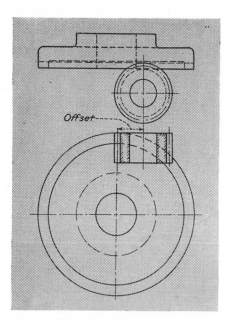

Fig. 5—The permissible amount of off-set of tapered gears with non-intersecting axis is limited by the taper angle. For face gears it is about one-fourth of the pitch diameter of the tapered member.

taper the more restricted the face width on the tapered member becomes. For example, for teeth having one degree of taper, the face width could be great before the 10 percent over-size and 5 percent undersize limitations were met. But if the angle of taper were near 90 deg, it would take a relatively narrow face to reach these limitations. This relationship is shown graphically in Fig. 3.

At 0 and 180 deg angles of taper there is no change in pressure angle across the face width of the gear (since the gear is a conventional external or internal gear in these cases) and hence

no limitation on face width. The variation in pressure angle across the face width of a tapered internal gear is not exactly the same as that of a tapered external gear, and for that reason, the maximum face widths of internal tapered gears for varying angles of taper is not the same. However, the same general rules apply for establishing the approximate diameter.

If it is necessary to use a face width greater than that possible by tapering one member only, the possibility of tapering both members should be investigated. If the diameters of gears involved are not extreme, a satisfactory contact can be obtained by properly relating cutter diameters. The tool vendor should be consulted in all cases.

Tapered Gears With Non-Intersecting Axis

So far in this article, consideration has been given only to tapered and face gears with intersecting axes. It is equally practicable to make tapered gears that operate with non-intersecting axes. There is a limitation to the amount of offset that can be employed, especially on high angle gears. On a 90 deg gear, as shown in Fig. 5, the maximum practicable offset is approximately one fourth of the pitch diameter of the tapered member. If too great an offset is employed, the teeth of the tapered member are badly mutilated, thus resulting in too great a loss of contact surface.

On low angles of taper the restriction of offset is different. The maximum offset in this case is equal to the sum of the pitch radii. A 90 deg tapered gear combination that is "on center" is shown at A in Fig. 6; position

B shows the maximum "off center" position. At this extreme, the taper effect actually disappears and the teeth of that member which are basically spur, take an appearance of being on a helix, the angle of which is the taper angle when the pinion is in position A. Intermediate positions of the pinion result in a combination of taper and apparent helix. If the maximum offset is employed with helical gears and the so called angle of taper is made 90 deg, these gears then become what is known as "hourglass gears."

The approximate large and small diameters of "offset" tapered gears can be calculated in the same way as for "on-center" gears, by merely taking the offset into consideration.

Let N = Number of teeth in tapered gear.
P = Theoretical diametral pitch of pinion.
ϕ = Taper angle.
e = Amount of offset (in.).

Then

Large diameter =
$$\sqrt{\left(\frac{1.1\,N \pm 2\cos\phi}{P}\right)^2 + e^2} \quad (13)$$

Small diameter =
$$\sqrt{\left(\frac{0.95\,N \pm 2\cos\phi}{P}\right)^2 + e^2} \quad (14)$$

It should be remembered that these diameters are approximate only.

The taper angle to which the gear blank should be turned is less than the shaft angle. This can be calculated as follows:

Tangent Blank Angle =
$$\frac{(\text{Large Diameter}-\text{Small Diameter})}{2 \times \text{face width}} \quad (15)$$

In taper gear drives it is possible to put the taper in either gear, whichever gives maximum benefit. If the gear ratio is relatively high and the maximum possible face width is desired, the taper should be put into the large member. This permits a greater change in diameter from end to end of the tapered gear for a given percentage of pitch diameter, thereby increasing the face width. Another influencing factor is the cutter; since the cutter for the tapered member must be about the size of the mating gear, one diameter might be practicable for the cutter whereas the other might not. For example, if a pair of gears were to be 5 inches and 10 inches in diameter, it would be better to put the taper on the 10 in. gear and use a 5 in. cutter than it would to put the taper on the 5 in. gear and use a 10 in. cutter.

There might be cases when it would be helpful in gear cutting or in assembling to put the taper in the smaller member. However, this should be avoided except in low taper angles. Face gears (90 deg taper angle) should always be larger than the mating pinions.

Fig. 6—A low shaft angle tapered gear combination that is at the on-center position (A) and at the maximum off-center position (B).

Fig. 1—Top view of a 20 tooth on-center face gear. Note how teeth are trimmed at inner edges.

Face Gear Design Factors

VICTOR FRANCIS and JOSEPH SILVAGI

Rules for determining the maximum practical outside diameter, the minimum practical inside diameter, and the face width of face gears. Data are included for calculating the principal dimensions of on-center face gears of different tooth pressure angles and gear tooth ratios.

THE FACE GEAR is a crown-type bevel gear that meshes at right angles with a conventional involute pinion. The pinion may be either spur or helical. The gears may have right-angle intersecting shafts forming a bevel gear set; or they may have right-angle non-intersecting shafts forming an off-set or hypoid arrangement.

Face gears have been commercially manufactured for many years. They have been used in textile machinery, truck engine governor drives, power mower transmissions, and gear cutting machinery, but the use of face gears has not been as widespread as their practicability warrants.

The pinion, being a conventional spur or helical gear, can be produced on a hobbing machine or on a gear shaper. The same shaper cutter can be used to produce both the pinion and the face gear.

On a conventional bevel gear of octoid tooth form, the face width theoretically can be increased outward indefinitely. Actually, the tool and the machine are the limiting factors. The pitch merely increases in proportion to the diameter, the pressure angle at the center of the tooth remains constant. Increasing the width of the face towards the apex of the gear is limited by the thickness of the tool and by side trimming of the teeth at the small end, which results from the tool being at the gear root angle and not at the theoretical cone angle.

On the face gear, however, the operating pressure angle increases towards the outside diameter, Fig. 1. The depth of the tooth remains constant. The maximum usable outside diameter is the diameter or a slightly larger diameter at which the teeth become pointed. The limiting inside diameter is the point where tooth trimming occurs, which is always outside the diameter where the operating pressure angle becomes zero. With a small number of teeth and a low gear ratio, it is possible to have the minor trimming diameter equal the pointed outside diameter; such a condition would leave no usable face width. The gear shown in Fig. 1 has a face width of one inch; if the trimmed portion on the inside and the pointed portion on the outside were removed, an effective face width of one-third of an inch would remain.

The physical appearance of face

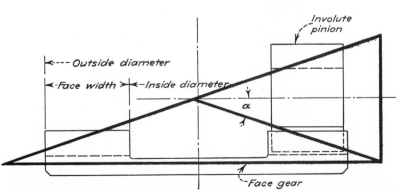

Fig. 2—Rolling pitch bodies for a spur gear and on-center face gear are cones.

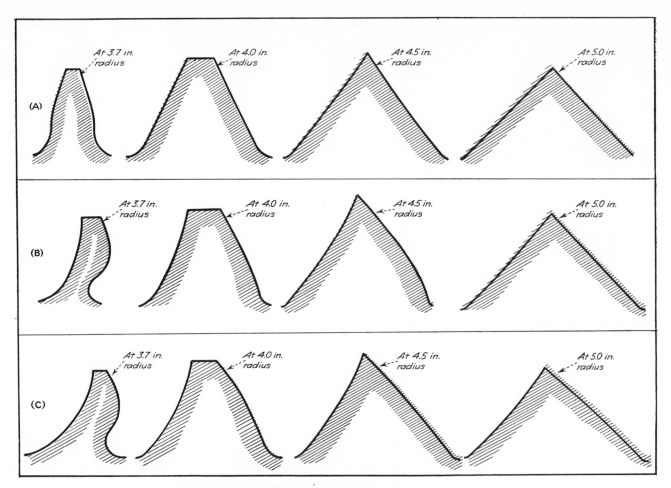

Fig. 4—Calculated profiles for a 1 diametral pitch, 8 tooth gear, taken at 3.7, 4.0, 4.5 and 5 in. radii. (A) On-center gear. (B) Offset gear. (C) Helical gear. Profiles do not show the effects of trimming.

Fig. 3—Top view of offset face gear. Interference of the cutting tool produced the trimming that can be seen on the inner or small end of the teeth.

gears, at first glance, resembles that of crown gears, and does not suggest that the pitch skeletons are cones. The pinion is made as a true cylindrical involute gear. The two gears when meshed with intersecting axes, Fig. 2, act as bevel gears and do have cones as their pitch surfaces or rolling bodies.

Since the operating pressure angle changes from the minor to the major diameters and the working depth of the two gears is constant, the instantaneous line of tooth contact is a diagonally curved line moving along the tooth as the face gear and the pinion roll together.

For ease of computation when designing the gears, it is sometimes assumed that the face gear tooth surface is composed of approximately straight lines. The surface, however, does not contain any theoretically straight lines. The line elements composing the tooth surface can be concave, convex, and even both, depending on the numbers of teeth and the gear ratio.

With pinion offset or helical conditions, the profile variations are more pronounced. In Fig. 3, notice how the teeth become trimmed at the small end as a result of radial crowding and how the teeth become pointed at the outside diameter because of the pressure angle increasing with increas-

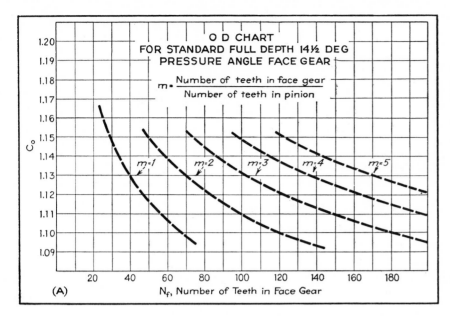

(A)

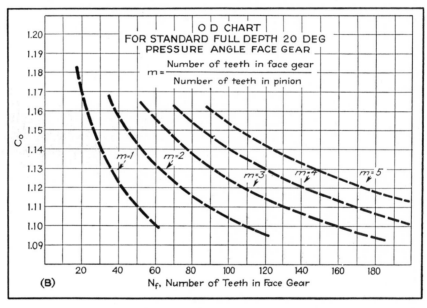

(B)

Fig. 5—Charts for obtaining the factor C_0 used in the formula for computing maximum usable outside diameter of on-center face gears. (A) Values for standard full depth $14\frac{1}{2}$ deg pressure angle. (B) Values for standard full depth 20 deg pressure angle. (C) Values for standard 25 deg pressure angle.

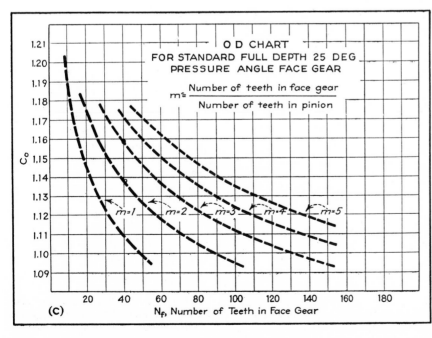

(C)

ing diameters. In Fig. 4 are shown profiles of the tooth forms at different diameters of the face gear for on-center, hypoid, and helical types. These profiles are for an extremely exaggerated case of an eight tooth face gear meshing with an eight tooth, 30 deg pressure angle pinion. This condition was chosen to show more clearly the curvatures of the profiles that are obtained. The effects of trimming or undercutting have been omitted in drawing these profiles.

In actual practice the gear ratio is seldom as low as 1.0, and the number of teeth in the face gear would never be as small as eight. A five diametral pitch set with 20 teeth in the pinion and 20 teeth in the gear would have about 0.005 in. concavity of the profile for an on-center type.

The face width of face gears is limited by the teeth becoming pointed at the outside diameter and trimmed or undercut at the inside diameter.

Profile curvatures are caused by numerous factors and the exact determination of the maximum outside diameter is mathematically difficult. To facilitate the design of spur, on-center face gears, charts for assisting the designer in determining the maximum outside diameter at which the face gear teeth become pointed are given in Fig. 5 for standard full depth $14\frac{1}{2}$, 20, and 25 deg pressure angle face gears. In most cases, the outside diameter will be designed smaller than this maximum diameter. Some gears, however, will have to be designed with a slightly larger outside diameter to obtain sufficient face width to carry the desired load. The tooth contact ratio at this increased diameter will still be greater than one but a large amount

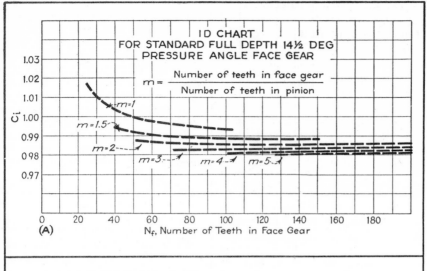

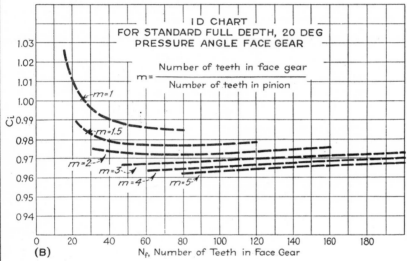

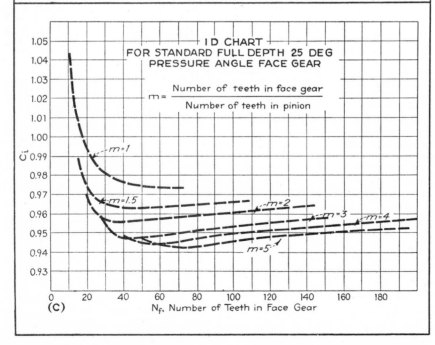

Fig. 6—Charts for obtaining the factor C_i used in the formula for computing the minimum usable inside diameter of on-center face gears. (A) Values for standard full depth $14\frac{1}{2}$ deg pressure angle. (B) Values for standard full depth 20 deg pressure angle. (C) Values for standard full depth 25 deg pressure angle.

of relative sliding will occur. Thus, if necessary, it is permissible to have pointed teeth for a small portion of the face width.

N_f = number of teeth in the face gear.
N_p = number of teeth in the pinion.
m = N_f/N_p.
D_{fo} = pitch diameter of the face gear and is equal to m times the pitch diameter of the pinion, in.
C_o = equals a constant from chart, Fig. 5, obtained from the intersection of the N_f ordinate with ratio m curve and reading horizontally to the left.

the pointed diameter is

Pointed diameter = $C_o D_{fo}$, in.

For determining the Minor Diameter where trimming occurs, the charts given in Fig. 6 will be found useful. They are based upon the establishment of the face gear pitch plane, Fig. 7, as the limit for the height of the trimmed portion.

To use the I D Charts, locate the intersection of N_f with the m ratio curve, and read across horizontally to obtain a value of C_i. Then

Minor dia of face gear = $C_i D_{fo}$

EXAMPLE 1. Find the minimum inside dia, maximum outside dia, and the face width of a 40 tooth face gear driven by a 20 tooth full depth standard tooth thickness, 2 diametral pitch, 20 deg pressure angle pinion.

m = 40/20 = 2.0
D_{fo} = (20/2) × 2 = 20

SOLUTION: Use chart in Fig. 6 (B). The intersection of the 40 tooth face gear ordinate N_f crosses the m curve 2 at the point where C_i has a value of 0.972.
Hence, the minimum ID is

$C_i D_{fo}$ = 0.972 × 20 = 19.44 in., approximately 19 7/16 in.

This dimension is the minimum diameter that may be used, since the chart is based on a diameter where trimming occurs at the middle of the tooth height. A slight amount of trimming may occur above or below the middle as shown in Fig. 7. This small amount of trimming, however, will not be detrimental to the quiet running qualities of the gears and allows the use of the maximum amount of profile.

To find the maximum outside diameter of the face gear, use the OD Chart Fig. 5(B). The intersection of the 40 tooth face gear ordinate crosses the m curve 2, at the point where C_o equals 1.159. Hence, the maximum OD is

$C_o D_{fo}$ = 1.59 × 20 = 23.18 in., approximately 23 3/16 in.
Face width = (23 3/16 − 19 7/16)/2 = 1 7/8 in.

EXAMPLE 2. Find the minimum inside diameter, maximum outside diameter and the face width of a 126 tooth face gear driven by a 36 tooth full depth standard tooth thickness, 10 diametral pitch, $14\frac{1}{2}$ degree angle pinion.

$$m = 126/36 = 3.5$$
$$D_{fg} = (36/10) \times 3.5 = 12.60$$

SOLUTION: From chart Fig. 6(A)

$$C_i = 0.981$$

Then

$$ID = C_i D_{fg} = 0.981 \times 12.60 = 12.36$$
in., approximately 12 3/8 in.

From chart Fig. 5(A)

$$C_o = 1.129$$

Then

$$OD = C_o D_{fg} = 1.129 \times 12.6 = 14.23 \text{ in.,}$$
approximately $14\frac{1}{4}$ in.

Face width $= (14\frac{1}{4} - 12\ 3/8)/2 = 15/16$ in.

If the face width as computed with chart values is too large and must be reduced, most of the un-needed face should be removed from the outer diameter and a smaller amount from the inner diameter, since small pressure angles transmit power more effectively than do large pressure angles.

On parallel axis gear sets, when undercutting would occur in a pinion of

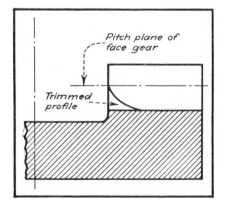

Fig. 7—Intersection of trimmed profile with pitch plane of face gear.

standard tooth form, the common method of avoiding the undercut is to make the pinion oversize. This solution, in effect, transforms the pinion into one of a slightly larger diametral pitch and a larger pressure angle. By operating the modified pinion with a gear made proportionally undersize, the original lower and superior operating pressure angles can be maintained.

The operating pressure angle of a face gear set, however, varies from a minimum value at the inside diameter to a maximum value at the outside diameter of the face gear. Because of this change of pressure angle the advantage of making a pinion oversize to avoid undercutting is lost. The better practice is to make the set with a higher basic pressure angle. Values of C_o and C_i for pressure angles lying

between those given on the charts can be accurately interpolated from the chart values.

EXAMPLE 3. Find the value of C_o for a 54 tooth face gear driven by an 18 tooth full depth standard tooth thickness 22 deg pressure angle pinion?

$$m = 54/18 = 3.0$$

SOLUTION: From charts Figs. 5(A) and (B).

For 20 deg, $C_o = 1.164$
For 25 deg, $C_o = 1.144$
$$\Delta C_o = 1.164 - 1.144 = 0.020$$
$$\Delta \phi = 25 - 20 = 5 \text{ deg}$$

For 22 deg pressure angle

$$C_o = 1.164 - (0.020 \times 2/5) = 1.1566$$

When compared with the conventional type of bevel gear, face gears have the following advantages:

1. The face gear can be checked for accuracy of tooth profile on a checking fixture.

2. The pinion of a face gear set can be adjusted axially at assembly without affecting contact or backlash. Thus, only one assembly distance has to be maintained accurately as compared with two assembly distances for conventional bevel gears. On the condition that its face remains in contact with the face gear, the pinion may be given any amount of axial movement. There is no axial thrust on the pinion bearings of an on-center gear set.

3. Pins or balls can be used for checking the size of face gears and the dimension can be computed when designing the gears.

4. The generation is positive with no trial and error settings required to obtain the conjugate profile.

5. Hypoid, or off-center face gears, are produced on the same machine and with the same cutter that are used to generate on-center face gears and the cylindrical involute pinions. Furthermore, the cost of the product is not increased by changing from on-center to hypoid face gears.

The disadvantages of face gears as compared with the conventional bevel gear are:

1. In some designs, the face width of face gears is limited by pointed teeth and trimming.

2. Shaper cutters for face gears are usually less versatile than the cutter blades used to produce conventional bevel gears.

Face gears are made by a continuous generating-shaping process on a Fellows Gear Shaper or similar type machine, Fig. 8, using a special fixture designed to put the face gear axis at right angles to the pinion axis. The pinion-type cutter is a duplicate of the mating pinion. If a twenty tooth pinion is to mesh with a sixty-eight tooth face gear, the shaper cutter for the face

gear should also have twenty teeth.

In practice, the cutter is made somewhat larger in diameter and possibly in number of teeth in order to use standard cutters, to obtain more latitude in assembly, or to achieve good production with the larger cutters. Loss of contact and profile modification occurs with this practice, depending on how much the cutter deviates from the mating pinion.

To produce optimum running face gears, a cutter with the same number of teeth as the pinion should be used and should be slightly larger in diameter. This condition must prevail throughout the life of the cutter to prevent binding of the pinion with the face gear at assembly. A side trimming attachment can be used on the generator to compensate for thinning the cutter tooth as it is reground. A cutter with a smaller number of teeth than the pinion should never be used.

The relative number of teeth determine the permissible variation from the ideal number of teeth that the cutter may have. For instance, if a twenty tooth pinion is to drive a twenty-four

Fig. 8—Shaping a 20 tooth on-center face gear with a 20 tooth, 20 deg cutter.

tooth face gear, the cutter should have twenty teeth to maintain good conjugate action. But if a twenty tooth pinion is to drive a hundred tooth face gear, the number of teeth in the cutter could vary from twenty to twenty-five, depending on the degree of quietness, cutter life, and production rate desired.

ACKNOWLEDGMENT: Photographs used to illustrate this article have been obtained through the courtesy of Atlas Gear Company.

Gear-force Analysis

TYLER G HICKS
Mechanical engineer, Hicksville, N Y

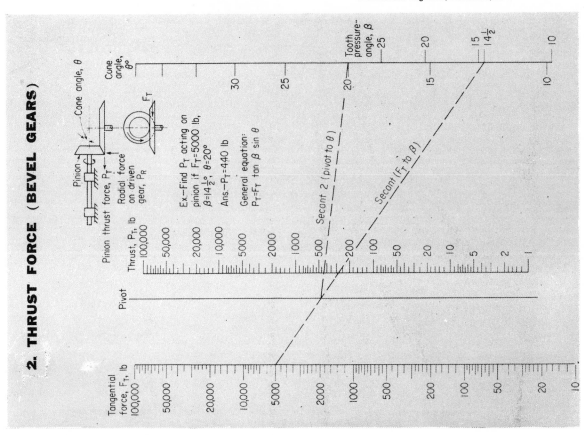

2. THRUST FORCE (BEVEL GEARS)

Ex.—Find P_T acting on pinion if $F_T = 5000$ lb, $\beta = 14\frac{1}{2}$, $\theta = 20°$.

Ans.—$P_T = 440$ lb

General equation:
$$P_T = F_T \tan \beta \sin \theta$$

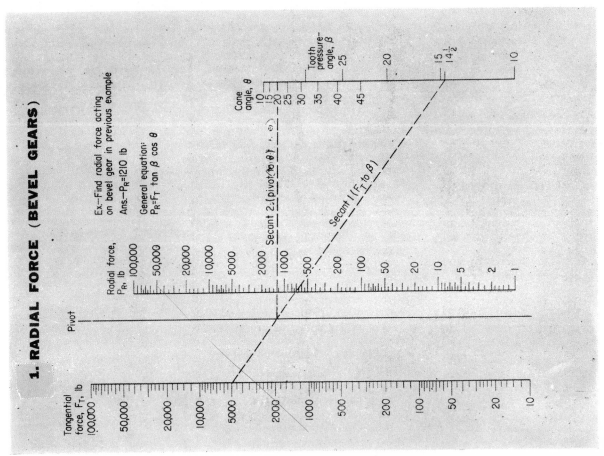

1. RADIAL FORCE (BEVEL GEARS)

Ex.—Find radial force acting on bevel gear in previous example

Ans.—$P_R = 1210$ lb

General equation:
$$P_R = F_T \tan \beta \cos \theta$$

3
PLANETARY GEAR SYSTEMS

Efficiency and speed-ratio formulas for

Planetary gear systems

Here are complete torque, speed, and power analyses, and final formulas ready for use

JOHN H. GLOVER, Product Design Engineer, Transmission and Chassis Div, Ford Motor Co, Detroit, Mich

THE remarkably high speed reductions that are produced in relatively small spaces by planetary gear systems fascinate gear designers. New gear arrangements are periodically developed—you might say invented—and it is surprising to note how similar many of them look when compared to each other. Yet, slight differences can cause one arrangement to have a much higher or lower speed-reduction ratio than another.

But speed reduction is not the only design criterion. Just as important is the overall efficiency of the gear system. This determines the amount of power that the gear system consumes (mostly in the form of heat), and hence the horsepower requirements of the motor that drives the planetary.

Most designers would find it difficult to calculate the overall efficiency of a complex planetary. And when they derive the efficiency formula they cannot be sure that the formula is correct. In fact, the types with multiple carriers and split power, such as Ex 8, p 98, seldom have been analyzed

This article, then, not only provides new easy-to-use "cook-book" formulas for overall efficiency and speed ratio, but also presents torque and angular speed values for the member-gears of the planetaries which are needed for the design of the gears themselves. Such torque and speed values are also needed for the design of clutches or fastenings to lock the reaction gears from turning (the gears that do not rotate in the assembly).

How to use the boxed examples

Schematic diagrams displaying the skeletal structure of the various gears and planet carrier hook ups (see the illustrations below, left, for an explanation of the diagrams). The schematics also show the torque flow and power flow through all the members of the gear assembly.

Formulas for the overall speed ratio, R. This is a function of the number of teeth of the gears (N_1, N_2, etc). Typical values for the number of teeth have been selected in the examples to illustrate how torque and power splits or combines while flowing through the gears.

Formulas for overall efficiency, E. These formulas require knowing only

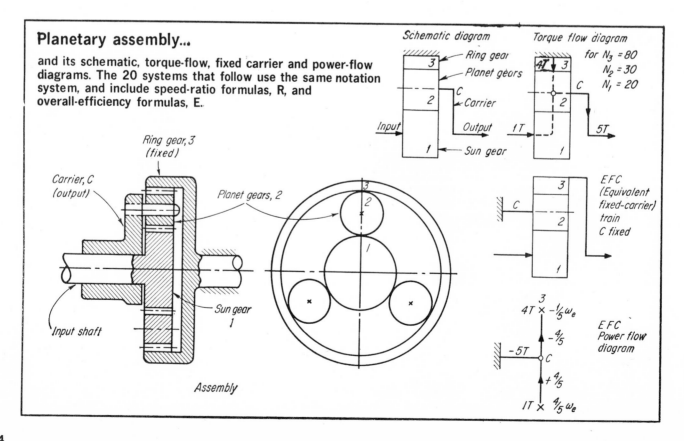

Planetary assembly...

and its schematic, torque-flow, fixed carrier and power-flow diagrams. The 20 systems that follow use the same notation system, and include speed-ratio formulas, R, and overall-efficiency formulas, E.

the number of teeth and the efficiency between meshing teeth. For example, $\epsilon_{1,4}$ is the mesh efficiency between gears 1 and 4. You can assume a mesh efficiency of $\epsilon = 0.98$ (98%) for each external mesh and $\epsilon = 0.99$ for each internal mesh, in keeping with common practice. Because power losses are always *positive* in sign, the terms in the efficiency equations which constitute losses are frequently enclosed in bars to indicate that an absolute (positive) value is always to be used.

Table of torque and speed values for the various members of the system. Each member is listed in the first column. The letter C stands for carrier (see list of symbols at right); the numbers represent gears.

The second column in the table shows the role that each member plays in the gear set. It may be the input or output member; it may be a reaction member, as is the case with the ring gear (3) in Example 1; or it may be an idler gear with no influence on the gear ratio.

The third column gives the torque acting on each member in relative terms. The torque on the input is set at 1, idlers have zero torque, and the torque on the other members is usually in terms of the speed ratio, R.

Hence in Example 1, the torque on the output member is equal to $-R$, or for the number of teeth shown, $R = \frac{1}{5}$, and the output torque will be $\frac{1}{5}$ the input torque. The reaction member takes up $\frac{4}{5}$ of the input torque. Input speed is assigned 1.

The fourth column gives the absolute angular speed, ω, of the member around its own axis (which includes the speed imparted to the member by a moving carrier).

The fifth column gives the angular speed of tooth engagement, ω_ϵ, which is its absolute speed minus the speed of the carrier.

By multiplying the T and ω values you obtain the power transmitted through a member. Also, by multiplying the T and ω_ϵ values you obtain the power circulating in its tooth engagement (if it is a gear).

The boxed systems also display an "equivalent fixed carrier" power-flow diagram. By assuming that the carrier is fixed you quickly see a non-dimensional, relative-value flow of torque, speed, and power, starting with assigned values of unity for input torque and input angular speed.

SYMBOLS

C = carrier (also called "spider")—a non-gear member of a gear train whose rotation affects gear ratio
N = number of teeth
R = overall speed reduction ratio = ω_i/ω_o
ω = angular speed
T = torque
ϵ = efficiency between meshing teeth
E = overall planetary gear mesh efficiency

Use of double subscripts
ϵ_{13} = mesh efficiency between gears *1* and *3*

Subscripts
o = output
i = input
r = reaction
c = carrier
e = equivalent
1, 2, 3, etc.— gears in a train

1. Simple planetary system -- input to carrier

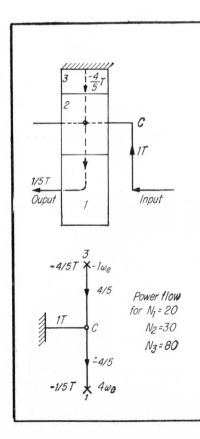

Power flow for $N_1 = 20$ $N_2 = 30$ $N_3 = 80$

Speed-ratio formula

$$R = \frac{1}{1 + \frac{N_3}{N_1}}$$

Efficiency formula

$$E = \frac{1}{1 + \left[\frac{N_3}{N_1 + N_3}\left(\frac{1}{\epsilon_{32}\,\epsilon_{21}} - 1\right)\right]}$$

Member	Role	T	ω	ω_e
C	Input	1	1	–
1	Output	$-R$	1/5	$\omega - \omega_c$
		$-1/5$	5	4
3	Reaction	$R-1$	0	$\omega - \omega_c$
		$-4/5$		-1
2	Idler	0	–	$m_{21}\,\omega_{e1}$
				$-8/3$

2. Simple planetary system -- input to sun

(See p 101 for other planetary inversions.)

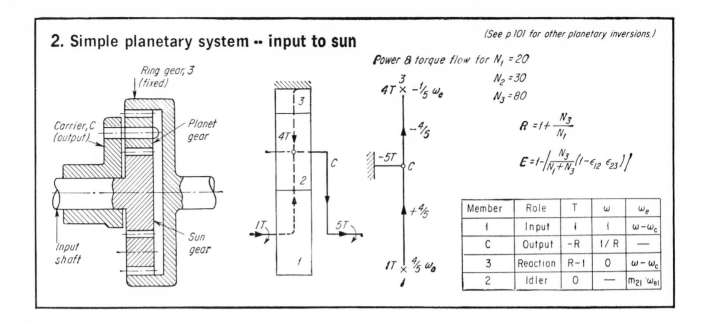

Ring gear, 3 (fixed)

Carrier, C (output)

Planet gear

Input shaft

Sun gear

Power & torque flow for $N_1 = 20$
$N_2 = 30$
$N_3 = 80$

3
$4T \times -\frac{1}{5}\omega_e$

$-\frac{4}{5}$

$-5T$ C

$+\frac{4}{5}$

$1T \times \frac{4}{5}\omega_e$
1

$$R = 1 + \frac{N_3}{N_1}$$

$$E = 1 - \left| \frac{N_3}{N_1 + N_3}(1 - \epsilon_{12}\,\epsilon_{23}) \right|$$

Member	Role	T	ω	ω_e
1	Input	1	1	$\omega - \omega_c$
C	Output	$-R$	$1/R$	—
3	Reaction	$R-1$	0	$\omega - \omega_c$
2	Idler	0	—	$m_{21}\,\omega_{e1}$

3. Simple planetary system -- with carrier fixed

Torque flow

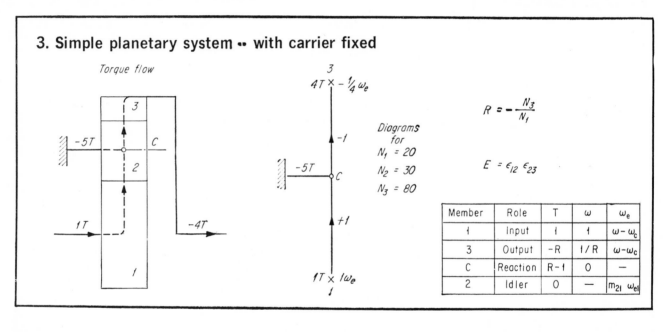

3
$4T \times -\frac{1}{4}\omega_e$

-1

$-5T$ C

Diagrams for
$N_1 = 20$
$N_2 = 30$
$N_3 = 80$

$+1$

$1T \times 1\omega_e$
1

$$R = -\frac{N_3}{N_1}$$

$$E = \epsilon_{12}\,\epsilon_{23}$$

Member	Role	T	ω	ω_e
1	Input	1	1	$\omega - \omega_c$
3	Output	$-R$	$1/R$	$\omega - \omega_c$
C	Reaction	$R-1$	0	—
2	Idler	0	—	$m_{21}\,\omega_{e1}$

4. Compound planetary -- input to carrier

Torque flow

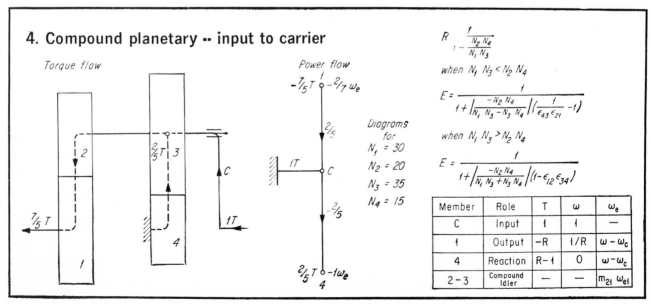

Power flow

$-\frac{7}{5}T \circ -\frac{2}{7}\omega_e$

$\frac{2}{5}$

$1T$ C

Diagrams for
$N_1 = 30$
$N_2 = 20$
$N_3 = 35$
$N_4 = 15$

$-\frac{2}{5}$

$\frac{2}{5}T \circ -1\omega_e$
4

$$R = \frac{1}{1 - \frac{N_2\,N_4}{N_1\,N_3}}$$

when $N_1\,N_3 < N_2\,N_4$

$$E = \frac{1}{1 + \left| \frac{-N_2\,N_4}{N_1\,N_3 - N_3\,N_4} \right| \left(\frac{1}{\epsilon_{43}\,\epsilon_{21}} - 1 \right)}$$

when $N_1\,N_3 > N_2\,N_4$

$$E = \frac{1}{1 + \left| \frac{-N_2\,N_4}{N_1\,N_3 + N_3\,N_4} \right| (1 - \epsilon_{12}\,\epsilon_{34})}$$

Member	Role	T	ω	ω_e
C	Input	1	1	—
1	Output	$-R$	$1/R$	$\omega - \omega_c$
4	Reaction	$R-1$	0	$\omega - \omega_c$
2-3	Compound Idler	—	—	$m_{21}\,\omega_{e1}$

5. Compound planetary -- input to sun

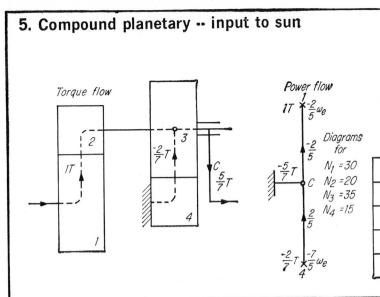

Torque flow

Power flow

$$R = 1 - \frac{N_2 N_4}{N_1 N_3}$$

When $N_1 N_3 < N_2 N_4$

$$E = 1 - \left|\frac{-N_2 N_4}{N_1 N_3 - N_2 N_4}\right| (1 - \epsilon_{12}\,\epsilon_{34})$$

When $N_1 N_3 > N_2 N_4$

$$E = 1 - \left|\frac{-N_2 N_4}{N_1 N_3 - N_2 N_4}\right| \left(\frac{1}{\epsilon_{43}\,\epsilon_{21}} - 1\right)$$

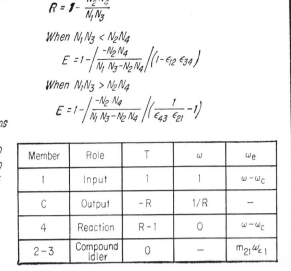

Diagrams for
$N_1 = 30$
$N_2 = 20$
$N_3 = 35$
$N_4 = 15$

Member	Role	T	ω	ω_e
1	Input	1	1	$\omega - \omega_c$
C	Output	$-R$	$1/R$	—
4	Reaction	$R-1$	0	$\omega - \omega_c$
2-3	Compound Idler	0	—	$m_{21}\,\omega_{e1}$

6. Double-eccentric planetary

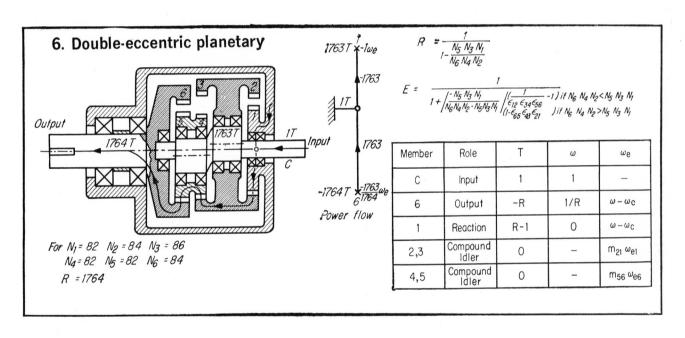

Output

Input

Power flow

$$R = \frac{1}{1 - \dfrac{N_5 N_3 N_1}{N_6 N_4 N_2}}$$

$$E = \frac{1}{1 + \left|\dfrac{-N_5 N_3 N_1}{N_6 N_4 N_2 - N_5 N_3 N_1}\right| \left(\begin{array}{l}\frac{1}{\epsilon_{12}\,\epsilon_{34}\,\epsilon_{56}} - 1 \; \text{if } N_6 N_4 N_2 < N_5 N_3 N_1 \\ (1 - \epsilon_{65}\,\epsilon_{43}\,\epsilon_{21}) \; \text{if } N_6 N_4 N_2 > N_5 N_3 N_1\end{array}\right)}$$

For $N_1 = 82$ $N_2 = 84$ $N_3 = 86$
$N_4 = 82$ $N_5 = 82$ $N_6 = 84$
$R = 1764$

Member	Role	T	ω	ω_e
C	Input	1	1	—
6	Output	$-R$	$1/R$	$\omega - \omega_c$
1	Reaction	$R-1$	0	$\omega - \omega_c$
2,3	Compound Idler	0	—	$m_{21}\,\omega_{e1}$
4,5	Compound Idler	0	—	$m_{56}\,\omega_{e6}$

7. Triple-planet planetary

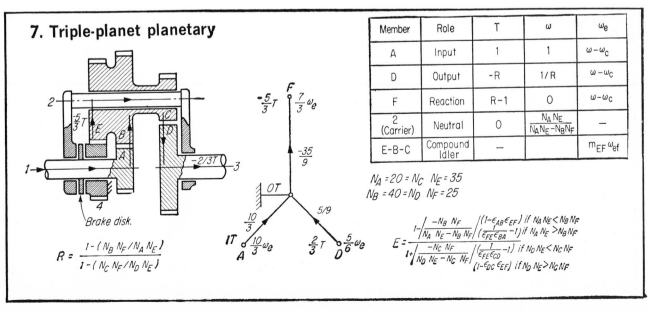

Brake disk.

Member	Role	T	ω	ω_e
A	Input	1	1	$\omega - \omega_c$
D	Output	$-R$	$1/R$	$\omega - \omega_c$
F	Reaction	$R-1$	0	$\omega - \omega_c$
2 (Carrier)	Neutral	0	$\dfrac{N_A N_E}{N_A N_E - N_B N_F}$	—
E-B-C	Compound Idler	—	—	$m_{EF}\,\omega_{ef}$

$N_A = 20 = N_C$ $N_E = 35$
$N_B = 40 = N_D$ $N_F = 25$

$$R = \frac{1 - (N_B N_F / N_A N_E)}{1 - (N_C N_F / N_D N_E)}$$

$$E = \frac{1 - \left|\dfrac{-N_B N_F}{N_A N_E - N_B N_F}\right| \begin{array}{l}(1 - \epsilon_{AB}\,\epsilon_{EF}) \; \text{if } N_A N_E < N_B N_F \\ (\frac{1}{\epsilon_{FE}\,\epsilon_{BA}} - 1) \; \text{if } N_A N_E > N_B N_F\end{array}}{1 + \left|\dfrac{-N_C N_F}{N_D N_E - N_C N_F}\right| \begin{array}{l}(\frac{1}{\epsilon_{FE}\,\epsilon_{CD}} - 1) \; \text{if } N_D N_E < N_C N_F \\ (1 - \epsilon_{DC}\,\epsilon_{EF}) \; \text{if } N_D N_E > N_C N_F\end{array}}$$

8. Triple ring-gear drive

$$R = \left(1 + \frac{N_3}{N_1}\right)\left[\left(1 + \frac{N_6}{N_4}\right)\left(-\frac{N_9}{N_7}\right) - \frac{N_6}{N_4}\right] - \frac{N_3}{N_1}$$

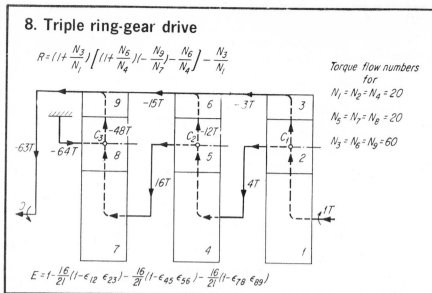

Torque flow numbers for

$N_1 = N_2 = N_4 = 20$

$N_5 = N_7 = N_8 = 20$

$N_3 = N_6 = N_9 = 60$

$$E = 1 - \frac{16}{21}(1 - \epsilon_{12}\,\epsilon_{23}) - \frac{16}{21}(1 - \epsilon_{45}\,\epsilon_{56}) - \frac{16}{21}(1 - \epsilon_{78}\,\epsilon_{89})$$

Member	Role	$\cdot T$	ω
1	Input	1	1
10	Output	$-R$	$1/R$
C_3	Reaction	$R-1$	0
3	Split output from 1	$-m_{13}$	ω_0
C_1		$-(1-m_{13})$	ω_4
4	Input to 2	$-T_{c1}$	$\omega_e + \omega_{c2}$

Member	Role	T	ω
6	Split output from 2	$-(1-m_{13})m_{45}$	ω_0
C_2		$T_{c1}(1-m_{46})$	ω_7
7	Input to (3)	$-T_{c2}$	ω_e
9	Output from (3)	$-T_7\,m_{79}$	ω_0
2	Idler	0	$m_{21}\,\omega_{e1}$
5	Idler	0	$m_{54}\,\omega_{e4}$
8	Idler	0	$m_{87}\,\omega_{e7}$

9. Minuteman-cover drive

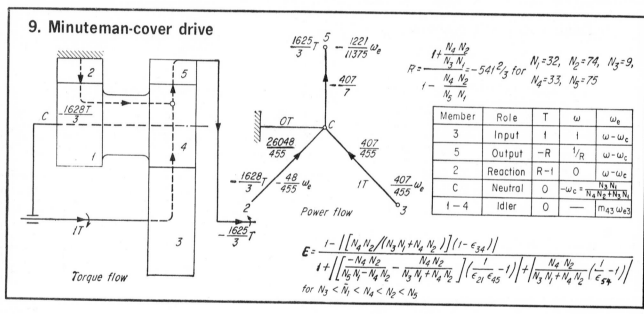

$$R' = \frac{1 + \dfrac{N_4 N_2}{N_3 N_1}}{1 - \dfrac{N_4 N_2}{N_5 N_1}} = -541\tfrac{2}{3} \text{ for } \quad N_1 = 32,\ N_2 = 74,\ N_3 = 9,\ N_4 = 33,\ N_5 = 75$$

Power flow

Torque flow

Member	Role	T	ω	ω_e
3	Input	1	1	$\omega - \omega_c$
5	Output	$-R$	$1/R$	$\omega - \omega_c$
2	Reaction	$R-1$	0	$\omega - \omega_c$
C	Neutral	0	$-\omega_c = \dfrac{N_3 N_1}{N_4 N_2 + N_3 N_1}$	
1 – 4	Idler	0	—	$m_{43}\,\omega_{e3}$

$$E = \frac{1 - \left|\left[N_4 N_2 /(N_3 N_1 + N_4 N_2)\right](1 - \epsilon_{34})\right|}{1 + \left|\left[\dfrac{-N_4 N_2}{N_5 N_1 - N_4 N_2} - \dfrac{N_4 N_2}{N_3 N_1 + N_4 N_2}\right]\left(\dfrac{1}{\epsilon_{21}\,\epsilon_{45}} - 1\right)\right| + \left|\dfrac{N_4 N_2}{N_3 N_1 + N_4 N_2}\left(\dfrac{1}{\epsilon_{54}} - 1\right)\right|}$$

for $N_3 < \bar{N}_1 < N_4 < N_2 < N_5$

Coupled planetary drives

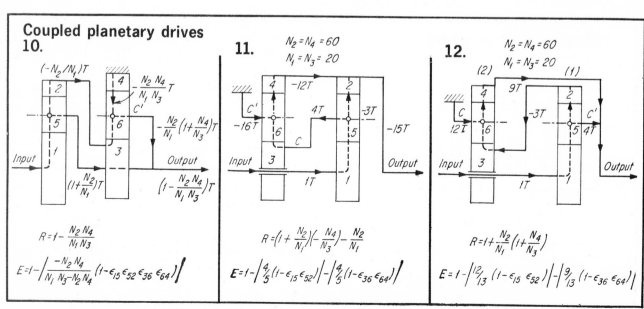

10.

$$R = 1 - \frac{N_2 N_4}{N_1 N_3}$$

$$E = 1 - \left|\frac{-N_2 N_4}{N_1 N_3 - N_2 N_4}(1 - \epsilon_{15}\,\epsilon_{52}\,\epsilon_{36}\,\epsilon_{64})\right|$$

11. $N_2 = N_4 = 60$; $N_1 = N_3 = 20$

$$R = \left(1 + \frac{N_2}{N_1}\right)\left(-\frac{N_4}{N_3}\right) - \frac{N_2}{N_1}$$

$$E = 1 - \left|\frac{4}{5}(1 - \epsilon_{15}\,\epsilon_{52})\right| - \left|\frac{4}{5}(1 - \epsilon_{36}\,\epsilon_{64})\right|$$

12. $N_2 = N_4 = 60$; $N_1 = N_3 = 20$

$$R = 1 + \frac{N_2}{N_1}\left(1 + \frac{N_4}{N_3}\right)$$

$$E = 1 - \left|\frac{12}{13}(1 - \epsilon_{15}\,\epsilon_{52})\right| - \left|\frac{9}{13}(1 - \epsilon_{36}\,\epsilon_{64})\right|$$

13. Coupled planetary drives (continued)

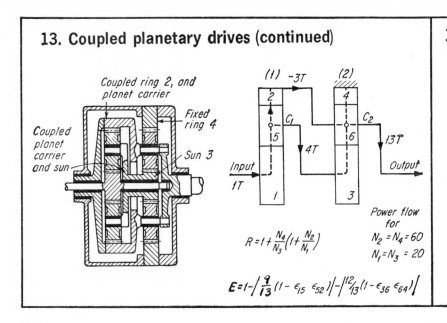

Coupled ring 2, and planet carrier

Coupled planet carrier and sun

Coupled planet carrier and sun

Fixed ring 4

Sun 3

(1) $-3T$ (2)

Input $1T$

C_1 C_2

$4T$

$13T$

Output

Power flow for
$N_2 = N_4 = 60$
$N_1 = N_3 = 20$

$$R = 1 + \frac{N_4}{N_3}\left(1 + \frac{N_2}{N_1}\right)$$

$$E = 1 - \left|\frac{9}{13}(1 - \epsilon_{15}\,\epsilon_{52})\right| - \left|\frac{12}{13}(1 - \epsilon_{36}\,\epsilon_{64})\right|$$

14. Differential drive

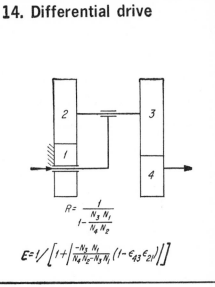

$$R = \frac{1}{1 - \frac{N_3\,N_1}{N_4\,N_2}}$$

$$E = 1 \Big/ \left[1 + \left|\frac{-N_3\,N_1}{N_4\,N_2 - N_3\,N_1}(1 - \epsilon_{43}\,\epsilon_{21})\right|\right]$$

Fixed differential drives
15.

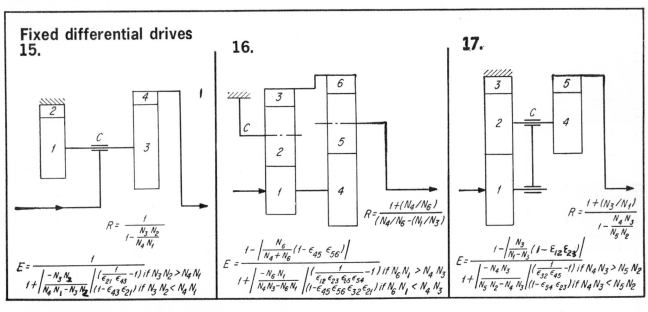

$$R = \frac{1}{1 - \frac{N_3\,N_2}{N_4\,N_1}}$$

$$E = \frac{1}{1 + \left|\frac{-N_3\,N_2}{N_4\,N_1 - N_3\,N_2}\right| \begin{vmatrix}(\frac{1}{\epsilon_{21}\epsilon_{43}} - 1) & \text{if } N_3\,N_2 > N_4\,N_1 \\ (1 - \epsilon_{43}\epsilon_{21}) & \text{if } N_3\,N_2 < N_4\,N_1\end{vmatrix}}$$

16.

$$R = \frac{1 + (N_4/N_6)}{(N_4/N_6) - (N_1/N_3)}$$

$$E = \frac{1 - \left|\frac{N_6}{N_4 + N_6}(1 - \epsilon_{45}\,\epsilon_{56})\right|}{1 + \left|\frac{-N_6\,N_1}{N_4\,N_3 - N_6\,N_1}\right| \begin{vmatrix}(\frac{1}{\epsilon_{12}\epsilon_{23}\epsilon_{65}\epsilon_{54}} - 1) & \text{if } N_6\,N_1 > N_4\,N_3 \\ (1 - \epsilon_{45}\epsilon_{56}\epsilon_{32}\epsilon_{21}) & \text{if } N_6\,N_1 < N_4\,N_3\end{vmatrix}}$$

17.

$$R = \frac{1 + (N_3/N_1)}{1 - \frac{N_4\,N_3}{N_5\,N_2}}$$

$$E = \frac{1 - \left|\frac{N_3}{N_1 - N_3}(1 - \epsilon_{12}\,\epsilon_{23})\right|}{1 + \left|\frac{-N_4\,N_3}{N_5\,N_2 - N_4\,N_3}\right| \begin{vmatrix}(\frac{1}{\epsilon_{32}\epsilon_{45}} - 1) & \text{if } N_4\,N_3 > N_5\,N_2 \\ (1 - \epsilon_{54}\epsilon_{23}) & \text{if } N_4\,N_3 < N_5\,N_2\end{vmatrix}}$$

---

18. Fordomatic transmission planetary

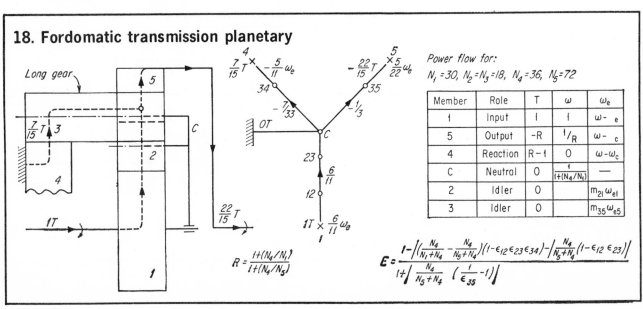

Long gear

$\frac{7}{15}T$

$\frac{22}{15}T$

$1T$

$\frac{4}{7/15}T$ $-\frac{5}{11}\omega_e$

34 $-\frac{7}{33}$

$0T$ C $-\frac{1}{3}$

23 $\frac{6}{11}$

12

$1T \times \frac{6}{11}\omega_e$

$-\frac{22}{15}T \times \frac{5}{22}\omega_e$

35

Power flow for:
$N_1 = 30,\ N_2 = N_3 = 18,\ N_4 = 36,\ N_5 = 72$

Member	Role	T	ω	ω_e
1	Input	1	1	$\omega - e$
5	Output	$-R$	$1/R$	$\omega - c$
4	Reaction	$R-1$	0	$\omega - \omega_c$
C	Neutral	0	$\frac{1}{1+(N_4/N_1)}$	—
2	Idler	0		$m_{21}\,\omega_{e1}$
3	Idler	0		$m_{35}\,\omega_{e5}$

$$R = \frac{1 + (N_4/N_1)}{1 + (N_4/N_5)}$$

$$E = \frac{1 - \left|\left(\frac{N_4}{N_1 + N_4} - \frac{N_4}{N_5 + N_4}\right)(1 - \epsilon_{12}\,\epsilon_{23}\,\epsilon_{34})\right| - \left|\frac{N_4}{N_5 + N_4}(1 - \epsilon_{12}\,\epsilon_{23})\right|}{1 + \left|\frac{N_4}{N_5 + N_4}\left(\frac{1}{\epsilon_{35}} - 1\right)\right|}$$

19. GM Hydramatic planetary

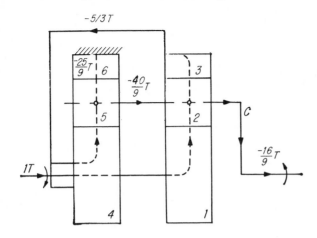

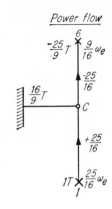

$N_1 = 45 = N_4$

$N_2 = 15 = N_5$

$N_3 = 75 = N_6$

$$R = -\frac{N_3}{N_1}\left(1 + \frac{N_6}{N_4}\right) + \left(1 + \frac{N_3}{N_1}\right)$$

$$E = 1 - \left|\frac{-N_3 N_6}{N_1 N_4 - N_3 N_6}(1 - \epsilon_{12}\,\epsilon_{23}\,\epsilon_{45}\,\epsilon_{56})\right|$$

20. Tractor transmission planetary

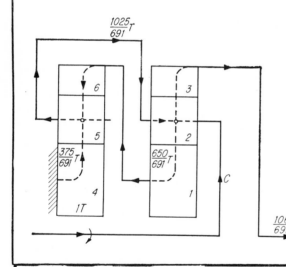

Power flow

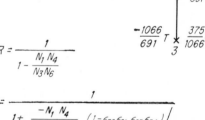

$N_1 = 50$

$N_2 = 17 = N_5$

$N_3 = 82$

$N_4 = 45$

$N_6 = 78$

$$R = \frac{1}{1 - \dfrac{N_1 N_4}{N_3 N_6}}$$

$$E = \frac{1}{1 + \dfrac{-N_1 N_4}{N_3 N_6 - N_1 N_4}(1 - \epsilon_{32}\,\epsilon_{21}\,\epsilon_{65}\,\epsilon_{54})}$$

21. More basic variations

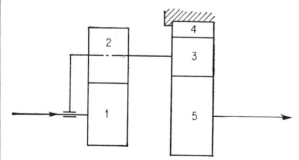

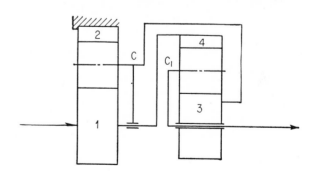

Gears 2 and 3 rotate together

$$R = \frac{1 + \dfrac{N_2 N_4}{N_1 N_3}}{1 + \dfrac{N_4}{N_5}}$$

$$R = \frac{1 + (N_2/N_1)}{1 + \dfrac{N_2/N_1}{1 + (N_3/N_4)}}$$

22. Simple planetaries and inversions

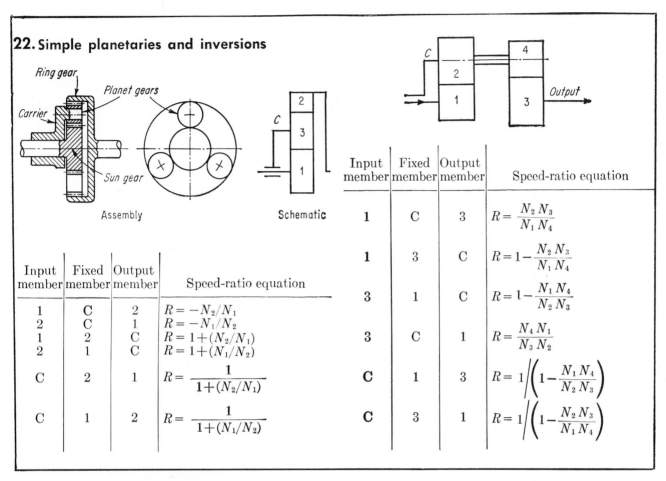

Assembly   Schematic

Input member	Fixed member	Output member	Speed-ratio equation
1	C	2	$R = -N_2/N_1$
2	C	1	$R = -N_1/N_2$
1	2	C	$R = 1 + (N_2/N_1)$
2	1	C	$R = 1 + (N_1/N_2)$
C	2	1	$R = \dfrac{1}{1 + (N_2/N_1)}$
C	1	2	$R = \dfrac{1}{1 + (N_1/N_2)}$

Input member	Fixed member	Output member	Speed-ratio equation
1	C	3	$R = \dfrac{N_2 N_3}{N_1 N_4}$
1	3	C	$R = 1 - \dfrac{N_2 N_3}{N_1 N_4}$
3	1	C	$R = 1 - \dfrac{N_1 N_4}{N_2 N_3}$
3	C	1	$R = \dfrac{N_4 N_1}{N_3 N_2}$
C	1	3	$R = 1 \left/ \left(1 - \dfrac{N_1 N_4}{N_2 N_3}\right)\right.$
C	3	1	$R = 1 \left/ \left(1 - \dfrac{N_2 N_3}{N_1 N_4}\right)\right.$

23. Daimler preselective drive

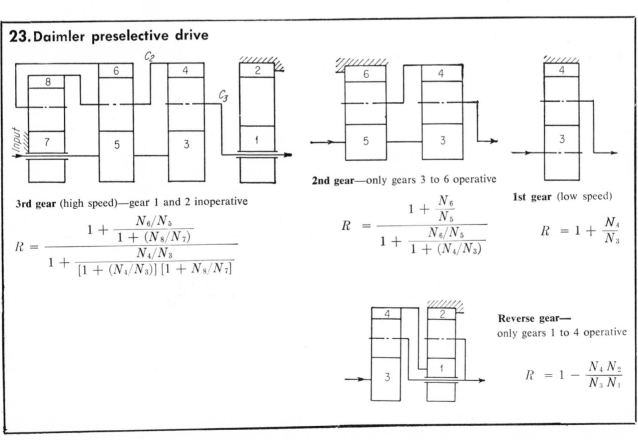

3rd gear (high speed)—gear 1 and 2 inoperative

$$R = \frac{1 + \dfrac{N_6/N_5}{1 + (N_8/N_7)}}{1 + \dfrac{N_4/N_3}{[1 + (N_4/N_3)]\,[1 + N_8/N_7]}}$$

2nd gear—only gears 3 to 6 operative

$$R = \frac{1 + \dfrac{N_6}{N_5}}{1 + \dfrac{N_6/N_5}{1 + (N_4/N_3)}}$$

1st gear (low speed)

$$R = 1 + \frac{N_4}{N_3}$$

Reverse gear—
only gears 1 to 4 operative

$$R = 1 - \frac{N_4 N_2}{N_3 N_1}$$

24. Two-gear planetary drives

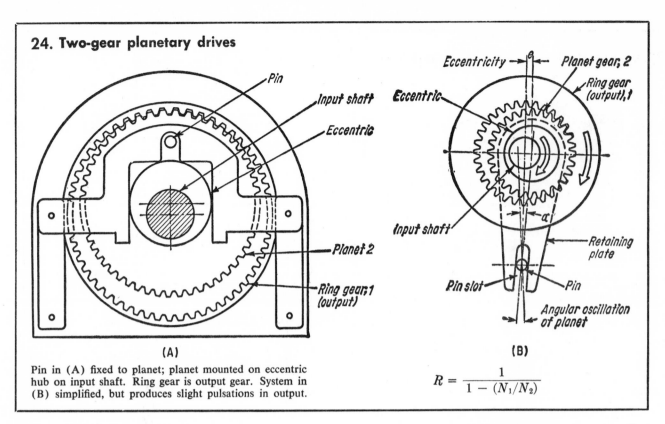

(A)

(B)

Pin in (A) fixed to planet; planet mounted on eccentric hub on input shaft. Ring gear is output gear. System in (B) simplified, but produces slight pulsations in output.

$$R = \frac{1}{1 - (N_1/N_2)}$$

25. Planocentric drive (General Electric Co.)

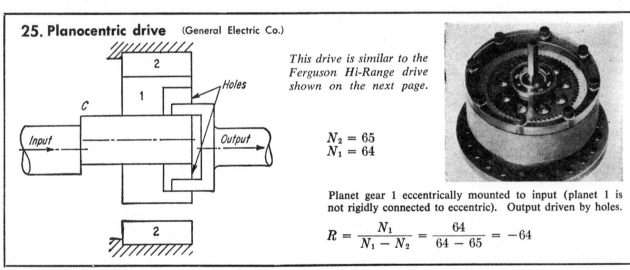

This drive is similar to the Ferguson Hi-Range drive shown on the next page.

$N_2 = 65$
$N_1 = 64$

Planet gear 1 eccentrically mounted to input (planet 1 is not rigidly connected to eccentric). Output driven by holes.

$$R = \frac{N_1}{N_1 - N_2} = \frac{64}{64 - 65} = -64$$

26. Wobble-gear drive

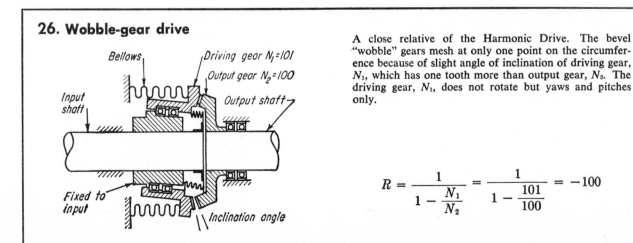

A close relative of the Harmonic Drive. The bevel "wobble" gears mesh at only one point on the circumference because of slight angle of inclination of driving gear, N_1, which has one tooth more than output gear, N_2. The driving gear, N_1, does not rotate but yaws and pitches only.

$$R = \frac{1}{1 - \dfrac{N_1}{N_2}} = \frac{1}{1 - \dfrac{101}{100}} = -100$$

27. Lycoming turbine drive

$$R = \left(1 + \frac{N_3}{N_2}\right)$$

$$\times \left(1 + \frac{N_4}{N_1}\right)$$

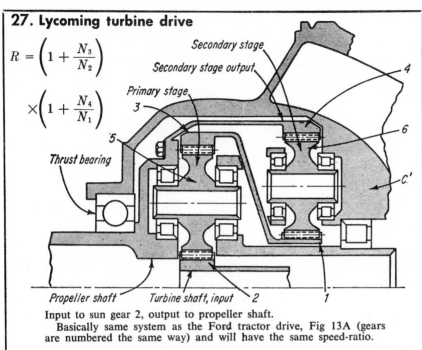

Secondary stage
Secondary stage output
Primary stage
Thrust bearing
Propeller shaft *Turbine shaft, input*

Input to sun gear 2, output to propeller shaft.

Basically same system as the **Ford** tractor drive, Fig 13A (gears are numbered the same way) and will have the same speed-ratio.

29. Compound spur-bevel gear drive

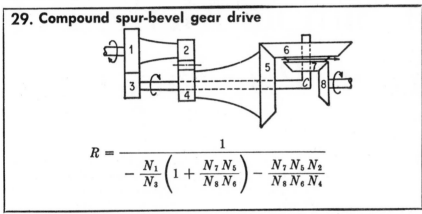

$$R = \frac{1}{-\frac{N_1}{N_3}\left(1 + \frac{N_7 N_5}{N_8 N_6}\right) - \frac{N_7 N_5 N_2}{N_8 N_6 N_4}}$$

30. Humpage's bevel gears

Output   *Input*

$$R = \frac{1 + \dfrac{N_5}{N_1}}{1 - \dfrac{N_3 N_5}{N_4 N_2}}$$

28. Harmonic drive

(United Shoe Machinery Corp.)

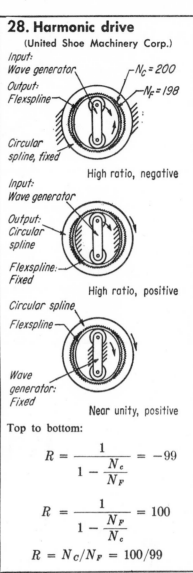

Input:
Wave generator $N_C = 200$
Output:
Flexspline $N_F = 198$

Circular spline, fixed

High ratio, negative

Input:
Wave generator
Output:
Circular spline
Flexspline:
Fixed

High ratio, positive

Circular spline
Flexspline

Wave generator:
Fixed

Near unity, positive

Top to bottom:

$$R = \frac{1}{1 - \dfrac{N_c}{N_F}} = -99$$

$$R = \frac{1}{1 - \dfrac{N_F}{N_c}} = 100$$

$$R = N_C / N_F = 100/99$$

31. Ferguson Hi-Range speed reducer. The pinion is driven by an eccentric cam on the input shaft and transmits its output through a set of pins on the

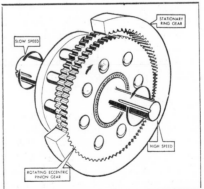

STATIONARY RING GEAR
SLOW SPEED
HIGH SPEED
ROTATING ECCENTRIC PINION GEAR

output shaft. The pins are inserted in oversize holes which permit the eccentric movement of the pinion. The unit can operate with a single pinion at low speeds but dual pinions are used for improved balance.

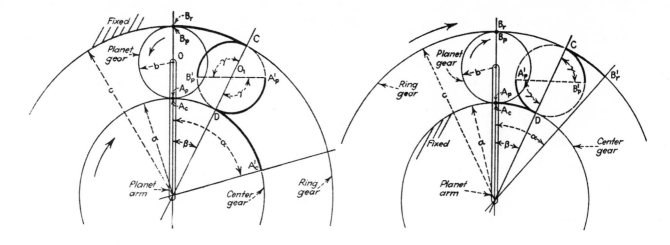

Fig. 1—Simple planetary train with fixed ring gear.   Fig. 2—Simple planetary train with fixed center gear.

A Simple Method of
Ratios in Planetary Gear

SIGMUND RAPPAPORT

THE COMMONLY USED METHOD of finding the ratio of input to output of a planetary gear train usually calls for considerable faith on the part of the analyst. A data table is developed by subjecting the train to a predetermined series of lockings, unlockings and rotations; then the velocity ratio is found in a specified column and row of the table. Little, if any qualitative understanding of the system is gained in the process.

In the method described below, called the "Method of Equal Arcs," the ratio is derived mathematically and the motion of any element in the train can be followed with respect to the other elements. Thus, the modifications required to produce a given effect is often obvious whereas the previous method may require many blind trial and error substitutions.

Four examples are given to illustrate the application of this new technique to both simple and compound gear trains. In all cases, however, the basic theory is the same. For example, Fig. 1 shows a simple planetary train in which the ring gear is fixed. The center gear and planet arm are movable; either can be driving or driven without affecting the analysis.

For the conditions shown, two points on the planet gear are in mesh; B_p mates with B_r which is on the ring gear, and A_p with A_c which is a point on the center gear.

Rotating the center gear through an arbitrary angle α, causes the planet arm to follow through some angle β, which must be evaluated. As a result of this rotation, the center

of the planet gear is at O', and points C and D are the new meshing points with the ring and center gears respectively. A'_p and B_p' are the new locations of the original points of contact.

Since the relative motion between the ring and planet gears is entirely a result of tooth action

$$\text{arc } B_p'C = \text{arc } B_rC \qquad (1)$$

Also,

$$\text{arc } A_p'D = \text{arc } DA_c' \qquad (2)$$

A_c' is the new location of point A_c when the center gear is rotated as described above.

Furthermore, it can be seen from the geometry that

$$\text{angle } B_p'O_1C = \text{angle } DO_1A_p' = \gamma \qquad (3)$$

Thus,

$$\text{arc } B_rC = \text{arc } B_p'C = \text{arc } DA_p' = \text{arc } DA_c' \qquad (4)$$

Let

a = pitch circle radius of the center gear
b = pitch circle radius of the planet gear
c = pitch circle radius of the ring gear,
α, β = angular displacement of center gear and planet arm, respectively.

It follows from Eq (4) that

$$\beta c = \gamma b = (\alpha - \beta) a \qquad (5)$$

Eq (5) is the general equation for the planetary gear train of Fig. 1. As indicated previously, it can be used to solve either of the two types of practical problems that exist: (1) finding the planet arm displacement β when the center gear is the input, its displacement α being given; or (2) finding the center gear displacement α when the planet arm is the input, its displacement β being given.

For type (1), $\beta = \alpha \dfrac{1}{1 + c/a}$ \qquad (6)

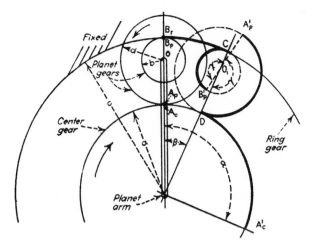

Fig. 3—Compound planetary train with fixed ring gear.　　Fig. 4—Compound planetary train with fixed outer ring gear.

Determining Trains

and for (2)

$$\alpha = \beta\left(1+\frac{c}{a}\right) \qquad (7)$$

The velocity ratio of input to output will, of course, be either α/β or β/α, depending on which element is the input.

CASE II—FIXED CENTER GEAR

In Fig. 2, the ring gear is rotated through the arbitrary angle α, since the center gear is stationary. This carries point B_r to B_r' and the planet gear arm moves through some angle β. As before

$$\text{arc } A_\bullet D = \text{arc } A_p'D = \text{arc } B_r'C = \text{arc } CB_r' \qquad (8)$$

Therefore,

$$\beta a = \gamma b = (\alpha - \beta)\,c \qquad (9)$$

For a ring gear input of α, the planet arm output from Eq (9) is

$$\beta = \alpha\,\frac{1}{1+a/c} \qquad (10)$$

Similarly, for a planet arm input of β

$$\alpha = \beta\,(1+a/c) \qquad (11)$$

Also, solving Eqs (9) and (10) simultaneously

$$\gamma/\alpha = \frac{a}{b}\left(\frac{c}{a+c}\right) \qquad (12)$$

This ratio γ/α is equal to the number of revolutions of the planet gear about its own shaft, per revolution of the ring gear.

CASE III—COMPOUND PLANETARY TRAINS

In Fig. 3 the ring gear is fixed; the center gear is rotated through the arbitrary angle α which moves point $A_\bullet$ to the

position $A'_\bullet$; the planet arm follows through angle β. Then, adjacent arcs being equal in length,

$$\text{arc } B_r C = \text{arc } B_p'C \text{ and} \qquad (13a)$$
$$\text{arc } A_p'D = \text{arc } DA'_\bullet \qquad (13b)$$

Also:

$$\beta c = \gamma b, \text{ and} \qquad (14a)$$
$$\gamma d = (\alpha - \beta)\,a \qquad (14b)$$

The velocity ratios can be obtained by solving Eqs (14a) and (14b) simultaneously. For example, if the center gear has a given input equal to α, the output of the train is

$$\beta = \alpha\left(\frac{1}{1+\dfrac{cd}{ab}}\right) \qquad (15)$$

Similarly, if the arm is the input, β being known, then

$$\alpha = \beta\left(1+\frac{cd}{ab}\right) \qquad (16)$$

The velocity ratio α/β or β/α can be developed from either of the latter two equations.

CASE IV

This compound train differs from that of the previous example in that the larger planet now engages the outer ring gear which is fixed, and the center gear is replaced by an inner ring gear with internal teeth to mesh with the small planet.

Rotating the planet arm so that point O moved to O_1 carries point A on the inner ring gear to some point A' while the larger planet rolls up on the outer, fixed ring gear. From Fig. 4,

$$\text{arc } B_r C = \text{arc } B_p'C \text{ and} \qquad (17a)$$
$$\text{arc } DA_p' = \text{arc } DA' \qquad (17b)$$

Therefore,

$$\beta c = \gamma d \text{ and} \qquad (18a)$$
$$\gamma b = (\beta - \alpha)\,a \qquad (18b)$$

From Eqs (18) the velocity ratio can be determined as

$$\beta/\alpha = \frac{1}{1-\dfrac{bc}{ad}} \qquad (19)$$

The four preceding examples illustrate the principles, but should not be interpeted as the only types of problems that can be solved. The "Method of Equal Arcs" can be applied to any epicyclic gear train.

Planetary Gear Train Design

Method for calculating planetary gear ratios which can be applied to determine the usefulness of any planetary configurations. Criteria to help evaluate the potential of a planetary gear train schematic. Information on pinion spacing, gear noise, bearing design.

OLIVER K. KELLEY
Chief Engineer
Buick Motor Division
General Motors Corporation

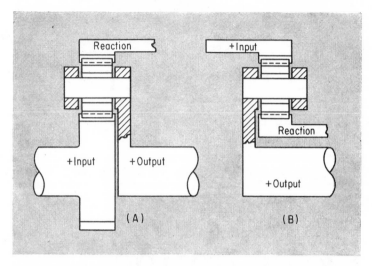

Fig. 1—Single, simple planetary gear system. (A) Where the sun gear is driving and the ring gear is the reaction element, reduction ratios of 2.8/1 or greater are obtainable. (B) When the ring gear is driving and the sun gear is the reaction member, the trouble point of small pinions show up when reduction ratio of 1.55/1 or more are attempted. Between 1.55/1 and 2.8/1 compound planetaries are needed.

Compact design, more efficient gearing and bearing design, easier to make quiet, and more suitable for designs of uninterrupted power shifting are the major advantages of planetary gear train over the countershaft gear train designed for the same function. To take advantage of these qualities, however, a choice from a wide variety of compounding systems must be made and some special gear and bearing problems solved.

SELECTING THE GEAR REDUCTION

A single simple planetary gear system, Fig. 1, will supply reduction gear ratios between 1.35/1 and 1.55/1 when the ring gear is driving and the sun gear is the reaction member. A 2/1 reduction ratio is impossible with a simple planetary gear system because it would require zero pitch diameter pinions.

Hence, the compound planetary gear system. The impossible 2/1 reduction ratio is achieved in several ways, Fig. 2; by dividing the work between the planetary sets, or using a regenerative system.

Both designs are useful. The first is a good way of obtaining two different reduction ratios through use of either the first ring gear or the first carrier as the reaction member, thus getting optimum efficiency. The second design obains a single reduction ratio and approximately

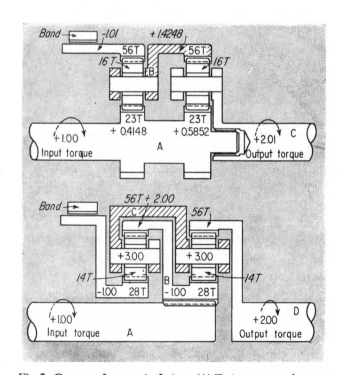

Fig. 2—Compound gear train designs. (A) Train uses one planetary set to reduce the full-speed gear action of the second set, thus also reducing the final drive ratio of the second set. (B) Left-side ring gear input planetary drives the right-side planet carrier at a reduction ratio. Planet pinions of the right-side carrier split the carrier torque between the output ring gear and the original input shaft to increase the input torque artificially.

equal reverse ratio by using either the first sun gear or the carrier combination as a reaction member, resulting in reasonable efficiency.

When nearly 2/1 ratios are sought, the Ravigneaux compound planetary design, Fig. 3, is a solution and has the structural advantage of a single carrier. Since the same tooth load prevails throughout the train, the reaction gear tooth load equals the driving gear tooth load and the reaction torque is directly proportional to the number of teeth in the reaction gear and input gear.

Where less than 2/1 reduction ratio is involved the reaction gear is smaller than the input gear; where more than a 2/1 ratio is required the opposite is true. Design is widely used as it is one of the better ways of keeping pitch line velocities down, thus reducing noise.

SYSTEM ANALYSIS

As the arrangements of compound systems become more complicated a rapid method of determining, (1) the reduction ratios available from a given design, and (2) whether the parts will go together, is needed. The following steps, shown in Fig. 4, will solve the first problem.

1. Hold any member of a gear train and turn another member one turn. Record the speed for each rotating member of the gear train when the speed of the held member is zero. The resulting values disclose the speeds of all the members of a gear train.

2. Choose any of the rotating members for the next possible stationary member and add an equal quantity of opposite sign to its speed value. This makes its speed

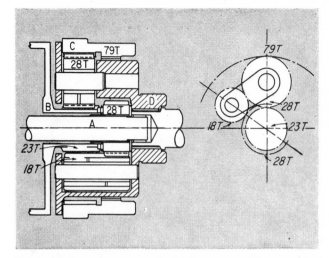

Fig. 3—Ravigneaux compound planetary gears. There are two sun gears, three long pinions, three short pinions, and a ring gear which is used for reverse reaction. The 28-tooth sun gear is the input, the 23-tooth sun is the forward reduction reaction gear, the 79-tooth planet ring gear is reverse reaction gear, and the planet carrier is the output member. Both forward and reverse reduction ratios are 1.82/1.

value zero. Then, add this same quantity to all the other speed values. The original zero value member has a speed and all values are changed.

3. Repeat this procedure until every member has been equated to zero, while the other members have changed their speeds to corresponding new values.

Fig. 4—Analysis of double planetary gear and a Ravigneaux gear train. Reciprocals would give twelve overdrive ratios.

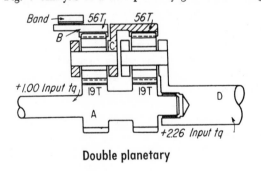

Double planetary

		A	B	C	D	Step 4 Reduction Ratios Forward	Reverse
Step 1		+1.00	−0.393=19/56	0	+0.2533=19/75	A/D=3.947 to 1	A/B=2.947 to 1
		−0.2533	−0.2533	−0.2533	−0.2533		B/D=1.34 to 1
Step 2		+0.7467	−0.5920	−0.2533	0	B/C=2.337 to 1	A/B=1.261 to 1
		+0.5920	+0.5920	+0.5920	+0.5920		A/C=2.947 to 1
Step 3		+1.3387	0	+0.3387	+0.5920	A/C=3.947 to 1	
		−1.3387	−1.3387	−1.3387	−1.3387	A/D=2.26 to 1	
						D/C=1.747 to 1	
Step 3a		0	−1.3387	−1.00	−0.767	B/C=1.3387 to 1	
						C/D=1.3387 to 1	
						B/D=1.793 to 1	

		A	B	C	D	Step 4 Reduction Ratios Forward	Reverse
Step 1		+1.00	−1.2174=28/23	+0.3544=28/79	0	A/C=2.821 to 1	B/C=3.435 to 1
		−0.3544	−0.3544	−0.3544	−0.3544		B/A=1.2174 to 1
Step 2		+0.6456	−1.5718	0	−0.3544	B/D=4.435 to 1	B/A=2.434 to 1
		+1.5718	+1.5718	+1.5718	+1.5718		A/D=1.821 to 1
Step 3		+2.2174	0	+1.5718	+1.2174	A/C=1.410 to 1	
		−2.2174	−2.2174	−2.2174	−2.2174	A/D=1.821 to 1	
						C/D=1.291 to 1	
Step 3a		0	−2.2174	−0.6456	−1.00	B/C=3.435 to 1	
						B/D=2.2174 to 1	
						D/C=1.549 to 1	

Ravigneaux

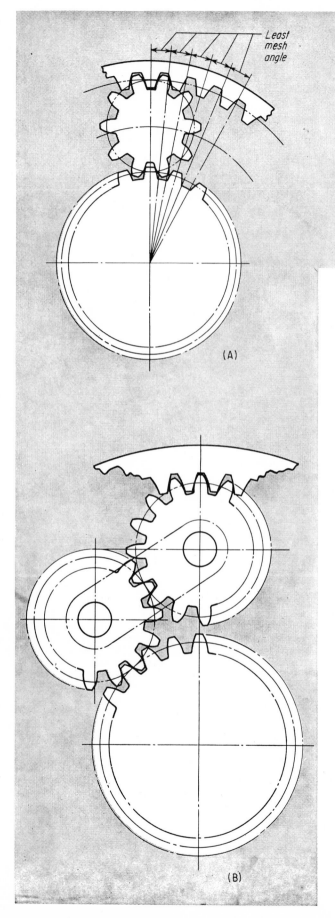

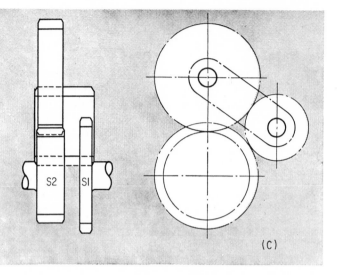

Fig. 5—Least mesh angle of (A) simple planetary gear, (B) double pinion gear set, (C) long and short pinion meshing with two sun gears. The pinions must mesh in increments of the least mesh angle to assemble both the sun and ring gears.

4. Now examine the potential speed relations of the compound gear train for usable forward and reverse ratios. The full potential of the gear train is revealed quickly by this method.

To fit the planetary gear train together several rules must be observed. First, if standard or only slightly modified addendum-dedendum tooth proportions are used, the pinion will not fit if the number of teeth in the sun is odd and in the ring is even, or vice versa. They both must be either odd or even. Each simple planetary gear train has a number of places for the pinions where they will fit into the sun gear mesh and the ring gear mesh simultaneously. This number is equal to the sum of the teeth of the sun and ring gears.

The least mesh angle is shown in Fig. 5(A) for a simple planetary gear set and is $360°/R + S$, where R is the number of teeth in the ring gear and S is the number of teeth in the sun gear. To have evenly spaced pinions, the angle between them must equal $360°/N$, where N is the number of pinions. Then, for evenly spaced pinions, this angle must be an integral multiple of the least mesh angle or

$$(360°/N)/(360°/R + S) = \text{whole number}$$

or

$$k = (R + S)/N$$

For a double pinion gearset, Fig. 5(B), the least mesh angle is $360°/R - S$, and for evenly spaced pinions the equation is

$$k = (R - S)/N$$

For a long and short pinion meshing with two sun gears, Fig. 5(C), the least mesh angle is $360°/(S_1 + S_2)$ and for evenly spaced pinions the equation is

$$k = (S_1 + S_2)/N$$

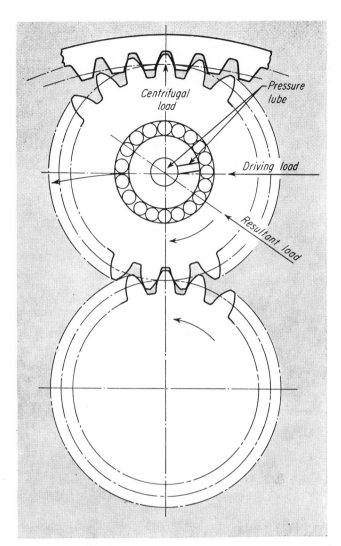

Fig. 6—Centrifugal load on pinion gear. Needles lose rotational speed because of rubbing friction when not under load. When forced to the load-carrying part of bearing, high rotational acceleration causes further wear. Problem is usually found in heavy-duty high-speed equipment.

GEAR NOISE

Gear noise is proportional to the square of the meshing pitch line velocity. Everything else being equal, the greatest single help comes from reduced pitch line velocities. This is the advantage of planetary gears. They offer many choices of designing a gear train for most needs that still give low pitch line velocities and are easier to make quiet in operation.

Helix gear planetary designs which depend on the helix overlap for tooth engagement are often noisier when carrying light loads only. This is caused by slight misalignment of opinions, which tends to cause gear edge loading only, thus losing all help from the helical overlap. Load application on the pinion teeth at both the sun and ring gear contacts will deflect the structure sufficiently to establish contact across the face of the gear, thus bringing into full advantage the helical overlap.

BEARING DESIGN

The pinion gear presents a bearing problem in a simple planetary system as the sun and ring gears do not carry a driving load. The pinion gears usually use needle bearings to transmit loads to the pinion carrier. These needles must resist destructive end-thrust and centrifugal forces.

The relatively flexible long bearing needle tries to roll on a deflected-at-the-middle shaft and also on the inner surface of a rigid pinion bore. This does not give true rolling action. Also, the high helix angle thrust loads, especially in the larger pinions, impose a cocking couple on the bearing. This shifts the load to a spiral pattern on the sides of the deflected shaft. The resultant spiral pattern can produce high end thrust on the needles. Hardened steel thrust washers on the pinion shafts between the pinion and the carrier are generally adequate to resist this thrust. Needle-bearing thrust-washers are often used to increase the transmission gear train thrust capacity.

Helical gears in planetaries produce equal and opposite thrust loads on the sun gear and the ring gear. In the case of a small diameter driving sun gear, this thrust load can produce a real problem. To cope with this problem, end thrust washers are used which have small needle bearings mounted in the washer. The short needle bearings are caged radially to absorb the end thrust loads of the driving and driven gears.

High centrifugal forces in planet carriers can cause severe rubbing action between the needles in the pinion bearing. Fig. 6 shows that this force makes the needles bunch together during the unloaded portion of their trip around the races. The resultant rubbing action causes each needle to lose its rpm. When the needle enters the actual working part of the path, it must accelerate instantaneously to the full rotational speed of the bearing. This causes slippage, wear, and heat. The solution is the use of caged needles and extreme lubrication.

An example of this technique is a Ravigneaux gear set, Fig. 3, with the following number of teeth. Ring gear (R) = 79, front sun gear (S_1) = 23, rear sun gear (S_2) = 28.

The Front gears form a simple planetary set and
$$k = (R + S_1)/N = (79 + 23)/3 = 37$$
The Rear sun gear, the two pinions, and the ring gear are a double pinion set and
$$k = (R - S_2)/N = (79 - 28)/3 = 17$$
The two sun gears and the pinions are another gear set to consider and
$$k = (S_1 + S_2)/N = (23 + 28)/3 = 17$$
All these combinations result in whole numbers which means the transmission will assemble and the pinions are equally spaced. Without such a check, it is difficult to determine if an interference exists.

volume requirements of

EPICYCLIC GEAR SYSTEMS

Equations for volume of spur or helical trains don't apply to epicyclic systems, where power flows through several planets. The author therefore modifies the equations—to provide a systematic analysis of volume, power capacity, speed ratios and stresses.

ROBERT J WILLIS JR
Lightweight gear application engineer
General Electric Co, West Lynn, Mass

For epicyclic gearing we need more than the standard equations that guide the preliminary design of spur or helical gears. As written, such equations do not allow for the multiple power paths in the planetary or star gear systems. But they can be adapted for the purpose. The following two equations, found in D. W. Dudley's book, *Practical Gear Design*, serve as a starting point:

$$C^2F = \frac{31,500\,Q}{K}; \text{ Where } K = \frac{W_t}{Fd_p}\left[\frac{m_g + 1}{m_g}\right] \qquad (1)$$

$$Q = \frac{\text{horsepower}}{\text{pinion rpm}} \times \frac{(m_g + 1)^3}{m_g} \qquad (2)$$

Substituting for Q in Eq (1)

$$C^2F = \frac{31,500}{K} \times \frac{\text{hp}}{n_p} \times \frac{(m_g + 1)^3}{m_g} \qquad (3)$$

Since, by definition,

$$C = \tfrac{1}{2}(d_p + d_p m_g),$$

$$C^2 = \left[\frac{d_p}{2}(m_g + 1)\right]^2$$

Substituting C^2 in Eq (3) gives

$$Fd_p{}^2 = \frac{126,000 \times \text{hp}}{Kn_p} \times \left[\frac{m_g + 1}{m_g}\right] \qquad (4)$$

The various factors required to solve Eq (4) are normally known, or can be made readily available in normal practice. The solution indicates that for a given set of design conditions a specific rotor volume is required—the volume of a cylinder formed by the active gear-face width and the diameter of the pitch circle.

Once rotor volume has been determined by Eq (4) for a specific set of design conditions, it must be kept. It is possible to vary the relationship of face width to diameter so long as their product remains constant. For high-speed precision gearing, however, it is not good practice to allow the face width to become too large with respect to gear dia. This can cause misalignment and bending, and lead to load-carrying problems. It is well to limit the face width to 60-70% of the gear dia, unless a double-helical set is used. In the latter case the individual faces should not exceed the 70% value.

Epicyclic vs parallel-shaft ratings for 6000 hp, 7.15:1 ratio

Ratings	Epicyclic gear	Double-helical parallel-shaft precision gear
Weight, lb	7500	15,000
Height, in.	51½	71
Length, in.	51	56
Width, in.	53	93
Offset, in.	none	35.475
Volume, cu ft	81	215
WR² of gear	2600	15,150
Losses, hp	110	130
Face width, in.	7.25	16
Oil capacity, gal	60	35
Pitch-line velocity, fps	140	326
k factor	295	101

Planetary gear for pulling 95-ton blast-resistant lid to cover and uncover underground Minuteman missiles. Schematic on p 98.

In practice, Eq (4) must be altered to account for the effect of the number of planets designated by b—Eq (5) below. (The possible number of planets in an epicyclic gear is further explained by the author in his PRODUCT ENGINEERING article, Nov 24 '58, p 57.)

$$Fd_s^2 = \frac{126,000 \text{ hp}}{b\,K\,n_s}\left[\frac{m_s + 1}{m_s}\right] \qquad (5)$$

For a simple planetary gear

$$m_s = \frac{M_o - 2}{2}$$

and for a star gear

$$m_s = \frac{M_o - 1}{2}$$

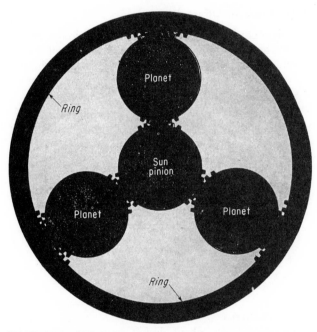

STAR GEARING is a variation of the basic planetary gear system. Geometry of these two types is the same; difference is in the ratio of output to input.

Other basic equations used are:

$$W_t = \frac{126,000 \text{ hp}}{b\,n_s\,d_s} \qquad (6)$$

$$U = \frac{W_t\,P_d}{F \cos \psi} \qquad (7)$$

Eq (7) applies to both straight spur and helical gear designs. For spur teeth, where helical angle is zero, $\cos \psi$ equals 1.0.

The reasons for using helical teeth in epicyclic gearings are the same as in normal gearing practice:

• Helical gears run more smoothly at high rpm.

• Helical gears tend to dampen disturbing dynamic or vibration frequencies.

• Double helical gears will cancel end thrusts and permit free floating of the sun pinion.

SYMBOLS

b — Number of planets	M_o — $(d_R/d_s) + 1$, overall epicyclic ratio for planetary gear
C — Center distance, in.	
d_p — Pitch diameter of planet gear, in.	n_s — Number of sun teeth
d_R — Pitch diameter of ring gear, in.	n_p — Number of planet teeth
d_s — Pitch diameter of sun pinion, in.	P_d — Diametral pitch
F — Face width, in.	S_c — Compressive stress, psi
m_y — Speed ratio. $(m_g + 1)^3/m_g$ is ratio factor	S_t — Tensile stress, psi
	S_t' — Tensile stress for reverse loading, psi
m_s — d_p/d_s, ratio of planet to sun pinion	U — Unit load, lb/in. of face width
	W_t — Tangential force, lb

When using double helical gearing, you must split the ring gear to allow for assembly.

Putting Equations to Work

The following steps, using arbitrary but representative figures, show how to get preliminary design data from these equations.

1. Solve for rotor volume, using Eq (5) and the calculated value of m_s. Suppose the answer is $Fd_s^2 = 25$.

2. Solve Eq (6) for tangential driving load. Since the sun-pinion diameter is one of the unknowns in the general problem, no numerical value is substituted for d_s. The result might be $W_t = 18,000/d_s$.

3. Choose a value of unit load. For high-speed high-power precision gear sets, a value of 12,000 to 13,000 lb/in. will result in reasonable root stresses for limited-life designs. Table below shows typical unit loads based on 100,000,000 cycles with reversed bending load accounted for. Substitute the value found for W_t above into Eq (7) and solve for Fd_s. For our example:

$$Fd_s = 1.05\,P_d$$

TABLE OF UNIT LOADS

Material (Carbon Steels)	Hardness, R_c	U
Furnace Hardened & Machined	33–38	6000–8000
Carburized Ground Fillets	58–63	9000
Flame or Induction Hardened	48–53	9000
Carburized Unground Fillets	58–63	12,000
Carburized, Special Design With Rigid Quality Control	60 min	15,000

4. There are now two equations with the factors F and d_s available in useful form. By solving these two equations for F and setting them equal to one another:

$$\frac{25}{d_s^2} = \frac{1.05\,P_d}{d_s} \text{ or}$$

$$d_s = \frac{23.809}{P_d}$$

5. A table as shown in sample problem later is now made, listing various values for d_s and P_d.

At this point engineering judgment will guide the designer to the best combination. For example, a specific application may have size limitations, yet the value of diametral pitch which results in the proper-size gear set may not be what the designer considers ideal. Some considered maneuvering with the variables is necessary to arrive at the required design objectives—minimum weight and size.

Taking the above basic equations a step further in our derivation by substituting the W_t in Eq (6) into Eq (7) gives:

$$U = \frac{126,000 \text{ hp } P_d}{b \, n_s F \, d_s} \tag{8}$$

or

$$F d_s = \frac{126,000 \text{ hp } P_d}{b \, n_s \, U} \tag{9}$$

By solving for F in Eq (5) and (9), and setting results equal, the following fundamental equation for d_s is obtained:

$$d_s = \frac{U}{K \, P_d} \left[\frac{m_s + 1}{m_s} \right] \tag{10}$$

It must be noted that in epicyclic gearing the planet gear is, in effect, an idler; therefore, the root stress of the sun pinion-planet mesh must not exceed 70% of the max allowable root stress to allow for reversed loading. It should also be realized that the number of cycles, even for a short design life, will be relatively large under normal circumstances, with the result that the stress is at or beyond the knee of the fatigue-life curve. In slower-speed gear systems this might be considered design for an infinite life. In normal practice—that is, high-speed power-transmission gearing—an epicyclic-gear set is classified as precision gearing where there has been careful selection of materials and supervision of manufacturing methods and heat-treating. This is especially true where minimum size and weight are design objectives. For such conditions, the maximum allowable root stress is approximately 54,000 psi. The following equation closely approximates the root stress:

$$S_t \cong \frac{W_t \, P_d \, (3)}{F} = 54,000 \tag{11}$$

Where reversed loading is encountered, as in the planet gears, it is customary to use as a limit 70% of the maximum allowable root stress:

$$S'_t = \frac{(0.7) \, W_t \, P_d \, (3)}{F} = 37,800 \tag{12}$$

The value of root stress is directly proportional to unit load (Eq. 7). Therefore, value of unit load for this method is 12,300 lb/in. of face width—and this only when a gear set with very high performance is desired. Actually, gear sets have successfully transmitted approximately 30 hp/lb of structural weight. (Planetary-gear feasibility designs for aircraft made by the author, using this procedure, indicate this figure could be as high as 34 hp/lb.) This weight includes casing structure, bearings, lubrication and oil-scavenge means, but does not include accessory gearing or the lubrication and scavenge pump.

Final Check Points

In applying the above procedure check these points:

• For spur gearing, limit face width of the sun pinion-planet mesh to 60-70% of the diameter of the smallest gear.

• The planet gears are in stress reversal. Be sure the limiting value of root stress is treated accordingly. For spur-gear root stress use Eq (12).

• Check the design for Hertz compressive stress to assure that the limiting value is not exceeded. For 20° spur gear, Hertz stress is $S_c = 5712 \sqrt{K}$.

• Check scoring factor or lubricant flash-point temperature. Kelley's method has been found reasonable to apply by actual test determination.

Application to Offset Gears

This method may also be used for the design of offset gears where the value of center distance is not fixed or given. When used for such a purpose, the term b (number planets) is dropped from the equations developed. Also, the factor affecting root stress to account for reverse loading is discarded unless the gear arrangement contemplated has an idler. The following equations are applicable:

$$F d_p^2 = \frac{126,000 \times \text{hp}}{K \, n_p} \left[\frac{m_g + 1}{m_g} \right]$$

$$W_t = \frac{126,000 \times \text{hp}}{n_p \, d_p}$$

$$U = \frac{W_t \, P_d}{F} = \frac{126,000 \times \text{hp} \times P_d}{n_p \, F \, d_p}$$

$$F d_p = \frac{126,000 \times \text{hp} \times P_d}{U \times n_p}$$

$$d_p = \frac{U}{K \, P_d} \left[\frac{m_g + 1}{m_g} \right]$$

SAMPLE PROBLEM I

To illustrate the above design procedure, consider the following example:

Data Given	Data Assumed
Planetary Gear	$K = 1000$
$M_o = 5$	$U = 12,300$
hp = 3000	
$n_s = 6000$	
$b = 4$	
$m_s = \dfrac{M_o - 2}{2} = 1.5$ (for planetary gear)	

Solution:

$$F d_s^2 = \frac{126,000 \times 3000}{4 \times 1000 \times 6000} \left[\frac{1.5 + 1}{1.5} \right] = 26.25$$

$$W_t = \frac{126,000 \times 3000}{4 \times 6000 \times d_s} = \frac{15,750}{d_s}$$

$$U = 12,300 = \frac{15,750 \, P_d}{F \, d_s}; \text{ therefore}$$

$$F d_s = 1.28 \, P_d; \text{ solving for } F \text{ and equating}$$

$$\frac{26.25}{d_s^2} = \frac{1.28 \, P_d}{d_s}$$

$$d_s = 20.5 / P_d$$

By choosing various values for d_s and P_d the following table can be constructed.

TABLE FOR EXAMPLE 1				
P_d	6	6.12	7	8
d_s	3.416	3.35	2.93	2.565
Fd_s	7.67	7.83	8.95	10.23
F	2.242	2.34	3.06	3.99
d_p	5.125	5.02	4.40	3.85
d_r	17.08	16.75	14.67	12.84
F/d_s	0.658	0.700	1.045	1.554
W_t	4610	4700	5380	6140
S_t	37,000	36,900	36,900	36,900
S_c		181,800		

Why construct the table when an immediate, specific solution for the variables is possible by substituting $0.6d_s$ or $0.7d_s$ for F after the value of Fd_s^2 was established? Simply because the table indicates the degree of change in each of the variables with constant increment changes of Pd_s and therefore presents a broader picture.

SAMPLE PROBLEM 2

Data Given	Data Assumed
Planetary Gear	$K = 500$
$M_o = 5$	$U = 15,000$
hp $= 3000$	
$n_s = 12,000$	
$b = 4$	
$m_s = \dfrac{M_o - 2}{2} = 1.5$	

Solution

$$Fd_s^2 = \frac{126,000 \times 3000}{4 \times 500 \times 12,000}\left(\frac{1.5 + 1}{1.5}\right) = 26.25$$

$$W_t = \frac{126,000 \times 3000}{4 \times 12,000 \times d_s} = \frac{7880}{d_s}$$

$$U = 15,000 = \frac{7880\, P_d}{Fd_s}; \text{ therefore}$$

$Fd_s = 0.525\, P_d$; solving for F and equating

$$\frac{26.25}{d_s^2} = \frac{0.525\, P_d}{d_s}$$

$$d_s = 50/P_d$$

Again, by using various values for d_s and P_d in the above equations the following table can be constructed.

TABLE FOR EXAMPLE 2									
P_d	7	8	9	10	11	12	13	14	15
d_s	7.15	6.25	5.56	5.0	4.55	4.17	3.85	3.57	3.33
Fd_s	3.68	4.20	4.72	5.25	5.77	6.30	6.82	7.35	7.87
F	0.515	0.672	0.849	1.05	1.27	1.51	1.77	2.06	2.36
d_p	10.71	9.36	8.34	7.50	6.82	6.25	5.77	5.35	5.00
d_R	28.60	25.00	22.24	20.00	18.20	16.68	15.40	14.28	13.32
W_t	1102	1260	1415	1575	1732	1890	2045	2210	2370
S_t					64,500				
S_c					128,000				

Analysis

Let us assume that the casing must not exceed 16 in. diameter. A study of the above table indicates:

- The root stress exceeds the max allowable value of 54,000 found in Eq (11).
- The sun pinion is the smallest gear and should have 24 or 25 teeth for good design.
- To fit a 16-in.-dia housing, a reasonable starting point is a ring gear with 13.32 in. pitch dia.
- A ratio of $F/d_s = 0.7$ for the sun pinion should be checked.

Procedure

To choose the most feasible design parameters from the above table proceed as follows:

$$P_d = \frac{n_s}{d_s} = \frac{25}{3.33} = 7.5$$

Check d_s when $F = 0.7\, d_s$.

$$Fd_s^2 = 28.25 = 0.7\, d_s^3$$

$d_s = 3.35$; therefore choice of $3.33 = d_s$ is good.

Check face width by limiting the value of $S_t = 54,000$

$$54,000 = \frac{2370 \times 7.5 \times 3}{0.7 \times F}$$

$$F = 1.411; \text{ Use } 1.45$$

Check F/d_s

$$F/d_s = \frac{1.45}{3.33} = 0.436 < 0.7$$

Check unit load

$$U = \frac{2370 \times 7.5}{1.45} = 12,258$$

Summing up:

$U = 12,250$
$S_t = 54,000$
$P_d = 7.5$
$d_s = 3.33$
$d_p = 5.00$
$d_R = 13.32$
$F = 1.45$
$W_t = 2370$
$S_c = 128,000$

Formula for epicyclic gear train of large reduction

M. F. SPOTTS, Northwestern Technological Institute

WHEN large speed reductions are required, the epicyclic train shown in Fig. 1 can be effectively employed. Input shaft A is keyed to gear 1 which turns gear 2 as well as the right side of the case through gear 3. Output shaft B is attached to the arm that carries gear 5 which in turn meshes with gear 4 and the case at gear 6.

Fig. 2 shows a front view of the gears. Radius OC of gear 3 was originally vertical but has been rotated to the position shown. In so doing arcs CD, DE, FG and GH are all equal to each other. If the rotation of shaft A is denoted by θ_a then angle COD, turned by gear 3, is $(N_1/N_3)\,\theta_a$, where N_1 is number of teeth in gear 1 and N_3 is number of teeth in gear 3. Arc GH subtends angle $(N_1/N_2)\,\theta_a$.

Starting points C and J for gears 3 and 6 were originally vertical. Terminal points H and P for gears 2 and 4 were also originally in the vertical. Rotation θ_b of the arm must therefore be an amount sufficient to make arcs JK, KL, MN, and NP equal to each other.

If θ_c represents the rotation of gear 5, then the angle subtended by arc NP is equal to $(N_5/N_4)\,\theta_c$. Arc JK subtends angle $(N_5/N_6)\,\theta_c$.

The following equation can now be written:

$$\angle COH = \left(\frac{N_5}{N_6} + \frac{N_5}{N_4}\right)\theta_c = \left(\frac{N_1}{N_3} + \frac{N_1}{N_2}\right)\theta_a$$

or

$$N_5\left(\frac{N_4 + N_6}{N_4 N_6}\right)\theta_c = N_1\left(\frac{N_2 + N_3}{N_2 N_3}\right)\theta_a$$

then
$$\theta_c = \frac{N_1 N_4 N_6}{N_2 N_3 N_5}\left(\frac{N_2 + N_3}{N_4 + N_6}\right)\theta_a \qquad (1)$$

Rotation θ_b of the arm as well as the shaft B is equal to

$$\theta_b = \frac{N_1}{N_3}\theta_a - \frac{N_5}{N_6}\theta_c$$

when the value of θ_c from Eq (1) is substituted, the result can be reduced to

$$\frac{\theta_b}{\theta_a} = \frac{N_1(N_2 N_6 - N_3 N_4)}{N_2 N_3 (N_4 + N_6)} \qquad (2)$$

EXAMPLE. Find the velocity ratio for the train if $N_1 = 12$, $N_2 = 51$, $N_3 = 76$, $N_4 = 49$, $N_5 = 12$, and $N_6 = 73$. SOLUTION. Substitution in Eq (2) gives

$$\frac{\theta_b}{\theta_a}\frac{12(51 \times 73 - 76 \times 49)}{51 \times 76(49 + 73)} = -\frac{1}{39,406}$$

An arrangement of gears like Fig. 1 is thus capable of giving very large reductions. It should be noted that these values for the numbers of teeth require non-standard pitch circle radii and non-standard thicknesses of teeth for gears 1, 2, and 3.

If gear 2 in the above example has 52 teeth, all gears will have standard proportions, but the ratio is then reduced to approximately 1/558.04.

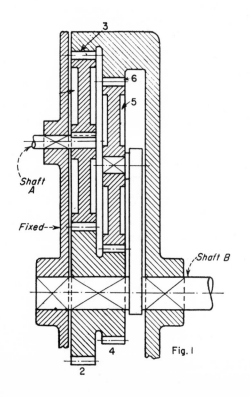

Fig. 1

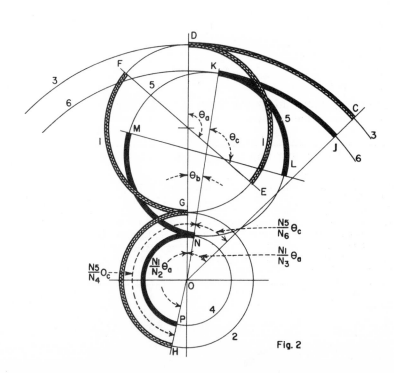

Fig. 2

novel concept gives
RECORD GEAR RATIOS

By pairing helical with spur gears in an otherwise standard planetary system, the author's method provides fantastically high speed-reductions.

SIGMUND RAPPAPORT

Here's a modification that will greatly increase the speed-reduction capabilities of an ordinary planetary gear system. It involves no drastic changes—just substitute a selected pair of helical gears for one of the two pairs of spur gears in the planetary train.

Though the gear assembly remains the same in size and weight, this simple change can easily boost the reduction ratio from, say, 1:21 to 1:1600. But any epicyclic gear train has this limitation—at high reduction ratios it will multiply torque to a maximum of 40:1, as will be shown later. This, however, does not prohibit its usefulness for low-power, high-reduction applications.

Where the Difference Lies

Capabilities of the spur-helical system are explained by the same simple equation that shows the conditions in a conventional four-spur planetary system. The latter, illustrated at the right, is "epicyclic" because its planet gears roll around the outside of center gears (with a "hypocyclic" type, the planet rolls on the inside of an internal ring gear). Gear a is fastened to the frame, the input shaft revolves the cage, and the size differences between the two gear sets (a and b; c and d) causes the output gear to revolve. The angular displacement (or velocity) ratio, r, between output and input is

$$\frac{\text{output displacement}}{\text{input displacement}} = r = 1 - \frac{ac}{bd} \qquad (1)$$

where symbols a, b, c and d can be considered as the number of teeth, or pitch diameters, of the respective gears.

A condition here is that center distance between gears a and b must equal center distance between gears c and d. Also, for a high reduction, the fraction in the equation should come close to unity without reaching it.

Author's version substitutes helical gears for one of the two sets of spur gears and boosts reduction ratio from 1:21 to 1:1600.

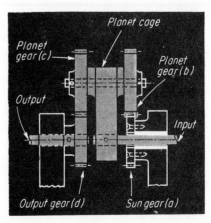

Conventional planetary gear train uses two sets of spur gears for speed reduction. A train like this (analyzed in article) gives only a 1:21 ratio.

(Unity occurs when both sets are identical i.e. a = d, and b = c. The cage then "idles" around the output gear without driving it).

Suppose space considerations limit the gears to not more than 41 teeth—what would be the highest reduction achievable? Mathematical analysis reveals that a, b, c and d should differ from one another by as little as possible, which yields, as possible teeth numbers, 39 and 40 in addition to the given 41. By maintaining $a + b = c + d$, the best that can be done is $a = 40$, $b = 41$, $c = 40$, $d = 41$, resulting in the ratio:

$$r = 1 - \frac{(40)\,(40)}{(41)\,(41)} = \frac{81}{1681} \approx \frac{1}{21}$$

which is not a very high reduction.

A higher reduction is possible if $b = c$. Thus: $a = 40$, $c = 41$, $d = 41$, resulting in the ratio:

$$r = 1 - \frac{(40)\,(41)}{(41)\,(41)} = \frac{1}{41}$$

This will mean, however, that $a + b$ does not equal $c + d$. Designers usually overcome this inequality in one of two ways:

• Insertion of idler gears between gears a and b, and between c and d. The idlers are carried around by the planet arm and their positions are determined by their size (which is immaterial).

• Cutting the teeth of the gear pair whose teeth sum is the larger, to a greater depth and making their OD accordingly smaller. This amounts to modifying their pitch—making it somewhat finer.

Neither method can be recommended. The first makes the drive heavier, less efficient and more expensive; the second is bad practice because it results in a nonstandard tooth form.

Spur-helical Method

Unequal sets of gears can be used if one set consists of helical gears. Here's how this affects the speed ratio:

There is a difference in pitch diameters between spur and helical gears. For spur, $PD = n/DP$, while for helical, $PD = n/DP \cos \beta$, where PD = pitch dia, DP = diametral pitch, n = number of teeth, β = helix angle.

This means helical gears can be employed as the gear pair with the smaller sum of teeth if β is determined from

$$\cos \beta = \frac{a + b}{c + d} \qquad (2)$$

This increases the pitch diameter of the helical gears by a factor of $1/\cos \beta$, and causes the center distance between gears a and b to become exactly equal to the center distance between c and d. Thus, the method escapes the disadvantages of extra gears or nonstandard tooth form, and is far superior. Let us see how high a speed reduction is possible in exactly the same space as the 1:21 or 1:41 drives.

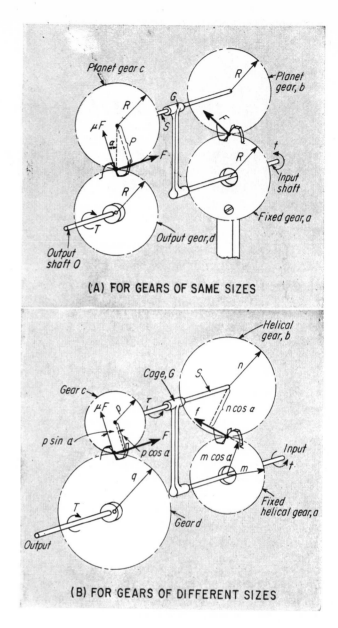

(A) FOR GEARS OF SAME SIZES

(B) FOR GEARS OF DIFFERENT SIZES

Among the 144 possible planet assembly combinations of gears with 39, 40 and 41 teeth, the one resulting in the highest ratio is found with $a = 39$, $b = 40$, $c = 41$, $d = 40$. Thus:

$$r = 1 - \frac{(39)\,(41)}{(40)\,(40)} = 1 - \frac{(40-1)\,(40+1)}{40^2} = 1 - \frac{1600-1}{1600} = \frac{1}{1600}$$

Here, gears a and b are helical, with a helix angle of $12°45'$. (A model of such a drive is shown in photo on page 67. The small pinion is the input, the pulley the output, and the output shaft makes one revolution when the input shaft turns 1600 times. Generally, the ratio becomes: $r = 1/n^2$, when one planet and the output gear each have n teeth, the fixed sun gear has $n - 1$ teeth and the second planet has $n + 1$ teeth. Thus, for a gear ratio of 1:3600, $r = 1/(60)^2$, and $a = 59$, $b = 60$, $c = 61$, $d = 60$, $\beta = 10°25'$.

Friction Losses

If there were no friction losses between the gear teeth and at the shaft bearings, ratio between input and

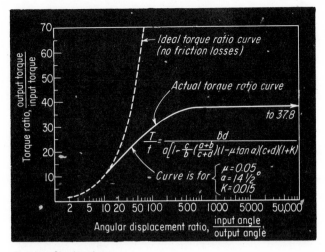

Torque limitation . . .
of epicyclic gear trains—curve shows tapering off to a 37.8 torque ratio.

output torques would be the reciprocal of the displacement ratio. But actually a large part of the input torque (or, to be exact, input work) is used up in overcoming friction.

To find the effect of friction in any epicyclic gear train, assume first all gears are of equal size. This results in a displacement ratio of $1/\infty$ or zero—if the input shaft is turned, the output shaft will remain at rest. For a frictionless gear system, the smallest input torque would be sufficient to turn the input shaft, no matter what load is applied to the output shaft. But actually the following will happen if a torque load T is applied to shaft O—see (A) on opposite page. It results in a tooth pressure force F normal to the tooth flank, thus creating a friction force equal to μF, where μ is the friction coefficient. Two torques, FP and $-\mu FP \tan a$ then act on gear c, where R is the pitch radius, a is the pressure angle and $P = R \cos a$. The friction torque created in the planet bearing G is proportional to the torque on the planet shaft S. The torque between c and d is counteracted by the torque between the fixed gear a and b; but to turn the input shaft the following torques must be overcome:

Friction torque between c and $d = \mu F P \tan a$
Friction torque between a and $b = \mu F P \tan a$
Friction torque at G $\qquad = K (2\mu F P \tan a)$

adding up to $2 \mu FP \tan a (1 + K)$, where K is the friction loss coefficient for the planet bearings.

Thus, the torque t necessary just to turn the input shaft without delivering any useful work (because the output rests) is $t = T (1 + K) 2\mu \tan a$. Thus:

$$\frac{T}{t} = \frac{1}{2 \mu \tan \alpha (1 + K)} \tag{3}$$

For typical constants of $\mu = 0.05$, $K = 0.015$ and pressure angle $a = 14\frac{1}{2}°$, the ratio between output and input torque $T/t = 37.8$. This condition is independent of size of the four gears, providing they are all the same size.

For the general case of gears of different sizes—see sketch (B)—with gears having pitch radii of m, n, p and q, the torque τ which acts on the planet shaft S and is produced by the output torque T and transmitted

back to the input is: $\tau = Fp \cos a - \mu Fp \sin a.$

This torque produces a tooth pressure force f of $\tau/n \cos a$ between gears a and b, hence

$$fn \cos \alpha = Fp (\cos \alpha - \mu \sin \alpha) \tag{4}$$

and

$$f = F(p/n)(1 - \mu \tan \alpha) \tag{5}$$

The input torque t therefore must supply $(F - f) \times (m + n)$ plus the friction torque created in the planet bearings G, which is proportional to the applied torque. Hence

$$\tau = (F - f) (m + n) (1 + K) (m/n) \tag{6}$$

where K is the friction loss coefficient in the planet bearings and the last term is caused by the ratio between sun gear and first planet.

From Eq (4), (5) and (6), and keeping in mind that T is proportional to F:

$$\frac{T}{t} = \frac{nq}{m \left[1 - \dfrac{p}{n} (1 - \mu \tan \alpha) \right] (m + n) (1 + K)} \tag{7}$$

It is more convenient to calculate with numbers of teeth than with pitch radii. This requires taking into account that $m + n = p + q$, even if $a + b$ is not equal to $c + d$. Thus, if m and n are the helical gears and a and b their respective numbers of teeth, both a and b must be divided by $\cos \beta$. But $\cos \beta = (a + b)/(c + d)$ and Eq (7) becomes:

$$\frac{T}{t} = \frac{bd}{a \left[1 - \left(\dfrac{c}{b} \right) \left(\dfrac{a+b}{c+d} \right) (1 - \mu \tan \alpha) \right] (c+d) (1+K)} \tag{8}$$

When all gears are of equal size, $a = b = c = d$, and Eq (8) degenerates into Eq (3). Thus the torque ratio approaches a constant value when the reduction ratio approaches $1/\infty$. The semilogarithmic graph on this page shows clearly how the curve approaches the limiting value asymptotically.

Test Results

Tests performed on a 1:1600 model showed that for input torque ranging from 1 in.-gm to 520 in.-gm the torque ratio ranged from 1:39 to 1:37.8, which agreed quite well with the theoretically derived value of 1/37.8.

How does this compare with torque ratios obtainable by an ordinary spur gear train of, say, 5 meshes? Assuming a 94% efficiency per mesh, the torque ratio would be: $(1:1600) \times 0.94^5 = 1:1150$. But 10 gears are needed for the 5-mesh drive.

In backlash conditions, the planetary system is slightly better than an ordinary gear train. Assuming a 0.002-in. linear backlash in all meshes and 32 diametral pitch, angular backlash is 8 minutes on the output shaft of the gear train, whereas corresponding backlash of the planetary drive is 6 minutes.

New multi-planet gears equalize their own loads

Unique arrangements of self-shifting levers and
fulcrums distribute the load among planetaries until all
pull equally, eliminating an old design drawback

Planetary gear systems, more popular in the 1930s than today, are getting a new lease on life. Their disadvantage of load imbalance is being overcome by ingenious devices developed in Europe and only now attracting attention in the U.S. Their advantages of compactness and high load capacity have never been doubted.

Two-gear and three-gear units worked out a few years ago by Felix Fritsch of Simmering-Graz-Pauker, a Viennese gear manufacturer, have had marked success in Europe. Now Fritsch has extended his system of eccentrically mounted gears and interconnected links to four-gear planetaries that operate without the old problem of load imbalance.

Fritsch's design ideas have been brought to the attention of American manufacturers by G. L. Dannehower of High Precision Products Co., Westfield, N. J. The two- and three-gear planetaries have been available in the U.S. through a licensee, Cincinnati Gear Co.

Appeal of planetaries. In comparison with equivalent parallel-shaft gearing, planetary systems have much to offer, including higher load capacities, lower tooth loadings, lower pitch-line velocities, higher efficiency, better lubrication, and an in-line arrangement of input and output shafts.

As the size comparison above shows, the more planetaries you put in a planetary gear system, the more compact the system becomes. Weight is also reduced. For the same power rating and speed ratio, a four-planet cluster has only half the overall diameter of a single-planet gear train. Size for size, it can transmit five times as much

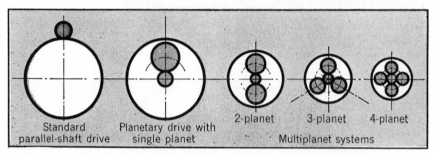

Adding planets makes the drive unit more compact (left to right). But this approach to reducing overall size creates problems unless planets are balanced.

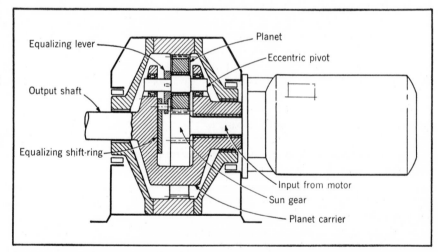

In an actual planetary application, the balancing system is built in like this. Cross-sectional view is of a motor's speed-reduction gear system.

torque as a parallel-shaft gear.

The key drawback has been that when more than one planet is used, load distribution among the planets is unbalanced unless each gear is perfectly concentric, of precisely the same size, and without any variations in tooth geometry and mounting dimensions. Imbalance can then tear the gear set apart.

The balancing system. In Fritsch's design, the self-equalizing arrangement of eccentrics and levers com-

pensates for any deviations. The use of eccentric pins is common to all the balancing systems in the drawings on the next page. Each pivot is connected with a radial lever that ends with a roller. The planet wheels are free to rotate on these eccentric pins, which are mounted on the planet carrier.

A four-planet system could be in equilibrium only when all eight tooth-contact points between the sun pinion, the planet gears, and the

Four linkage systems solve problem of balancing loads automatically in gears employing two or more planets.

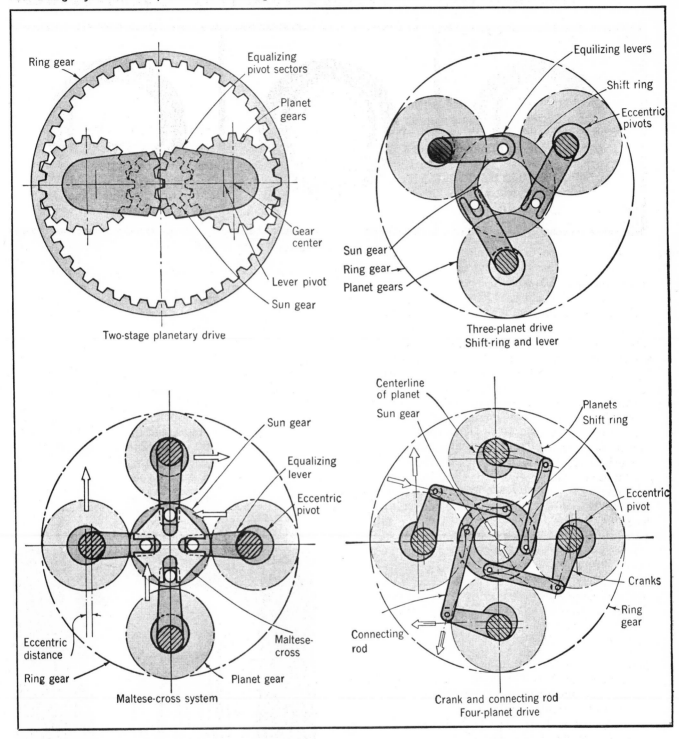

Two-stage planetary drive

Three-planet drive
Shift-ring and lever

Maltese-cross system

Crank and connecting rod
Four-planet drive

ring gear are transmitting the same magnitude of force. In actual operation, certain planets would tend to lag because of manufacturing inaccuracies, leaving one planet ahead of the others and forced to take on more of the load.

In the Maltese-cross equalizing system, for example, this extra force creates a coupling moment, causing the eccentric pin to pivot. With the help of the attached lever, this action shunts some of the driving load away from the lead planet to the floating Maltese cross. The cross adjusts itself by shifting and twisting until all four forces acting upon it are once more equal.

In a two-planet arrangement (top left drawing), the planets are mounted opposite each other, with a pair of meshing gear sectors in-stead of lever linkage. These gear segments are connected so any rotation of one eccentric produces an equal but opposite rotation in the other to equalize the loading on the planets.

In the three-planet design (top right drawing), the ends of the levers bear against a floating ring that adjusts its position until the three forces acting upon it are equal.

the Harmonic Drive... HIGH-RATIO GEARING

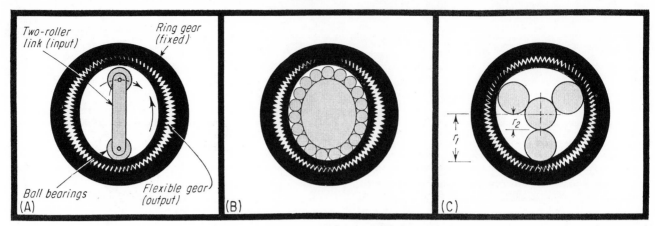

(A) Two-roller link (input) Ring gear (fixed) Ball bearings Flexible gear (output)

(B)

(C)

THREE VERSIONS OF DRIVE. Flexible gear is deflected in (A) by two-roller link; (B) by elliptical cam rolling within ballbearings; (C) by planetary-gear system for still-higher speed change.

How the Drive Works

The rotary version has a ring gear with internal teeth mating with a flexible gear with external teeth—see (A) in previous illustration. These teeth are straight-sided, and both gears have the same circular pitch, hence the areas of engagement are in full mesh. But the flexible inner gear has fewer teeth than the outside gear and therefore its pitch circle is smaller. Third member of the drive is a link with two rollers which rotates within the flexible gear, causing it to mesh with the ring gear progressively at diametrically opposite points. This propagates a traveling strain, or deflection wave in the flexible gear—hence, United Shoe's tradename, Harmonic Drive. If motion of the center link or "wave generator" is clockwise, and the ring gear is held fixed, the flexible gear will rotate (or "roll") counterclockwise at a slower rate, with constant angular velocity.

Teeth are stationary where in mesh, thus acting as splines in full contact. Movement of the flexible, driven member is confined to that area where teeth are disengaged. Each rotation of the center link moves the flexible gear a distance equal to the tooth differential between the two gears. Thus, speed ratio between center link (input) and flexible gear (output) is

$$\frac{V_o}{V_i} = \frac{N_f - N_r}{N_f}$$

where V_o = output velocity, rpm; V_i = input velocity, rpm; N_r = number of teeth (or pitch dia) of the ring gear; N_f = number of teeth (or pitch dia if permitted to take its full circular form) of the flexible gear.

For a drive with, say, 180 teeth in the ring gear and 178 teeth in the flexible gear

$$\frac{V_o}{V_i} = \frac{178 - 180}{178} = -\frac{1}{89}$$

Negative sign indicates that the input and output move in opposite directions. Actually, any one of the three basic parts can be held fixed and the other two used interchangeably as input and output.

The planetary-roller configuration in (C) permits better balance than with a two-roller link. In addition, the over-all gear ratio is increased by the planetary reduction. If driven by the sun roller,

$$\frac{V_o}{V_i} = \frac{N_f - N_r}{N_f} \left[\frac{r_2}{r_1 + r_2} \right]$$

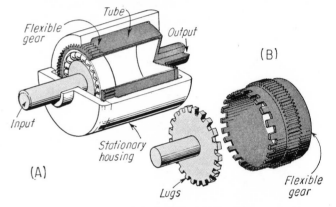

(A) Flexible gear Tube Output Input Stationary housing Lugs (B) Flexible gear

TWO METHODS OF COUPLING
flexible gear to shafts:
(A) by means of tubing; (B) with lugs.

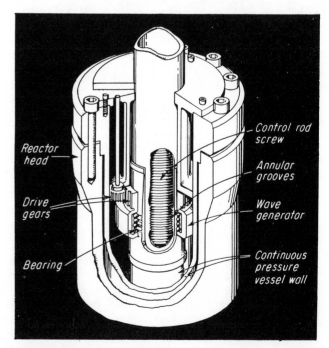

Reactor head Drive gears Bearing Control rod screw Annular grooves Wave generator Continuous pressure vessel wall

ROTARY-TO-LINEAR VERSION moves the control-rod linearly in this nuclear reactor head without need for mechanical contact through the sealed inner tube.

Coupling to Flexible Member

The flexible gear can be extended in the form of a tube (or cup) made of one or two parts, as shown in (A) at left. Flexibility of the tube permits the gear portion to be deflected as if it were a simple ring. Because deflection is small, the tube length need not be large. In addition, the inside space can be utilized to contain a motor, clutch, etc.

Circumferential lugs, (B), give another method of coupling; however, because of the deflection in the flexible gear, only relatively few of the lugs are in actual contact at any one time. This tends to limit the load capacity to the load-bearing capacity of those lugs in actual contact. Also, as the lugs slide radially in and out under a relatively great load, they are likely to wear locally at the point of contact.

Rotary-to-linear Mechanisms

The rotary-linear version of the Harmonic Drive can actuate a rod positively through entirely sealed walls—desirable for applications requiring absolute containment of pressure, vacuum, liquids or contaminants. The drive can provide positive mechanical control without the need for seals and glands.

In the illustration, left, of a nuclear reactor head, the control-rod screw is actuated by rotation of the drive gears without need for a device which penetrates through the inner, continuous-pressure vessel. Inner wall of the flexible tube has annular (zero lead) grooves. The control rod has double-lead threads. The wave generator is elliptical in shape, and when rotated by the drive gears, flexes the inner tube with its annular grooves, causing the minor-axis regions of engagement to progress around the lead screw and produce translation of the control-rod screw—without causing it to rotate.

Features Claimed by Inventor

Inventor and patentee of the drive is Dr C W Musser, a United Shoe consultant.

Dr Musser points out that the new drive avoids many characteristic difficulties of the conventional gear systems. Some distinctive features of the new drive are:

Large Percentage of Teeth in Contact. Usually 55% of them are in active engagement. This offers a much higher torque capacity than conventional gearing, which usually has only 2 to 6% of its teeth in full contact.

Large Torque Capabilities. Calculations by Dr Musser indicate a surprisingly large torque-producing capacity. For example, take a drive with pitch dia of driven gear = 4 in.; 2-lobe contact; coefficient of friction = 0.05; gear-reduction ratio of 1/100; and made of steel with a shear strength of 50,000 psi. Output torque capacity here would be 126,000 lb.-in., and tooth-contact pressure for this torque would be less than 22,000 psi. If the input were driven at 1800 rpm, the power output would be 36 hp. But a gear of this size would not have sufficient thermal capacity, even with cooling, to deliver an output of this amount continuously. However, if used for the transmission of 1 hp, it would have an overload or shock-resistance safety factor of 36, as compared with the safety factor—on the order of only 2 or 3—of many conventional gearing.

FIVE GEAR SYSTEMS — A COMPARISON (all systems producing 1 hp at 18 rpm)					
	Epi-cyclic	Her-ring-bone	Single worm	Heli-cal worm	Har-monic drive
Speed ratio	97.4	96.2	108	100	100
Efficiency, %	85	85	40	78	82
Number of gears	13	4	2	4	2
No. of bearings	17	6	6	6	2
Pitch-line velocity, ft/min.	1,500	1,500	1,500	1,500	18
Tooth-sliding velocity, ft/min.	2,500	2,500	1,500	2,500	28
Tooth-contact pressure, psi	50,000	50,000	5,000	50,000	605
Teeth in contact, %	7	5	2	3	50
Safety factor	3	2	2	2	36
Height, in.	13	14	23	16	6
Length, in.	15	20	19	17	6
Width, in.	13	10	14	10	6
Cubic volume, in.	2,500	2,800	6,100	2,700	216
Weight, lb.	246	280	230	205	30
Tooth contact	Line	Line	Line	Line	Surface
Quiet operation	No	Yes	Yes	Yes	Yes
Balanced Forces	Yes	No	No	No	Yes

Ratios from 10:1 to 1,000,000:1. For ratios above 200:1 the planetary wave generator will do. For ratios from about 1,000:1 to 1,000,000:1, a dual drive (a drive within a drive) can be employed.

High Mechanical Efficiency. Because of low friction losses a drive with a ratio of 100:1 has an over-all efficiency between 69 and 96%, depending on workmanship and lubrication. Increasing the ratio does not decrease the efficiency as markedly as is the case with standard gearing: a drive with a 400:1 ratio will be 80% efficient.

Adjustable Freedom from Backlash. By making the inner member capable of adjusting the deflection, backlash can be removed from a gear system by increasing the deflection to the point where the crest of the wave is radially deflected farther into the mating tooth spaces—until the teeth at each side come into contact.

Low Pitch-line Velocity. There are only two gears involved in the entire gear train under normal circumstances, and one of these is stationary with zero pitch-line velocity. The other has a rotational speed equal to that of the output shaft. Since the gear ratios are relatively large in many cases, the output rotational speeds are relatively small, usually of the order of 10 to 100 rpm.

Optimum Tooth Form

Tooth size, shape and differential greatly influence the percentage of teeth which are in engagement. In fact, for properly shaped teeth, 100% of the teeth are in contact, but in various degrees of engagement, as illustrated on next page. Proceeding from base line, which is the disengaged position, one side of the ring-gear teeth will become progressively more engaged with one side of the flexible gear teeth.

At point of 45° revolution, the teeth will engage 50% of the deflection; at 90° revolution they will be fully engaged. Beyond 90°, the teeth will become progres-

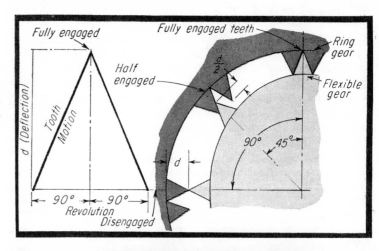

THEORETICAL TOOTH ACTION for two-roller version of drive has 100% of teeth in contact in various degrees of engagement.

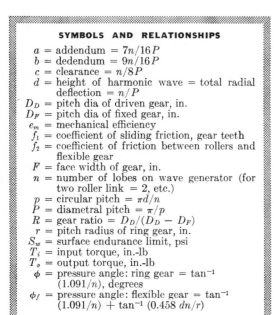

SYMBOLS AND RELATIONSHIPS

a = addendum = $7n/16P$

b = dedendum = $9n/16P$

c = clearance = $n/8P$

d = height of harmonic wave = total radial deflection = n/P

D_D = pitch dia of driven gear, in.

D_F = pitch dia of fixed gear, in.

e_m = mechanical efficiency

f_1 = coefficient of sliding friction, gear teeth

f_2 = coefficient of friction between rollers and flexible gear

F = face width of gear, in.

n = number of lobes on wave generator (for two roller link = 2, etc.)

p = circular pitch = $\pi d/n$

P = diametral pitch = π/p

R = gear ratio = $D_D/(D_D - D_F)$

r = pitch radius of ring gear, in.

S_w = surface endurance limit, psi

T_i = input torque, in.-lb

T_o = output torque, in.-lb

ϕ = pressure angle: ring gear = $\tan^{-1}(1.091/n)$, degrees

ϕ_f = pressure angle: flexible gear = $\tan^{-1}(1.091/n) + \tan^{-1}(0.458\,dn/r)$

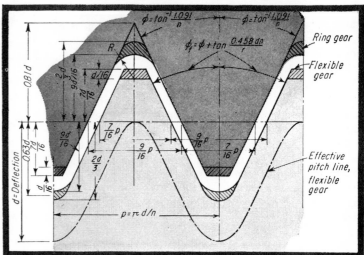

PRACTICAL TOOTH FORM makes allowances for dimensional tolerances and has slightly larger pressure angle for flexible gear than for ring gear. Approximately 55% of teeth are always in engagement, as against 2 to 6% for standard gears.

sively less engaged—here, however, the opposite side of the teeth are in contact. Thus, for the same direction of drive, successively opposite sides of the same tooth engage as the teeth advance.

Results of tests and calculations show that the straight-sided tooth forms, illustrated above, are the most practical. Note that the pressure angle for the flexible gear is slightly greater than for the ring gear. This compensates for the angularity of the flexible gear with the pitch-circle circumference during tooth contact.

Design Formulas

Because interaction between the teeth is dissimilar to that of conventional gearing, standard gear formulas are not applicable. Dr Musser points out that with the Harmonic Drive there is surface contact rather than line contact in the mating teeth; also, teeth cannot be analyzed as cantilever beams because of the pressure-angle relation-

ship to tooth size. Here are the design formulas relating torques, stresses and efficiencies (see list of symbols above):

Output-torque capacity

$$T_o = \frac{\pi r^2 F S_w \cos \phi}{4 \sin \phi}$$

Mechanical efficiency

$$e_m = \frac{(1 - f_1 \tan \phi)(1 - 0.458\,dnf_2/r)}{R(\tan \phi + f_1)(f_2 + 0.458\,dn/r)}$$

Output-input relationship

$$T_i = \frac{T_o}{Re_m}$$

Input-torque capacity

$$T_i = T_o \left[\frac{(\tan \phi + f_i)(f_2 + 0.458\,dn/r)}{(1 - f_i \tan \phi)(1 - 0.458\,dnf_2/r)} \right]$$

4
MODIFIED-TOOTH GEAR SYSTEMS

New tooth shape taking over?

Design of Novikov gears

Here's the first design article in this country giving the data on tooth geometry, load capacity, and manufacturing techniques.

Note: Novikov-type gears are similar to those developed by E. Wildhaber in the early 1920s (see p 135).

NICHOLAS CHIRONIS

THERE is a growing possibility that helical gears with circular-arc teeth will one day be widely substituted for involute gears in high-load and medium-load applications where speeds are not excessive. In Europe, debate among experts on the merits of these two types has recently intensified; opinion seems to be moving towards a reconsideration of circu-lar-arc gears. It is remarkable, therefore, that no research on this design has been published in the United States, although it is unofficially reported that several US firms have been carrying out experimental programs.

Although British and American patents on circular-arc gear systems have been on file as far back as the early 1920s, it is the Russians, with their "Novikov" system, who now lead the way. Their first public announcement was an academic paper in the April 1958 issue of *Vestnik Mashinostroy-eniya* (see References at the conclusion of this article). This paper not only gave the geometric relationships, it also revealed that Soviet research was being implemented by experiments in more than 50 factories and

1 . . HELICAL CIRCULAR-ARC GEARS employed in final stage of a triple-reduction gear box (left); close-up shown at right. The gears, which are being manufactured by AEI under the trade name Circarc, are of circular-arc profile in the plane normal to the helix. Most Novikov gears have circular-arc profiles in the transverse plane.

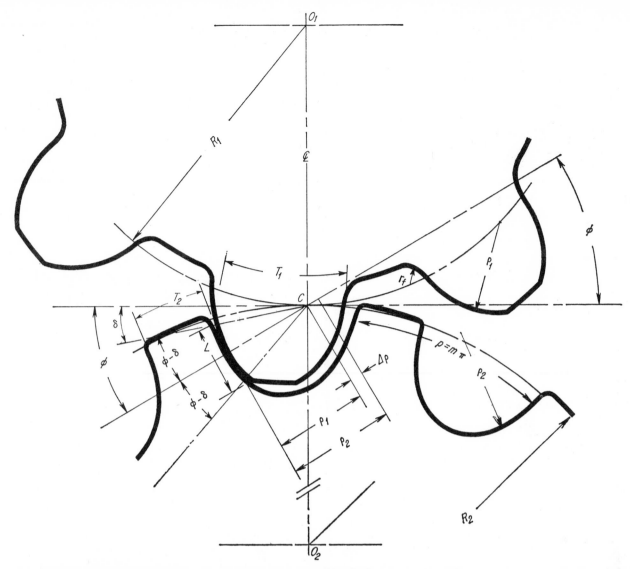

2 . . NOVIKOV GEAR TEETH with convex-concave tooth profiles consisting of circular arcs in the transverse section. The difference in radii, $\Delta\rho$, is usually small to obtain maximum load capacity, but ρ_2 can be much larger than ρ_1 (infinite if desired).

engineering institutes. The report stated that tests showed the new tooth form could carry three to four times the load of conventional involute gearing of the same size.

Other Russian papers soon followed. By December 1959, at the Leningrad technical conference on methods of reducing size and weight of gearing, the emphasis was on Novikov gears. Representatives of 123 Soviet institutes learned that comparative tests, conducted with automotive and agricultural-machinery gear boxes, had confirmed the higher strength and wear qualities of the new gearing. The conference decision was to give priority to developing the necessary production tools and establishing a system of standard pitches and tolerances for Novikov gears.

Subsequent Soviet reports indicate

that bevel and worm gears are being produced with Novikov teeth and that the new gearing is being considered for use in turbo-prop engines, helicopters and VTO aircraft, and marine drives. Japanese researchers report that the new principle is applicable to skew gearing.

In England, several companies have started serious investigations, and one —the AEI (Associated Electrical Industries Ltd, Rugby, England)—now markets a circular-arc type of gearing under the trade name Circarc, Fig 1. Germany also has been closely following the Soviet developments, and the topic of Novikov gearing dominated the 1960 International Conference on Gearing held at Essen.

In comparison with involute gearing, Novikov gears are reported to offer three advantages:

• They can take three to five times the load on the tooth flanks (tooth profiles) without detrimental pitting or wear. Any process applied to involute gearing for increasing the load capacity, such as hardening or soft-nitriding, can also be applied to Novikov gears.
• They retain lubricants between mating teeth more easily and form a thick oil film, perhaps as much as ten times thicker.
• They are somewhat more efficient in power transmission.

On the other hand, some reports state that Novikov gears are noisier than involute gearing and more sensitive to center-distance variations.

The Novikov principle

Novelty of the system lies in the fact that the tooth profiles consist of

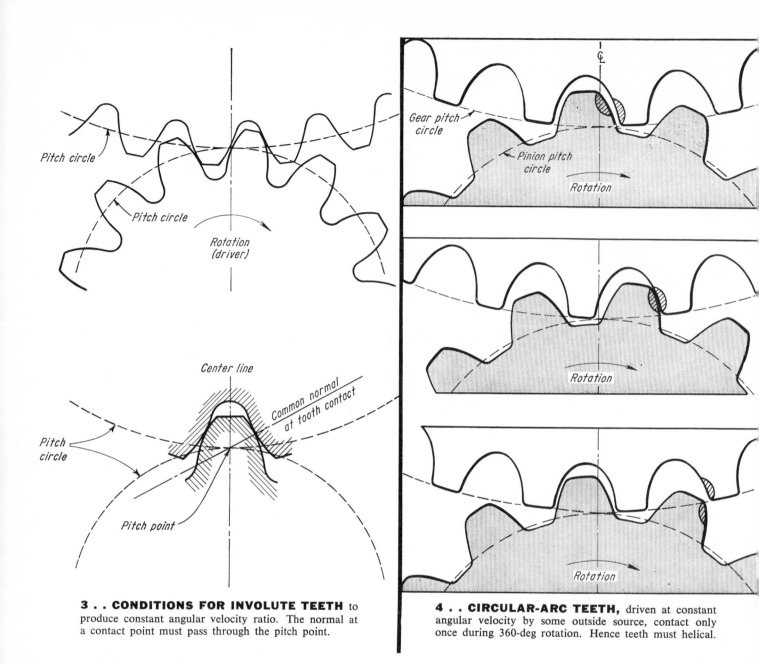

3 . . CONDITIONS FOR INVOLUTE TEETH to produce constant angular velocity ratio. The normal at a contact point must pass through the pitch point.

4 . . CIRCULAR-ARC TEETH, driven at constant angular velocity by some outside source, contact only once during 360-deg rotation. Hence teeth must helical.

circular arcs, usually in the transverse (frontal) plane, Fig 2. The tooth profiles of conventional gearing are shaped to an involute curve. (Sometimes, for mechanisms such as clocks and watches, the teeth are shaped to cycloidal curves.) Gear teeth based upon the involute curves present convex surfaces, one to the other, when in mutual engagement.

In the Novikov system, one member of the gear pair—usually the driving pinion—is made with teeth of a convex curve, or circular arc. The mating gear has teeth with a concave surface nearly conforming to the circular arc of the pinion. Thus the tooth shapes conform, or *envelope* one another.

The question comes up whether such teeth can provide a constant velocity ratio during rotation. With involute spur gears, Fig 3, the teeth make contact along a straight line across their width. Position of this contact line changes during rotation, shifting from near the root of the driving tooth to its tip. For constant velocity ratio, or *conjugate* action, the common normal of the tooth profiles at their point of contact must, in all positions of the contact, pass through a fixed point on the line-of-centers called the *pitch point*. The two pitch circles from the gear centers always pass through the pitch point, and the gears transmit motion as though they were a pair of friction contact wheels.

With Novikov gears, the common normal passes through the pitch point during only one position of rotation. Hence this principle is not applicable to spur and straight-bevel gears. Con-

jugate action can be achieved only by helical action, Fig 4. Assume that the two gears are driven at uniform speed at the proper ratio by some outside means. The surfaces of the teeth will mate and contact one another during only one point of rotation (in a given plane of rotation). If, however, the gear teeth are cut helical to their axis, Fig 5, the gears are equivalent to a whole series of thin spur gears, each displaced slightly from its neighbors. The gears are in contact in only one point for each width, w, the extent to which any tooth is twisted over a distance between two teeth, p. Thus, as the gears rotate, the contact point (or points, depending on the total face width, F) travels across the teeth; the pressure angle (angle ϕ in Fig. 2) remains constant, and conjugate action is obtained.

Elliptical area of contact

At first glance the theoretical point contact of Novikov teeth seems inferior to the line contact of involute teeth. In actual application, however, an elliptical area of contact results, Fig 5, which moves across the width of the teeth. This area is due to the close conformity of the mating teeth, the deformation of the contact point under load, and the wearing-in that results when the gears are operated for a time.

It has been established by Hertz and others that when two cylinders are in contact with one another parallel to their axes, both the area of contact due to deformation and the resulting surface stress depend on their relative radii of curvature, other things being equal. The area of contact is proportional to the face width and relative radii of curvature. Thus for spur gears in contact:

$$\text{Contact area} \propto \frac{F}{\dfrac{1}{\rho_1} + \dfrac{1}{\rho_2}}$$

For internal gears, or concave-convex gears:

$$\text{Contact area} \propto \frac{F}{\dfrac{1}{\rho_1} - \dfrac{1}{\rho_2}}$$

Thus, the contact height of circular-arc gears, for equal radii of curvature, is invariably much greater than for involute gears. It is this factor that is the basis for the high load capacity of Novikov gearing.

Novikov variations

Novikov gearing can take three different forms, Fig 6. All three are drawn to scale to produce the same transmission ratio. The first form, in which the smaller gear (pinion) has an all-addendum convex profile and the larger gear an all-dedendum concave profile, is the one employed in most applications.

The second form, with all-dedendum pinion, results in a smaller pinion than that in the first form. It is perhaps the least desirable of the three because of smaller tooth height and number of pinion teeth.

The third form is a combination of the other two and has good potential for future application. The addenda of both the pinion and gear teeth have convex profiles, and the dedenda concave profiles. Thus the same cutting tool is suitable for both pinion and gear.

Other circular-arc systems

Novikov's system agrees in all essential features with a circular-tooth form

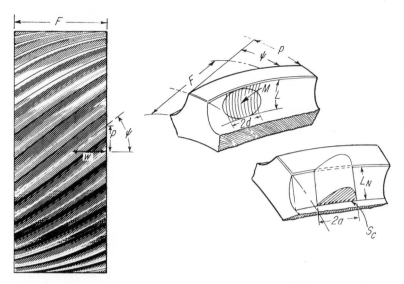

5 .. HELICAL NOVIKOV GEAR (left) has only one area of contact for each multiple of width *w*. Contact is theoretically at a point, but is actually an elliptical area (center) which moves across teeth from side to side as gear rotates. Pressure distribution at tooth shown at right.

system patented in 1923 by Ernest Wildhaber, who was then with Gleason Works and is now a practicing gear consultant in Rochester, NY. The main difference between the two systems is that most Novikov gears today are of circular-arc tooth form in the tranverse plane, whereas in the Wildhaber system the teeth are usually of circular-arc form in the normal plane (the plane perpendicular to the inclination of the gear teeth). Wildhaber's system permits tooth flanks to be accurately ground with wheels having circular contour in the axial section.

Novikov, however, may have considered such a modification. (M. L. Novikov, Dr Tech Sci, died in 1956; his work was published by his colleagues). Soviet engineers say that Novikov proposed that in mass production the profiles be generated by cutters having circular-arc contours in the normal section; he predicted that performance and load capacity would not be impaired. There are some indications that this modification is now being employed by the Russians. The Circarc gears, Fig 1, produced by AEI do have circular-arc contours in the normal section.

The Wildhaber system was developed soon after the British VBB (Vickers-Bostock-Bramley) tooth form, Fig 7, which has been used since the early 1920s in ship gearing and mine hoists. However, the American gear industry seems to have lost interest in the Wildhaber system when the VBB system failed to take hold in the US. It has been reported that the VBB system is particularly sensitive to variations in center distance (this may also be the case for Novikov gears, as will be discussed later) but since the flank load capacity of the VBB system is approximately three times greater than that of the 20-deg involute system, it is still not clear why this system did not become prominent, even in England.

Design Factors
Profile radii

The profile radius of curvature of the pinion, ρ_1, is generally made equal to the reciprocal of the pitch (see list of symbols, p 130):

$$\rho_1 = \frac{1}{P_d} = m \qquad (1)$$

A difference between the radii of the profiles of concave and convex teeth ($\Delta\rho = \rho_2 - \rho_1$ in Fig 2) is necessary to prevent the point of contact from reaching the upper edge of the concave teeth because of various inaccuracies and deformities. A positive deviation of the distance between the centers, or deflection of the shafts, may occur under load. If the point of contact shifts towards the edge of the convex tooth there will be a reduction in the load capacity. It is therefore generally agreed that ρ_2 should exceed ρ_1 by an amount

continued, next page

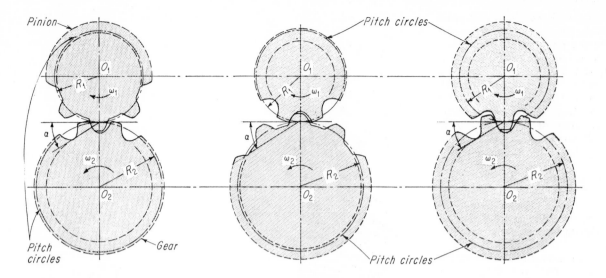

6 . . THREE FORMS of Novikov helical gears. All addendum pinion (left) is most popular form. Here the pinion has the convex tooth form, the gear the concave tooth form. The system with all-dedendum pinion (center) is the least popular form. The addendum-dedendum pinion (right) has good potential be-cause the same cutting tool can be used for both the pinion and gear. Variations in center distance do not affect constancy of transmission ratio.

$$\rho_2 \geqq 1.05 \text{ to } 1.2\rho_1 \qquad (2)$$

Face width

To ensure that the load is not carried at the extreme ends of the face width, there must be an overlap ratio greater than unity. In other words, the face width F, in Fig 5, must be appreciably greater than the axial pitch. Specifically:

$$F = (1.1 \text{ to } 1.2)\left(\frac{\pi}{P_d \tan \psi}\right) \quad (3)$$

Tooth thickness

The flexure strengths of pinion and gear are not equal for equal tooth thicknesses; tests have shown that the concave profile is stronger than the convex one. To equilize the flexure strength, the thickness of the teeth, T_1 and T_2, in Fig 2 has been designed with the following ratio at the Zdar Engineering Works, USSR:

$$T_1/T_2 = 1.5 \qquad (4a)$$

The tooth thickness of the pinion is made approximately

$$T_1 = 1.85/P_d \qquad (4b)$$

Pressure angle

A pressure angle of $\phi = 30$ deg is recommended by the Zdar Works. Other sources quote pressure angles varying from 20 to 30 deg. Soviet sources indicate that when the pressure angle is increased from 20 to 30 deg, load capacity increases 40% and the loading in the bearings 8%. Increasing the pressure angle also allows a reduction in the difference of radii ($\Delta\rho$) by a factor of nearly 2. Therefore it is best to employ a high pressure angle. However, if the pressure angle is made higher than 30 deg, the teeth become too pointed.

Helix angle

Recommendations for helix angle, ψ, vary from 10 to 30 deg. It will be shown later, however, that increasing the helix angle much above the lower recommended values greatly reduces the load capacity. The Zdar Works has been using a helix angle of 11 deg.

Decreasing the helix angle also increases the required face width of the gear (with a zero-degree helix angle you need an infinite face width). But increases in width are only proportional to cot ψ.

Load capacity

The flank load capacity of the Novikov tooth form has been computed by Gustav Niemann of Munich, Germany, and presented at the International Conference on Gearing, October 1960, Essen, Germany (see References). The following equations are based on his analysis. A high flank load capacity means higher pitting resistance and higher surface durability.

The contact height (L in Fig 2) is

$$L = 2 \sin (\phi - \delta)\rho_1 \qquad (5)$$

where δ is the clearance angle required by the hobbing operation. Angle δ is generally chosen as 5 deg.

The line of contact in the normal section, L_N, then equals

$$L_N = L \frac{\sin \phi}{\sin \phi_N} \qquad (6)$$

where the value for sin ϕ_N can be computed from the following relationship (subscripts N refer to values in the normal plane):

$$\tan \phi_N = \tan \phi \cos \psi \qquad (7)$$

The lengthwise radii of curvature in the normal section, Fig 8, are equal to

$$\rho_{1N} = \frac{R_1(1 + \tan \psi \cos \phi)^{3/2}}{\tan^2 \psi \sin \phi \left(1 + \dfrac{R_1}{\rho_1} \sin \phi\right)} \qquad (8)$$

$$\rho_{2N} = \frac{R_2(1 + \tan \psi \cos \phi)^{3/2}}{\tan^2 \psi \sin \phi \left(1 - \dfrac{R_2}{\rho_2} \sin \phi\right)} \qquad (9)$$

The relative radius of lengthwise curvature, ρ_N, in the normal plane is

$$\rho_N = \frac{1}{\dfrac{1}{\rho_{1N}} + \dfrac{1}{\rho_{2N}}} \qquad (10)$$

When Eq 8 and 9 are substituted into Eq 10 and symbols m_σ = reduction ratio = R_2/R_1, and $D_1 = 2R_1$ are

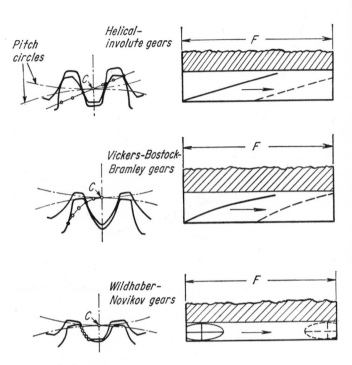

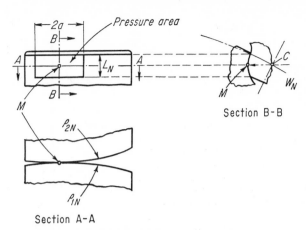

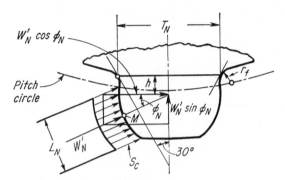

8 . . RADII OF CURVATURE, Novikov gears.

9 . . SURFACE-LOADING analysis.

7 . . COMPARISON of three helical gear systems. Line of pressure is straight line for involute, curved line for VBB gearing, and elliptical area for the W-N gearing.

employed, the value for ρ_N becomes

$$\rho_N = \frac{D_1}{2}\frac{(1+\tan\psi\cos\phi)^{3/2}}{\tan^2\psi\sin\phi} \times \left(\frac{m_G}{m_G+1}\right) \quad (11)$$

The face contact ratio m_F is equal to (it should be greater than unity):

$$m_F = \frac{Fp_d\tan\psi}{\pi} \quad (12)$$

The normal tooth load W_N, is

$$W_N = \frac{W_t}{m_F{'}\cos\phi\cos\psi} \quad (13)$$

where W_t is the tangential force at the pitch diameter, and m'_F is the integral portion of m_F (ie, when $m_F = 1.2$, $m'_F = 1.0$).

The contact rolling stress (Hertzian stress), S_c, now can be predicted by means of the following equation:

$$S_c = \left(\frac{W_NE}{5.72\,\rho_N L_N}\right)^{1/2} \quad (14)$$

A low predicted value in comparison to other types of gearing is an indication that the tooth flanks (tooth profiles) can accept high tangential loading.

Comparison of load capacities

The relative flank load capacity as a function of helix angle, Fig 10, have been computed by Niemann for Novikov and VBB gearing and for involute gearing with pressure angles of 20 and 28 deg.

Calculations for the Novikov gears were based on the previous equations. By keeping the contact stress, S_c, constant in the calculations for the involute gears, an index, in percent, was obtained relative to the 20-deg involute gearing.

For all curves, therefore, a permissible contact stress was used of $S_c = 76.5$ kg/mm² (109,000 lb/in.²) for tempered steel lubricated with mineral oil. Other constants were: $E = 2.1 \times 10^4$ kg/mm² (29.8×10^6 lb/in.²); reduction ratio $m_G = 3$; circumferential velocity $V = 10$ m/sec (32.8 ft/sec); number of teeth in pinion $= 30$. For the Novikov gears: face contact ratio

$m_F = 1.2$; pressure angle $\phi = 30$ deg; and $\delta = 5$ deg.

Referring to the Soviet experimental data in Fig 10, Niemann says that it is not possible to establish conclusively the relative load capacity because the Soviet publications from which his information was drawn do not state the gear materials, pressure angle, profile tolerances, and operating data. However, he assumed good design and workmanship and applied his equations to obtain the plot shown. This curve proves to be lower than that of the calculated values. Niemann theorizes that in the Soviet experiments the gearing was built with low pressure angles and helix angles, and that a larger pressure angle in that range would lift the curve to that of the calculated value.

In any event, the curves in Fig 10 show that:

● Flank load capacity (which is an indication of the wear resistance) of the Novikov gears quickly decreases with increasing helix angle, while that of the involute and VBB profiles in-

Symbols

Subscript 1 refers to gear with convex teeth (usually the pinion), subscript 2 to gear with concave teeth, subscript N to values in normal plane.

C = pitch point
d = semi-width of contact area, in.
D = pitch diameter, in.
E = Young's modulus, psi
F = total face width, in.
h = load height above tooth root, in.
K = Lloyd's factor
L = height of contact area, in.
m_F = face contact width, in.
m_G = reduction ratio = R_2/R_1
M = midpoint of contact
p = circular pitch = $\pi D/N$
P_d = diametral pitch = DN
r_f = fillet radius, in.
R = pitch radius, in.
S = stress, psi
T = circular tooth thickness, in.
W_N = normal tooth load, lb
W_t = tangential load at pitch dia, lb
w = minimum permissible width, in.
δ = clearance angle to facilitate hobbing, deg
μ = shearing stress factor
ρ = profile radius of curvature, in.
ρ_N = lengthwise radius of curvature
ϕ = pressure angle in transverse plane, deg
ψ = helix angle, deg
ω = angular velocity

creases with increasing helix angle.

• Novikov gearing attains an exceedingly high load capacity, particularly at low helix angles. For $\psi = 11$ deg —the value previously recommended —the capacity of Novikov gears is more than five times that of the 20-deg involute.

Niemann also concluded from tests that involute gearing (and undoubtedly Novikov gearing) can have a multifold increase in load capacity by means of soft-nitriding.

Beam strength of teeth

The shear and bending stresses acting on the tooth flank (tooth profile) can be computed with the following equations based on those developed by Niemann. There is some debate as to the validity of these equations. However, they do show the influence of the helix angle.

The maximum normal tooth force, W_N', per inch of face width (see Fig 9) is

$$W_N' = S_C L_N \qquad (15)$$

continued, next page

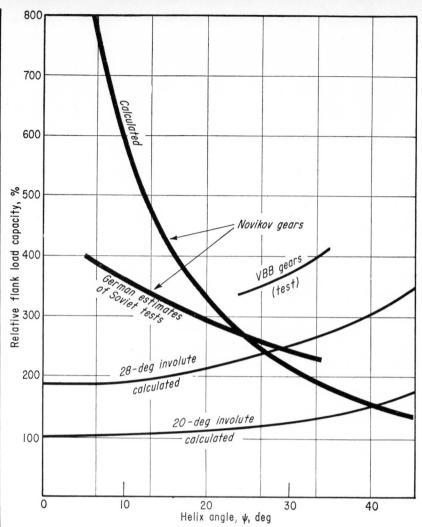

10 . . RELATIVE LOAD CAPACITIES of three types of gears.

11 . . RELATIVE BEAM strength of involute and Novikov gears.

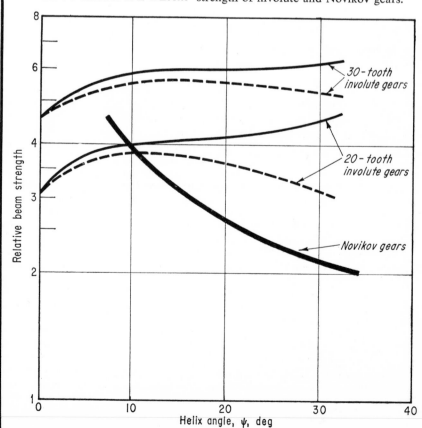

where S_c is determined from Eq 14, and L_N from Eq 6.

The mean shearing stress is

$$S_s = \frac{W_N{}' \cos \phi_N}{T_N} \qquad (16)$$

The tensile bending stress is

$$S_t = \frac{6hW_N{}' \cos \phi_N}{T_N{}^2} - \frac{W_N{}' \sin \phi_N}{T_N} \qquad (17)$$

where h is the load height above the root of the tooth, as shown in Fig 9.

The combined or "equivalent" stress value, taking into consideration both the shear and bending stresses, then is

$$S_e = [S_t{}^2 + (\mu S_s)^2]^{1/2} \qquad (18)$$

where μ is a German shearing stress factor which takes the tooth form into consideration ($\mu = 2.0$ for Novikov gears and 2.5 for involute gears).

An examination of Eq 18 will show that the equivalent stress increases with increasing helix angle. Hence, the beam strength of the Novikov tooth profile decreases with increasing helix angle. This is illustrated in Fig 11, in which Niemann estimates that the beam strength of equivalent involute gears surpasses that of the Novikov gears at the higher helix angles. Tooth failures in unhardened gears, however, are usually due to pitting and wear of the tooth flanks.

Power vs face width

As previously stated, decreasing the helix angle increases the load capacity—but also increases the required face width of the gearing. By assuming a 1.2 face contact ratio, W. J. Davies of the Aero Division of Rolls-Royce Ltd has come up with some striking results, illustrated in Fig 12 (See References). The designs are all for 14:43 teeth, with 3.25-in. centers, at 6100:1986 rpm. A face contact ratio of 1.2 is maintained by increasing the face width to make it proportional to cot ψ.

Using a contact stress of 116,000 lb/in.², Davies has estimated that the change from 30-deg helix to 10-deg helix increases the power capacity from 67 hp to 555 hp (8.3 times) but at the expense of a face width increase from 0.75 in. to 2.46 in. (3.3 times). Hence, when the horsepower per inch of face width is evaluated (see graph in Fig 12), Novikov gears with smaller helix angles are superior. Main restriction in this area may be the space (in width) available to the designer.

Center distance variations

Changes in the distance between shaft centerlines of a pair of Novikov

TABLE .. Effect of center distance variations on tooth failures.

TEST NO.	ΔCD, IN.	INITIAL POSITION (UNLOADED)	NO. OF CYCLES (GEAR) × 10⁶	FINAL CONDITION OF TEETH (ALL DAMAGE OCCURRED ON CONCAVE TEETH)
1 2 3	zero		28.6 23 20	Limited chipping in vicinity of tooth tip
4 5 6	+0.012		3.9 15.3 19.1	Fracture of tooth Fracture of tooth Chipping of tooth tip
7 8	+0.006		20.5 20.5	
9 10	−0.012		22.5 22.5	Chipping of tooth tip
11 12	−0.006		20.4 20.4	No traces of deterioration

gears have no effect on the constancy of the transmission ratio. An increase in the center distance leads to an increase in pressure angle—but the constant transmission ratio is still maintained.

There are references in European literature, however, that suggest that Novikov gearing may be sensitive to inaccuracies in the center distance because an increase in the center distance moves the contact area toward the tip of the concave tooth, with a consequent increase in flank pressure and root stress. This center-distance sensitivity is often cited as the main disadvantage of Novikov gearing.

To avoid tip contact, however, Soviet literature recommends that negative tolerances for the center distance be employed to decrease the distance between centers (this is feasible with Novikov gearing because the gears are easily "run in" or self-lapped under light load.) Also, it is recommended that tolerances for the profile radii of convex and concave teeth should both be the same in sign, preferably negative.

In an interesting series of experiments on the effect of center-distance changes, a Soviet researcher, I. N. Grishel (see References), concluded

that the maximum permissible center-distance tolerance is 0.3 mm (0.0118 in.), which is rather loose from the American standpoint. This is encouraging, as tighter CD tolerances should avoid any detrimental drop in load capacity.

The table above summarizes Grishel's results. In general, the experiments were conducted with the following values of ΔCD (change in center distance): ΔCD = +0.012 in., +0.006 in., −0.012 in., −0.006 in. The number of cycles to failure for gears with substantially no CD variation ranged from 20 × 10⁶ to 28.6 × 10⁶ cycles. With the negative tolerances, the working-in period proceeded much faster than with the positive tolerances. Also, with positive tolerances, no contact of the working profiles over the whole height of the tooth was obtained, even with a protracted period under full load during 20 × 10⁶ cycles.

With ΔCD = +0.012 (referring to the table) chipping of the working zone of profile in the vicinity of the outside diameter of concave-teeth gear develops very fast. In addition to this, because of the short length of the contact line and unsatisfactory working-in of the teeth, high contact stresses orig-

131

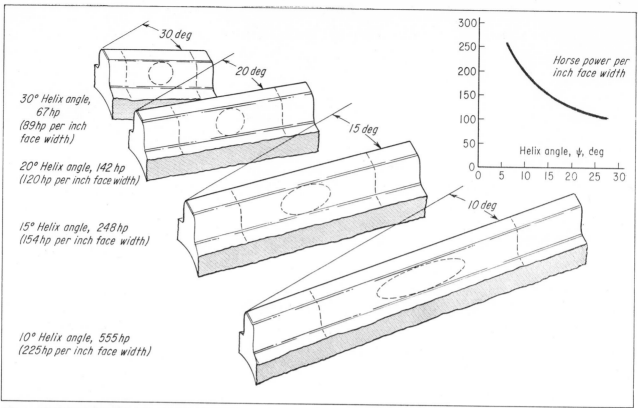

12 . . HORSEPOWER VS FACE WIDTH for Novikov gears. Decreasing the helix angle increases both width and horsepower per inch of face width.

inate on the working surfaces, causing progressive chipping of the concave profiles on the top of the teeth and, eventually, fatigue cracks resulting in tooth fracture.

With $\Delta CD = +0.006$, chipping of concave teeth occurred over a larger zone. Chipping did not show progressive development.

With $\Delta CD = +0.012$ and $+0.006$ there were no traces of deterioration of the convex teeth.

With $\Delta CD = -0.012$ there was no deterioration of the concave teeth but some chipping of the convex teeth, mainly in the zone of initial contact.

With $\Delta CD = -0.006$ there were no traces of deterioration on both the concave and the convex teeth.

Variations in center distance had no effect on the friction losses in the gearing. These losses decrease with increasing operation.

Lubrication and efficiency

In contrast to involute teeth the rate of slip on the tooth profiles in the normal plane is the same for every point of contact. Many European experimenters found that the quality of the tooth flanks is readily improved by a lapping run under load.

The rate of slip, V_s, is

$$V_s = V \rho_1 \left(\frac{1}{R_1} + \frac{1}{R_2} \right)$$

where V = circumferential velocity at the pitch circle.

In comparison with the involute system, successful lubrication appears to be somewhat easier. Novikov gears develop an oil-film thickness several times that of involute gears. There are two reasons for this: 1) The close conformity of the surfaces ensures that the pressure between teeth is distributed over a relatively large area of oil film. 2) The rapid movement of an oil wedge formed by the contact area moving across the teeth from side to side in the gear produces a strong pumping or hydrodynamic action.

Regarding the magnitude of the oil-film thickness, one Russian experimenter has suggested that it might be as much as ten times greater than that of involute gears. However, tests conducted by AEI on their Circarc gears showed a certain amount of burnishing of the surface markings, though not as much as on involute gears. Surface-finish amplitude of the tested gears was about 130 microin.;

hence AEI estimates the film thickness as being three to four times that of involute gears.

Although direct measurements of oil-film thickness have not been possible, electrical resistance measurements have been used to give an indication of relative oil-film properties. The increase in electrical resistance between teeth with time is given in Fig 13. The slope of such a line is a measure of the speed of running-in. The Circarc gear shows considerable variations of this slope at different loads, and involute gears tested on the same rig have also shown a small decrease in slope with increasing loads, Fig 14. The improvement in load-carrying capacity is well illustrated by this graph.

Lloyd's K factor is determined from the equation

$$K = \frac{W_t}{F D_p} \left(\frac{m_G + 1}{m_G} \right)$$

where W_t = tangential tooth load, lb; F = face width; D_p = pinion pitch dia; and m_G = gear ratio.

According to Russian experiments, the efficiency of Novikov gears is substantially higher, with losses up to half that of involute gearing, perhaps

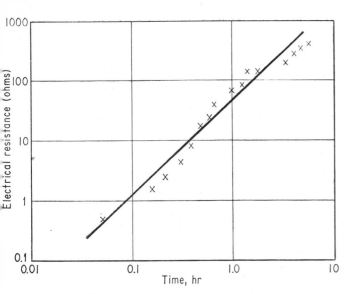

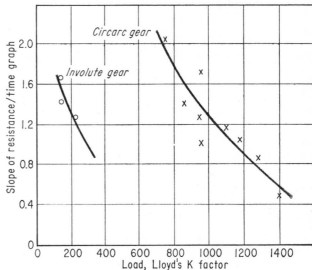

13 . . RESISTANCE-TIME GRAPH for Circarc gears. Resistance is proportional to oil-film thickness. Curve shows value of running-in circular-arc gears.

14 . . COMPARISON of running-in characteristics of involute and Circarc gears. The curves also show the higher load capacity of circular-arc gearing.

because of the thick hydrodynamic film.

Gear noise

The running tests carried out with Novikov and involute gears in the same transmission case at the Gear and Transmission Design Research Center, Munich Polytechnic Institute, demonstrated higher sound pressures and decibel levels for the Novikov system at all speeds, Fig 15, even though the Novikov gears had previously been carefully lapped.

However, when measurements were taken by AEI with involute gears in one gearbox and Circarc gears in the other, the transmitted power being the same, similar noise levels were recorded for both. When higher powers were transmitted in the Circarc gears, the noise level increased. The two gear systems do, however, show consider-able differences in the character of the noise produced. The fact that the noise spectrum of involute gears is quite complex is well known; a typical main component analysis by AEI is shown in Fig 16. A similar analysis taken from a Circarc test is also shown; in this case the main components are the tooth contact frequency and harmonics. Because the spectrum is simpler, AEI suggests that it should be easier to find the cause of the noise and find remedial measures.

Some engineers have doubted the suitability of Novikov gears for high-speed applications. So far there is insufficient data on this matter, although the possible higher noise levels may lead to some restrictions.

Circular arcs in normal plane

As previously mentioned, gears can be made with profiles that are of cir-cular arcs in a section normal to the inclination of the teeth. Thus the tooth flanks of this Wildhaber type of gearing can be accurately ground with wheels having a circular contour in the axial section. Load capacities are approximately equal to those of the Novikov system, and mathematical relations may be simpler. Wildhaber presented an analysis of this system at the 1960 Essen gear conference (see References).

Hobbing tools

Tools for manufacturing circular-arc gears can be divided into two groups, hobbing tools, Fig 17, and shaping tools. In involute gearing, the straight-flank rack profile is the basis for all hobbing tools (rack cutter, flank cutter of hob). It is independent of number of teeth, helix angle, and tooth thickness.

17 . . GEAR HOBS for manufacture of Circarc gearing. Photo at right shows hobbing of gear.

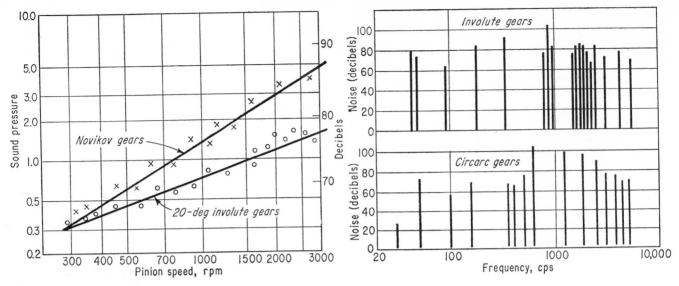

15 . . NOISE LEVEL as function of rotational speed for Novikov and 20-deg involute gears mounted in same transmission case and sound chamber (German tests).

16 . . SPECTRA of main noise components for Circarc gears and involute gears (English tests). Circarc gears are noisier only when more highly loaded.

A hobbing tool designed for a given helical circular-arc gear, however, can be used for cutting a different gear only if, in addition to the pitch, it has the same helix angle, tooth thickness, and profile radius—unless it is circular in the normal section. It is then suitable for any pitch diameter and helix angle.

Rack cutter: The rack cutter with concave teeth for cutting convex teeth in the pinion is shown in Fig 18 (from H. Winter and J. Looman; see References). The rack cutter for cutting concave teeth in the gear is its precise opposite, with convex tooth shape. The cutting edges of the cutter are circular in both cases and lie in transverse plane of the gear.

With helical circular-arc gears and their corresponding "circular-arc rack," the tooth on the work is not enveloped as time goes on, as is the case with involute gearing, in which the hobbing action takes place in small steps. The circular cutting edge coin-

cides with the frontal circular arc of the work-gear tooth only in a particular position, so that in this particular position the cutter tooth cuts the frontal arc, as it were, all at once on the last cut from base to crest. For the neighboring, parallel frontal planes to be cut in the same manner, the gear must rotate onward continuously as the cutter traverses the width.

Flank cutter: Teeth of a flank cutter must, if conversions are to be avoided, have a cutting outline of the proper matching profile—in other words, a circular arc in frontal section with same tooth radii and corresponding gap and tooth widths as the gear. The flank cutter has helical teeth. Unlike the rack cutters with their infinite pitch radius, the lateral flanks of the cutter teeth must be undercut very little, if at all.

Hobs: The hob form can be calculated exactly by Matthieu's method; in many cases, however, it is possible to reach an approximation by regard-

ing the rack profile in normal section as the hob profile. In normal section at the angle ψ, the rack has the shape—depending on gear type—of a circle or ellipse.

Shaping tools

On small-production or job quantities, it is more economical to use side or end-mill cutters. The profiles of these tools can be exactly calculated and accurately produced. In using contour grinding wheel or a side-milling cutter, Fig 19, the helical circular-arc gear, mounted between centers, is fed helically under the rotating cutter. An end-mill cutter is theoretically a special case of a side-mill cutter. It remains only to take account of the fact that the cutter axis now coincides with the y-axis (the vertical axis). End mill profiles resemble those of side mills.

Circular-arc gears with different number of teeth, but with the same helix angle, and cut with the same

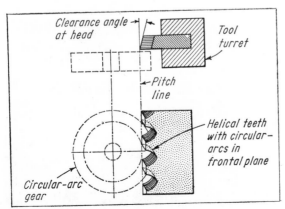

18 . . RACK CUTTER with concave circular profiles for cutting pinion with convex profiles.

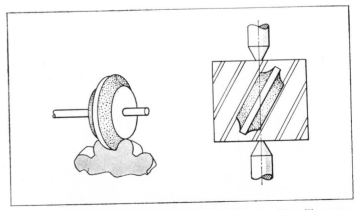

19 . . CONTOUR GRINDING WHEEL or side-mill cutter. Gear is fed helically under the rotating cutter.

cutter, will mesh correctly. This permits a system of standardization for both design and manufacture. The gears can be cut on standard production equipment and, since the hobs and shaping tools can be made with the same techniques employed for involute-gear tools, it should be economically feasible to produce circular-arc gears in this country.

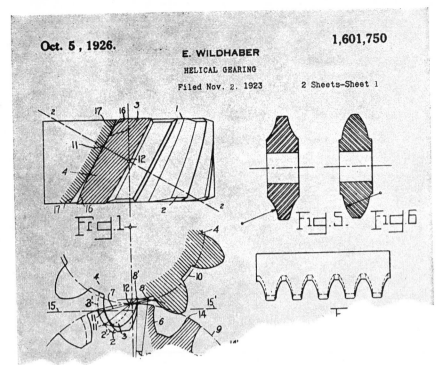

Oct. 5, 1926.

E. WILDHABER

HELICAL GEARING

Filed Nov. 2, 1923

1,601,750

2 Sheets-Sheet 1

Fig.1

Fig.5 Fig.6

According to Ernest Wildhaber, who received the US patent in 1926 on a circular tooth form, "All the characteristics of the Novikov gearing are completely anticipated by my patent. My gearing never had a real test here, although a pair of gears was made in the 1920's. It would be wrong to dismiss the Russian development as old and known."

Acknowledgements:

Our grateful acknowledgement of the valuable comments and technical evaluations in reviewing the text to:
Allan H. Candee, Gear Consultant, Rochester, NY.
Wells Coleman, Chief, Gear Analysis, Gleason Works, Rochester, NY.

E. J. Wellauer, Director, Research and Development, Falk Corp, Milwaukee, Wis.
C. F. Wells, Manager, Gear Engineering Dept, and **B. A. Shotter,** AEI Research Laboratory, Rugby, England.
Ernest Wildhaber, Gear Consultant, Rochester, NY.

Bruce W. Kelley, Staff Engineer, Research Dept, Caterpillar Tractor Co, Peoria, Ill.
George W. Michalec, Manager, General Engineering Dept, GPL Div, General Precision Inc, Pleasantville, NY.

REFERENCES

American

The Novikov Gearing System: Review of Soviet and Western Literature 1957-60. AID report 60-3. This report can be obtained from the Air Information Div, Washington, DC.

Russian

M. L. Novikov: USSR Patent No. 109,750 (1956).
R. V. Fedyakin and V. A. Chesnokov: The Novikov Gear Tooth System. *Vestnik Mashinostroyeniya,* April 1958. (In translation.)
R. V. Fedyakin and V. A. Chesnokov: Design of Tooth Gears of M. L. Novikov. *Vestnik Mashinostroyeniya,* May 1958. (In translation.)
I. N. Grishel: The Influence of Center Distance Errors on the Loading Capacity of Novikov Gears. *Vestnik Mashinostroyeniya,* May 1959.
F. S. Diktyar: Hobs for Cutting Novikov Gears. *Vestnik Mashinostroyeniya,* September 1959. (In translation.)
K. Jonas and A. Holemar: Novikov Gear Teeth (a summary of performance tests at the Zdar Engineering Works, USSR) *Technical Digest,* April 1962. (Published in English in Prague, Czechoslovakia.)

German

The following papers were presented at the International Conference on Gearing, October 1960, Essen, Germany, and published in the *VDI Berichte,* No. 47, 1961.
G. Niemann: Novikov Gear System and Other Special Gear Systems for High Load Carrying Capacity.
E. Wildhaber: Gears With Circular Tooth Profile Similar to the Novikov System.
H. Winter and J. Looman: Tools for making Helical Circular-Arc Spur Gears.
T. Matsuyama: Extension of Novikov's Gearing on Skew Gears.
W. H. Harrison: Load Tests on Gears With Teeth to the Wildhaber Novikov Design Basis.

English

W. J. Davies: Novikov Gearing. *Machinery,* Jan 13, 1960.
C. F. Wells and B. A. Shotter: The Development of 'Circarc' Gearing. *AEI Engineering,* March-April 1962.

Here's how to design full and semi-
Recess action gears

Chances are that some of your gear applications will function better with recess action shapes in place of the standard proportions

ELIOT K. BUCKINGHAM, Buckingham Associates Inc, Springfield, Vt

RECENT tests have shown that recess-action gears will wear longer, and operate smoother, with lower friction and less noise, than equivalent gears with standard proportion teeth. This holds true for the helical-worm types of recess-action gears (bevel, spiral, etc) as well as the spur type.

Recess-action gears (or, more simply, *RA gears*) are actually quite common, but they are known by different names. *Long-and-short addendum gears,* for example, are one form of RA gear. Such gears were usually employed to avoid interference and not for long-wearing qualities. Smaller pinions, also, are often designed with enlarged teeth to avoid undercut, as is the case with enlarged-tooth pinions. And if you go back far enough, you will find that old *clock gears* were actually RA gears.

As will be shown later, RA gears involve a change in the contact conditions between engaging teeth. All or most of the contact occurs during the recess portion, or *recess action*, of the line of contact; thus they avoid or reduce the detrimental effect of the *approach action.* This results in lower starting and running torques and produces higher operating efficiencies.

Contrary to some misconceptions, RA gears can be machined with standard hobs and cutters, and on standard hobbing machines. Their tooth form is an involute based on the same rack system as for the standard involutes, and they use the same center distance. Thus a pair of standard spur gears can be simply removed and replaced with a pair of RA gears (Fig 1)—without any other change in hardware.

Manufacturers in general find no difficulty in machining RA gears. You need merely specify the outer diameter D_o, the base diameter D_b, the pitch diameter D, and the tooth thickness t at D. The equations for finding these and other important geometric relationships are in Table I, p 140, for both full and semi-recess gears. The manufacturer also needs to know the measurements over pins. This he can compute from the data in Table I (you should be able to do it too), but as the calculations are involved and lengthy, the author programmed the steps in a computer and obtained overpin measurements to four decimal places (Tables II to V).

What limitations do RA gears have?

The first question that comes to mind is, how does the tooth strength (or beam strength) of RA gears compare to that of the standard involutes? The answer is that they are about the same—the driving gear of the RA system is somewhat stronger than an equivalent standard gear, and the follower gear is somewhat weaker. Hence, tooth strength seldom affects choice one way or another.

But an RA gear must be designed to operate either as a driver or as a follower. This is the main limitation of RA gearing. There's no problem with having an RA pinion drive a follower in either direction—in other words, the motor can be reversed without reducing the efficiency of the drive. Also, the gears can be used as a speed-increasing drive as well as a speed-reducing drive. But the power flow in the system must always be in the same direction, in this case from pinion to gear, not from gear to pinion. For example, RA gears cannot be employed for hoisting because when the brake is released to lower the weight, the weight will begin to drive the RA gears. This will cause binding in the tooth contact area with resulting high friction and wear.

This limitation also means that an idler cannot be used in an RA gear system (with standard center distances) because such a gear functions both as a follower and as a driver.

Who's using RA gears

Many critical drives have been built with recess action gears, including:
- Indexing drives for large telescopes; for example, the Mt Palomar 200-in. telescope employs RA gears for several of its drives
- Radar units for the dew-line stations
- High-speed torpedo drives rotating up to 100,000 rpm
- Servo gear heads
- Precision mechanical timers

Some of the companies that are successfully employing recess-action gears include:
- **Cone Automatic Machine Co,** Windsor, Vt, which is using this form in two locations on its latest bar-and-chucking machines (see photos, opposite page).
- **United Shoe Machinery Co,** Beverly, Mass, which is using RA gearing on main power drives. In the past, the company was spending several thousand dollars every year replacing or repairing worn-out gear drives. Since switching over, it has discovered that many of the drives that have been in operation for years are in just as good condition as when they left the factory.
- **Veeder-Root,** Hartford, Conn.—They discovered that RA gears have

text continued, page 139

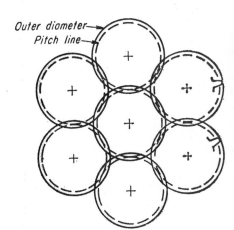

Outer diameter
Pitch line

Standard involute gears

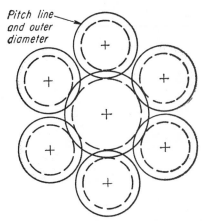

Pitch line
and outer
diameter

Full recess action gears

1 .. Full recess action gears (RA gears) on a six-spindle bar-and-chucking automatic machine (top right). Specifications call for 1 : 1 ratio from the input gear to the six spindles which must be equally spaced around the bull gear. With standard involute gears (top), all gears have the same outer diameter and thus have to be staggered to avoid interference. Full RA gears (bottom) enabled Cone Automatic Machine Co to fit all gears in same plane which considerably simplified the mounting design and also resulted in improved quietness, longer gear life and increased rigidity. The RA gears still provide a 1 : 1 ratio even though the bull gear is now larger than the spindle gears. All center distances remain the same, so it is easy to replace one type for another. The spindle carrier and gear-drive arrangement are shown at right. RA gears are also used in a second location (not shown) to transfer the high-speed power from the input shaft clutches to the control shafts that govern the machine functions during its "idle" portion of the work cycle.

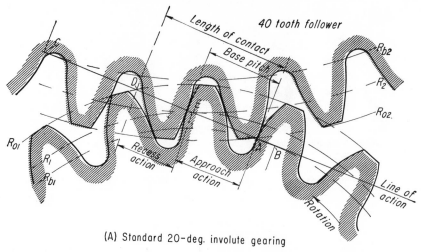

(A) Standard 20-deg. involute gearing

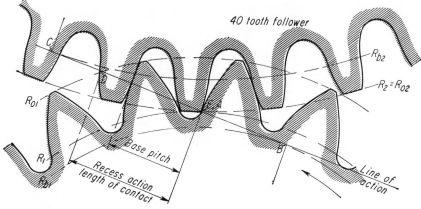

(B) Full recess gear system

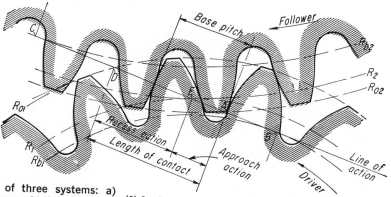

(C) Semi recess gear system

2 . . Comparison of three systems: a) standard involute gears, b) full RA gears, and c) semi-RA gears. All are 20-deg pressure angle spur gears with a 20-tooth pinion (driver) and a 40-tooth follower. Note the directions of rotation. Teeth of the standard form first make contact at A; detrimental wear occurs during the entire approach action from A to E. Recess action E to D, on the other hand, is of a beneficial nature. In full RA system, all tooth contact occurs during recess action. But because the length of contact has been reduced, the minimum number of teeth permissible for the pinion has gone up to 20 teeth or more. Semi-RA gears may be the best compromise as they reduce approach action yet increase contact length.

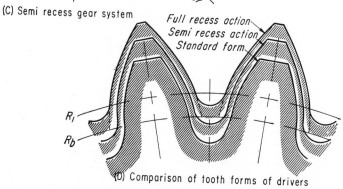

(D) Comparison of tooth forms of drivers

long operating life, especially when die cast, in applications where standard gears failed after short operation.

• **Servotron Corp,** South Berwick, Maine, is using RA gears in gear heads for space applications (see photo, p 89).

• **Servo-Dynamics Corp,** Somersworth, NH. Noise was a problem with its small ¾ to 1½ in. dia servo-gear heads. After building one unit with RA gears, the reduction in noise (with no changes in tolerances or class of gear) was so noticeable that the company now sticks to this form exclusively.

• **Cramer Control Div, Giannini Controls Corp,** Saybrook, Conn, which uses RA gears for many of its special control units, and in standard units where noise and life have been a problem. It has a large inventory of standard form gears plus a wide variety of designs, so the production control difficulties have made impractical the complete switching to this form to date, although it is being considered.

• **Bendix Corp,** Teterboro, N J. Some engineers here tend to select RA gears when performance is critical.

• **Other users** include Pacific Pump, Instron Corp, Moore Business Forms, DeLaval-Holroyd Inc, Whitin Machine Works, Welding Engineers Inc, and International Harvester.

Why reduce the approach action?

Tooth contact on spur gears is a line contact across the face of the tooth, and the motion is a combination of sliding and rolling. Many attempts have been made to reduce the rate of sliding, but this is not as important as the *direction* of the sliding.

Let's look at the type of contact action present in a pair of standard involute gears in contact (Fig 2A). Initial contact takes place at *A* (note direction of rotation) where the follower tooth begins to dig into the flank of the driver tooth. As contact progresses from *A* to *E* there is approach action. Recess action occurs from *E* until contact ceases at *D*.

Approach action is a sliding in. It is a detrimental scraping action, which tends to wear away the surfaces of the gears. The friction is high (Fig 3), causing scuffing and, later, noise. And the rougher the surface finish, the worse these conditions will be.

On all except hardened steel gears, there is a plastic flow of materials toward the root of the tooth of the driver and toward the pitch line of the follower. This tends to develop a hollow at the pitch line of the driver and build up a high spot on the pitch line of the follower, which increases the error in action, the dynamic loads, and the rate of wear, and sometimes destroys the teeth.

The change in direction of sliding at the pitch line also tends to break down the oil film, which increases the problems of lubrication and, in turn, increases friction. All in all, the approach action should be reduced as

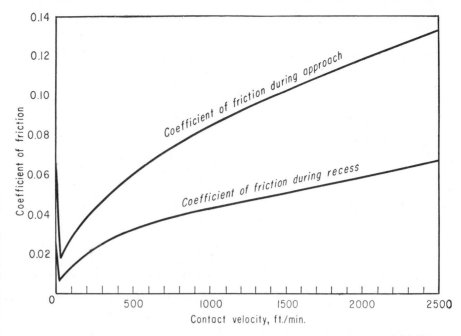

3 . . The reason recess action gears wear longer: The coefficient of friction during the recess action is half that during the approach action. Also, the magnitude of this difference increases significantly with increase in speed. The action is polishing for recess action, scraping for approach.

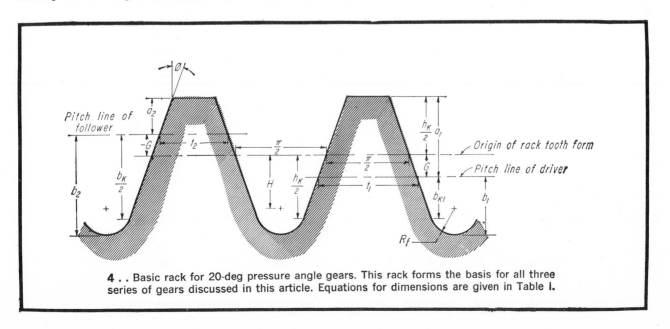

4 . . Basic rack for 20-deg pressure angle gears. This rack forms the basis for all three series of gears discussed in this article. Equations for dimensions are given in Table I.

much as the other factors in the gear design will allow.

Why recess action is beneficial

Recess action is a sliding out. Friction is lower and in a direction to help rotation. Nature of the action is beneficial, tending to cold work the surfaces, smooth out the rough spots, and work-harden the contact area in the process. This in turn increases the surface endurance limits of the materials and the load capacity of the pair. The lubrication problem is not as critical. With RA gearing it is frequently possible to run the surfaces dry after the gears have been run in. An inhibitor is recommended to keep the gear teeth from sticking together. Also, a dry film lubricant becomes feasible in many cases where a standard gear would require a wet type.

Some interesting tests have been conducted in England on the surface endurance of metals. Test rolls of the metals were driven by means of eccentric gears. Under these conditions of changing surface speeds there is pure rolling between them at two diametrically opposite elements of the test cylinders. In effect these are two pitch lines, because it is only at the pitch line contact of two mating gear teeth that there is momentary pure rolling. In these tests, the direction of sliding is reversed at the pitch lines, exactly as would occur with gear teeth.

The result was that the surfaces broke down on the areas of approach action. In no case was surface failure evident on the areas of recess action. This does not mean, necessarily, that there never would be failure on these recess areas. It does indicate, however, that the approach action is much more destructive than the recess action.

Wooden clock gears were full recess action. But some experts had noted the reduced wear. Willis Milham writes in *Time and Timekeepers* (MacMillan Co, 1941) that as the tooth of a wheel meets the tooth of a pinion, the teeth roll or slide on each other, engaging deeper and deeper until the line of centers is passed. They then draw apart until completely disengaged. He noted that the friction and wear due to entering is much greater than the friction and wear of leaving, and recommended that the teeth should not meet until at or beyond the line of centers.

It is interesting to note that the Novikov gearing system (see "Design

TABLE Ia..GENERAL VALUES FOR GEARS OF ANY PRESSURE ANGLE

Term	Symbol	Value
Addendum	a	$G + h_k/2$
Dedendum, total	b	$(h_k/2) + H + R_t - a$
Dedendum, working	b_k	$(h_k/2) - G$
Distance to rack origin	G	See Fig. 4
Distance to center of fillet radius	H	$\dfrac{h_k}{2} - \left[\dfrac{\pi}{4} - \dfrac{h_k}{2}\tan\phi\right]\tan\phi$
Outside diameter	D_o	$N + 2a$
Radius, fillet	R_t	$\left[\dfrac{\pi}{4} - \dfrac{h_k\tan\phi}{2}\right]\dfrac{1}{\cos\phi}$
Root radius	R_r	$(N/2) - b$
Tooth depth, total	h_t	$a + b$
Tooth depth, working	h_k	as specified
Tooth thickness at R	t	$(\pi/2) + 2G\tan\phi$

TABLE I..GEOMETRY VALUES FOR STANDARD, AND RA GEARS
(Pressure angle φ=20 deg for all cases)

Term	Symbol	Standard Form	Semi-RA gears Driver	Semi-RA gears Follower	Full RA gears Driver	Full RA gears Follower
Addendum	a	1.00/P	1.50/P	0.50/P	2.00/P	0
Base radius	R_b	$R\cos\phi$	$R_1\cos\phi$	$R_2\cos\phi$	$R_1\cos\phi$	$R_2\cos\phi$
Base pitch	p_b	$P\cos\phi$	$P\cos\phi$	$P\cos\phi$	$P\cos\phi$	$P\cos\phi$
Circular pitch	P	π/P	π/P	π/P	π/P	π/P
Dedendum	b	1.296/P	0.796/P	1.796/P	0.296/P	2.296/P
Diametral pitch	P	N/D	N/D	N/D	N/D	N/D
Outer radius	R_o	$\dfrac{N+2}{2P}$	$\dfrac{N+3}{2P}$	$\dfrac{N+1}{2P}$	$\dfrac{N+4}{2P}$	$\dfrac{N}{2P}$
Outer diameter	D_o	$\dfrac{N+2}{P}$	$\dfrac{N+3}{P}$	$\dfrac{N+1}{P}$	$\dfrac{N+4}{P}$	$\dfrac{N}{P}$
Pitch diameter	D	N/P	N/P	N/P	N/P	N/P
Pitch radius	R	D/2	D/2	D/2	D/2	D/2
Inside radius (internal)	R_i	$\dfrac{N-2}{2P}$	none	$\dfrac{N-1}{2P}$	none	none
Fillet radius	R_t	0.44847/P	0.44847/P	0.44847/P	0.44847/P	0.44847/P
Root radius	R_r	$\dfrac{N-2.592}{2P}$	$\dfrac{N-1.592}{2P}$	$\dfrac{N-3.592}{2P}$	$\dfrac{N-.592}{2P}$	$\dfrac{N-4.592}{2P}$
Tooth thickness (at pitch radius, R)	t	1.5708/P	1.9348/P	1.2068/P	2.2987/P	0.84286/P

Note: Subscript **1** relates to driver; subscript **2** relates to follower

of Novikov Gears," p 124) is also all recess action gearing.

The full RA gear system

The term "recess action gears" is used here for any gears where the amount of recess action is substantially more than the amount of approach action. The two series of RA gears discussed here—the full RA and the semi RA—are based on the same rack as for the 20-deg standard, full-depth involute gears, Fig 4 (stub forms can also be employed). Because the rack is a gear of infinite diameter, and is also the form of the hob used to generate the gears, the same hob may be used for all three series.

Standard gears place the pitch line in the center of the working depth of the tooth. This gives a tooth thickness at the pitch line for 1-P (one diametral pitch) gears of 1.5700 for both driver and follower, and an outside diameter of $N + 2$ (Table I).

For full recess action, however, all contact must take place after the line of centers (Fig 2B). To obtain this condition, the outside diameter of the follower is made equal to the pitch diameter. Referring again to Fig 4, if the arc tooth thickness at the center of the rack tooth equals $\pi/2$, then the thickness at the outside diameter equals 0.8429. This will be the arc tooth thickness at the pitch line of all followers in this series for 1 P.

If the gears are to operate on the same center distance as a pair of standard form gears, then 1 P the sum of the arc tooth thicknesses on the pitch line must be equal to π. This gives a tooth thickness at the pitch line of the driver of 2.2987, and is the thickness of a tooth of a gear with an outside diameter equal to $N + 4$. All that is done in effect is to reduce the size of the follower, and increase the size of the driver an equivalent amount.

Limitations on number of teeth

For a pair of gears to operate effectively, there must be adequate contact ratio, there should not be excessive undercut, and the teeth should not be pointed. This imposes the following restrictions on a series of full RA gears:

Maintain 20 or more teeth in the

5 . . Large standard-form gear (essentially a rack) with three smaller gears, all with the same OD, but with different number of teeth. The top gear is a full RA driver with 10 teeth, the one on left with 11 teeth is a semi RA driver, and the one on right is a standard 12-tooth pinion.

N	y	M/P	Mr	N	y	M/P	Mr
TABLE II. . SEMI-RECESS ACTION DRIVER Pin measurements, M, are over 1.92/P dia wires							
10	.232	13.5403	0.942	55	.279	58.9393	1.272
11	.237	14.4386	0.962	56	.279	59.9659	1.274
12	.242	15.5920	0.980	57	.279	60.9467	1.277
13	.246	16.5073	0.997	58	.279	61.9723	1.280
14	.249	17.6352	1.012	59	.280	62.9537	1.282
15	.252	18.5627	1.027				
16	.254	19.6721	1.040	60	.280	63.9784	1.285
17	.256	20.6085	1.053	61	.280	64.9604	1.287
18	.258	21.7041	1.065	62	.280	65.9841	1.289
19	.260	22.6475	1.076	63	.280	66.9667	1.292
				64	.280	67.9896	1.294
20	.261	23.7321	1.087	65	.280	68.9727	1.296
21	.263	24.6811	1.097	66	.281	69.9949	1.298
22	.264	25.7570	1.106	67	.281	70.9784	1.300
23	.265	26.7105	1.115	68	.281	71.9998	1.302
24	.266	27.7792	1.124	69	.281	72.9838	1.304
25	.267	28.7365	1.132				
26	.268	29.7993	1.139	70	.281	74.0046	1.306
27	.269	30.7597	1.147	71	.281	74.9890	1.308
28	.269	31.8174	1.154	72	.281	76.0092	1.310
29	.270	32.7806	1.161	73	.281	76.9940	1.312
				74	.281	78.0135	1.314
30	.271	33.8340	1.167	75	.282	78.9988	1.316
31	.271	34.7995	1.173	76	.282	80.0177	1.317
32	.272	35.8491	1.179	77	.282	81.0033	1.319
33	.272	36.8167	1.185	78	.282	82.0217	1.321
34	.273	37.8631	1.191	79	.282	83.0077	1.322
35	.273	38.8824	1.196				
36	.274	39.8759	1.201	80	.282	84.0256	1.324
37	.274	40.8469	1.206	81	.282	85.0118	1.325
38	.275	41.8879	1.211	82	.282	86.0293	1.327
39	.275	42.8603	1.215	83	.282	87.0159	1.328
				84	.282	88.0329	1.330
40	.275	43.8990	1.220	85	.282	89.0197	1.331
41	.276	44.8727	1.224	86	.282	90.0363	1.333
42	.276	45.9093	1.228	87	.283	91.0234	1.334
43	.276	46.8842	1.232	88	.283	92.0396	1.335
44	.276	47.9189	1.236	89	.283	93.0270	1.337
45	.277	48.8949	1.240				
46	.277	49.9280	1.243	90	.283	94.0428	1.338
47	.277	50.9049	1.247	91	.283	95.0304	1.339
48	.277	51.9365	1.250	92	.283	96.0458	1.341
49	.278	52.9143	1.254	93	.283	97.0338	1.342
				94	.283	98.0488	1.343
50	.278	53.9445	1.257	95	.283	99.0370	1.344
51	.278	54.9232	1.260	96	.283	100.0517	1.345
52	.278	55.9520	1.263	97	.283	101.0400	1.347
53	.279	56.9315	1.266	98	.283	102.0544	1.348
54	.279	57.9592	1.269	99	.283	103.0430	1.349
				100	.283	104.0571	1.350

driver, and 27 or more teeth in the follower. For equivalent standard involute gears, the minimums are 12 or more teeth in the driver, and 15 or more teeth in the follower.

The semi-RA gear system

Because of the rather high minimums in the numbers of teeth that RA gears require, we frequently turn to a compromise system—the semi-RA gears (Fig 2C)—which provides most of the benefits of recess action, and yet permits 10 or more teeth in the driver, and 20 or more in the follower. The semi-RA system uses the same rack as before and is listed in Table I.

Overpin measurements for RA gears

When involute gears are hobbed, the machinist generally starts with a gear blank which has been turned down to the proper outer diameter. The hobs then cut into the blank until they reach the proper tooth depth. This depth can be determined by several inspection methods, one of which is the overpin measurement.

In this method, the desired tooth thickness, t, is translated into a distance, M, across two pins of a specific diameter. These are placed diametrically opposite into the teeth (for even number of teeth), and one tooth away from the diametrically opposite position (for odd number of teeth). These data are given in Tables II to IV for gears with a diametral pitch of $P = 1$. For gears with other diametral pitches, divide the tabulated values of M by the pitch (see design problem below). The tables include Lewis strength of factors y, and the length of approach M_a or recess M_r. There are separate tables for drivers and followers, and for semi-RA and full RA gears.

Sample design problem

The problem below is divided into two sections: calculations of the gear geometry (size), and predictions of wear and strength.

Design of a pair of 64-pitch gears, 4:1 reduction ratio, 15 teeth on the driver and 60 on the follower. The gears to be 20 deg, full depth form, with $\frac{3}{16}$-in. face width. Material: steel, 300 Bhn with bending endurance limit of 75,000 psi. Determine the torque capacity of the drive.

Because there are less than 20 teeth in the driver, use the semi-recess action form.

Driver design

Pitch dia $= N/P = 15/64 = 0.2343$ (from Table I)
Outer dia $= (N+3)/P = 18/64 = 0.2812$ (from Table I)
Measurement over wire from Table II (using a wire size $= 1.92/64 = 0.030$) for $N = 15$:
$M = 18.5627/64 = 0.29004$

Follower design

Pitch dia $= 60/64 = 0.9375$ (from Table I)
Outer dia $= (N+1)/P = 61/64 = 0.9531$
Measurement over 0.030 wire (same as driver) from Table III:
$M = 62.2126/64 = 0.97207$
Center distance
$C = (D_1 + D_2)/2 = (0.2343 + 0.9375)/2 = 0.5859$ in. This is center distance for tight mesh. See AGMA standards for backlash, etc.
Contact ratio $= M_a + M_r = 1.027 + 0.467 = 1.494$

Load capacity of drive

$W_w =$ Limiting maximum wear load, lb $= D_1 FKQ$
$F =$ face width $= 0.1875$ in. (given)
$K =$ wear load factor $= 196$ (from *Manual of Gear Design, Vol II*
$Q =$ ratio factor $= 2N_2/(N_1+N_2) = 2(60)/(15+60) = 1.60$
$W_w = 0.2343 (0.1875) (196) (1.6) = 13.77$ lb

Beam strength of teeth

$W_s =$ max beam load, lb $= spFy$
$s =$ bending endurance limit $= 75,000$ psi (given)
$p =$ circular pitch $= \pi/P = \pi/64 = 0.04909$
$y =$ Lewis y-factor $= 0.252$ for driver (Table II, for $N = 15$) and 0.242 for follower (Table III, for $N = 60$)

TABLE III..SEMI-RECESS ACTION FOLLOWER
Pin measurements, M, are over 1.92/P dia wires

N	y	M/P	Ma	N	y	M/P	Ma
20	.178	22.2035	0.428	60	.242	62.2126	0.467
21	.182	23.1445	0.430	61	.243	63.1924	0.468
22	.185	24.2046	0.432	62	.244	64.2128	0.468
23	.189	25.1508	0.434	63	.244	65.1932	0.468
24	.192	26.2056	0.436	64	.245	66.2129	0.469
25	.194	27.1561	0.438	65	.245	67.1940	0.469
26	.197	28.2064	0.440	66	.246	68.2131	0.470
27	.200	29.1607	0.441	67	.247	69.1947	0.470
28	.202	30.2072	0.443	68	.247	70.2132	0.470
29	.204	31.1646	0.444	69	.248	71.1953	0.471
30	.207	32.2078	0.446	70	.248	72.2134	0.471
31	.209	33.1680	0.447	71	.249	73.1960	0.471
32	.211	34.2084	0.448	72	.249	74.2135	0.471
33	.212	35.1710	0.449	73	.250	75.1966	0.472
34	.214	36.2089	0.450	74	.250	76.2136	0.472
35	.216	37.1736	0.451	75	.251	77.1971	0.472
36	.218	38.2094	0.452	76	.251	78.2137	0.473
37	.219	39.1760	0.453	77	.252	79.1977	0.473
38	.221	40.2098	0.454	78	.252	80.2138	0.473
39	.222	41.1781	0.455	79	.252	81.1982	0.473
40	.223	42.2102	0.456	80	.253	82.2139	0.474
41	.225	43.1800	0.457	81	.253	83.1987	0.474
42	.226	44.2105	0.457	82	.254	84.2140	0.474
43	.227	45.1818	0.458	83	.254	85.1991	0.474
44	.228	46.2108	0.459	84	.254	86.2141	0.474
45	.229	47.1834	0.460	85	.255	87.1996	0.475
46	.230	48.2111	0.460	86	.255	88.2142	0.475
47	.232	49.1848	0.461	87	.255	89.2000	0.475
48	.233	50.2114	0.461	88	.256	90.2143	0.475
49	.234	51.1862	0.462	89	.256	91.2004	0.476
50	.234	52.2116	0.463	90	.256	92.2143	0.476
51	.235	53.1874	0.463	91	.257	93.2008	0.476
52	.236	54.2119	0.464	92	.257	94.2144	0.476
53	.237	55.1886	0.464	93	.257	95.2011	0.476
54	.238	56.2121	0.465	94	.258	96.2145	0.476
55	.239	57.1896	0.465	95	.258	97.2015	0.477
56	.239	58.2123	0.466	96	.258	98.2146	0.477
57	.240	59.1906	0.466	97	.259	99.2018	0.477
58	.241	60.2125	0.467	98	.259	100.2146	0.477
59	.242	61.1915	0.467	99	.259	101.2022	0.477
				100	.259	102.2147	0.477

Because the y value for the follower is lower, the follower will be limiting member:

$$W_s = 75{,}000 \ (0.04909) \ (0.1875) \\ (0.242) = 167.1 \text{ lb}$$

The wear load is lower than the tooth strength; hence this determines the capacity of the drive:

Maximum output torque

$$T = W_w D_2/2 = 13.77 \ (0.9375/2) \\ = 6.45 \text{ in.} - \text{lb}$$

6 . . Servo gear head by Servotron Corp is designed with RA gears so as to operate with little or no lubrication under high operational loading. The ¾-in. dia unit transmits 75 in.-oz torque and holds 250 in.-oz statically.

		TABLE IV. . FULL RECESS ACTION DRIVER						
		Pin measurements, M, are over 1.92/P dia wires					
N	**y**	**M/P**	**Mr**	**N**	**y**	**M/P**	**Mr**
20	.296	24.3798	1.369	**60**	.293	64.7736	1.653
21	.296	25.3336	1.383	61	.293	65.7574	1.657
22	.296	26.4177	1.396	62	.293	66.7834	1.660
23	.296	27.3757	1.408	63	.293	67.7677	1.664
24	.296	28.4519	1.420	64	.293	68.7928	1.667
25	.296	29.4134	1.431	65	.293	69.7775	1.670
26	.296	30.4829	1.442	66	.292	70.8018	1.674
27	.296	31.4474	1.452	67	.292	71.7869	1.677
28	.296	32.5113	1.462	68	.292	72.8103	1.680
29	.295	33.4782	1.472	69	.292	73.7959	1.683
30	.295	34.5373	1.481	**70**	.292	74.8186	1.686
31	.295	35.5064	1.490	71	.292	75.8045	1.689
32	.295	36.5614	1.498	72	.292	76.8264	1.692
33	.295	37.5323	1.506	73	.292	77.8127	1.695
34	.295	38.5836	1.514	74	.292	78.8340	1.698
35	.295	39.5562	1.522	75	.292	79.8206	1.700
36	.295	40.6043	1.529	76	.292	80.8413	1.703
37	.295	41.5783	1.536	77	.292	81.8282	1.706
38	.294	42.6235	1.543	78	.292	82.8483	1.708
39	.294	43.5989	1.550	79	.292	83.8355	1.711
40	.294	44.6415	1.556	**80**	.292	84.8550	1.713
41	.294	45.6180	1.563	81	.292	85.8425	1.715
42	.294	46.6584	1.569	82	.292	86.8615	1.718
43	.294	47.6359	1.575	83	.292	87.8493	1.720
44	.294	48.6743	1.580	84	.292	88.8678	1.722
45	.294	49.6527	1.586	85	.292	89.8558	1.725
46	.294	50.6892	1.591	86	.292	90.8738	1.727
47	.294	51.6685	1.596	87	.292	91.8621	1.729
48	.294	52.7033	1.601	88	.292	92.8796	1.731
49	.293	53.6834	1.606	89	.292	93.8681	1.733
50	.293	54.7166	1.611	**90**	.292	94.8852	1.735
51	.293	55.6974	1.616	91	.292	95.8740	1.737
52	.293	56.7292	1.620	92	.292	96.8907	1.739
53	.293	57.7107	1.625	93	.292	97.8796	1.741
54	.293	58.7411	1.629	94	.292	98.8959	1.743
55	.293	59.7233	1.633	95	.291	99.8851	1.745
56	.293	60.7525	1.638	96	.291	100.9010	1.747
57	.293	61.7352	1.642	97	.291	101.8904	1.749
58	.293	62.7633	1.645	98	.291	102.9059	1.751
59	.293	63.7466	1.649	99	.291	103.8955	1.752
				100	.291	104.9107	1.754

		TABLE V. . FULL RECESS ACTION FOLLOWER			
		Pin measurements, M, are over 1.92/P dia wires			
N	**y**	**M/P**	**N**	**y**	**M/P**
27	.158	28.0715	64	.222	65.1847
28	.162	29.1214	65	.223	66.1665
29	.165	30.0850	66	.224	67.1858
			67	.225	68.1681
30	.168	31.1309	68	.225	69.1869
31	.171	32.0962	69	.226	70.1697
32	.173	33.1386			
33	.176	34.1055	**70**	.227	71.1879
34	.178	35.1450	71	.228	72.1712
35	.181	36.1134	72	.228	73.1888
36	.183	37.1505	73	.229	74.1726
37	.185	38.1202	74	.230	75.1897
38	.187	39.1551	75	.231	76.1739
39	.189	40.1262	76	.231	77.1906
			77	.232	78.1751
40	.191	41.1592	78	.233	79.1913
41	.193	42.1315	79	.233	80.1762
42	.195	43.1627			
43	.196	44.1362	**80**	.234	81.1921
44	.198	45.1659	81	.234	82.1773
45	.200	46.1403	82	.235	83.1928
46	.201	47.1687	83	.236	84.1784
47	.203	48.1441	84	.236	85.1935
48	.204	49.1712	85	.237	86.1794
49	.206	50.1475	86	.237	87.1941
			87	.238	88.1803
50	.207	51.1734	88	.238	89.1947
51	.208	52.1506	89	.239	90.1812
52	.210	53.1755			
53	.211	54.1535	**90**	.239	91.1952
54	.212	55.1773	91	.240	92.1821
55	.213	56.1561	92	.240	93.1958
56	.214	57.1791	93	.241	94.1829
57	.215	58.1585	94	.241	95.1963
58	.216	59.1806	95	.242	96.1837
59	.217	60.1607	96	.242	97.1968
			97	.242	98.1844
60	.218	61.1821	98	.243	99.1973
61	.219	62.1628	99	.243	100.1851
62	.220	63.1834			
63	.221	64.1647	**100**	.244	101.1977

For miniaturized gear systems
Enlarged-tooth pinions

Here's how to increase the number of teeth
in contact and thereby avoid undercut
when you want fewer – but larger – pinion teeth.

RICHARD L THOEN, staff engineer, General Mills Inc, Minneapolis

MANY of the gearing problems associated with miniaturization can be avoided by resorting to pinions with small number of teeth. But the trend today is in the other direction—toward specifying ever finer pitches. It may surprise most gear designers to know that we have generated—with conventional hobs—pinions having as few as four teeth; and, with special form cutters, pinions having as few as three teeth (Fig 1 and 2). These fewer teeth result in larger, stronger tooth forms, higher contact ratios (number of teeth in mesh), and fewer of the manufacturing and inspection problems inherent to the ultra-fine pitches.

At present, if you follow the standard procedure for modifying pinions you will not be able to come up with optimum results. By "optimum" we mean pinions that yield the maximum contact ratio. In fact, before the contact ratio can be determined, it is necessary to compute the amount of undercut—a rather laborious computation.

The graphs presented here give a clear picture of the variables that influence the optimum tooth proportions of modified pinions. It will be shown, for example, that decreasing the pressure angle actually increases the contact ratio of enlarged pinions—a favorable result, and contrary to popular opinion. Included in this analysis are proposed tooth proportions of a generating rack (or hob) which, it is hoped, will be standardized by the industry for the production of pinions of all pitches in the fine-pitch category — 20 diametral pitch (DP) and finer. This hob is one of the many standard hobs now employed in the industry—hence, in many cases, special tooling costs will be unnecessary.

Ease of fabrication

The pinions and gears in Fig 1 were generated with the same hob. The hob also generates the pinion outside radii. This offers all the manufacturing and inspection advantage of top hobbing; but, unlike topped gears, the outside radii are not dependent upon the dimensions and tolerances of a topping hob. Also, because of their relatively exposed tooth surfaces, the pinions can be checked for tooth-thickness by span-gaging with a conventional micrometer, and measured for lead error with a simple setup consisting of a surface plate, shop centers and a comparator.

Maximum contact ratio

Key advantage of this modification system is that the pinions yield the maximum contact ratio consistent with the pinion tooth-thickness tolerance and hob-addendum dimensions. Although the system is applicable to pinions with 21 teeth and fewer, it is best for pinions in the range of 4 to 12 teeth.

The gear addendums are enlarged—they will vary from 1.0 (standard) to 1.5. This dimension, which is the addendum of a 1 DP tooth, will increase as the number of pinion teeth decrease. The gear-tooth thickness is standard ($\pi/2$), with bilateral tolerances, which minimizes inventories of stock gears, master gears, measuring pins and cutting tools.

Because of the enlarged pinion, the center distance should be increased to avoid reducing the thickness of the mating gear, Fig 3. This illustration shows an 8-tooth pinion meshing with two gears, one of standard tooth thickness at enlarged center-distance, and another of reduced tooth thickness at standard center-distance. In general, the standard tooth-thickness system is more desirable than the standard center-distance system. The hob proportions will be described later.

The manner in which contact ratio varies with enlargement is shown in Fig 4. The tooth forms are for 9-, 18- and 36-tooth gears, all drawn to the same diametral pitch (1 DP, for example). The generating rack has a 1.0 dedendum, a 1.6 addendum and a 20° pressure angle. The gear enlargements correspond to rack offsets of zero (standard center-distance), 1.0 and 2.0. At zero offset all three gears are topped, the 9- and 18-tooth gears are undercut, and the 36-tooth gear is not undercut. At an offset of 1.0 the conditions are the same except that the 9-tooth gear is pointed and the 18-tooth gear is not undercut. At an offset of 2.0 all three gears are pointed and are not undercut.

Referring to the 9-tooth gears, the contact ratio (with a rack) is less for the condition of zero offset than for 1.0 offset. The detailed manner in which this contact ratio varies, plus that for the 18 and 36-tooth gears, is shown in Fig 5. The maximum contact ratio occurs at the transition point between undercut and no undercut. (The 36-tooth curve peaks at a negative offset.) Mathematically, this point occurs at an offset equal to

$$\Delta C = a_R - \frac{N}{2} \sin^2 \phi_s$$

where ΔC = rack offset, a_R = generating rack addendum, N = number of teeth and ϕ_s = standard pressure angle.

The maximum contact ratio increases with rack addendum (a_R) and number of teeth, Fig 6. Only the peaks of curves, such as in Fig 5 for a 1.6 addendum, are plotted in Fig 6. Compare, for example, the peak of the 9-tooth curve in Fig 5 with the value for the 9-tooth curve at 1.6 addendum in Fig 6. With standard pinions, the contact ratio decreases as the rack addendum increases (a semifinishing hob produces more undercut than a finishing hob). Thus, you can enlarge pinions (to avoid undercut) and obtain higher contact ratios by employing a long-addendum cutting tool. In fact, the 3-tooth pinion (Fig 2) requires a cutting tool that is almost pointed.

Because the maximum contact ratio increases as the tip width of the cutting tool decreases, Fig 6 is replotted in terms of rack tip width, Fig 7. To show also how the maximum contact ratio varies with pressure angle, only the 4-tooth curve of Fig 6 has been plotted in Fig 7. Compare, for example, the 4-tooth curve at zero tip width (2.16 rack addendum) in Fig 6 with the 20° curve at zero tip width in Fig 7.

Importance of tip width

Contrary to common opinion, Fig 7 shows that the maximum contact ratio of enlarged pinions varies inversely with pressure angle. As an illustration, a tooth modification system based on 14½°, 0.4 tip width offers slightly greater possibilities than one based on 25°, 0.4 tip width. In short, from the standpoint of maximum contact ratio, a cutting tool should have a long addendum and low pressure angle.

Thus, the system for modifying pinions should be based on tip width, (which is inversely proportional to the rack addendum). However, the tip widths of present-day cutting tools are not constant—they vary with the diametral pitch. For instance, Fig 8 shows how the active addendums of 20°, non-topping, Class A hobs vary over the range of 20 to 120 DP. The upper curve represents a sharp-cornered semi-finishing hob and is dependent upon the 0.002-in. clearance constant, tooth-thickness tolerance, and tooth-thinning allowance. The effect of tooth thinning has been included because many semifinishing hobs are used as finishing hobs. The lower curve represents a round-cornered finishing hob, and is dependent upon the 0.002 constant, tooth-thickness tolerance, tip-radius allowance,

1 ENLARGED-TOOTH PINIONS — one with only 4 teeth — and gears generated with the same hob.

2 A 3-TOOTH PINION and rack formed with special form cutters illustrated.

3 ENLARGED-TOOTH PINION can be meshed with gear at enlarged center-distance (left), or at standard center-distance (right). Gear at left has standard tooth-thickness; tooth thickness for gear at right is greatly reduced.

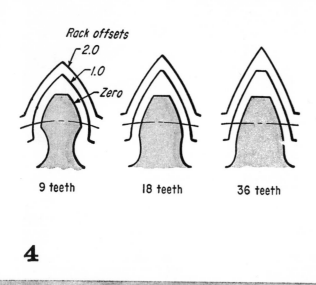

Rack offsets
2.0
1.0
Zero

9 teeth   18 teeth   36 teeth

4

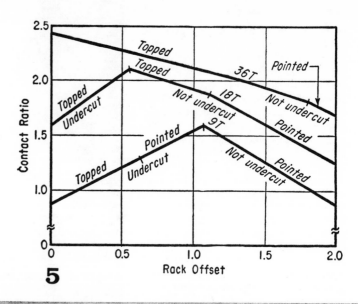

5

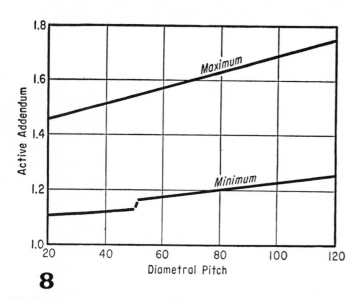

8

4   **EFFECT OF ENLARGEMENT** on tooth form is seen here.

5   **CONTACT RATIO** and tooth form vary with offset of generating rack. A high contact ratio is key design goal.

6   **RACK ADDENDUM** — its influence on the maximum contact ratio.

7   **THE RATIO** of maximum contact versus rack tip width.

8   **ADDENDUM VARIATION** on fine-pitch hobs employed in gear industry. This variation can be avoided.

9   **UNDERCUT** on 20 and 120 DP standard 12-tooth pinions.

10   **TOOTH PROPORTIONS** of author's generating rack to produce modified pinions and standard gears.

and tip-radius tolerance. The values for tooth thinning and tip radius (which are slightly different from standard) were taken from a hob manufacturer's catalog.

In the application of pinions with small tooth numbers, the contact ratios usually are small and so it is important to know the generated radial dimensions, namely, the generated root radius, inside form radius, outside form radius and outside radius. The determination of radial dimensions is time-consuming, especially the inside form radius when the pinion is undercut. (Actually, there is relatively little need for modified pinions with 12 teeth and more. The undercut that occurs on a standard pinion is usually

advantageous in that the undesirable portion of the involute near the base circle is removed, thereby permitting a more generous tolerance on the outside diameter of the mating gear, and roll testing with a standard master gear.) From a designer's standpoint, therefore, variation of hob addendum with pitch is an inconvenience in that the generated root and inside form radii cannot be determined once and for all. Instead, each pitch must be treated separately. To illustrate this, Fig 9 shows two 12-tooth standard 20° pinions generated with sharp-cornered semifinishing hobs, one with a 20 DP hob (1.46 addendum) and the other with a 120 DP hob (1.75 addendum). Allowing for the standard 0.1

outside radius tolerance, the contact ratios (with a rack) are 1.06 for 20 DP and 0.97 for 120 DP. Further, these contact ratios would be less if conventional tolerances for pinion tooth thickness and top land runout (which also vary with pitch) were applied.

It is best if the tooth proportions of the generating rack do not vary with pitch. We have been employing the generating rack (hob) shown in Fig 10 for modified pinions and standard gears. It is independent of pitch, and as desired, the tip width is narrow but not narrower than the generally regarded practical limit. In the range of 20 to 146 DP the addendum tolerance (1.6—1.5 = 0.1) is greater than the addendum tolerance for Class

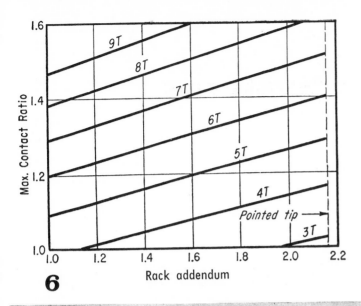

6

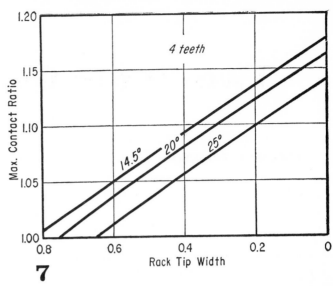

7

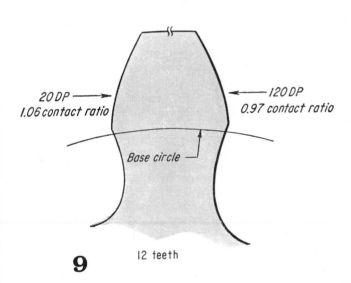

9

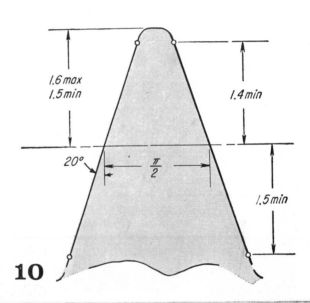

10

A hobs. The tip and root forms are not specified.

Two of the hobs standardized by the gear industry have these properties—the addendum limits of 64 DP semifinishing and 150 DP finishing Class A hobs lie between 1.5 and 1.6. In adjacent pitches, it is possible to find hobs with addendums that measure from 1.5 to 1.6, namely, semifinishing hobs in the range of 24 to 96 DP and finishing hobs in the range of 120 to 200 DP.

With a generating rack that is independent of pitch, it becomes practicable to tabulate the dimensions of modified pinions and of standard pinions with undercut. We have prepared such tables with the aid of digital computers. Only a few tables need be constructed, say 3 to 5 based on tooth-thickness tolerances, not a series of tables based on combinations of pinion-tooth thickness and diametral pitch.

For all practical purposes the fillet curve generated by a hob is the same as the fillet curve generated by a rack; therefore, tables based on a generating rack are applicable to hobbed pinions. The fillet curve generated by a circular shaper cutter differs in that it depends also upon the number of cutter teeth and the amount of required sharpening. Consequently, as in any modification system, pinions that are produced with circular shaper cutters will require special treatment.

Two-Tooth Gear Systems

SIGMUND RAPPAPORT

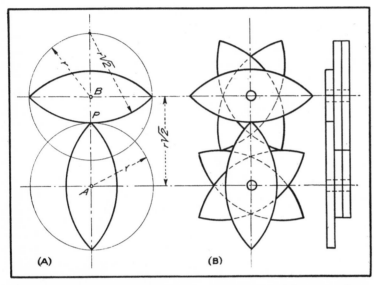

Fig. 1—Diagram (A) shows location of contact point P in two coplanar pitch circles. Three laminations (B) are used for constrained motion.

THE PROFILE OF A GEAR TOOTH is determined by the laws of kinematics under the assumption that two coplanar pitch circles roll, without sliding, around each other in opposite directions at equal circumferential speed. But what happens to the gear system if this assumption is modified by the condition that the two planes must rotate at equal speed in the same direction?

Applying the laws of relative motion in the diagram, Fig. 1, the point P is located at distance r from the rotating center A of one plane. Relative to the second plane, which rotates around center B, point P describes a circle having the radius $r\sqrt{2}$.

But the two gears generated using radius $r\sqrt{2}$ do not make up a constrained kinematic chain because the movement of one gear does not completely control the movement of the other. To make one gear precisely define the movement of the other at all times, each gear must consist of a set of at least three two-tooth gear laminations. The latter are spaced at equal angles around the gear center as shown in Fig. 2.

Four of these laminated two-tooth gears mounted in mesh with their centers at the corners of a square fulfill the original condition as stated above. If one of the four gears is rotated, the other three will turn in the same direction. Similarly, five or more of these gears arranged, in mesh, around a circle would increase the number of centers.

Although there is an inherent high friction loss due to the constant sliding of the tooth tip of one gear along the face of the other, which results in very low coefficiency of the gear system, there are undoubtedly possible applications for this type of gear mechanism. It is not suitable for use as a power transmission element, but the pocket formed between the teeth as they rotate could conceivably be utilized for metering semi-solid materials.

Wright Machinery Co.

Fig. 2—Four or more gears equally spaced on a circle have same rotation.

What you can do with NONSTANDARD SPUR GEARS

Altering pressure angles offers increased freedom in selecting center distances, gear ratios and speeds, and no special manufacturing tools are required.

EDWARD A BRASS,
Chief, preliminary mechanical design
Lycoming Div, AVCO Corp, Stratford, Conn

■ Often precise gear-ratio specifications make it difficult, or even impossible, to produce a workable planetary gear combination with standard spur gears. This is particularly true where space limitations must also be considered. Special center distances or gear ratios of ordinary gear combinations are equally impossible with standard spur gears.

Nonstandard spur gears can overcome such limitations and offer other advantages as well. Low operating pressure angles offer better backlash control, less noise, less shock, and higher contact ratios. High operating pressure angles provide increased load-carrying capacity.

Pressure-angle variation is the key

Standard spur gears operate with either a 14½, or a 20 deg pressure angle. But two gears manufactured for standard operating pressure angle and center distance can be operated at nonstandard pressure angle with corresponding center distance, providing proper tooth engagement and backlash are maintained. In a planetary gear system (see diagram) the mathematical relationship between pressure angle, center distance, diametral pitch, and number of gear teeth is:

$$C = \frac{N_s + N_p}{2P} \frac{\cos \phi}{\cos \phi_e} \qquad (1)$$

SYMBOLS

C — Center distance, in.
N — Number of gear teeth
P — Diametral pitch
R — Overall gear ratio
ϕ — Manufactured pressure angle corresponding to manufactured diametral pitch.
ϕ_e — Operating pressure angle of the external mesh.
ϕ_i — Operating pressure angle of the internal mesh.
ϕ_1 — Operating pressure angle of primary mesh in double planetary.
ϕ_2 — Operating pressure angle of secondary mesh in double planetary.

Subscripts

p — Planet gear
r — Ring gear
s — Sun gear

$$C = \frac{N_r - N_s}{2P} \frac{\cos \phi}{\cos \phi_i} \qquad (2)$$

For a simple gear and pinion combination the center distance is the same as in Eq (1), which with Eq (2), establishes the center distance for planetary gears.

When standard spur gears are used, the pressure angle is the same for both meshing gears. This makes the right-

continued, next page

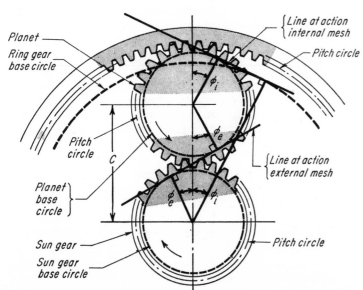

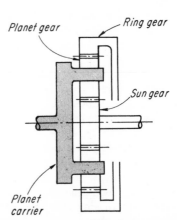

GEOMETRIC RELATIONSHIP (left) of operating pressure angles and center distances for a single planetary gear system (above) is the basis for effective use of nonstandard spur gears.

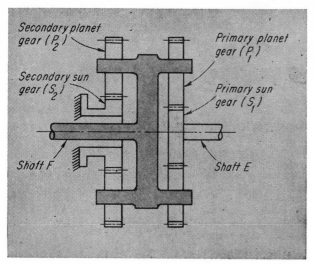

NONSTANDARD spur gears in a double planetary gear system can boost the reduction ratio from 1:21 to 1:1600.

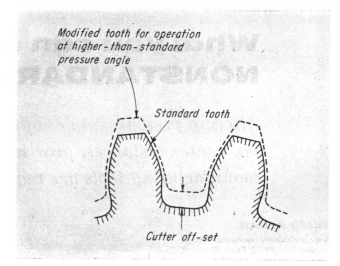

STANDARD tools can be used with predetermined cutter offset to generate nonstandard spur gears. Cutter offset shown will produce a modified tooth for operation at higher-than-standard pressure angle.

hand factor in the above equations cancel out so that:

$$C = \frac{N_s + N_p}{2P} \qquad (3)$$

and

$$C = \frac{N_r - N_s}{2P} \qquad (4)$$

This limits the freedom with which center distance C can be varied. Likewise, the gear ratio cannot be changed without varying the center distance.

By varying the pressure angle to suit particular design conditions nonstandard spur gears offer:

- Precise gear ratios and optimum meshing teeth combinations in a planetary system.
- Precise center distance for a given gear ratio.
- Very high speed ratios in a planetary system.
- Flexibility in choosing numbers of gear teeth in order to space planets equally.
- Flexibility in choosing numbers of gear teeth in order to de-phase the ratio of planets to the number of teeth in sun and ring gears.
- Operating pressure angle that will give the best compromise between high and low operating angles.

Precise gear ratios in a planetary system with standard spur gears are limited to the number of gear teeth that give a whole number for the planet gear teeth. For example, if a ring gear with 87 teeth and a sun gear with 28 teeth are specified, it would be impossible to find a planetary gear that would fit between them, because it would require $29\frac{1}{2}$ teeth. With nonstandard spur gears, however, it is possible to use 28, 29, 30, or 31 teeth, simply by adjusting the operating pressure angles. In this case 28- and 30-tooth planets are not good choices, because of the common factor with the 28-tooth sun gear. The 29 tooth planet likewise is not a good choice because of the common factor with the 87-tooth ring gear. So this narrows the choice to the 31-tooth planet.

Eq (1) and (2) show that three operating pressure angles are involved. If the gears are selected with a manufactured pressure angle of 20 deg and a corresponding diametral pitch of 10, and if the operating pressure angle for the

planet gear external mesh is assumed to be 18 deg, the remaining angle ϕ_i can be calculated as follows:

$$C = \frac{28 + 31}{2\,(10)} \frac{\cos 20°}{\cos 18°} = 2.92 \text{ in.}$$

Substituting this value of C in equation (2) gives:

$$\phi_i = \cos^{-1} \left[\frac{(87 - 31) \cos 20°}{2\,(10)\,(2.92)} \right] = 25.49°$$

Precise center distances can be maintained by merely varying the pressure angles, and this can be done without changing the gear ratios. This is particularly advantageous when errors in center distance need to be corrected without changing gear ratios. Also, there are occasions when precise center distances are more important than the specified gear ratios. Nonstandard spur gears could fulfill such requirements, and still give more exact gear ratios than is possible with standard spur gears.

Flexibility in choosing gear teeth numbers is also desirable in a planetary system in order to space the planets equally. This condition requires that the sum of the teeth in the ring gear and sun gear be divisible by the number of planets. This is a requirement that is not always easily fulfilled with standard spur gears without sacrificing desired gear ratios.

For example, assume that an approximate ratio can be satisfied with 142 teeth in the ring gear and 28 teeth in the sun gear. If three planets are to be used, then for equally spaced planets, the ring gear must have, as the closest approximation, 143 teeth; so that $(143 + 28)/3 = 57$. With nonstandard gears you could use 143 teeth, even though the tentative number of teeth in the planet is $57\frac{1}{2}$.

Flexibility to de-phase the gear meshes is sometimes a problem that can be solved with nonstandard spur gears. This is a case where the number of planets must have no common factor with either the number of teeth in the sun gear, or the number of teeth in the ring gear. For example, in a planetary system with 33 teeth in the sun gear, 123 teeth in the ring gear and 3 planets, the unit could be de-phased by changing the sun gear to 31 and the ring gear to 116. At the same time the original gear ratio can be maintained approximately.

Choosing the best operating pressure angle is sometimes an advantage by itself. In standard spur gears with either $14\frac{1}{2}$ or 20-deg pressure angles it is a dilemma to choose between the advantages of low and high operating pressure angles. For power gearing of an operating pressure angle of about 25 deg is usually preferred. But in a planetary transmission this high angle is an advantage only at the external mesh since the load-carrying capacity of the internal mesh is inherently higher than the external mesh.

Usually, the rim of the ring gear is subjected to high bending stresses set up by the separating components of the planet gear-tooth loading. For a given value of the tangential tooth load, the bending stresses in the ring gear rim are directly proportional to the tangent of the operating pressure angle. This means that the stress in the ring gear rim increases by about 35% in going from 17 to 25-deg operating pressure angle.

Here is the procedure for designing a planetary system with operating pressure angles of about 17 deg for the internal meshes, and 25 deg for the external meshes. Assume that preliminary specifications call for a ring gear with 108 teeth, a sun gear with 42 teeth, and a diametral pitch of 10 corresponding to a pressure angle of 20 deg, then calculate:

Step (1) $N_p = \dfrac{N_r - N_s}{2} = \dfrac{108 - 42}{2} = 33$

Step (2) $C = \dfrac{N_p + N_s}{2P} \dfrac{\cos 20}{\cos 20} = \dfrac{33 + 42}{2(10)} = 3.75$ in.

Starting with these values, the final adjustments can be made to obtain the desired operating pressure angles as follows:

Step (3) $\phi_e = \cos^{-1}\left[\dfrac{N_p + N_s}{2P} \times \dfrac{\cos 20}{C} \right]$

$\phi_i = \cos^{-1}\left[\dfrac{N_r - N_s}{2P} \times \dfrac{\cos 20}{C} \right]$

First varying N_p, then C, and maintaining N_p and N_s the above equations give the following combinations:

N_p	C	ϕ_e	ϕ_i
33	3.750	20	20
32	3.750	22.00	17.78
31	3.750	23.84	15.26
31	3.760	24.19	15.81
31	3.775	24.69	16.59
31	**3.784**	**24.99**	**17.04**

This shows that the last combination in the above table is the closest that could be obtained for this example to take advantage of high operating pressure angle in the external meshes and low operating pressure angle in the internal meshes.

Very high speed ratios can be obtained with nonstand-

ard spur gears in a double planetary system shown here. Mr. S. Rappaport showed in his article (PE—Jun 22 '59, p 65) how this trick can also be performed with a combination of spur and helical gears.

For standard spur gears the key relationships in this unit are:

$$R = \cfrac{1}{\left[1 - \cfrac{N_{p1}}{N_{p2}} \cfrac{N_{s2}}{N_{s1}} \right]} \tag{5}$$

$$N_{s1} + N_{p1} = N_{s2} + N_{p2} \tag{6}$$

$$C = \frac{N_{s1} + N_{p1}}{2P} \frac{\cos \phi}{\cos \phi_1} \tag{7}$$

$$\phi_2 = \cos^{-1}\left[\frac{N_{s2} + N_{p2}}{2P} \frac{\cos \phi}{C} \right] \tag{8}$$

Eq (5) shows that very high ratios can be obtained by making the quantity $(N_{p1} N_{s2})/(N_{p2} N_{s1})$ approach unity. However, (6) shows that for standard spur gears the sums of teeth in the sun and planet of each planetary unit must be equal. On the other hand, with nonstandard gears by varying the pressure angle, Eq (7) and (8) can be used to arrive at suitable proportions. For example, a double planetary system with standard spur gears $N_{s1} = 41$, $N_{p1} = 40$, $N_{s2} = 40$, $N_{p2} = 41$ will give an approximate overall ratio:

$$R = \cfrac{1}{\left[1 - \cfrac{40 \times 40}{41 \times 41} \right]} = 21$$

But a double planetary system with nonstandard spur gears, $N_{s1} = 40$, $N_{p1} = 39$, $N_{s2} = 40$, $N_{p2} = 41$ will give, an approximate overall ratio:

$$R = \cfrac{1}{\left[1 - \cfrac{39 \times 41}{40 \times 40} \right]} = 1600$$

Typical operating pressure angles for this system can be determined from Eq (7) and (8). In this case they could be $\phi = 20°$, $\phi_1 = 23°$, $\phi_2 = 19.2°$ for $C = 4.040$ in.

Spur gears for such a double planetary system have an advantage over helical gears in that they do not introduce extraneous axial thrust loads.

What about the disadvantages?

Nonstandard spur gears are essentially custom designed to fulfill particular needs, and this usually means that they cost slightly more. However, properly designed, they can be made with standard tools in the same way as standard spur gears. The main difference is the degree of cutter offset required for the particular operating pressure angles and center distances selected to maintain the desired backlash (see diagram above, left). Several systematic methods are available for determining cutter offset and for proportioning the cutter offset (tooth thicknesses) to optimize the bending strength of the gear teeth.

Charts for Designing Long-Short Addendum Spur Gears

WAYNE H. BOOKMILLER

Gear Engineering Division, General Electric Company

ENLARGEMENT OF PINIONS to avoid undercutting the teeth or to increase their strength, or both, has become standard practice.

The gear that mates with an enlarged pinion must be mounted at a greater center distance or must be reduced in diameter. The latter method is usually preferred.

Gear sets made with long addendum (enlarged) pinions and short addendum gears can be designed so the beam strengths of gear and pinion are either equal or related by a fixed ratio. The family of curves in Fig. 1 shows the addenda necessary to have the same form factor Y for the pinion and gear. The actual addenda are found by dividing the curve values by the diametral pitch.

Since the Y factor is based on the worst application of load, and a stress concentration factor K is included, the Lewis Equation as now used is; where

S = Root stress, psi
W_T = Tangential driving pressure, lb
P_D = Diametral pitch
F = Face width, in.
K = Stress concentration factor
Y = Form factor of tooth

$$S = \frac{W_T \times P_D \times K}{F \times Y} \qquad (1)$$

The form factor Y and the stress concentration factor K, for use in Eq. (1), are found on Figs. 2 and 3 respectively. These curves apply to long-short addendum gear sets designed for equal Y factor pinions and gears.

As an example of the use of Eq. (1) and Figs. 1, 2 and

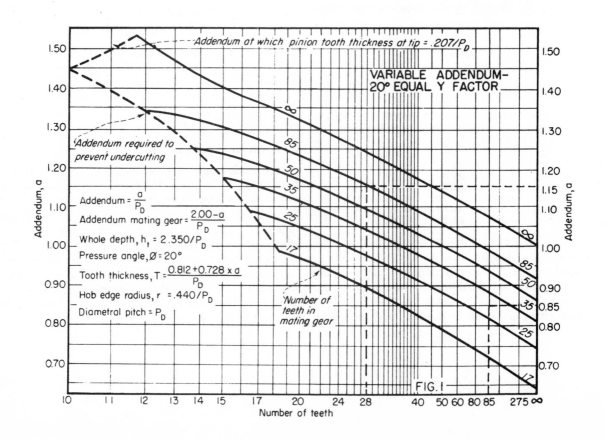

Addendum at which pinion tooth thickness at tip = .207/P_D

VARIABLE ADDENDUM—
20° EQUAL Y FACTOR

Addendum required to prevent undercutting

Addendum = $\frac{a}{P_D}$

Addendum mating gear = $\frac{2.00-a}{P_D}$

Whole depth, h_t = 2.350/P_D

Pressure angle, $\varnothing$ = 20°

Tooth thickness, $T = \frac{0.812+0.728 \times a}{P_D}$

Hob edge radius, r = .440/P_D

Diametral pitch = P_D

Number of teeth in mating gear

FIG. 1

Number of teeth

Addendum, a

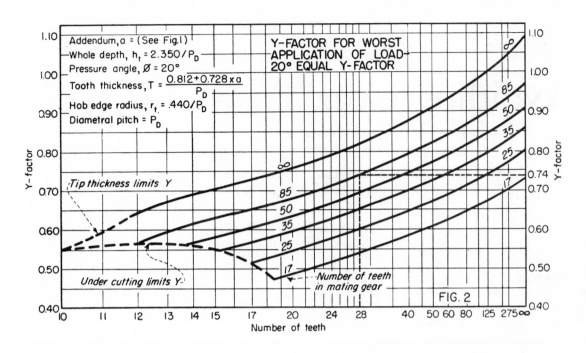

FIG. 2

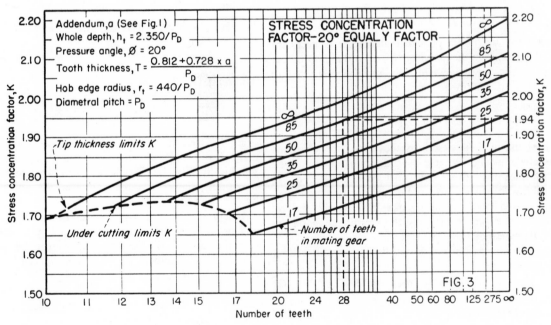

FIG. 3

3, consider the following gear set:

	Pinion	Gear
Number of teeth	28	85
Diametral pitch	20	20
Pressure Angle, deg	20	20
Pitch diameter, in.	1.400	4.250
Speed, rpm	20,000	6,600
Face width, in.	$\frac{1}{2}$	$\frac{1}{2}$
Horsepower	15	.....

Pinion addendum $= \dfrac{a}{P_D} = \dfrac{1.150}{20} = 0.0575$ in.

Pinion tooth thickness $= \dfrac{0.812 + 0.728 \times 1.15}{20} = 0.0825$ in.

Gear addendum $= \dfrac{a}{P_D} = \dfrac{0.850}{20} = 0.0425$ in.

Gear tooth thickness $= \dfrac{0.812 + 0.728 \times 0.850}{20} = 0.0715$ in.

$W_T = \dfrac{15 \times 33,000 \times 12}{20,000 \times 2\pi \times 0.7} = 67.5$ lb

$Y = 0.740$ (Fig. 2) $K = 1.940$ (Fig. 3)

$S = \dfrac{W_T \times P_D \times K}{F \times Y} = \dfrac{67.5 \times 20 \times 1.94}{0.500 \times 0.740} = 7,080$ psi, for pinion

The gear root stress can be calculated in a similar manner and will be approximately equal to the pinion root stress.

Avoiding Gear Tooth Interference

Simple, direct method of calculating maximum diametral pitch that can be used without undercutting or using non-standard tooth forms. Also useful in evaluating the strength advantages of pressure angles larger than $14\frac{1}{2}$ deg for a given application.

REINER J. AUMAN
Kerns Manufacturing Company

TWO METHODS are commonly used for checking gear teeth for possible interference. One is a graphical technique, using a layout drawn several times size. The other is analytical; it involves solving an equation like the following (see Fig. 3 for notation):

$$\left(\frac{d_o}{2}\right)^2 = \left(\frac{d_1 \cos \beta}{2}\right)^2 + \left(\frac{d_1+d_2}{2}\right)^2 \sin^2 \beta \qquad (1)$$

Having the maximum outside diameter, it is necessary to find the maximum diametral pitch (or minimum number of teeth). The following equation is frequently used to calculate the maximum diametral pitch:

$$p = \frac{2}{d_o - d_1}$$

Both of the methods described above tend to be tedious and usually must be repeated several times for any given problem since they solve only one set of conditions. If standard involute gears are to be used, a direct method can be derived that will also make it relatively easy to investigate the effects of changing the pressure angle on the outside diameters of the gears.

Using trigonometric substitutions, Eq (1) can be written as follows:

$$d_o{}^2 = d_1{}^2 + 2d_1 d_2 \sin^2 \beta + d_2{}^2 \sin^2 \beta \qquad (2)$$

Taking the maximum addendum as $1/p$, the maximum outside diameter becomes

$$d_o = \left(d_1 + \frac{2}{p}\right) \qquad (3)$$

Substituting Eq (3) into Eq (2) and m for d_2/d_1,

$$p_{max} = \frac{2}{d_1[\sqrt{1+(2m+m^2)\sin^2\beta} - 1]} \qquad (4)$$

This equation solves for the maximum diametral pitch directly. For further simplification, Eq (2) can be modified for solution by substituting n_1/p for d_1 and n_2/p for d_2:

$$n_{1\ min} = \frac{n_2{}^2 \sin^2 \beta - 4}{4 - 2n_2 \sin^2 \beta} \qquad (5)$$

Expressing the righthand side of this equation by k, a function of pressure angle, Eq (5) becomes

$$n_{1\ min} = k$$

or,

$$\frac{n_{1\ min}}{n_{2\ min}} = \frac{k}{n_{2\ min}} = \frac{d_1}{d_2} = \frac{1}{m} \qquad (6)$$

This states that $\dfrac{k}{n_{2\ min}}$ must be equal to the gear ratio. However, both gears have the same diametral pitch

$$p = \frac{n_1}{d_1} = \frac{n_2}{d_2}$$

Thus, Eq (6) becomes

$$p_{max} = \frac{k}{d_1} = \frac{n_{2\ min}}{d_1 m} \qquad (7)$$

Eq (7) relates all the necessary variables at the extreme condition before undercutting and can be represented graphically, Fig. 1. This chart can be used to evaluate the maximum diametral pitch and also the effect on pitch by changing the pressure angle.

It should be noted that the numbers along the abscessa and ordinate of the lefthand chart in Fig. 1 merely represent the ratio of the gear train. They can be either pitch diameters or numbers of teeth and can be multiples of the original values as long as their ratio is unchanged.

EXAMPLE: Gear set with $d_1 = 8.00$ in., $d_2 = 2.00$ in., and pressure angle $\beta = 14\frac{1}{2}$ deg is given.

SOLUTION: Drawing line III from the origin through the intersection of lines I and II, line IV horizontally through the intersection of line III and the $14\frac{1}{2}$ deg curve, and line V vertically through the intersection of line IV and the $d_1 = 8$ line, the maximum diametral pitch is 14.3.

Since, by definition the maximum limit of $d_2/d_1 = 1.0$, equal sized gears can have the maximum allowable pitch for any particular case where center distances are fixed.

A 1:1 gear ratio is very common, particularly with pump gears. Thus, Fig. 2 plots this ratio based on Eq (4) with $m = 1.0$:

$$p_{max} = \frac{2}{d[\sqrt{1 + 3\sin^2\beta} - 1]} \qquad (8)$$

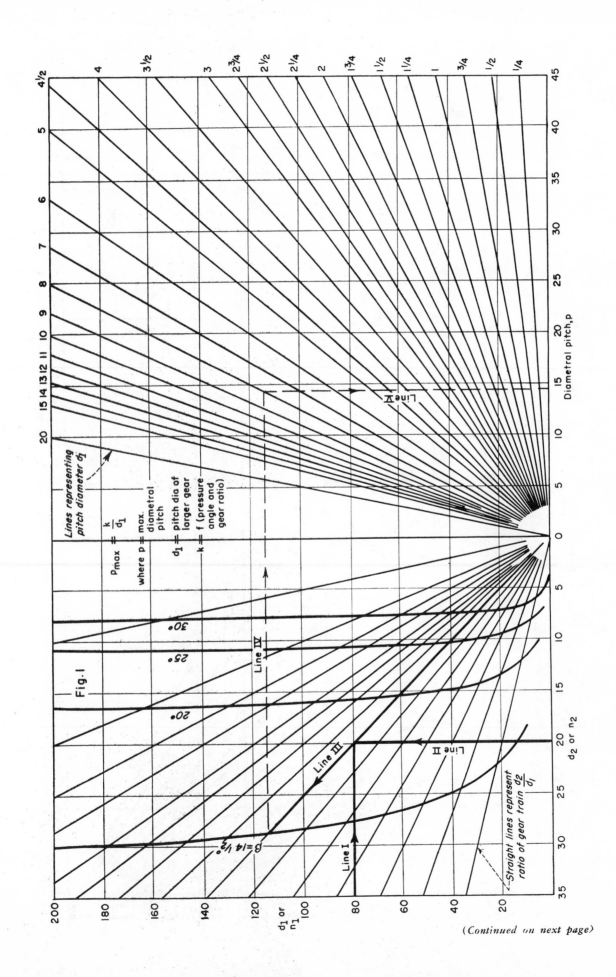

Fig. 1

Lines representing pitch diameter d_1

$p_{max} = \dfrac{k}{d_1}$

where p = max. diametral pitch

d_1 = pitch dia of larger gear

k = f (pressure angle and gear ratio)

Diametral pitch, p

d_2 or n_2

d_1 or n_1

Straight lines represent ratio of gear train $\dfrac{d_2}{d_1}$

$\beta = 14\frac{1}{2}°$

Line I

Line II

Line III

Line IV

Line V

(Continued on next page)

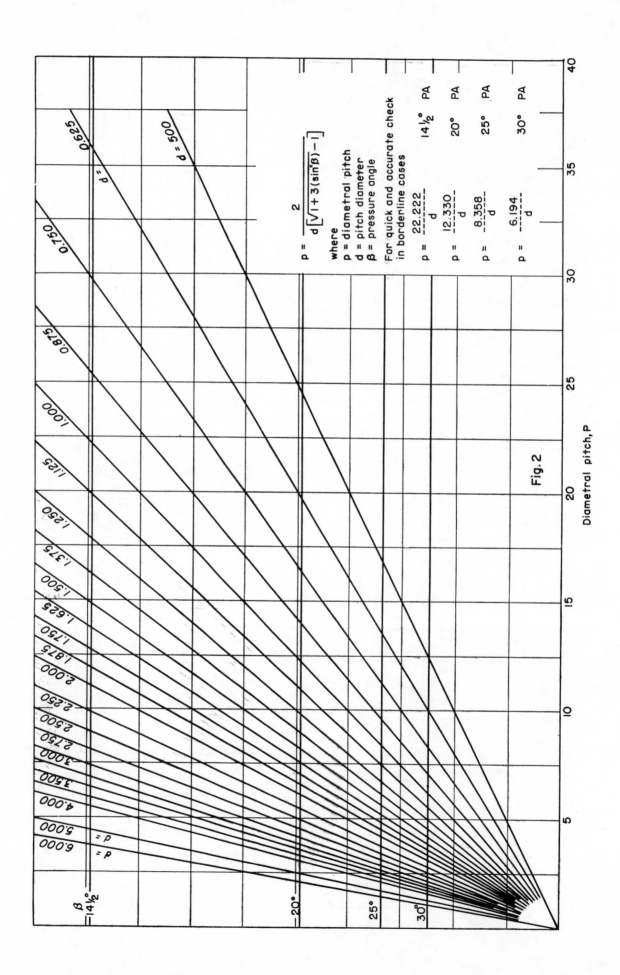

$$p = \dfrac{2}{d\left[\sqrt{1+3(\sin^2\beta)}-1\right]}$$

where

p = diametral pitch
d = pitch diameter
β = pressure angle

For quick and accurate check
in borderline cases

$p = \dfrac{22.222}{d}$	14½°	PA
$p = \dfrac{12.330}{d}$	20°	PA
$p = \dfrac{8.358}{d}$	25°	PA
$p = \dfrac{6.194}{d}$	30°	PA

Fig. 2

Diametral pitch, P

5

NONCIRCULAR GEAR SYSTEMS

DESIGN GUIDE...WHEN YOU NEED
NONCIRCULAR GEARS

Gears in this category are not new—but generally are not understood, or employed very frequently. Yet they offer distinct advantages over linkages and cams. Here's a discussion most types of noncircular gears—also equations and tables giving their basic characteristics, and three design problems putting the equations to work

B. BLOOMFIELD, computer division
The Fellows Gear Shaper Co.

The popularity of noncircular gears has grown steadily in recent years mainly because of their usefulness in mechanical computing devices. Previously, there had been a steady—but slow—demand for them in automatic machinery. They were held back, not so much by manufacturing cost as by the lack of sufficient design knowledge about them. Even now, few handbooks or texts on mechanical design include information on such gears—and then too briefly to instill confidence.

Noncircular gears generally cost more than competitive devices such as linkages and cams. But with the development of modern production methods, such as the tape-controlled gear shaper, cost has gone down considerably. Also, in comparison with linkages, noncircular gears are more compact and balanced—and can be more easily balanced. These are important considerations in high-speed machinery. Further, the gears can produce continuous, unidirectional cyclic motion—a point in their favor when compared with cams. The disadvantage of cams is that they offer only reciprocating motion.

Applications can be classed into two groups:

• Where only an over-all change in angular velocity of the driven member is required: quick-return drives, intermittent mechanisms as in printing presses, planers, shears, winding machines, automatic-feed machines.

• Where precise, nonlinear functions must be generated, as in computing machines for extracting roots of numbers, raising numbers to any power, generating trigonometric and logarithmic functions.

TYPES OF NONCIRCULAR GEARS

It is always possible to design a special-shaped gear to roll and mesh properly with a gear of any shape—sole requirement is that distance between the two axes must be constant. However, the pitch line of the mating gear may turn out to be an open curve, and the gears can be rotated only for a portion of a revolution—as with two logarithmic-spiral gears (illustrated in Fig 1).

True elliptical gears can only be made to mesh properly if they are twins, and if they are rotated about their focal points. However, gears resembling ellipses can be generated from a basic ellipse. These "higher-order" ellipses (see Fig 2) can be meshed in various interesting combinations to rotate about centers A, B, C or D. For example, two 2nd-order elliptical gears can be meshed to rotate about their geometric center; however, they will produce two complete speed cycles per revolution. Difference in contour between a basic ellipse and a 2nd-order ellipse is usually very slight. Note also that the 4th-order "ellipses"

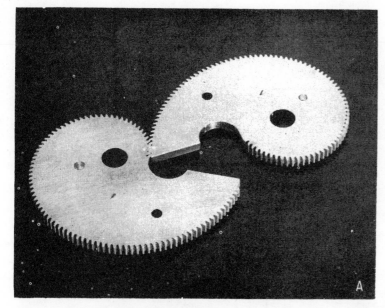

1 **LOGARITHMIC SPIRAL GEARS** in (A) are open curved, usually employed in computing devices. Elliptical-shape gears (B) are closed curved, frequently found in automatic machinery. Special-shape gears (C) offer wider range of velocity and acceleration characteristics.

resemble square gears (this explains why the square gears, sometimes found as ornaments on tie clasps, illustrated in Fig 3, actually work).

A circular gear, mounted eccentrically, can roll properly only with specially derived curves (shown in Fig 4). One of the curves, however, closely resembles an ellipse. For proper mesh, it must have twice as many teeth as the eccentric gear. When the radiis r, and eccentricity, e, are known, the major semi-axis of the elliptical-shape gear becomes 2r + e, and the minor 2r − e. Note also that one of the gears in this group must have internal teeth to roll with the eccentric gear. Actually, it is possible to generate internal-tooth shapes to rotate with noncircular gears of any shape (but, again, the curves may be of the open type).

Noncircular gears can also be designed to roll with special-shaped racks (shown in Fig 5). Combinations include: an elliptical gear and a sinusoid-like rack (a 3rd order ellipse is illustrated but any of the elliptical rolling curves can be used in its place—main advantage is that when the ellipse rolls, its axis of rotation moves along a straight line); and a logarithmic spiral and straight rack (the rack, however, must be inclined to its direction of motion by the angle of the spiral).

DESIGN EQUATIONS

Equations for noncircular gears are shown here in functional form for three common design requirements. They are valid for any noncircular gear pair. Symbols are defined in the box on the next page:

CASE I—Polar equation of one curve and center distance

continued, next page

2 Basic and High-order Elliptical Gear Combinations

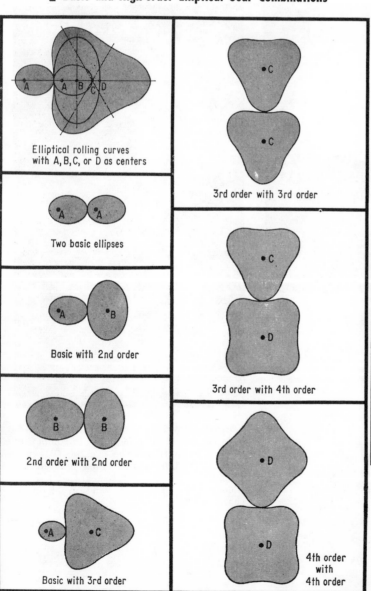

Elliptical rolling curves with A, B, C, or D as centers

Two basic ellipses

Basic with 2nd order

2nd order with 2nd order

Basic with 3rd order

3rd order with 3rd order

3rd order with 4th order

4th order with 4th order

3 SQUARE GEARS ON TIE CLASP seem to defy basic kinematic laws, are actually a takeoff on a pair of 4th order ellipses.

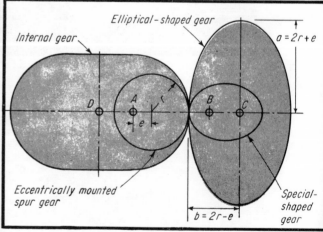

Internal gear
Elliptical-shaped gear
$a = 2r + e$
Eccentrically mounted spur gear
$b = 2r - e$
Special-shaped gear
e

4 ECCENTRIC SPUR GEAR rotating about point **A**, will mesh properly with any of the three gears shown with centers at points **B**, **C** and **D**.

160

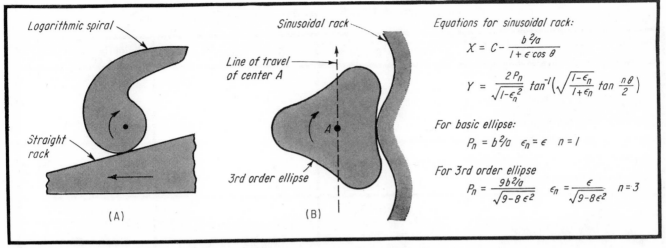

Equations for sinusoidal rack:

$$X = C - \frac{b^2/a}{1 + \epsilon \cos \theta}$$

$$Y = \frac{2 P_n}{\sqrt{1 - \epsilon_n^2}} \tan^{-1}\left(\sqrt{\frac{1 - \epsilon_n}{1 + \epsilon_n}} \tan \frac{n\theta}{2}\right)$$

For basic ellipse:

$$P_n = b^2/a \quad \epsilon_n = \epsilon \quad n = 1$$

For 3rd order ellipse

$$P_n = \frac{9b^2/a}{\sqrt{9 - 8\epsilon^2}} \quad \epsilon_n = \frac{\epsilon}{\sqrt{9 - 8\epsilon^2}} \quad n = 3$$

5 RACK AND GEAR COMBINATIONS are possible with noncircular gears. Straight rack for logarithmic spiral (A) must move obliquely; center of 3rd order ellipse (B) follows straight line.

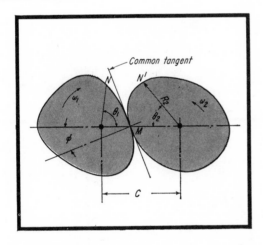

Symbols

a = semi-major axis of ellipse
b = semi-minor axis of ellipse
C = center distance (see above sketch)
ϵ = eccentricity of an ellipse = $\sqrt{1 - (b/a)^2}$
e = eccentricity of an eccentrically mounted spur gear
N = number of teeth
P = diametral pitch
r_c = radius of curvature
R = active pitch radius
S = length of periphery of pitch circle
X, Y = rectangular coordinates
θ = polar angle to R
ϕ = angle of obliquity
ω = angular velocity
$f(\theta), F(\theta), G(\theta)$ = various functions of θ
$f'(\theta), F'(\theta), G'(\theta)$ = first derivatives of functions of θ

are known; to find the polar equation of the mating gear:

$$R_1 = f(\theta_1)$$
$$R_2 = C - f(\theta_1)$$
$$\theta_2 = -\theta_1 + C \int \frac{d\theta_1}{C - f(\theta_1)}$$

CASE II—Relationship between angular rotation of two members and center distance are known; to find polar equations of both members:

$$\theta_2 = F(\theta_1)$$
$$R_1 = \frac{C F'(\theta_1)}{1 + F'(\theta_1)}$$
$$R_2 = C - R_1 = \frac{C}{1 + F'(\theta_1)}$$

CASE III—Relationship between angular velocities of two members and center distance are known; to find polar equations of both members:

$$\omega_2 = \omega_1 G(\theta_1)$$
$$R_1 = \frac{C G(\theta_1)}{1 + G(\theta_1)}$$
$$R_2 = C - R_1$$
$$\theta_2 = \int G(\theta_1) d\theta_1$$

Velocity equations and characteristics of five types of noncircular gears are listed in the table on the next page.

CHECKING FOR CLOSED CURVES

Gears can be quickly analyzed to determine whether their pitch curves are open or closed by means of the following equations:

In Case I, if $R = f(\theta) = f(\theta + 2N\pi)$, the pitch curve is closed.

In Case II, if $\theta_1 = F(\theta_2)$ and $F(\theta_0) = 0$, the curve is closed when the equation $F(\theta_0 + 2\pi/N_1) = 2\pi/N_2$ can be satisfied by substituting integers or rational fractions for N_1 and N_2. If fractions must be used to solve this

Type	Comments	Basic equations	Velocity equations $\omega_1 = $ constant
Two ellipses rotating about foci	Gears are identical. Comparatively easy to manufacture. Used for quick-return mechanisms, printing presses, automatic machinery	$R = \dfrac{b^2}{a[1 + \epsilon \cos\theta]}$ $\epsilon = $ eccentricity $= \sqrt{1 - \left(\dfrac{b}{a}\right)^2}$ $a = \frac{1}{2}$ major axis $b = \frac{1}{2}$ minor axis	$\omega_2 = \omega_1 \left[\dfrac{r^2 + 1 + (r^2 - 1)\cos\theta_2}{2r}\right]$ where $r = \dfrac{R\,max}{R\,min}$
2nd Order elliptical gears rotating about their geometric centers	Gears are identical. Geometric properties well known. Better balanced than true elliptical gears. Used where two complete speed cycles are required for one revolution.	$R = \dfrac{2ab}{(a + b) - (a - b)\cos 2\theta}$ $C = a + b$ $a = $ maximum radius $b = $ minimum radius	$\omega_2 = \omega_1 \left[\dfrac{r + 1 + (r^2 - 1)\cos 2\theta_2}{2r}\right]$ where $r = \dfrac{a}{b}$
Eccentric circular gear rotating with its conjugate	Standard spur gear can be employed as the eccentric. Mating gear has special shape	$R_1 = e \cos\theta_1 + \sqrt{a^2 - e^2 \sin^2\theta}$ $\theta_2 = -\theta_1 + C \displaystyle\int \dfrac{d\theta}{C - e\cos\theta_1 - \sqrt{a^2 - e^2 \sin^2\theta_1}}$ $R_2 = C - R_1$	$\dfrac{\omega_2}{\omega_1} = -\dfrac{R_1}{R_2}$
Logarithmic spiral gears	Gears can be identical although can be used in conbinations to give variety of functions. Must be open gears	$R_1 = A e^{k\theta_1}$ $R_2 = C - R_1$ $\quad = A e^{k\theta_2}$ $\theta_2 = \dfrac{1}{k} \log (C - Ae)^{k\theta_1}$ $e = $ natural log base	$\dfrac{\omega_2}{\omega_1} = \dfrac{A e^{k\theta_1}}{C - A e^{k\theta_1}}$
Sine-function gears	For producing angular displacement proportional to sine of input angle. Must be open gears	$\theta_2 = \sin^{-1}(k\theta_1)$ $R_2 = \dfrac{C}{1 + k\cos\theta_1}$ $R_1 = C - R_2$ $\quad = \dfrac{C k\cos\theta_1}{1 + k\cos\theta_1}$	$\dfrac{\omega_2}{\omega_1} = k\cos\theta_1$

equation, the curve will have double points (intersect itself), which is, of course, an undesirable condition.

In Case III, if $\theta_2 = \int G(\theta_1)\, d\theta_1$, let $G(\theta_1)\, d\theta_1 = F(\theta_1)$, and use the same method as for Case II, with the subscripts reversed.

With some gear sets, the mating gear will be a closed curve only if the correct center distance is employed. This distance can be found from the equation:

$$4\pi = \int_0^{2\pi} \frac{d\theta_1}{C - f(\theta_1)}$$

ANGLE OF OBLIQUITY

It is also desirable to avoid too great an angle between the common normal to the mating pitch curves at the point of tangency and the line of centers (called the angle of obliquity, ϕ). This angle is equal to

$$\phi = \tan^{-1}\left(\frac{f'(\theta)}{f(\theta)}\right)$$

where $f'(\theta)$ is the first derivative of $f(\theta)$. This angle should not exceed 45°, to prevent the teeth from pulling out of engagement.

It is often necessary to know if the pitch curve contains a reverse portion. This condition exists if, in the equation for the angle of obliquity ϕ_1, there is more than one value of θ for each value of ϕ during any π-range of θ (or any 180° range of θ).

DETERMINING NUMBER OF TEETH

As with circular gears, it is desirable to know if the teeth of noncircular gears are in danger of being undercut. According to AGMA standards, to avoid danger of undercutting when the outside diameter of a gear is $(N + 2)/P$, the minimum number of teeth should be 18.

For a smaller number of teeth, the outside diameter can be made oversize in accordance with AGMA standards. This principle can be applied to noncircular gears by determining the instantaneous radius of curvature of the pitch curve and multiplying it by twice the diametral pitch. If the answer is less than 18, there is danger of undercut. However, this does not usually become serious until the number is 12 or less.

The radius of curvature, r_c, can be determined from the following equation where $R = f(\theta)$ is the polar equation of the pitch curve, and R' and R'' are first and second derivatives of R:

$$r_c = \frac{[R^2 + (R')^2]^{3/2}}{R^2 - R(R'') + 2(R')^2}$$

All of the above properties (open or closed gears, angle of obliquity, reverse curvature, radius of curvature) can be determined from a carefully made, large-scale layout of the pitch curve. When the formula is unknown and the pitch curves have been determined empirically, layout is the only method available. However, even when the polar equation is known, the equation for angle of obliquity and radius of curvature may be hard to solve and a layout will determine these values to sufficient accuracy.

Number of teeth in a noncircular gear is determined as follows. Find the length of the periphery of the pitch curve. If it is a closed gear, employ the equation:

$$S = \int_0^{\theta = 2\pi} [(dR)^2 + (Rd\theta)^2]^{1/2}$$

If this is too hard to solve, the length can usually be approximated from a careful layout. For elliptical gears, the length can be approximated from the equation:

$$S = \pi \sqrt{2(a^2 + b^2) - \frac{(a - b)^2}{2.2}}$$

For open gears, the teeth number is usually not very important because the gears do not make a complete revolution. For closed gears, the number equals SP/π and

should, of course, be integral. Also, if the mating gears are identical (elliptical, for example), the teeth number should be odd to insure proper meshing.

The diametral pitch in the formula for number of teeth can be varied slightly to provide the proper number. This, in effect, will make the gears slightly oversize or undersize, depending on whether the pitch is increased or decreased. If this change is under 10%, the flexibility of the involute system will take care of it.

THE MANUFACTURING TECHNIQUES

Simplest method—but also slowest and most expensive —of making noncircular gears involves the division of the gear into several segments, each of which can be considered as approximating an arc of a circle. Each segment is then cut separately on a gear shaper, so set up that the work rotates about the instantaneous center of the segment. Rotation is geared for that segment. This involves danger of considerable error at the junction of two segments.

Perhaps the most accurate way of cutting noncircular gears is the tape-controlled gear shaper. The center distance between cutter and work, and the angular velocity of the work spindle, is varied in accordance with the instructions contained on a previously prepared tape, which can be paper, plastic, film or magnetic. The tape delivers its instructions to the machine through a special control unit.

Such a unit, coupled to a Fellows gear shaper, has been developed and used by Frederick Cunningham, Stamford, Conn. He has made the gears illustrated on p 158. Norden Laboratory Division of Norden-Ketay, and a few others are also using this technique. The method is especially suitable for making masters for the production gears.

Another way is to mount a special cam on a gear shaper, as illustrated in Fig. 6, to control the center distance of cutter and work spindles. Cutter is held in an eccentric adapter or bushing to vary the linear velocity.

The gear shaper can also act as a copying machine by having a master gear and pinion regulate the velocity ratio and center distance of the cutter and work. This method allows using a standard machine with a special work-holding fixture. The master pinion is a duplicate of the cutter. The master noncircular gear can be generated on the same fixture, using a cam and follower in place of the noncircular gear and pinion. To insure correct rolling action during the cutting, steel tapes are wound around the cam and follower, and the diameter of each is made small enough to allow for thickness of the tape.

This method has proved successful, although master gear and pinion are usually preferred for production. It also lends itself to using a rack-shaped cutter, (see Fig. 7) with a rack or bar in place of the master pinion or follower. The fixtures are set up on a Fellows 7-Type Gear Shaper. Three motions are imparted to the work.

continued, next page

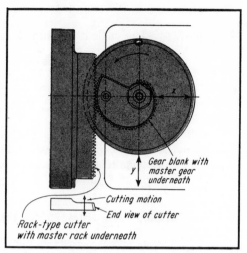

7 RACK-TYPE CUTTER is operated by master gear and rack directly underneath the work and cutter. As the slide is moved in **y** direction, the master rack rotates the work-holding outfit through the master gear, and at the same time moves the work in **x** direction. Thus, three motions are imparted to the work while the rack-type cutter is reciprocating in a direction perpendicular to both **x** and **y** to cut the teeth.

6 CAM-CONTROLLED FELLOWS GEAR SHAPER varies the center distance between work and reciprocating cutter. Cutter is held in eccentric adapter.

Sample Calculations

PROBLEM I—Polar coordinate of one curve is given as:
$R_1 = 2 + \cos \theta_1$.

Find the equation of the mating gear and correct center distance.

From the equations for Case I:

$$R_2 = C - (2 + \cos \theta_1)$$

Since

$$dR_1 = - \sin \theta_1$$

Then

$$\theta_2 = - \theta_1 + C \int \frac{d\theta_1}{C - (2 + \cos \theta_1)}$$

$$\theta_2 = - \theta_1 + \frac{C}{\sqrt{C^2 - 4C + 3}} \tan^{-1} \left[\frac{\sqrt{C^2 - 4C + 3} \sin \theta_1}{(C - 2) \cos \theta_1 - 1} \right]$$

The center distance at which the mating gear will be a closed curve is

$$4\pi = \int_0^{2\pi} \frac{d\theta_1}{C - (2 + \cos \theta_1)}$$

$$4\pi = \frac{2\pi C}{\sqrt{(C - 2)^2 - 1}}$$

$$C = (2/3) (4 + \sqrt{7})$$

This value can be substituted into the equations above

VARIABLE-FREQUENCY OSCILLATOR designed by Arma Div of American Bosch Arma for operational ground equipment serving missiles and space vehicles. Specifically, the noncircular gears convert linear rotation of an input shaft, θ_1, to the nonlinear function: $\theta_2 = (636.43\theta_1)/(\theta_1 + 306.43)$.

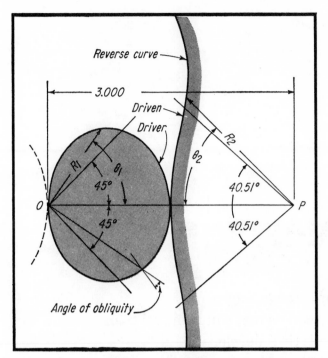

Reverse curve

3.000

Driven
Driver

R_2

R_1

θ_1

45°

45°

θ_2

40.51°

40.51°

O

P

Angle of obliquity

CONVERSION OF SINE FUNCTION to a linear function ($\sin \theta_1 = \theta_2$) by means of special gears rotating about points **O** and **P**. Details pertain to sample problem II.

for R_2 and θ_2. These two equations then define the pitch curve for the mating gear.

Angle of obliquity is

$$\tan \phi = \frac{\sin \theta_1}{2 + \cos \theta_1}$$

This value is never over 45°, and indicates no reversal of the pitch curve because ϕ is single-valued during any multiple of π-range (or 180° range) of θ_1.

Length of the periphery is

$$S = \int_0^{2\pi} \sqrt{\sin^2\theta_1 + (2 + \cos\theta_1)^2}\, d\theta$$

By trigonometric identities, this equation can be converted to

$$S = 6 \int_0^{\pi/2} \sqrt{1 - \frac{8}{9}\sin^2\left(\frac{\theta_1}{2}\right)}\, d\left(\frac{\theta_1}{2}\right)$$

which is in the elliptical-integral form of

$$S = 6 \int \sqrt{1 - k^2 \sin^2 x}\, dx$$

Looking up the value of this integral for $k = \sqrt{8/9}$ in a table such as *Handbook of Mathematical Tables and Formulas*, by Burington (p 279) gives

$$S = 6.68$$

Then, assuming the circular pitch, P, to be 16:

$$N = SP/\pi = 34.04$$

Therefore $N = 34$

PROBLEM II—It is desired to linearize the function $\sin \theta$. Thus: $\sin \theta_1 = \theta_2$
where θ_2 is in radians. Final shape of the gears will be similar to those drawn above.

Given center distance = 3 in.; driver oscillates between ±45°. From the equations for Case II:

$$\theta_2 = F(\theta_1) = \sin \theta_1$$

$$R_1 = \frac{C\, F'(\theta_1)}{1 + F'(\theta_1)} = \frac{3\cos\theta_1}{1 + \cos\theta_1} = \frac{3}{1 + \sec\theta_1}$$

$$R_2 = \frac{C}{1 + F'(\theta_1)} = \frac{3}{1 + \cos\theta_1} = \frac{3}{1 + \sqrt{1 - \theta_2^2}}$$

Although R_1 is a closed gear—because $f(\theta_1) = f(\theta_1 + 2N\pi)$—$R_2$ is open and is discontinuous when $\theta_2 = 1$. The gears will, however, operate satisfactorily within the range desired. Inspection indicates that the formulas for length of curve, radius of curvature, and angle of obliquity are not subject to ready solution, and the curves are therefore laid out; see illustration at left. Length of the driver curve between ±45° is approximately 2.29 in. Using 48 pitch (0.0655 circular pitch) there will be 35 teeth in this portion of the curve. Approximate min radius of curvature from the layout is 0.8 in., giving 76 teeth, which will present no undercutting problem.

The layout shows no reverse curve in the usable portions of the two curves, and the angle of obliquity is always less than 45°. The driver has double points (crosses itself) at $\theta_1 = 90°$. Driven is discontinuous at $\theta_2 = 57.3$.

PROBLEM III—Desired velocity relationship between two shafts is $\omega_1 = \omega_2\, B \cos \theta_1$ where B is a constant. Required angular displacement of θ_2 is from +45° to −45°. From equations for Case III:

$$G(\theta_1) = B \cos \theta_1$$

$$R_1 = \frac{C\, B \cos \theta_1}{1 + B \cos \theta_1}$$

$$\theta_2 = \int B \cos \theta_1 d\theta_1 = B \sin \theta_1$$

$$R_2 = \frac{C}{1 + B \cos \theta_1} = \frac{C}{1 + \sqrt{B^2 - \theta_2^2}}$$

Equation for R_1 is a closed curve if $B < 1$ (although the equation may cross itself); but R_2 is discontinuous when $\theta_2 = B$. However the curves are valid for the portion specified (±45°).

Angle of obliquity is:

$$\tan \phi = (B - 1)\left[\frac{\sin\theta_2 + B\sin\theta_2\cos\theta_2}{\cos\theta_2(1 + B\cos\theta_2)}\right]$$

Here, ϕ is less than 45° and single valued for the range given, indicating no reversals. Also, since $\phi = 0$ when $\theta_2 = 0$, there can be no great change of curvature during the interval 0 to 45°. A layout will verify this, and for the purposes of this combination, a comparatively fine pitch (32 or 48) can be used.

ELLIPTICAL GEARS
for
CYCLIC SPEED VARIATIONS

Like other noncircular gearing they offer advantages for
high-speed operation. The fact that they can be made in identical twin
pairs gives them a special advantage. Article develops handy
design equations, works out two sample problems.

SIGMUND RAPPAPORT, *kinematician*
Ford Instrument Co, Div of Sperry Rand Corp, and
Adjunct professor, Polytechnic Institute of Brooklyn

Though you can't get intermittent motion out of a set of elliptical gears, their popularity is growing. This is because they meet the need for speeding up automatic machinery. These speeds are becoming too high to permit an intermittent mechanism to come to a full stop. This is why today's preference is for a varying-speed drive that slows down periodically. If such cyclic motion is practical, the elliptical type has this recommendation—smooth cyclic motion with a minimum of sudden changes in velocity and imbalance. Many other types of noncircular gears can give the same results, but the elliptical types are simpler. They can be made as twin pairs, and their contour can be approximated by a series of circular arcs. Both advantages reduce cost of manufacture.

Design Equations

The twin principle simplifies design. It is illustrated with the pair of identical ellipses drawn at right. They rotate in opposite directions: one around its focus F_1; the other around corresponding focus F'_1. When the gears rotate through angles θ_1 and θ_2, respectively, all points on arc MN come in successive contact with points of the equally long and identically shaped arc MP. Basic rule is that $R_1 + R_2$ must be constant. From symmetry, $F_1'P$ (which is R_2) equals F_2N; therefore distance $F_1N + F_2N$ is constant, which is the definition of an ellipse.

From this same condition the instantaneous angular velocity $d\theta_2 dt$ of the output gear easily be found if the constant input velocity ω is given. The instantaneous velocity ratio $(d\theta_2/dt) : (\omega)$ is determined by the inverse ratio of the instantaneous radii vectors:

$$\frac{d\theta_2}{dt} = \omega \frac{R_1}{R_2}$$

From the polar equation of the ellipse,

$$R_1 = \frac{a(1 - \epsilon^2)}{1 + \epsilon \cos \theta_1}$$

where ϵ (the numerical eccentricity of the ellipse) is defined as

$$\epsilon = \frac{\sqrt{a^2 - b^2}}{a}$$

with a as major, and b as minor semi-axis. This equation can also be written as

$$\epsilon = \frac{e}{a}$$

where e is distance from focus to center of the ellipse.

Symbols

a = major semi-axis
b = minor semi-axis
ϵ = numerical eccentricity
e = distance from focal point to geometric center of ellipse
F_1 = focal point and center of rotation of ellipse
F_2 = focal point of ellipse
K = ratio of velocity change of output gear
N = number of teeth
R = instantaneous radius
θ = instantaneous angle of rotation
ρ = radius for approximating the contour of ellipse
ω = constant angular velocity of input
Subscripts *1* and *2* relate to input and output elliptical gear, respectively, unless otherwise stated.

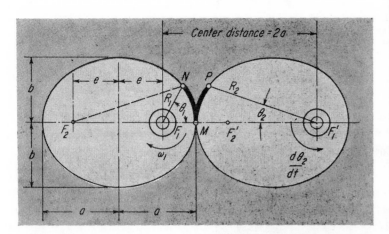

TWO IDENTICAL ELLIPSES rotating around foci F_1 and F_1', will have equal arcs MN and MP in successive contact.

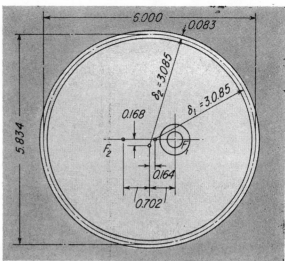

DIMENSIONS OF ONE OF THE ELLIPSES for producing over-all velocity change of 2.6 per cycle.

From $R_1 + R_2 = 2a$, is it found that

$$R_2 = 2a - \frac{a(1 - \epsilon^2)}{1 + \epsilon \cos \theta_1}$$

$$R_2 = a \left[\frac{1 + 2\epsilon \cos \theta_1 + \epsilon^2}{1 + \epsilon \cos \theta_1} \right]$$

or

$$\frac{d\theta_2}{dt} = \omega \left[\frac{1 - \epsilon^2}{1 + \epsilon^2 + 2\epsilon \cos \theta_1} \right]$$

The extremes occur at $\theta_1 = 0$ and $180°$, with $\cos \theta_1 = +1$ and $\cos \theta_1 = -1$. Thus, by substituting for ϵ, the minimum velocity at $\theta_1 = 0$ is:

$$\frac{d\theta_2}{dt_{(min)}} = \omega \left[\frac{a - e}{a + e} \right]$$

and, for $\theta_1 = 180°$:

$$\frac{d\theta_2}{dt_{(max)}} = \omega \left[\frac{a + e}{a - e} \right]$$

Maximum and minimum angular-output velocities are the reciprocals of each other. If K is the ratio of maximum to minimum, then

$$K = \frac{d\theta_2}{dt_{(max)}} \bigg/ \frac{d\theta_2}{dt_{(min)}}$$

$$K = \left(\frac{a + e}{a - e} \right)^2$$

It is good practice to keep K under 5 to insure smooth running without "whip." Design usually starts with the choice of center distance $2a$; in other words, with the size of the gears. If the major semi-axis a is given, and a certain over-all ratio K is desired, the minor semi-axis b can be found from

$$b = \frac{2aK^{1/4}}{1 + K^{1/2}}$$

Sample Calculations

PROBLEM I—Design for an approximate 2.8:1 velocity change

GIVEN: Exact center distance $2a = 6$ in.

Approximate over-all ratio $K = 2.8$
Preferred diametral pitch (DP) = 12
Thus, $a = 3$, and

$$b = \frac{6(2.8)^{1/4}}{1 + (2.8)^{1/2}} = 2.905$$

Circumference of the pitch ellipse is approximated closely enough by $(a + b)\pi$, if, as in most applications, K does not much exceed the value of 4; that is, if the ellipse is not too flat. The number of teeth, N, of a given diametral pitch that can be accommodated on this circumference is the same that can be placed on an ordinary circular spur gear of the diameter $(a + b)$. Thus,

$$a + b = N/DP$$

and

$$N = (3 + 2.905)(12) = 70.86$$

If an elliptical gear has an odd number of teeth, a space on one end of the major axis will be opposed on the other end by a tooth. This allows making the two mating gears identical—and cutting them simultaneously. Thus, choosing the nearest odd integer, the number of teeth becomes 71. Values for b and K must be adjusted:

$$a + b = 71/12 = 5.917$$
$$b = 2.917$$

Dimension e is

$$e = \sqrt{9 - 2.917^2} = 0.702$$

$$K = \left[\frac{3.702}{0.298} \right]^2 = 2.6$$

All dimensions determining the pitch ellipse are now known (see illustration above).

Outline of the gear blank is not an ellipse, but a curve equidistant from the pitch ellipse. The distance equals the addendum (in this case $1/12$ in. = 0.083). It is often sufficient to treat this outline as an aproximation and compose it of circular arcs, the radii of which are:

$$\rho_1 = \frac{b^2}{a} = 2.836$$

$$\rho_2 = \frac{a^2}{b} = 3.085$$

The addendum must be added to the radii to obtain

CIGARET MACHINE (in photo) employs elliptical-gear set designel by author for cyclic operation of tobacco cutter.

the radii of the blank. The larger radius can be turned or milled; the smaller radius is milled. Adjoining portions of the arcs can be blended in by hand filing. Teeth are cut with a form cutter, one space at a time, while the blank is held in a fixture which is rotatable around the centers of the radii. The starting cut is put on the major axis. Or any of the methods suggested in the previous article on noncircular gears can be applied.

PROBLEM II—Design for a given displacement ratio

The tobacco transport on a small European-design machine for mouthpiece cigarets (F Lerner Automatic Machinery, Vienna, Austria) had an intermittent drive to stop the transport while the cutter went through the tobacco. It was found sufficient merely to slow down the tobacco speed when the cutter whips through it, and replacing the intermittent drive by a pair of elliptical gears (see photo above) gave an appreciable increase in output speed.

Requirements: When the driver turns through 120°, the driven gear shall turn approximately 180° and shall complete its full turn when the driver goes through its own remaining 240° (see diagram). Center distance is 5 in. (hence a = 2.5); DP = 10.

It follows from the geometry that

$$\overline{F_1F_2} = 2e = \tfrac{1}{2}\overline{F_1S}$$
$$\overline{F_1S} = 4e = 4\sqrt{a^2 - b^2}$$

and from the fact that the angle $F_2F_1S = 60°$

$$\overline{F_2S} = 2\sqrt{3(a^2 - b^2)}$$

But the ordinate in the focus of an ellipse is

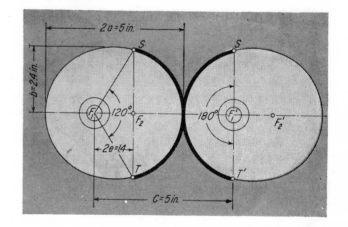

$$\overline{F_2S} = b^2/a$$

Therefore

$$b^4 = 12a^2(a^2 - b^2)$$

Substituting for a = 2.5 and solving for b gives

$$b = 2.408$$

Diameter of a spur gear which has the same circumference is

$$a + b = 4.908$$

from which the number of teeth is

$$N = 10(4.908) = 49.08$$

Nearest odd integer is 49. Again adjusting b gives, b = 2.4. Thus

$$e = \sqrt{2.5^2 - 2.4^2} = 0.700$$

From this all other dimensions are determined.

For a prescribed variable output, try
Twin eccentric gears

A pair of eccentrically mounted standard gears can compensate for nonlinearity in instruments. These new equations simplify calculations.

S V MIANO, design supervisor, Eclipse-Pioneer Div, Bendix Corp, Teterboro, NJ

TWO circular gears mounted eccentrically on shafts will mesh correctly when operated—but they will not run "true". The backlash between mating teeth and the number of teeth in contact will vary within each cycle and you will obtain a cyclically varying output. This characteristic of eccentric gears, however, offers a means of compensating for the nonlinearity in the relationship between variables. Eccentric gears are now employed in instruments which give barometric pressure as a function of altitude (a nonlinear relationship), in potentiometers and synchros to vary the output as a function of angular displacement, and in output shafts of limited-travel gear trains to compensate for the accumulated eccentricities of the intermediate gears.

A cyclic output can also be obtained by means of cams, four-bar linkages and noncircular gears. However, it is usually less expensive to employ standard spur gears with bores offset to a predetermined eccentricity (the eccentricity in most of these applications is usually small). Also an eccentric-gear system is more rigid—and frequently more accurate—than systems employing cams or linkages.

A new mathematical approach is given here which simplifies the determination of the eccentricity required to produce a desired nonlinear angular displacement of the output shaft for a given rotation of the input shaft. The typical system, Fig 1, employs two identical gears of known pitch diameters and eccentricity ϵ. The gears are in mesh so that their centers of rotation (eccentric centers) E_1 and E_2 are a fixed distance D apart. At the start of rotation, the geometric centers of the gears, C_1 and C_2, lie in the common center line X-X. Center distance D is slightly greater than the standard center distance of the gears when in intimate mesh (its value will be determined later).

When gear 1 is rotated clockwise about center E_1, its geometric center C_1 is displaced through angle θ to position C'_1. As the gears are always in mesh, geometric center C_2 of gear 2 assumes position C'_2 after its counter-clockwise displacement about eccentric center E_2. A line joining the displaced geometric centers C'_1 and C'_2 assumes an inclination of α degrees relative to the starting common center line X-X. Line E_2M is constructed parallel to the line between C'_1 and C'_2. Radial line $E_1C'_1$ is extended to M.

From the geometry of the figure, when gear 1 is rotated clockwise through angle θ relative to the original common center line X-X it has also rotated through angle θ-α relative to the new common center line $C'_1C'_2$, and consequently through angle θ-2α degrees. Thus, as gear 1 rotates at a constant angular velocity, gear 2 lags by a varying angle 2α which ranges from a minimum of zero at the starting position to a maximum when θ is $90 + \alpha$, (Fig 2). The derivation of equations involving the various pertinent quantities follows from the geometry of Fig 1 and 2.

Determination of lag angle

Consider triangle E_1E_2M in Fig 1. Line E_1N is constructed perpendicular to line E_2M.

$$E_1 N = D \sin \alpha = 2 \, \epsilon \sin (\theta - \alpha)$$

SYMBOLS

A = constant, dimensionless	R_b = base-circle radius of gears, in.
c = minimum contact ratio	T_o = circular thickness at standard pitch radius, in.
C_1, C_2 = geometric centers of gears 1 and 2, respectively	T_1, T_2 = circular thickness at points of closest and farthest mesh, respectively, in.
D = distance between centers of rotation (or between gear-shaft centerlines), in.	
E_1, E_2 = centers of rotation	ϵ = eccentricity (distance between actual and geometric gear centers), in.
p = circular pitch	2α = lag angle (lag between input and output gears), deg
PD = pitch diameter, in.	θ = input rotation, deg
r_1, r_2 = operating pitch radii, in.	ϕ = pressure angle, deg
R_o = outside radius of gears, in.	ρ = distance between geometric centers, in.

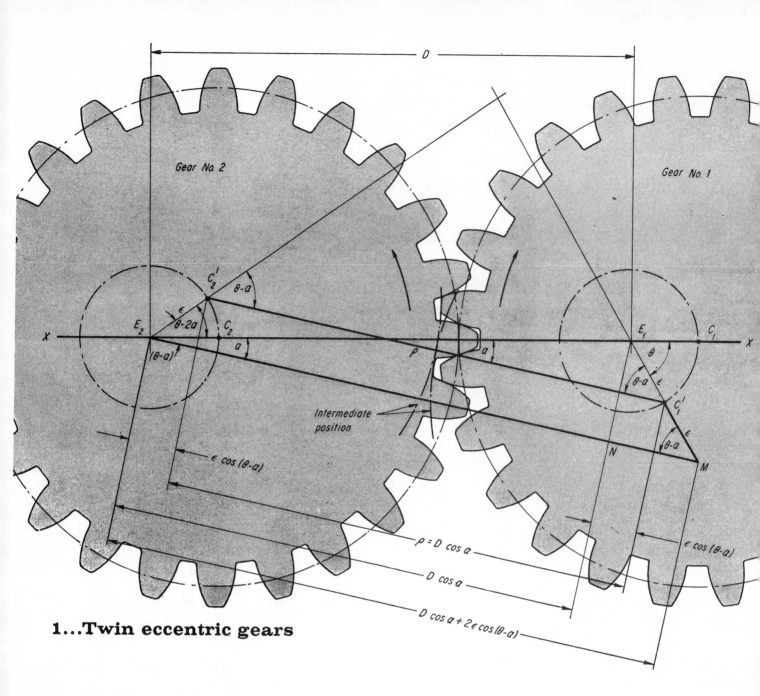

1...Twin eccentric gears

By trigonometric expansion

$$D \sin \alpha = 2 \epsilon \sin \theta \cos \alpha - 2 \epsilon \cos \theta \sin \alpha$$

Thus

$$\frac{\sin \alpha}{\cos \alpha} = \frac{2 \epsilon \sin \theta}{D + 2 \epsilon \cos \theta}$$

or

$$\tan \alpha = \frac{\sin \theta}{\dfrac{D}{2 \epsilon} + \cos \theta}$$

Therefore, the angle that gear 2 lags gear 1 is

$$2 \alpha = 2 \tan^{-1} \left[\frac{\sin \theta}{\dfrac{D}{2 \epsilon} + \cos \theta} \right]$$

Angle α is **minimum** ($\alpha = 0$) when $\theta = 0$. The maximum value of α is

obtained by differentiating the previous equation

$$\frac{d\alpha}{d\theta} = \frac{d}{d\theta} \left[\tan^{-1} \frac{\sin \theta}{\dfrac{D}{2 \epsilon} + \cos \theta} \right]$$

and setting the derivative equal to zero:

$$\frac{d\alpha}{d\theta} = \frac{\dfrac{D}{2 \epsilon} (\cos \theta) + 1}{\left(\dfrac{D}{2 \epsilon} \right)^2 + \left(\dfrac{D}{\epsilon} \right) \cos \theta + 1} = 0$$

Thus, α is a maximum when the numerator of the above equation equals zero:

$$\frac{D}{2 \epsilon} (\cos \theta) + 1 = 0$$

or when $\quad \cos \theta = -\dfrac{2 \epsilon}{D}$

Referring to Fig 2, α is maximum when line $E_2 M$, which is numerically equal to the geometric center distance of the gears, is at right angles with line $E_1 M$, which is numerically equal to twice the eccentricity. From Fig 2, maximum α also occurs when

$$\sin \alpha = \frac{2 \epsilon}{D}$$

and when maximum $\alpha = \theta - 90°$. Thus, amount of lag (2α) between input and output gears will occur when

$$\theta = \tfrac{1}{2} (2 \alpha + 180°)$$

Rate of change of angle

The rate of change of angle α is represented by the first derivative of α, as given by the equation for $d\alpha/d\theta$.

At any angular position of the input gear, it can be determined at what rate the output gear is lagging in angular displacement. For instance, when the input gear is at the datum angular displacement position ($\theta = 0$), the angle of lag is zero, but the rate of change of α is maximum. Thus, $\cos \theta = 1$, and

$$\left(\frac{d\alpha}{d\theta}\right)_{max} = \frac{1}{\left(\frac{D}{2\epsilon}\right) + 1}$$

When the input gear displacement increases from $\theta = 0$ to θ equal to an angle whose cosine is $-2\epsilon/D$, $\cos \theta$ decreases from 1 to $-2\epsilon/D$, and the rate of change of the output gear decreases until $\cos \theta = -2\epsilon/D$ with the result that $d\alpha/d\theta = 0$. Thus the rate of displacement of the output gear has ceased to decrease and α has attained its maximum value.

Between $\theta =$ the angle whose cosine is $-2\epsilon/D$, and $\theta = 180$, angle α decreases from maximum to zero while its rate of change increases from zero to its maximum rate of

$$\frac{1}{\frac{D}{2\epsilon} + 1}$$

Variable geometric center distance

As a pair of eccentric gears rotate through a cycle, their respective teeth in mesh appoach each other and then separate. The operating pitch diameters of the two gears simultaneously decrease and then increase during the rotational process.

It is important to know the conditions of maximum and minimum operating pitch diameters because the former is the condition of the teeth tightly in mesh, while the latter is the condition of maximum backlash. To control the limits of these two ex-

tremes it is necessary to control the variable distance between the geometric centers of the gears. This distance is denoted by ρ.

Referring again to Fig 1, the instantaneous center distance ρ is the distance between geometric centers C_1' and C_2'. An equation for ρ in terms of α is obtained as follows:

Length of line $E_2N = D \cos \alpha$. If, from this length the distance $\epsilon \cos (\theta - \alpha)$ at gear 2 is subtracted, and the same distance $\epsilon \cos (\theta - \alpha)$ at gear 1 is added, the result is length $C_2'C_1'$, or the instantaneous center distance ρ. Thus

$$\rho = D \cos \alpha$$

When α is zero, that is, when both the geometric and eccentric centers are in line, $\cos \alpha = 1$. At this point ρ is a max and is equal to: $\rho_{max} = D$.

When α is maximum ($\cos \alpha$ is minimum), ρ is minimum. Therefore, disregarding provision for backlash, minimum ρ is the sum of the pitch radii of the two gears, or the standard center distance.

If the gears are cut so that the circular tooth thickness is one half the circular pitch, and the operating centers are separated by an amount F to provide for manufacturing errors, then

$$\rho_{min} = \frac{PD_1 + PD_2}{2} + F$$

Contact ratio

Because an intimate mesh occurs only when variable center distance ρ is minimum, it follows that when ρ increases to a maximum value of D there is a separation of the gears accompanied by a decrease in contact ratio (the number of teeth in contact). The limit of permissible separation is a function of the minimum contact ratio

desired for proper gear transmission. It is interesting to note that at maximum separation of the gears, a pair of 80-pitch precision gears of 0.7500-in. pitch diameter and as much as 0.0312-in. eccentricity, still operate with a contact ratio considerably better than 1.2—the generally accepted minimum contact ratio. This indicates that in a wide range of applications of eccentric gears the question of reduced contact ratio is not critical.

Backlash

The geometric center distance ρ varies from a minimum $PD_1/2 + PD_2/2$ (standard center distance) to D as a maximum. It follows that as ρ approaches D in length and the gears separate, the operating pitch radii and the pressure angle increase while the circular tooth thickness at the point of action decreases. The result is an increase in backlash. Just how critical this backlash is depends upon the accuracy demanded. The introduction of suitable antibacklash devices will partly compensate for this backlash. The equations for computing the amount of backlash at the point of maximum separation between gears are given in the example that follows.

NUMERICAL EXAMPLE

Fixed distance and eccentricity

It is required that a nonlinear output shaft lag an input shaft from a minimum of zero degrees at the zero and 180° positions of the input shaft, to a maximum of 2.8918 deg. Determine the distance between the eccentric centers, D, the amount of eccentricity necessary, ϵ, and the lag between input and output at intermediate points of rotation.

Two 80-diametral-pitch eccentric

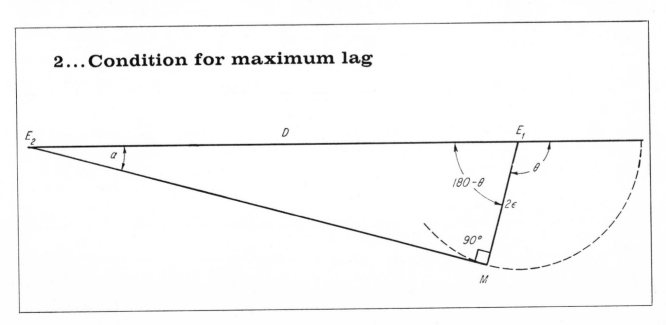

2...Condition for maximum lag

3...Shift in output symmetry by varying eccentricity.

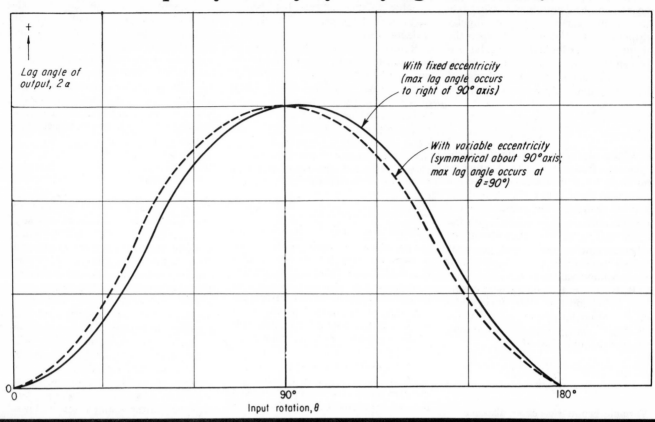

gears of 0.7500-in. pitch are arbitrarily chosen to operate at 0.7528-in. center distance at their closest position.

From the data given, the max lag angle 2α is equal to 2.8918 (all angles are in degrees).

$$\text{Maximum } \alpha = 1.4459$$
$$\tan \alpha = 0.025,24$$
$$\cos \alpha = 0.999,68$$
$$\sin \alpha = 0.025,23$$
$$\rho = D \cos \alpha = 0.7528$$

Then
$$D = 0.7530$$

This is the fixed center distance between the operating eccentric centers. The distance is $0.7528 - 0.7500 = 0.0028$ greater than the standard center distance.

The max lag angle will occur when the input gear is displaced an amount

$$\theta = \tfrac{1}{2}(2.8918 + 180) = 91.4459$$

The required eccentricity is

$$\cos (91.4459) = -\frac{2\,\epsilon}{0.7530}$$

$$2\,\epsilon = 0.7530\,(0.025,23) = 0.0190$$

$$\epsilon = 0.0095.$$

Values of the lag angle of the output shaft may be calculated for any displacement of the input shaft by substituting in the equation

$$\tan \alpha = \frac{\sin \theta}{\dfrac{0.7530}{0.0190} + \cos \theta}$$

Minimum contact ratio

Find the minimum contact ratio for the previous gear pair when:

ϕ_o = pressure angle at standard center distance, $= 20°$ ($\cos 20° = 0.939,69$; INV $20° = 0.014,90$)
R_o = outside radius of gears $= 0.3875$
R_b = base circle radii of gears $= 0.3524$
r_o = standard pitch radii $= 0.3750$
D = center distance $= 0.7530$
p = circular pitch $= 0.039,27$
ϕ_1 = pressure angle when teeth are at closest mesh (when $\rho_{min}=0.7528$)
ϕ_2 = pressure angle at when teeth are farthest apart (when $\rho_{max} = D$)
r_1 = operating pitch radius at $\rho_{min} = \tfrac{1}{2}\,\rho_{min} = 0.3764$
r_2 = operating pitch radius at $\rho_{max} = \tfrac{1}{2}\,\rho_{max} = 0.3765$

$$\cos \phi_2 = \frac{r_o}{r_2} \cos (\phi_o)$$

$$\cos \phi_2 = \frac{0.7500\,(0.939,69)}{0.7530} = 0.935,95$$

$$\phi_2 = 20.6180° \quad \sin \phi_2 = 0.352,14$$
INV $\phi_2 = 0.016,38$

The formula for contact ratio is:

$$c = \frac{2\,\sqrt{R_o - R_b} - D \sin \phi_2}{p \cos \phi_1}$$

$$c = \frac{2\,(0.161,77) - 0.261,58}{(0.039,27)\,(0.939,69)} = 1.6$$

Amount of additional backlash

If the input gear is reversed, and an antibacklash device is not employed, it is desirable to know the amount of backlash when taking readings from an instrument using the eccentric gears. The amount of backlash may be applied as a correction upon the lag angle at any operating center distance ρ. For instance, in going from maximum α to $\alpha = 0$, the operating center distance ρ increases from 0.7528 to 0.7530. Tooth action takes place at a reduced circular thickness, and backlash increases as a function of the tooth thickness where the line of contact crosses the involute.

T_o = circular thickness at standard pitch radius $= p/2$ $= 0.019,63$
T_1 = circular thickness at 0.3764 pitch radius
T_2 = circular thickness at 0.3765 rad.

$$\cos \phi_1 = \frac{r_o}{r_1} \cos \phi_o$$

$$\cos \phi_1 = \frac{0.7500}{0.7528}\,(0.939,69) = 0.936,20$$

$$\phi_1 = 20.5775; \quad \text{INV } \phi_1 = 0.016,28$$

The circular thicknesses at points of minimum and maximum mesh will be:

$$T_1 = 2\,r_1 \left(\frac{T_o}{2\,r_o} + \text{INV } \phi_o - \text{INV } \phi_1 \right)$$

$$T_1 = 2\,(0.3764) \left(\frac{0.019,63}{0.7500} + \right.$$

$$\left. 0.014,90 - 0.016,28 \right) = 0.018,66$$

Differences in output—fixed vs variable eccentricity

$\theta°$	$2\alpha_1$ (WITH VARIABLE ECCENTRICITY)	$2\alpha_2$ (WITH FIXED ECCENTRICITY)	DIFFERENCE IN DECIMAL OF A DEGREE	DIFFERENCE IN DEGREES
0	0	0	0	0
10	0.5022	0.4900	0.0122	0° 0′ 44″
20	0.9892	0.9660	0.0232	0° 1′ 24″
30	1.4460	1.4148	0.0312	0° 1′ 52″
40	1.8590	1.8232	0.0358	0° 1′ 9″
50	2.2154	2.1794	0.0364	0° 2′ 11″
60	2.5044	2.4724	0.0324	0° 1′ 57″
70	2.7174	2.6932	0.0242	0° 1′ 27″
80	2.8478	2.8344	0.0134	0° 0′ 48″
90	2.8918	2.8908	0.0010	0° 0′ 4″
91.4459	2.8908	2.8918	−0.0010	−0° 0′ 4″
100	2.8478	2.8594	−0.0116	−0° 0′ 42″
110	2.7174	2.7402	−0.0228	−0° 1′ 22″
120	2.5044	2.5356	−0.0312	−0° 1′ 52″
130	2.2154	2.2512	−0.0358	−0° 1′ 9″
140	1.8590	1.8950	−0.0360	−0° 2′ 10″
150	1.4460	1.4780	−0.0320	−0° 1′ 55″
160	0.9892	1.0128	−0.0236	−0° 1′ 25″
170	0.5022	0.5148	−0.0126	−0° 0′ 45″
180	0	0	0	0

$$T_2 = 2\,(0.3765)\left(\frac{0.919,63}{0.7500} + 0.014,90 - 0.016,38\right) = 0.018,59$$

The increase in backlash is 0.018,66 − 0.018,59 = 0.000,07 in. This is equivalent to an angle (or play in the gears) of 0° 1′ 20″.

Variable eccentricity

Unlike a sinusoidal curve which is in the form of $y = A \sin \theta$, the eccentric gear curve

$$\tan \alpha = \frac{\sin \theta}{(D/2\,\epsilon) + \cos \theta}$$

$$= \left[\frac{1}{\dfrac{D}{2\,\epsilon} + \cos \theta}\right] \sin \theta$$

is not symmetrical about the axis through the node, Fig 3. The reason for this is that the coefficient of $\sin \theta$, composed of the terms in the bracket, is not constant because of the variable, $\cos \theta$.

We have employed a mechanical device to vary the eccentricity ϵ at a rate relative to $\cos \theta$ so that the denominator remains constant. The resulting curve is then symmetrical (as shown in Fig 3). Referring to the previous example where maximum angle $\alpha = 1.4459°$, the use of a me-

chanical device results in a symmetrical curve, $\tan \alpha = A \sin \theta$, by letting maximum α occur at input angle $\theta = 90°$. At this point, $\sin \theta$ is maximum and equal to one. Then

$$\tan 1.4459 = A \sin 90°$$

and

$$A = \frac{\tan 1.4459}{1} = 0.025,24$$

The equation of the symmetrical curve is then

$$\tan \alpha = 0.025,24 \sin \theta$$

This is a more direct relationship between input and lag than is obtained from the case where ϵ is fixed. Hence the variable-center system has potential use in instrumentation.

Values of angle α have been calculated for 10° intervals in angle θ, (see table). This table shows values of the lag angle, $2\alpha_2$, as determined from the nonsymmetrical eccentric gear equation $\tan \alpha = \sin \theta/(\cos \theta + D/2\epsilon)$, and values of $2\alpha_1$, as determined from the symmetrical curve equation $\tan \alpha = 0.025,24 \sin \theta$. (Subscripts 1 and 2 refer to the symmetrical and nonsymmetrical case, respectively.) Also shown are the differences between the two sets of values. All values are in degrees.

The equation for varying the eccentricity to maintain a constant coefficient of $\sin \theta$ can be obtained by set-

ting the bracketed coefficient:

$$\frac{1}{\left(\dfrac{D}{2\,\epsilon}\right) + \cos \theta}$$

equal to the predetermined constant, A. Thus

$$\frac{1}{\left(\dfrac{D}{2\,\epsilon}\right) + \cos \theta} = A$$

and

$$\epsilon = \frac{A\,D}{2\,(1 - A \cos \theta)}$$

Hence, for the conditions previously given:

$$\epsilon = \frac{0.009,50}{1 - 0.025,24 \cos \theta}$$

This equation gives the rate that the gear centers of both gears must travel with respect to their geometric centers, during rotation of the input gear.

6
GEAR SPECIFICATIONS, INSPECTION, AND TESTS

NEW AGMA CLASSIFICATION SYSTEM FOR GEARS . . .

. . . widens quality range—classifies materials

what it offers

✓ Covers spur, helical and herringbone gears
✓ Sets up 16 new quality classes
✓ Expands backlash classes of fine-pitch gears
✓ Establishes new backlash tolerances for coarse-pitch gears
✓ Assigns code numbers to gear materials and heat treatment
✓ Gives quality requirements for many industries and applications

NICHOLAS P CHIRONIS

■ Here is the all-inclusive system for specifying gears that designers have been waiting for. It brings together data on fine- and coarse-pitch gears, and establishes a new quality classification system applicable to a wide span of precision —from "commercial" to "ultraprecision" types of gears. These terms have been obsoleted, and 16 classes of quality may be specified—the higher the number, the higher the quality.

The new AGMA system also classifies, by means of a letter-number code system, popular gear materials and their heat treatment. In addition, backlash specifications for fine-pitch gears are classified in steps from A to E, as against the previous A to D system. Thus by combining the various classification numbers, the designer has a simple and easy-to-use code for specifying quality, backlash, material and heat treatment of a gear. This will go a long way toward building a better communication bridge between gear user and manufacturer.

Its scope

The new system covers spur, helical and herringbone gears. In many cases, it gives separate data for coarse-pitch gears; from 0.5 P to 19.99 P (P = diametral pitch); and fine-pitch gears, from 20 P to 200 P.

Four areas are covered: quality classifications of gear teeth; backlash classifications relating to the amount of tooth-thinning to be built into the gear; materials and heat-treatment classifications; typical quality requirements of applications.

At the end of this article are examples of how to combine the various codes to call out a complete set of specifications.

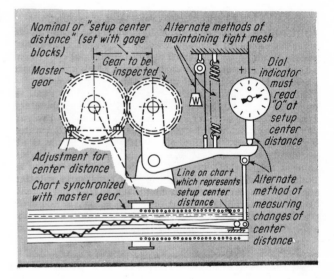

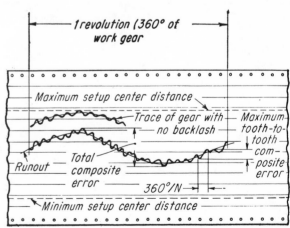

GEAR-ROLLING is the preferred way to inspect quality of fine-pitch gears. The new quality specifications give permissible tooth-to-tooth and total composite errors.

Quality numbers—fine-pitch gears

Quality numbers refer to the degree of precision applied to making the gear teeth. The new system eliminates the confusing terms "commercial" and "precision." Table 1 shows both the former and present quality designations.

Tolerances for the new quality designations, Table 2, are given in terms of:

Tooth-to-tooth composite tolerance. This is the allowable center distance variation as the gear is rotated in tight mesh with a master gear for one complete revolution of the gear. It includes effects of variations in circular pitch, thickness and profile.

Total composite tolerance. This is the allowable total variation in center distance as the gear is rotated with the master. It includes the effect of all tooth and runout (gear eccentricity) variations.

Both terms are illustrated at left. The gear to be inspected is held in tight mesh with a master gear in a gear-rolling fixture. Variations in the center distance are recorded on a chart (or observed on a dial indicator), and then compared with the acceptable deviations (tolerances) in Table 2. A perfect gear would produce a straight line on the chart—there would be no changes in center distance.

At present, there has been no need to classify fine-pitch gears with tolerances wider than those for Quality 5. Table 2 starts with this grade.

Quality numbers—coarse-pitch gears

The new system lists tolerances for quality numbers from 3 to 15. Quality 16 has been left open for future needs. Included are tolerances for runout, pitch, profile and lead. Also, for quality numbers from 8 to 12, the total-composite tolerance is given. If this is specified, it should

text continued, next page

IS IT MANUAL OR STANDARD?

■ Although AGMA cautiously refers to the new system as a manual, to all practical purposes it will function as a standard. The two previous standards it obsoletes were, in fact, so designated by AGMA.

Why the reluctance now? Primarily because some of the new quality classifications call for extremely tight tolerances. This is bound to be troublesome for some manufacturers. The AGMA is a looser organization than that of the technical societies. If a manufacturer does not approve a standard he may pull out of the organization. This has happened before. Technical committees setting up standards have learned to proceed slowly. This classification system is a big step. By referring to it as a manual, the manufacturer is presented with a guide post rather than a directive.

But the American Standards Association (ASA) has indicated interest in adopting the new system as an ASA standard. ASA has done this previously with other AGMA publications. Thus, unless a change is made in nomenclature, the AGMA may find itself offering the gear designer a "manual," while the ASA is offering the same contents as a "standard." ■

WHY THE CHANGE?

■ In many instrument and control applications, gears have become the limiting factor of over-all system accuracy. Until now, the AGMA (American Gear Manufacturers Association) standard on fine-pitch gears (AGMA 236.04) provided for only seven quality classes—four classes of "commercial" gears, and three of "precision" gears. The standard did not include material specifications, nor did the quality classes satisfy the need for more precise gears in specialized applications.

Then, in a PRODUCT ENGINEERING article (Dec 22 '58, p 46), Lou Martin, well-known gear expert and member of several AGMA technical committees, proposed a classification system for quality of fine-pitch gears that could easily be extended in the ultraprecision direction. At that time he envisioned a 15-class system with tolerances that decrease logarithmically in preferred-number steps. He employed the terms "commercial," "precision" and "ultraprecision"—but suggested that these terms be dropped in favor of a code system to range from 1 to 15. The article raised objections from some gear manufacturers that to establish tolerances and assign code numbers to ultraprecise gears (a grade of quality not in the standards at that time) would encourage gear designers to specify such gears—and the gear industry had sufficient difficulty mass-producing Precision 3 gears (the best listed).

But it was only a matter of time. In Oct '59, an AGMA committee worked up a new standard, containing a 16-class quality range, and applicable to coarse-pitch as well as fine-pitch gears. It was routed to company members, but ran into resistance from some of the fine-pitch gear manufacturers. Main complaint: it failed to go far enough in eliminating gear specification problems. Many of the suggestions that resulted, including requests to improve the backlash classifications, were incorporated into it, and reaction last month at the AGMA meeting in Hot Springs, Va, was favorable. The system is now issued as the "Gear Classification Manual (AGMA 390.01) for Spur, Helical and Herringbone Gears." It obsoletes the two previous standards: AGMA 236.04 "Inspection of Fine-Pitch Gears," and AGMA 231.02 "Inspection of Coarse-Pitch and Helical Gears." However, portions of AGMA 236.04 dealing with inspection procedure are being retained under a new number, AGMA 236.05. ■

TABLE 1—FORMER AND PRESENT QUALITY DESIGNATIONS FOR FINE-PITCH GEARS

Former designation	Present Quality No.
Commercial 1	5 or 6
Commercial 2	6 or 7
Commercial 3	8
Commercial 4	9
Precision 1	10 or 11
Precision 2	12
Precision 3	13 or 14

Table 2—Fine-pitch gear tolerances (in inches)

AGMA Quality No.	No. of Teeth, T and Pitch Diameter	Diametral Pitch Range	Tooth-to-tooth Composite Tolerance	Total Composite Tolerance
5	Up to 20 T (inclusive)	20 to 80	0.0037	0.0052
	Over 20 T; to 1.999 in.	20 to 32	0.0027	0.0052
	Over 20 T; 2 to 3.999 in.	20 to 24	0.0027	0.0061
	Over 20 T; 4 in. and over	20	0.0027	0.0072
6	Up to 20 T	20 to 200	0.0027	0.0037
	Over 20 T; to 1.999 in.	20 to 48	0.0019	0.0037
	Over 20 T; 2 to 3.999 in.	20 to 32	0.0019	0.0044
	Over 20 T; 4 in. and over	20 to 24	0.0019	0.0052
7	Up to 20 T	20 to 200	0.0019	0.0027
	Over 20 T to 1.999 in.	20 to 100	0.0014	0.0027
	Over 20 T; 2 to 3.999 in.	20 to 48	0.0014	0.0032
	Over 20 T; 4 in. and over	20 to 40	0.0014	0.0037
8	Up to 20 T	20 to 200	0.0014	0.0019
	Over 20 T; to 1.999 in.	20 to 200	0.0010	0.0019
	Over 20 T; 2 to 3.999 in.	20 to 100	0.0010	0.0023
	Over 20 T; 4 in. and over	20 to 64	0.0010	0.0027
9	Up to 20 T	20 to 200	0.0010	0.0014
	Over 20 T; to 1.999 in.	20 to 200	0.0007	0.0014
	Over 20 T; 2 to 3.999 in.	20 to 200	0.0007	0.0016
	Over 20 T; 4 in. and over	20 to 120	0.0007	0.0019
10	Up to 20 T	20 to 200	0.0007	0.0010
	Over 20 T; to 1.999 in.	20 to 200	0.0005	0.0010
	Over 20 T; 2 to 3.999 in.	20 to 200	0.0005	0.0012
	Over 20 T; 4 in. and over	20 to 200	0.0005	0.0014
11	Up to 20 T	20 to 200	0.0005	0.0007
	Over 20 T; to 1.999 in.	20 to 200	0.0004	0.0007
	Over 20 T; 2 to 3.999 in.	20 to 200	0.0004	0.0009
	Over 20 T; 4 in. and over	20 to 200	0.0004	0.0010
12	Up to 20 T	20 to 200	0.0004	0.0005
	Over 20 T; to 3.999 in.	20 to 200	0.0003	0.0005
	Over 20 T; 2 to 3.999 in.	20 to 200	0.0003	0.0006
	Over 20 T; 4 in. and over	20 to 200	0.0003	0.0007
13	Up to 20 T	20 to 200	0.0003	0.0004
	Over 20 T; to 1.999 in.	20 to 200	0.0002	0.0004
	Over 20 T; 2 to 3.999 in.	20 to 200	0.0002	0.0004
	Over 20 T; 4 in. and over	20 to 200	0.0002	0.0005
14	Over 20 T	20 to 200	0.00019	0.00027
	Over 20 T; to 1.999 in.	20 to 200	0.00014	0.00027
	Over 20 T; 2 to 3.999 in.	20 to 200	0.00014	0.00032
	Over 20 T; 4 in. and over	20 to 200	0.00014	0.00037
15	Up to 20 T	20 to 200	0.00014	0.00019
	Over 20 T; to 1.999 in.	20 to 200	0.00010	0.00019
	Over 20 T; 2 to 3.999 in.	20 to 200	0.00010	0.00023
	Over 20 T; 4 in. and over	20 to 200	0.00010	0.00027
16	Up to 20 T	20 to 200	0.00010	0.00014
	Over 20 T; to 1.999 in.	20 to 200	0.00007	0.00014
	Over 20 T; 2 to 3.999 in.	20 to 200	0.00007	0.00016
	Over 20 T; 4 in. and over	20 to 200	0.00007	0.00019

generally be in lieu of the other tolerances because a composite check is sufficient to tie down the gear-tooth errors. However, the coarser pitch gears, and those of the lower quality number, are not checked by means of a rolling fixture; hence the need for the other tolerances.

The key tolerances for qualities 8 to 15 for coarse-pitch gears are shown in Table 3.

Why the need for backlash

Backlash is the amount by which the width of a tooth space exceeds the thickness of the engaging tooth on the pitch circles. Backlash can be caused by machining variations in the center distance between gears, play in bearings, deflection under load, gear-tooth wear, etc.

Backlash is also "built-in" the gears by deliberately

continued on page 180

Table 3—Coarse-pitch gear tolerances (in ten-thousandths of an Inch)

AGMA Quality Number	Normal Diametral Pitch	Total Composite Tolerance — Pitch Diameter (inches)						
		1	3	6	12	25	50	100
8	2				80	94	111	135
	4		46	52	58	68	79	93
	8	35	38	42	46	52	60	70
	16–19.99	27	30	34	37	44	50	58
9	2				57	66	78	95
	4		33	37	42	48	54	66
	8		27	30	33	38	43	50
	16–19.99	19	22	24	27	31	36	42
10	2				40	48	58	69
	4		23	26	29	34	39	47
	8		19	21	24	26	30	35
	16–19.99	14	15	17	19	22	26	30
11	4		17	18	21	24	28	34
	8		14	15	17	19	22	26
	16–19.99	10	11	12	14	16	18	21
12	4		12	13	15	17	20	25
	8		10	11	12	13	15	18
	16–19.99	7	8	9	10	11	13	16

AGMA Quality Number	Normal Diametral Pitch	Pitch Tolerance — Pitch Diameter (Inches)						
		1½	3	6	12	25	50	100
13	2			2½	2¾	2¾	3¼	3½
	4		2	2	2½	2½	2¾	3
	8	1¾	1¾	2	2	2¼	2½	2¾
	16–19.99	1½	1½	1¾	1¾	1¾	2	2¼
14	2			1¾	1¾	2	2¼	2½
	4		1½	1½	1½	1¾	2	2¼
	8	1½	1½	1½	1½	1½	1¾	2
	16–19.99	1½	1½	1½	1½	1½	1½	1¾
15	2			1½	1½	1½	1½	1¾
	4		1½	1½	1½	1½	1½	1½
	8	1½	1½	1½	1½	1½	1½	1½
	16–19.99	1½	1½	1½	1½	1½	1½	1½

Table 4—Backlash tolerances for fine-pitch gearing (in inches, measured in normal plane)

Backlash Designation	Normal Diametral-pitch Range	Tooth Thinning to Obtain Backlash (per gear)	Resulting Approximate Backlash (per pair)
A	20 thru 45	0.0020	0.0040 to 0.0080
	46 thru 70	0.0015	0.0030 to 0.0070
	71 thru 90	0.0010	0.0020 to 0.0055
	91 thru 200	0.0008	0.0015 to 0.0030
B	20 thru 60	0.0010	0.0020 to 0.0040
	61 thru 120	0.0008	0.0015 to 0.0030
	121 thru 200	0.0005	0.0010 to 0.0020
C	20 thru 60	0.0005	0.0010 to 0.0020
	61 thru 120	0.0004	0.0008 to 0.0015
	121 thru 200	0.0003	0.0006 to 0.0010
D	20 thru 60	0.0003	0.0006 to 0.0010
	61 thru 120	0.0002	0.0004 to 0.0007
	121 thru 200	0.0001	0.0002 to 0.0004
E	All Pitches	Do not provide tooth thinning	0.0000 to 0.0002

Table 5—Backlash tolerances for coarse-pitch gearing

(in inches, measured in normal plane)

Center Distance	Normal Diametral Pitches				
	0.5-1.99	2-3.49	3.5-5.99	6-9.99	10-19.99
Up to 5.........	....	....	....	....	0 005–.015
Over 5 to 10...	....	....	....	0.010–.020	0.010–.020
Over 10 to 20..	....	....	0.020–.030	0.015–.025	0.010–.020
Over 20 to 30..	....	0.030–.040	0.025–.030	0.020–.030	....
Over 30 to 40..	0.040–.060	0.035–.045	0.030–.040	0.025–.035	....
Over 40 to 50..	0.050–.070	0.040–.055	0.035–.050	0.030–.040	....
Over 50 to 80..	0.060–.080	0.045–.065	0.040–.060	....	....
Over 80 to 100.	0.070–.095	0.050–.080	....	....	....
Over 100 to 120	0.080–.110	....	....	....	....

thinning the tooth thickness. This avoids a tight mesh because of incorrect mounting distance or because of gear expansion caused by an increase in the operating temperature. A tight mesh may result in objectionable gear noise, increased power losses, over-heating, rupture of the lubricant film, overloaded bearings and premature gear failures. There are gearing applications where a tight mesh (zero backlash) is required. On the other hand, specifying unnecessarily close backlash tolerances will increase the cost of the gears.

Fine-pitch backlash classifications

Two key changes in backlash in the new table, Table 4, from the previous standard are:

Backlash designations have been expanded from the previous A to D designations to the present A to E.

Previously, designation D called for no measurable backlash. Some gear manufacturers complained that although they could do better than C without too much difficulty, to achieve D greatly increased the cost of the gear. One manufacturer (*PE—June 6 '60, p 20*) came out with a CD backlash designation for gears with backlash between C and D.

The new system follows essentially this reasoning by shifting the previous D designation to E, and then adding new, in-between tolerances to D.

A new column has been added giving the amount of tooth-thinning built into the gear. Previously the table gave only the backlash in the normal plane.

Backlash does not affect the quality of a gear. Referring to the chart on p 68, the top trace has no backlash (backlash is measured at point of closest engagement). The second trace has a certain amount of backlash, resulting in a reduction in center distance. Both show the same total composite, and tooth-to-tooth deviations; hence both are rated the same, as far as quality goes.

Coarse-pitch backlash specifications

The backlash table for coarse-pitch gears, Table 5, is based on center-distance dimensions (indicative of gear size) rather than on gear pitch, as for the fine-pitch data, Table 4. In addition, the table gives only the normal-plane

backlash; there is no information on tooth-thinning. Also missing are code letters to easily indicate the desired backlash. This means that the coarse-pitch gear designer must specify the actual backlash dimension because the new AGMA system specifically states that "definite backlash tolerances on coarse-pitch gearing shall be considered binding on the gear manufacturer only when agreed upon in writing" (a similar statement covers the fine-pitch gears).

Material and treatment

Code numbers have been assigned to a list of popular material and treatment combinations, Table 6. Nonmetallics are not included, but it is expected that the tables will be expanded as the need arises.

Although no attempt is made to relate material-treatment combinations to particular applications, these suggestions are offered by the AGMA:

Carbon steels, annealed from bar stock, forgings, or castings are usually satisfactory for pinion and gear when the loads are uniform, or moderate shock and the size of the gearing is not an important factor.

Alloy-steel lengthens the life expectancy of pinions. They may be used with cast iron gears, or with annealed, forged or cast steel gears where the ratio is about 6:1 or higher.

Alloy-steel pinions and gears, heat-treated, cut down space requirements by permitting smaller centers and face width.

Specify minimum hardness on steel pinions to be 40 Bhn higher than minimum hardness on the gear, for ratios from 2:1 to 8:1 when both are heat-treated. For ratios below 2:1 they are generally made to the same hardness.

For convenience of the heat treater, a range of 40 Bhn should be specified. For example, a pinion might be specified with a range from 265-305 Bhn and the mating gear 225-265 Bhn.

Where considerable impact loads exist, the use of alloys, plus lowering the hardness to a 50-56 Rc range, are recommended for carburized gears and pinions.

Where accelerated wear is encountered in service, heat

Table 6—Classification of materials

Designation Number	Treatment	Hardness Range	Designation Number	Treatment	Hardness Range
CARBON STEEL			N1-11	Quench and Temper	315 BHN Min.
U-1	Anneal	180 BHN	N1-13	"	350 BHN Min.
H-1	Normalize and Temper	210–250 BHN	N1-14	Flame-harden or Induction-harden	48 Rc Min.
H-2	Quench and Temper	225–265 BHN			
H-3	"	245–285 BHN	**STAINLESS**		
H-4	"	265–305 BHN	**(300 SERIES)**		
H-5	"	285–325 BHN	S-1	None Required	....
H-6	"	300–340 BHN			
F-1	Flame-harden	43 Rc Min.	**STAINLESS**		
F-2	Flame-harden	48 Rc Min.	**(400 SERIES)**		
F-3	"	52 Rc Min.	S-2	None Required	....
F-4	"	55 Rc Min.			
IN-1	Induction-harden	43 Rc Min.	**STAINLESS**		
IN-2	"	48 Rc Min.	**(410, 416, 440)**		
IN-3	"	52 Rc Min.	S-3	Quench and Temper Induction-harden or Bright-harden	as specified
IN-4	"	55 Rc Min.			
C-1	Cyanide	55 Rc Min.			
CN-1	Carbonitride	55 Rc Min.			
CH-1	Carburize	48 Rc Min.	**STAINLESS (440)**		
CH-2	"	50 Rc Min.	S-4	Harden and Temper Induction-harden or Bright-harden	as specified
CH-3	"	55 Rc Min.			
CH-4	"	58 Rc Min.			
CH-5	"	60 Rc Min.			
			TIN BRONZE CASTING		
ALLOY STEEL			B-1	As-cast	60 BHN Min. (500 Kg)
U-11	Anneal	180 BHN	**MANGANESE BRONZE**		
H-11	Normalize and Temper or Quench and Temper	225–265 BHN	B-2	As-cast	121 BHN Min.
			B-3	"	163 BHN Min.
H-12	Quench and Temper	245–285 BHN	B-4	"	187 BHN Min.
H-13	Quench and Temper	265–305 BHN			
H-14	"	285–325 BHN	**ALUMINUM BRONZE**		
H-15	"	300–340 BHN	B-5	As formed or Heat-treated	143 BHN Min.
H-16	"	335–375 BHN			
H-17	"	350–400 BHN	B-6	Heat-treated	179 BHN Min.
H-18	"	400–450 BHN (43–48 Rc)	**NICKEL ALUMINUM BRONZE**		
F-11	Flame-harden	43 Rc Min.	B-7	Heat treated	197 BHN Min.
F-12	"	48 Rc Min.			
F-13	"	52 Rc Min.	**ALUMINUM BAR (2017-T4)**		
F-14	"	55 Rc Min.	A-2	Heat-treated	105 BHN (500 Kg)
IN-11	Induction-harden	43 Rc Min.			
IN-12	"	48 Rc Min.	**ALUMINUM BAR (2024-T4)**		
IN-13	"	52 Rc Min.	A-4	Heat-treated	120 BHN (500 Kg)
IN-14	"	55 Rc Min.			
C-11	Cyanide	55 Rc Min.	**ALUMINUM SHEET (2017-T3)**		
CN-11	Carbonitride	55 Rc Min.	A-1	Heat-treated	....
N-11	Nitride	48 Rc Min.			
N-12	Nitride	64 Rc Min.	**ALUMINUM SHEET (2024-T3)**		
CH-11	Carburize	48 Rc Min.	A-3	Heat-treated	120 BHN (500 Kg)
CH-12	"	50 Rc Min.			
CH-13	"	55 Rc Min.	**ALUMINUM BAR OR SHEET (6061-T6)**		
CH-14	"	58 Rc Min.	A-5	Heat-treated	95 BHN (500 Kg)
CH-15	"	60 Rc Min.			
			HH BRASS		
NODULAR IRON			BR-1	As-rolled	....
N1-1	Anneal	180 BHN Min.			
N1-2	Anneal or Normalize and Temper	210 BHN Min.	**NONMETALLIC**		
			NM	....	....
N1-3	"	225 BHN Min.			
N1-4	"	245 BHN Min.			
N1-5	Quench and Temper	255 BHN Min.			
N1-6	"	265 BHN Min.			
N1-7	"	270 BHN Min.			
N1-8	"	280 BHN Min.			
N1-9	"	285 BHN Min.			
N1-10	"	300 BHN Min.			

treating to obtain higher hardness will, in most cases, help reduce wear. Gears are usually made from stainless steels, brass, bronze, aluminum and nonmetal materials for applications where the power transmitted is relatively light, such as cameras, electronic instruments, control and guidance systems, radar, and meters.

For small power-tool applications subject to severe service, gears are usually made from carbon steels and hardened to various ranges such as Ch-1 to Ch-4, or induction-hardened to IN-2 and IN-3, or alloy steels IN-11 and IN-12. In cases where more hardness is required, alloy steels are used, carburized to CH-11 and CH-14.

Recommended quality numbers

The AGMA has gathered the quality requirements of many industries and applications. These are grouped in Table 7 under fine-pitch and coarse-pitch applications. To keep cost down, use the lower quality number shown.

AGMA class numbers

When specifying a gear, call for the AGMA class number, obtained by combining the designations for quality, backlash and material. A typical class number for fine-pitch gear is 10C-A-4.

This specifies that the gear is to be made to quality 10, backlash designation **C**, and of aluminum bar 2024-T4, heat-treated to 120 Bhn.

For coarse-pitch gears, all gear-tooth tolerances need not be made to the same quality number. For example the class number:

8-CH-15 (except pitch-quality 10)

specifies quality 8 tolerances except for pitch, which is made to tigher (quality 10) tolerances. ■

Table 7—Quality numbers for gear applications

Fine-pitch applications	Quality Number	Fine-pitch applications	Quality Number
COMPUTING & ACCOUNTING MACHINES		**HOME APPLIANCES**	
Accounting-Billing	9–10	Blender	6–8
Adding Machine-Calculator	7–9	Mixer	7–9
Comptometer	6–8	Washing Machine	8–10
Computing	10–11	**PHOTOGRAPHIC EQUIPMENT**	
Data Processing	7–9	Commercial, 8–10; Aerial	10–12
Typewriter	8		
		SMALL POWER TOOLS	
INSTRUMENTS & CONTROLS		Bench Grinder	6–8
Accelerometer	10–12	Drills-Saws	7–9
Aircraft Instrument	12	Sander-Polisher	8–10
Computers	10–14		
Anti-Aircraft Detector	12	**MISCELLANEOUS**	
Automatic Pilot	9–11	Clocks	6
Radar-Sonar-Tuner	10–12	Counters	7–9
Recorder-Telemeter	10–12	Gages	8–10
Servo System Component	9–11	Motion-picture Equipment	8
Sound Detector	9	Pumps	5–7
Transmitter-Receiver	10–12	Vending Machines	6–7

Coarse-pitch Applications	Quality Number	Coarse-pitch Applications	Quality Number
AEROSPACE		**MARINE INDUSTRY**	
Control Gearing	7–11	Accessories	5–8
Engine Accessories	10–13	Small Propulsion	10–12
Small Engines	12–13		
		METAL WORKING	5–8
AGRICULTURE	5–7		
		OVERHEAD CRANES	5–6
AIR COMPRESSOR	10–11		
		PRECISION GEAR DRIVES	
CONSTRUCTION EQUIPMENT		Diesel Electric Gearing	8–9
Cranes, Open Gearing	3–6	High Speed Reels	8–9
Enclosed Gearing	6–8	Locomotive Timing Gears	9–10
Ditch Digger; Transmission	6–8	Pump Gears	8–9
		Turbine	9–10
ENGINES			
Combustion, Engine Accessories	10–12	**PRINTING INDUSTRY**	9–7
Timing Gearings	10–12		
Transmission	8–10	**PUMP INDUSTRY**	
		Liquid	10–12
FOUNDRY INDUSTRY	5–6	Rotary, vacuum	6–8
		Slush-duplex-triplex	6–8
MACHINE TOOL INDUSTRY			
Hand Motion	6–9	**RADAR AND MISSILE**	
Power Drives, 0–800 fpm	6–8	Antenna Elevating	8–10
800–2000 rpm	8–10	Data Gear	10–12
2000–4000 fpm	10–12	Ring Gear	9–12
Over 4000 fpm	12 & up	Rotating Drive	10–12
Accurate Indexing	12 & up		

EDITOR'S NOTE: Tables 1, 2, 4, 5 and 6 are published in entirety from corresponding tables in the Manual. Some changes have been made in styling. Also, the material designations in Table 6 have been reorganized to provide quicker reference to specific materials.

Tables 3 and 7 have been excerpted from more complete tables in the Manual. Copies of the Manual are obtainable from AGMA, No 1 Thomas Circle, N. W., Washington 5, DC, at $1.50 per copy.

Analytical geometry leads to accurate
Gear center coordinates

Converting center distances to rectangular coordinates
seems easy. Use trig, you'd say. But ever-so-slight
errors in finding angles cause excessive backlash

WILLIAM SCHWINT, Land Armament Dept, Sperry Gyroscope Co, Div of Sperry Rand Corp, Great Neck, NY

A common chore in gear-train design is to convert the center-to-center distances between gears to a rectangular-coordinate system. This is done to permit the shop to accurately machine the shaft centers by measuring from a common datum line.

Most designers faced with such a problem turn to trigonometry—and then run into a tedious process that still does not easily produce results of sufficient accuracy.

If minimum backlash is wanted, all decimals in the calculations must be carried to four places. The trig method then necessitates that all angles be computed in minutes and nearest seconds. Trig tables must be consulted and values interpolated. Any errors or deviations that creep in get carried through the entire sequence of calculations, and you are not sure of the results until the centers are machined and the gears mounted. At that point, any

Gear plate

Differential

Precision switch

Spring
return
mechanism

Instrument
counter

1. Range readout mechanism indicates range in meters and has the capability of adding or subtracting over two operating ranges of 0 to 5000, and 5000 to 10,000 meters. The mechanism contains three miniature differentials, two spring return mechanisms, two meter instrument counters, precision rotary switches, and numerous spur gears whose shaft centers are accurately located by means of the numerical method employed in this article.

deviations in the calculations show up as excessive back-lash.

The alternate method proposed here is the analytical geometry approach. You set up equations for circles with centers at the known shaft centers—usually, those of the input and output shaft. The equations are in the form of

$$x^2 + y^2 = r^2$$

and

$$(x - h)^2 + (y - k)^2 = r^2$$

By solving them as simultaneous equations you obtain directly the x and y coordinates of the gear center desired. No angles are involved as the unknowns are obtained with a mathematical approach that is direct.

The analytical geometry method is also easier to check for accuracy because the entire solution can be accomplished with the aid of a desk calculating machine. An example of both methods is given below so that they can be compared.

Trigonometric method

First one angle is found, then that angle is used to find a particular side of a triangle. This dimension in turn is used to find another angle, and again another side. If any error is made, the error is multiplied in the next sequence.

A typical application involves two gears and an idler to be mounted on a plate, Fig 3. The centers at A and C are the input and output shaft centers determined by other considerations. The location of the idler gear at B is based on the gear designer's computations of the desired speed-reduction ratio between A and C. Thus the following information is known:

• Rectangular coordinates of center C with respect to center A. This is given as $x = 1.81250$ and $y = -0.53125$ (all dimensions in inches).

2.

Precision-drilled plate
requires that distances between
centers be computed to
ten-thousandths of an inch.

3.

Trigonometric method
necessitates that all angles be
computed in seconds and is
not easily checked for errors.

• Gear center distances: $AB = 0.796875$ and $BC = 1.75000$.

Step 1. Construct a horizontal and vertical datum line through gear center A which now becomes a point of origin and is labeled $(0,0)$.

Step 2. Draw line AC and CD to form the right triangle ACD, and solve for angle θ and then for side AC.

$$\tan \theta = \frac{0.53125}{1.8125} = 0.293103$$

$$\theta = 16° 20' 10''$$

$$AC = \frac{0.53125}{0.281271} = 1.8887$$

Step 3.

$$\theta + \epsilon + 90° = 180°$$
$$\epsilon = 89° 59' 60'' - 16° 20' 10''$$
$$\epsilon = 73° 39' 50''$$

Step 4. Solve for angle α of triangle ABC by the law of cosines

$$a^2 = b^2 + c^2 - 2bc \cos \alpha$$

$$\cos \alpha = \frac{(1.88875)^2 + (0.796875)^2 - (1.75)^2}{2(1.88875)(0.796875)}$$

$$= 0.378663$$
$$\alpha = 67° 44' 55''$$

Step 5. Draw line BE to form the right triangle ABE and solve for angle λ

$$\lambda + \alpha + \epsilon = 180°$$
$$\lambda = 179° 59' 60'' - 67° 44' 55'' - 73° 39' 50''$$
$$\lambda = 38° 35' 15''$$

Step 6. Solve triangle ABE for BE and AE which are the desired x and y coordinates.

$$\phi + \lambda + 90° = 180°$$
$$\phi = 51° 24' 45''$$

Step 7. Solve for the y coordinate

$$y = (0.796875) \sin 51° 24' 45''$$
$$y = 0.62277 \text{ in.}$$

Step 8. Solve for the x coordinate

$$x = (0.796875) \cos 51° 24' 45''$$
$$x = 0.49729$$

Analytical geometry method

Here the same example is worked out by analytical geometry:

Step 1. Construct two arcs, at gear centers A and C, Fig 4, with the given radii of $AB = 0.796875$ and $BC = 1.75000$.

Step 2. Set up the equations of the two circles and solve for the unknowns. Two solutions result, but one solution is in the fourth quadrant, so is discarded:

At the origin $(0,0)$

$$x^2 + y^2 = (0.79687)^2$$
$$x^2 + y^2 = 0.63501 \tag{1}$$

At distance h and k from the origin

$$(x - 1.8125)^2 + [y - (-0.53125)]^2 = (1.750)^2$$
$$(x^2 - 3.625x + 3.2852) +$$
$$(y^2 + 1.0625y + 0.2822) = 3.0625 \tag{2}$$

Step 3. Solve for y in Eq 1

$$y = (0.63501 - x^2)^{1/2}$$

Step 4. Substitute y into Eq 2 and solve for the x coordinates.

$$12.6401x^2 - 7.32061x + 0.51782 = 0$$
$$x_1 = 0.49668 \quad \text{or} \quad x_2 = 0.08210$$

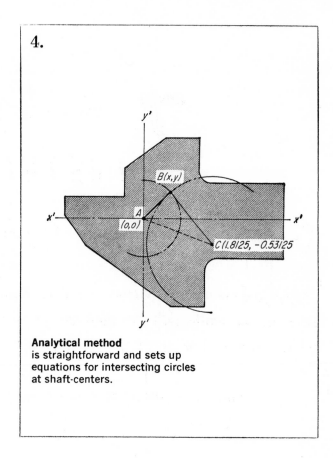

4.

Analytical method
is straightforward and sets up
equations for intersecting circles
at shaft-centers.

From Fig 4, we can see that the greater of the two values ($x_1 = 0.49668$) is the one we are looking for.

Step 5. Going back to Step 3 we solve for y:

$$y = [0.63501 - (0.49668)^2]^{1/2}$$
$$y = 0.62313$$

Both these values can be quickly checked with the aid of a computer. Additional x and y coordinates can be obtained by simply continuing with the same solution as shown above. The gear center most recently found becomes the new origin and another set of horizontal and vertical datum lines are constructed. After all coordinate dimensions have been calculated for all gear centers, one set of datum lines are established, Fig 5, and then all x and y dimensions are measured from this reference.

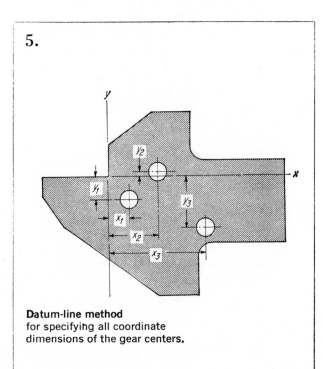

5.

Datum-line method
for specifying all coordinate
dimensions of the gear centers.

7 *Rules simplify* INSTRUMENT GEAR SPECIFICATIONS

Specifying too many requirements is time-consuming and expensive. This article evaluates such considerations as backlash and tolerances, and boils specifications down to a minimum.

WILLIAM A. WIEGAND, *President, Dynamic Gear Co., Inc.*

Gears for light loads and moderate duty—often called instrument gears—cover a wide pitch range and are made in many precision grades. Selection can be simplified by following a few simple rules and keeping the specifications to a minimum.

1 Select the finest pitch that will carry the load. This permits reducing over-all size of the gear train. Stocked gears are available to 120 pitch, and can be made to order to 200 pitch.

2 For minimum backlash, pick gears with a $14\frac{1}{2}°$ pressure angle. The backlash here is considerably less than with 20° pressure-angle gears of equal precision (see chart). The 20° gears carry more load and have a longer life, load on most instruments is light and a long cyclic life is not required.

A 0.0015-in. tolerance on the pitch diameter, for example, will result in greater backlash (0.0005 in.) with 20° gears than with $14\frac{1}{2}°$ gears (0.00038). Viewed another way, a backlash range from 0 to 0.0013 in. for $14\frac{1}{2}°$ gears would require a max tolerance range of 0.0005 in. (for example, pitch dia might be 1.000 in. plus 0 minus 0.0005). But for 20° gears, the tolerance range is cut to 0.00038 in. This may increase cost of the gear by 40%.

3 Call for Precision I or Commercial IV class gears whenever possible, avoiding Precision II unless you are sure the tight tolerances are required. Over-designing gear trains is a common fault. Sometimes it is deliber-

ately done to insure acceptability of prototypes—a later relaxation of tolerance is intended but, too often, never comes. Short target dates can also be responsible for needlessly tight tolerances; the designer assumes there will be insufficient time after receipt of gears for special fitting into the assembly. This makes both prototype and production models costlier than necessary. A Precision II gear can cost nearly 150% as much as its Commercial IV counterpart. For a typical gear in production quantities, here is how cost compares for the three classes of gears:

Commercial IV	$3
Precision I	3.54
Precision II	4.45

4 Specify bore tolerance, face run-out, and pitch-dia tolerance in line with AGMA requirements for the class. Gear drawings often call for over-loose tolerances on these dimensions, especially when the specifications include a drawing of the gear blank. A gear cannot be made more accurate than the gear blank on which it is cut.

Typical is the error of calling for a 0.001-in. tolerance on the bore for Precision I or II gears.

5 Obtain the max tolerance on OD by using the AGMA formula, 0.2/pitch. This gives max allowable variation below nominal OD. Applying the formula to a 48-pitch gear, for example, indicates acceptability of an OD tolerance that is 0.004 in. under nominal.

Of value also is the common practice of applying

AGMA Gear Tolerances

Type of Tolerance	Gear Class		
	Commercial IV	Precision I	Precision II
Tooth-to-tooth error........	0.0007	0.0004	0.0003
Tooth composite error......	0.0015	0.0010	0.0005
Plus tolerance on bore dia...	0.0008	0.0005	0.0003
Lateral (face) runout per in. of dia.................	0.0020	0.0015	0.0010
Minus tolerance on PD......	0.0020	0.0010	0.0005
Concentricity of OD to bore.	0.0015	0.0010	0.0005

Note: See p 177 for new class designations.

double the pitch dia tolerance to the OD. On a Precision I gear with −0.001-in. tolerance on the PD, a −0.002 in.-tolerance would be allowed on the OD; on Precision II with −0.0005 PD, the OD tolerance would be −0.001 in. For normal applications, however, such tolerances may be tighter than necessary, and it is better to apply a tolerance of −0.002 in., or the formula −0.2/P, whichever is the lesser figure. Where a gear is not top-hobbed, these tolerances should certainly be adequate.

6 Specify top-hobbing whenever concentricity of the OD to the PD is important. In this operation, PD and OD are generated simultaneously and the whole depth of tooth is governed by the hob or cutter. It is possible to hob a gear to PD size and find the OD slightly oversize because of slight errors in manufacturing or sharpening the hob.

If this occurs split the tolerance to permit a plus tolerance—that is, specify +0.001 in. or +0.1/P, whichever is the lesser (rather than specifying the −0.002 in. or −0.2/P).

Two methods for specifying pitch dia of instrument gears, and its tolerances, are the center-to-center distance specifications, and measurements over pins. In the center-distance specifications, tolerance is not applied directly to the PD, but to the variations in center distance as the gear is rolled in mesh with a hardened and ground master gear.

Because AGMA specifications tie down the permissible center-distance variations and backlash for the classes of gears, pitch dia should be specified on the drawing as being for reference only. For example, by specifying a Precision I, Class C, 48-pitch gear, the total permissible variation in center distance is 0.001 in., and the backlash is 0.001 to 0.002 in. These figures can be obtained from the AGMA Standard 236.04 for fine-pitch gears, or from "Methods for Specifying Precision of Spur Gears," Nov. '56, p. 135.

7 Keep drawing specifications of the gear teeth to a minimum. Unless gears are to be critically inspected, only the information in the required-data column of the simplified form (right) need be specified.

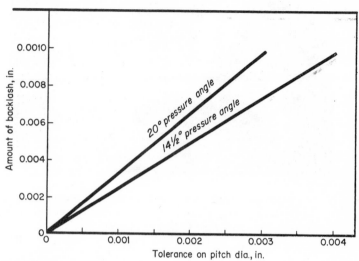

Backlash values . . .

for instrument gears with 14½° pressure angle are considerably less than those for 20° gears.

TOO MANY SPECIFICATIONS?

Only the gear-tooth information in the required-data column below need be specified. The optional data, usually included for inspection purposes only, are derived: (6) and (7) from standard gear formulas; (8) and (9) from reference tables; (10) through (16) from the AGMA specifications in (5).

Required Data

1. Number of teeth — *48*
2. Diametral pitch — *48*
3. Pressure angle — *14½°*
4. Pitch dia (ref.) — *1000*
5. AGMA spec — *Precision I, Class C*

Optional Data

6. Addendum — *0.0208*
7. Whole depth of tooth — *0.0478*
8. Measurement over wires — *1.0208*
9. Wire dia — *0.0360*
10. Center-to-center distance — *1.125 ± 0.0005 (with 1.5 in. master gear)*
11. Testing pressure — *22 oz.*
12. Tooth-to-tooth error — *0.0005*
13. Total composite error — *0.0010*
14. Plus tolerance on bore dia — *0.0005*
15. Lateral runout per in. of dia — *0.0015*
16. Concentricity of OD to bore — *0.0010*

a workable approach to
GEAR TOLERANCES

WILLIAM A WIEGAND, president
Dynamic Gear Co. Inc, Amityville, NY

Most problems —especially in the instrument field— stem from difficulty in understanding the center-distance specifications in the AGMA standards. What the gear designer really wants is simply this: to be able to specify the class of precision—say, "Precision II" gears—and then consult a table for (1) the exact distance he should locate the shaft centers on his product; (2) the tolerance on this center-to-center distance; and (3) the amount of backlash he can anticipate—because backlash, and sometimes lack of backlash ("binding" in a gear train) is one of his biggest headaches.

For example, suppose you are designing a gear train with two 1-in., 14½° pressure-angle, intermeshed gears. The nominal center-to-center distance is 1 in. If precision is a requirement and gears in the Precision I class are specified, you should locate the centers 1.0004 in. apart with a tolerance of plus 0.0004 in., minus 0.0000. This will result in a backlash play ranging from 0.00021 in. to 0.00145 in. which means an angular play or "lost motion" in one gear (holding the second gear stationary) of 0.024 to 0.165°. But nowhere in either the AGMA's present or Martin's proposed standard can this information be found. Consequently, many designers "play it safe" by purchasing the most precise gears that can be made in production—only to run into trouble anyway because of incorrect center-distance dimensions or tolerances.

THESE TABLES FOR WORKING DATA

This is why I have set up two tables (preceding page) which permit an engineer to specify a class of gear and immediately know what the center distance should be, along with resulting backlash. The AGMA specifications of total composite error (TCE) and tooth-to-tooth composite error (TTCE) are also given.

One table is for cases where a measurement over wires is specified; the other, where a center-distance check is specified. These tables could be as much a part of gearing standards as the bore tolerance is a function of a particular class of gear.

Gears specified to a measurement over wires (sometimes called pins) are more economical to produce and inspect than gears specified to center-distance measurements. This is because inspection with over-wires requires only a micrometer, a set of wires, and bench

centers to hold the gears.

Equipment for a center-distance check, on the other hand, is very costly, requires specially trained personnel to interpret the results, and can be found in relatively few locations other than the plants of gear-manufacturers and large users. But a tighter center distance can be safely specified (this is seen by comparing the two tables).

Our company is supplying vast quantities of Precision Class I and Precision Class II gears. To fabricate Precision Class III gears in production quantities and offer them at production prices, meticulous care must be taken not only in actual tooth cutting but also in manufacture of the blank. Presently we are producing these gears on the new Barber Coleman 2½-4 Hobbing Machine which has work spindle to hob spindle accuracy within 20 seconds. This in turn holds indexing error on a 2½-in.-dia gear to within 0.00014 in.

However, inspection of these parts remains a slow, tedious, cautious process because we are examining production gears that have about the same accuracy as the master with which they are being checked. Bearing this in mind, I cannot appreciate the need for Mr. Martin's "Ultraprecision" Classes which go as fine as 0.00001-in. TCE. Because I am a major supplier of fine-pitch gears (48 through 120 DP) people would expect my organization to supply some of these ultraprecision fine-pitch gears. But right now I would not know how to even handle an Ultraprecision Class II gear on a semi-mass production basis, let alone Class VII. Anything approaching such accuracies must be done as a laboratory job and I do not believe that we need such standards for the type of work carried on in modern labs. There would even be a problem in inspection. Precision Class III Master Gears are readily available and we have been advised by the suppliers that 0.0001-in. accuracy can be provided; but this accuracy would not be good enough to check Ultraprecision Classes IV to VII.

Probably one day these standards and specifications will be required, but they certainly appear premature now. For example, one might propose a neat, compact set of traffic laws to govern rocket travel, and few will deny the possibility of such laws being needed some day. But who would want to be burdened with them at present?

GEAR AND BACKLASH SPECIFICATION—A PRACTICAL APPROACH

(Values shown in bold type are only for two intermeshed gears of 1-in. pitch dia)

FOR MEASUREMENT OVER WIRES

Precision Class	TCE	TTCE	Measurement Over .036-in. Dia Wires, and Tolerance	Distance Between Shaft Centerlines			20° Pressure Angle		14½° Pressure Angle	
				Nominal	Additive	Final	Minimum Backlash	Maximum Backlash	Minimum Backlash	Maximum Backlash
I	.001	.0004	+.000 **1.0506** −.001	**1.000**	.0014	+.0004 **1.0014** −.0000	.00029	.00276	.00021	.00196
II	.0005	.0003	+.000 **1.0506** −.0005	**1.000**	.0008	+.0003 **1.0008** −.0000	.00022	.00153	.00016	.00109
III	.00025	.0002	+.000 **1.0506** −.0003	**1.000**	.0005	+.0002 **1.0005** −.0000	.00015	.00095	.00010	.00068

WHEN MESHED WITH A 1.500-DIA MASTER GEAR

Precision Class	TCE	TTCE	Center Distance Meshed with Master and Tolerance	Distance Between Shaft Centerlines			20° Pressure Angle		14½° Pressure Angle	
				Nominal	Additive	Final	Minimum Backlash	Maximum Backlash	Minimum Backlash	Maximum Backlash
I	.001	.0004	+.000 **1.250** −.001	**1.000**	.0004	+.0004 **1.0004** −.0000	.00029	.00204	.00021	.00145
II	.0005	.0003	+.000 **1.250** −.0005	**1.000**	.0003	+.0003 **1.0003** −.0000	.00022	.00116	.00016	.00083
III	.00025	.0002	+.000 **1.250** −.0003	**1.000**	.0002	+.0002 **1.0002** −.0000	.00015	.00073	.00010	.00052

Note: See p 177 for new class designations.

HOW TO USE ABOVE TABLES

The two tables given here are suggested as a quick guide—not as a standard—for gear trains where minimum backlash is desired. All values, with the exception of those in bold type, hold true for any size or pitch gear.

The values which vary are: (1) Measurement over wires. This varies with the size (pitch dia) of the gear and pitch. See *The New American Machinists Handbook*, pages 4-53 to 4-57, McGraw-Hill Book Co, NYC, for other values. The tolerance, however, remains constant for that particular class of gear. (2) Nominal center distance. This, of course, varies with size. (3) Modified center distance. This also varies with size; however, its tolerance remains constant for that class.

You can choose either table, depending on whether an over-wire check or center-distance check is specified. Select also gear class and pressure angle. The shaft-centerline column marked "Final" then directly gives dimension and tolerance between shaft centerline that you should specify on the drawing of the product or instrument. Backlash columns give range of backlash (max and min values) that can occur—not including, of course, play in bearings, etc.

Note that backlash values for 14½° gears are smaller than for 20° gears.

These values can be converted to angular play, θ, in the gears by using the equation $\theta = 360B/\pi D$, where $B = $ backlash, in., $D = $ pitch dia, in.

OBTAINING THE VALUES

Here is the way the values in above table were calculated. The example is for Precision Class I.

Measurement over wires—Min Backlash

Assume pitch dia = 1.000 in., which is largest permissible value. Then measurement over wires is 1.0506 (obtained from machinist's tables or calculated).

Each pitch dia can have a runout of 0.0005 over nominal due to TCE which is 0.001 for this class. Therefore, for two gears meshed at their highest point, the center distance is:

$$C.D. = \frac{1.0000}{2} + 0.0005 + \frac{1.0000}{2} + 0.0005 = 1.0010$$

The gears should be mounted at a distance equal to nominal C.D. + TCE + TTCE. A plus tolerance is assigned to the C.D.; therefore, for tightest condition, distance between shaft centerlines = 1.0014. Clearance at point of tightest mesh is equal to 0.0004, and min backlash is: B = 2 (ΔC.D.) (tan θ) = 0.00029 min.

Measurement over wires—Max Backlash

Assume each gear has min pitch dia of 0.999. Then measurement over wires is 1.0496.

$$C.D. = \frac{0.999}{2} - 0.0005 + \frac{0.999}{2} - 0.0005 = 0.998$$

The gears should be mounted at a C.D. as obtained before except tolerance is added. Therefore C.D. for mounting = 1.0018, ΔC.D. = 1.0018 − 0.998 = 0.0038 and B = 0.00276 max.

Center-distance check

If a 1.000 pitch dia gear is meshed with a master gear of 1.500 pitch dia, the nominal C.D. = 1.250. Because the tolerance on this mesh is −0.001, the C.D. with full TCE will vary from 1.2500 to 1.2490. With two gears, C.D. will vary from 1.000 to 0.998.

The gears should be mounted at a C.D. equal to nominal plus TTCE, with a plus-tolerance equal to TTCE. Thus, if the shaft distance is at a minimum (1.0004) and gears are meshed at highest points (C.D. = 1.0000), the ΔC.D. = 0.0004 and B = 0.00029 min.

For maximum backlash, the ΔC.D. = 1.0008 − 0.998 = 0.0028, which gives B = 0.00204 max.

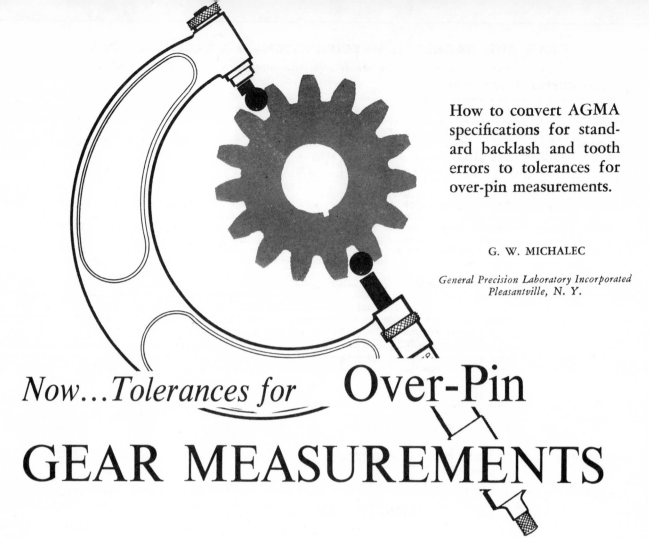

How to convert AGMA specifications for standard backlash and tooth errors to tolerances for over-pin measurements.

G. W. MICHALEC

General Precision Laboratory Incorporated
Pleasantville, N. Y.

Now...Tolerances for Over-Pin

GEAR MEASUREMENTS

Assuming proper tooth contour and spacing, the tooth thickness and gear concentricity are the important design specifications requiring tolerances to assure desired performance. Tooth width and concentricity combined can be checked with the variable-center-distance method described in an earlier article, "Methods for Specifying Precision of Spur Gears," November 1956 page 135. This article also discusses the limitations of the caliper method for direct measurement of tooth thickness and the dial indicator method for concentricity.

Over-pin measurement is another method for quickly measuring the tooth thickness. Two accurately ground pins of proper diameter are placed at opposite positions and the distance across the pins measured with a micrometer. For inspection of gears with odd number of teeth, the pins are placed in tooth spaces as nearly opposite as possible. The measurement is then compared with nominal values given in various tables, such as, "The New American Machinist Handbook," page 4-53, *McGraw-Hill Book Company*, New York; or the American Gear Manufacturers Association (AGMA), Standard 236.04. These tables, however, do not include tolerances. (See p 196 for overspin measurements.)

Over-pin measurements have these advantages: (1) Gears purchased in small quantities can be quickly inspected without need for special equipment. (2) Over-pin dimensions are easily specified on a gear drawing and just as easily interpreted by the inspection depart-

ment. (3) Tooth thickness of a gear can be checked during the machining process while the gear is still on the arbor.

Main limitations are that the method does not take into consideration the concentricity of the gear bore with the pitch circle nor the accuracy of the tooth profile.

Because of these limitations, AGMA specifications emphasize the use of the variable-center-distance method in which the test gear is rotated tightly in mesh with a master gear or rack of known accuracy. The resulting variations in the center distance are then recorded and analyzed. For this method, AGMA has established permissible tolerances for various classes of gears.

Although over-pin measurements are widely used, there are no standardized design tolerances for this method, thus limiting the use of this method in specifying gears. Many gear manufacturers have established their own tolerances, but this does not help the gear designer. Quite often the acceptance or rejection of a purchased gear is left to the discretion of the inspection department because there are no over-pin standards.

Solution is to set up a tolerance system for over-pin measurement that is equivalent to the AGMA tolerances in Tables I and II. But correlation between the two systems is not a simple matter. Tooth thickness directly affects backlash. Backlash, in turn, prevents jamming, over-loading and over-heating of gears. On the other hand, excessive backlash is objectionable if

TABLE I—PERMISSIBLE BACKLASH PER MESHED PAIR

From AGMA Standard 236.04—20° Pressure Angle

	Diametral Pitch	Backlash, In.
Class A	20 to 45	0.004 to 0.006
	46 to 70	0.003 to 0.005
	71 to 90	0.002 to 0.0035
Class B	20 to 60	0.002 to 0.004
	61 to 120	0.0015 to 0.003
	121 and finer	0.001 to 0.002
Class C	20 to 60	0.001 to 0.002
	61 to 120	0.0007 to 0.0015
	121 and finer	0.0005 to 0.001
Class D	No measurable backlash at any pitch	

Backlash values are understood to be backlash between two assembled gears at their tightest point of mesh. Backlash will be increased when the low points of runout are in contact.

Note: See p 177 for new class designations.

TABLE II—AGMA STANDARD FOR TOTAL AND TOOTH-TO-TOOTH COMPOSITE ERRORS

From AGMA Standard 236.04 for fine pitch gears

Class	Total Composite Error, in.	Tooth-to-Tooth Composite Error, in.
Commercial 1	0.006	0.002
2	0.004	0.0015
3	0.002	0.001
4	0.0015	0.0007
Precision 1	0.001	0.0004
2	0.0005	0.0003
3	0.00025	0.0002

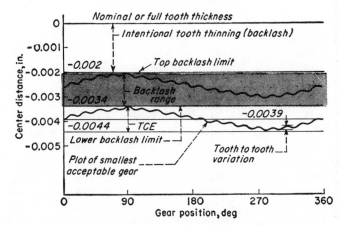

Fig. 1—Variations in center distance between a test gear and master gear. The AGMA standards specify the permissible backlash (in shade) and total composite error (TCE). Over-pin measurements cannot detect TCE; are difficult to correlate with AGMA standards.

the drive frequently reverses, if there is an over-running load or if high positional accuracy in the gear train is required.

The backlash information of Table I can be accurately converted into equivalent over-pin measurements by means of standard formulas. Unfortunately such a conversion is not correct because Table I states that the values are for the tightest point of mesh. Once the gears are mounted and rotated, there may be additional backlash because of eccentricity between the bore and pitch circle.

To account for eccentricity, AGMA has established classes of gears based on the Total Composite Error (TCE) as shown in Table II. A complete gear specification includes a number for the TCE quality and a letter for the backlash class, for example: Precision 1-B; Commercial 3-A. To insure accuracy, the effect of TCE must be analyzed and included in the correlation.

Comparison with AGMA Specifications

Typical variations in center distance, or "plot," of a gear with tooth-to-tooth errors and bore eccentricity are shown in Fig. 1. Assuming that a 48-pitch gear with a 20° pressure angle per AGMA Class Precision 1-A is desired, the permissible tolerances from Tables I and II are: TCE = 0.001; backlash per meshed pair = 0.003 to 0.005. Therefore, backlash per gear = 0.0015 to 0.0025.

Equivalent change in center distance for the upper limit of the backlash tolerance is

$$\Delta C.D = \frac{B}{2 \tan \phi} = \frac{0.0015}{2(.0364)} = 0.002$$

This establishes the top limit of the largest acceptable gear because all backlash values in Table I are for gears at their tightest point of mesh. Using the same formula, the lower limit is 0.0034. Fig. 1 shows the plots of two gears having a TCE equal to 0.001, with one gear at the upper level of the backlash level and the other at the lower level. In other words, the TCE specifications restrict the amplitude of the plot, and the backlash values establish the position of the plot with respect to the nominal datum line.

One problem in correlating this system to over-pin measurements is that the gear at the lower level can have a maximum variation in center distance of 0.0044 in. and still be acceptable. But because the over-pin measurement cannot detect TCE, it would be incorrect to establish tolerances based on the maximum variation. Therefore, the average value for the smallest acceptable gear, −0.0039 in this case, is used as the lowest limit for converting to over-pin measurements, and the total acceptable range is from −0.002 to −0.0039. This range is illustrated in Fig. 2 and the limits designated as the "theoretical" maximum and "theoretical" minimum.

There are still some gears, however, that may pass inspection by the over-pin method within these limits and still be rejected when inspected by AGMA standards. For example, referring again to Fig. 2, a gear at the top limit can also have some TCE which cannot be detected by over-pin gaging and part of its plot would fall above the −0.002 line. (*next page, please*)

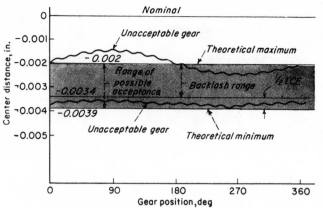

Fig. 2—Theoretical tolerance range (in shade) for over-pin measurements is equal to AGMA backlash range plus one-half TCE. However, certain gears unacceptable to AGMA standards may get through inspection.

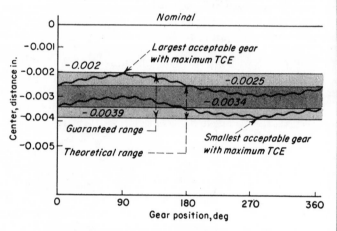

Fig. 3—Narrower range of tolerances (in white) than theoretical range can almost "guarantee" that a test gear is within AGMA specifications.

On the lower end of the size range, gears which fall just within the lower limit but have low TCE, would fail to pass the AGMA standards because the high points of their plots would not reach the minimum value of −0.0034 as specified by the backlash designation.

To avoid this situation, the limits should be reduced to −0.0025 and −0.0034; in other words, from the largest permissible value of the high point minus ½ maximum TCE, to the minimum value for the high point. This narrower range, shown graphically in Fig. 3, may then be considered as the "guaranteed" range for over-pin measurements. However, even this narrower range is valid only if the TCE does not exceed its limits, and since this cannot be a certainty, the literal meaning of "guarantee" means "highly probable."

All gears fulfilling the class requirements per guaranteed over-pin measurements will also fall within class limits on a center-distance check provided the TCE is within the allowable maximum. The converse is not true; all gears passing the center distance test will not fall within the guaranteed over-pin limits. But the exceptions are few and the guaranteed tolerances offer best standards.

text continued on page 195

TABLE III—VALUES OF

20-PITCH, T = 0.07854

Precision Class	T_a	T_b	T_c	T_d
1-A	0.07654	0.076176	0.07554	0.075176
1-B	0.07754	0.077176	0.07654	0.076176
1-C	0.07804	0.077676	0.07754	0.077176
1-D	0.07854	....	....	0.078176
2-A	0.07654	0.076358	0.07554	0.075358
2-B	0.07754	0.077358	0.07654	0.076358
2-C	0.07804	0.077858	0.07754	0.077358
2-D	0.07854	....	....	0.078358
3-A	0.07654	0.076449	0.07554	0.075449
3-B	0.07754	0.077449	0.07654	0.076449
3-C	0.07804	0.077949	0.07754	0.077449
3-D	0.07854	....	....	0.078449

24-PITCH, T = 0.06545

Precision Class	T_a	T_b	T_c	T_d
1-A	0.06345	0.063086	0.06245	0.062086
1-B	0.06445	0.064086	0.06345	0.063086
1-C	0.06495	0.064586	0.06445	0.064086
1-D	0.06545	....	....	0.065086
2-A	0.06345	0.063268	0.06245	0.062268
2-B	0.06445	0.064268	0.06345	0.063268
2-C	0.06495	0.064768	0.06445	0.064268
2-D	0.06545	....	....	0.065268
3-A	0.06345	0.063359	0.06245	0.062359
3-B	0.06445	0.064359	0.06345	0.063359
3-C	0.06495	0.064859	0.06445	0.064359
3-D	0.06545	....	....	0.065359

32-PITCH, T = 0.04909

Precision Class	T_a	T_b	T_c	T_d
1-A	0.04709	0.046726	0.04609	0.045726
1-B	0.04809	0.047726	0.04709	0.046726
1-C	0.04859	0.048226	0.04809	0.047726
1-D	0.04909	....	....	0.048726
2-A	0.04709	0.046908	0.04609	0.045908
2-B	0.04809	0.047908	0.04709	0.046908
2-C	0.04859	0.048408	0.04809	0.047908
2-D	0.04909	....	....	0.048908
3-A	0.04709	0.046999	0.04609	0.045999
3-B	0.04809	0.047999	0.04709	0.046999
3-C	0.04859	0.048499	0.04809	0.047999
3-D	0.04909	....	....	0.048999

TOOTH THICKNESS FOR COMPUTING OVER-PIN TOLERANCES

48-PITCH, T = 0.03272

Precision Class	T_a	T_b	T_c	T_d
1-A..........	0.03122	0.030856	0.03022	0.029856
1-B..........	0.03172	0.031356	0.03072	0.030356
1-C..........	0.03222	0.031856	0.03172	0.031356
1-D..........	0.03272	....	....	0.032356
2-A..........	0.03122	0.031038	0.03022	0.030038
2-B..........	0.03172	0.031538	0.03072	0.030538
2-C..........	0.03222	0.032038	0.03172	0.031538
2-D..........	0.03272	....	....	0.032538
3-A..........	0.03122	0.031129	0.03022	0.030129
3-B..... •	0.03172	0.031629	0.03072	0.030629
3-C..........	0.03222	0.032129	0.03172	0.031629
3-D..........	0.03272	....	....	0.032629

80-PITCH, T = 0.01964

Precision Class	T_a	T_b	T_c	T_d
1-A..........	0.01864	0.018276	0.01789	0.017526
1-B..........	0.01889	0.018526	0.01814	0.017776
1-C..........	0.01929	0.018926	0.01889	0.018526
1-D..........	0.01964	....	....	0.019276
2-A..........	0.01864	0.018458	0.01789	0.017708
2-B..........	0.01889	0.018708	0.01814	0.017958
2-C..........	0.01929	0.019108	0.01889	0.018708
2-D..........	0.01964	....	....	0.019458
3-A..........	0.01864	0.018549	0.01789	0.017799
3-B..........	0.01889	0.018799	0.01814	0.018049
3-C..........	0.01929	0.019199	0.01889	0.018799
3-D..........	0.01964	....	....	0.019549

64-PITCH, T = 0.02454

Precision Class	T_a	T_b	T_c	T_d
1-A..........	0.02304	0.022676	0.02204	0.021676
1-B..........	0.02379	0.023426	0.02304	0.022676
1-C..........	0.02419	0.023826	0.02379	0.023426
1-D..........	0.02454	....	....	0.024176
2-A..........	0.02304	0.022858	0.02204	0.021858
2-B..........	0.02379	0.023608	0.02304	0.022858
2-C..........	0.02419	0.024008	0.02379	0.023608
2-D..........	0.02454	....	....	0.024358
3-A..........	0.02304	0.022949	0.02204	0.021949
3-B..........	0.02379	0.023699	0.02304	0.022949
3-C..........	0.02419	0.024099	0.02379	0.023699
3-D..........	0.02454	....	....	0.024449

96-PITCH, T = 0.01636

Precision Class	T_a	T_b	T_c	T_d
1-A..........	....	....	....	....
1-B..........	0.01561	0.015246	0.01486	0.014496
1-C..........	0.01601	0.015646	0.01561	0.015246
1-D..........	0.01636	....	....	0.015996
2-A..........	....	....	....	....
2-B..........	0.01561	0.015428	0.01486	0.014678
2-C..........	0.01601	0.015828	0.01561	0.015428
2-D..........	0.01636	....	....	0.016178
3-A..........	....	....	....	....
3-B..........	0.01561	0.015519	0.01486	0.014769
3-C..........	0.01601	0.015919	0.01561	0.015519
3-D..........	0.01636	....	....	0.016269

72-PITCH, T = 0.02182

Precision Class	T_a	T_b	T_c	T_d
1-A..........	0.02082	0.020456	0.02007	0.019706
1-B..........	0.02107	0.020706	0.02032	0.019956
1-C..........	0.02147	0.021106	0.02107	0.020706
1-D..........	0.02182	....	....	0.021456
2-A..........	0.02082	0.020638	0.02007	0.019888
2-B..........	0.02107	0.020888	0.02032	0.020138
2-C..........	0.02147	0.021288	0.02107	0.020888
2-D..........	0.02182	....	....	0.021638
3-A..........	0.02082	0.020729	0.02007	0.019979
3-B..........	0.02107	0.020979	0.02032	0.020229
3-C..........	0.02147	0.021379	0.02107	0.020979
3-D..........	0.02182	....	....	0.021729

120-PITCH, T = 0.01309

Precision Class	T_a	T_b	T_c	T_d
1-A..........	....	....	....	....
1-B..........	0.01234	0.011976	0.01159	0.011226
1-C..........	0.01274	0.012376	0.01234	0.011976
1-D..........	0.01309	....	....	0.012726
2-A..........	....	....	....	....
2-B..........	0.01234	0.012158	0.01159	0.911408
2-C..........	0.01274	0.012558	0.01234	0.012158
2-D..........	0.01309	....	....	0.012908
3-A..........	....	....	....	....
3-B..........	0.01234	0.012249	0.01159	0.011499
3-C..........	0.01274	0.012649	0.01234	0.012249
3-D..........	0.01309	....	....	0.012999

SYMBOLS

a — Radius of pins inserted between teeth. Standard pins are based on the equation $a = \frac{1}{2}(1.728/P)$

B — Backlash per pair of gears as specified by AGMA letter class and listed in Table I.

C — Maximum total composite error (TCE) as specified by AGMA quality class and listed in Table II.

M — Over-pin measurements

N — Number of teeth

P — Diametral pitch

R — Standard pitch radius, $R = N/P$

R_b — Base circle radius, $R_b = R \cos 20°$

R_c — Radial distance to center of pin

T — Full tooth thickness at standard radius R, $T = 1.5708/P$

T_x — Desired tooth thickness at standard radius R.

ϕ — Pressure angle $= 20°$

ϕ_1 — Pressure angle at center of pin

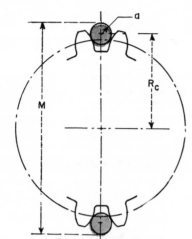

For even number of teeth

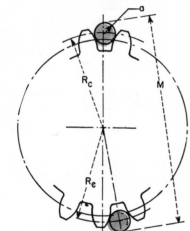

For odd number of teeth

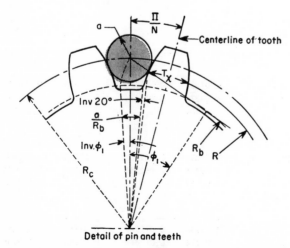

Detail of pin and teeth

Fig. 4—Geometry for computing over-pin measurements for gears with even and odd number of teeth.

No. of Teeth	Theoretical Maximum M_a	Guaranteed Maximum M_b	Guaranteed Minimum M_c	Theoretical Minimum M_d
5	0.2158	0.2151	0.2140	0.2133
6	0.2557	0.2550	0.2538	0.2530
7	0.2814	0.2807	0.2794	0.2787
8	0.3188	0.3180	0.3167	0.3159
9	0.3457	0.3450	0.3436	0.3429
10	0.3816	0.3809	0.3795	0.3787
11	0.4094	0.4086	0.4072	0.4064
12	0.4444	0.4436	0.4422	0.4414
13	0.4727	0.4719	0.4705	0.4697
14	0.5071	0.5063	0.5049	0.5040
15	0.5358	0.5350	0.5336	0.5327
16	0.5698	0.5690	0.5675	0.5667
17	0.5988	0.5980	0.5965	0.5956
18	0.6325	0.6316	0.6301	0.6293
19	0.6617	0.6608	0.6593	0.6585
20	0.6951	0.6942	0.6927	0.6918
21	0.7245	0.7237	0.7221	0.7213
22	0.7577	0.7568	0.7553	0.7544
23	0.7873	0.7864	0.7849	0.7840
24	0.8203	0.8194	a.8179	0.8170
25	0.8500	0.8491	0.8476	0.8467
26	0.8829	0.8820	0.8804	0.8795
27	0.9127	0.9118	0.9103	0.9094
25	0.9454	0.9445	0.9430	0.9421
29	0.9754	0.9745	0.9729	0.9720
30	1.0080	1.0071	1.0055	1.0046
31	1.0380	1.0371	1.0355	1.0346
32	1.0705	1.0696	1.0681	1.0672
33	1.1006	1.0997	1.0981	1.0972
34	1.1331	1.1322	1.1306	1.1297
35	1.1633	1.1623	1.1608	1.1598
36	1.1956	1.1947	1.1931	1.1922
37	1.2258	1.2249	1.2233	1.2224
38	1.2582	1.2573	1.2557	1.2547
39	1.2884	1.2875	1.2859	1.2850
40	1.3207	1.3198	1.3182	1.3173
41	1.3510	1.3501	1.3485	1.3476
42	1.3832	1.3823	1.3807	1.3798
43	1.4136	1.4127	1.4111	1.4101
44	1.4458	1.4449	1.4432	1.4423
45	1.4761	1.4753	1.4736	1.4727
46	1.5083	1.5074	1.5058	1.5048
47	1.5387	1.5378	1.5362	1.5353
48	1.5708	1.5699	1.5683	1.5673
49	1.6013	1.6004	1.5987	1.5978
50	1.6334	1.6324	1.6308	1.6299
51	1.6639	1.6629	1.6613	1.6604
52	1.6959	1.6950	1.6933	1.6924
53	1.7264	1.7255	1.7238	1.7229
54	1.7584	1.7575	1.7558	1.7549
55	1.7890	1.7880	1.7864	1.7854
56	1.8209	1.8200	1.8183	1.8174
57	1.8515	1.8506	1.8489	1.8480
58	1.8834	1.8825	1.8809	1.8799
59	1.9140	1.9131	1.9115	1.9105

Table IV—Over-pin Measurements for Various Values of Tooth Thicknesses (as obtained from a computer)
Class: AGMA Precision 1A
Pitch: 32
Pin Diameter = 0.054

Computing the Tolerance Ranges

Relationships between tooth thickness and over-pin measurements are based on the geometry of Fig. 4:
For odd-numbered teeth

$$M = 2\left[a + R_c \cos\left(\frac{90°}{N}\right) \right] \qquad (1)$$

For even-numbered teeth

$$M = 2[a + R_c] \qquad (2)$$

where

$$R_c = R\left(\frac{\cos 20°}{\cos \phi_1} \right) \qquad (3)$$

and the value of ϕ_1 is obtained from:

$$\text{inv } \phi_1 = \frac{T_x}{2R} + \frac{a}{R_b} - \frac{\pi}{N} + \text{inv } 20° \qquad (4)$$

By substituting the full tooth thickness T of a particular gear for T_x in Eq (4), a value for the involute of ϕ_1 is obtained, the angle of which can be found from a table of involutes. This will give the nominal value for M from Eq (1) or (2) which can be checked with existing tables of over-pin measurements.

Four Reduced Tooth Thicknesses

To obtain over-pin measurements that are equivalent to the AGMA standards, four reduced tooth thicknesses are substituted for T_x, values of which are obtained from

$T_a = \text{Theoretical maximum} = T - 1/2\, B_{min}$
$T_b = \text{Guaranteed maximum} = T - 1/2\, (B_{min} + 0.728C)$
$T_c = \text{Guaranteed minimum} = T - 1/2\, B_{max}$
$T_d = \text{Theoretical minimum} = T - 1/2\, (B_{max} + 0.728C)$

These values are substituted for T_x in Eq (4) to obtain the four limits of over-pin measurements: M_a, M_b, M_c, and M_d, listed in descending order of magnitude. The combination M_a and M_d constitute the upper and lower limits of the "theoretical" range and the combination M_b and M_c the "guaranteed" narrower range.

In using this data, one would be highly confident of a gear being correct per class if within the narrower limits. If the measurement only falls within the wider extreme limits, the gear is "probably" within class requirements. This can only be verified by rolling on a variable-center fixture to check if the TCE is within the permissible range for that class.

Tables for Computations

Values of tooth thicknesses T_a to T_d, to be used in computing the over-pin dimensions are listed by class in Table III. The actual dimensions M_a to M_d consist of hundreds of tables similar to the typical one shown in Table IV, and therefore are not presented here because of space limitations. This is because the diametral pitch cannot be eliminated from the equations. Although the AGMA tolerances are variable, they remain fixed over a range of pitches for a given gear class and it is not possible to condense the data in terms of one diametral pitch and divide by the pitch for other values. This results in a great number of possible tables when one considers the range of pitches and cross combinations of AGMA backlash and accuracy classes.

Complete tables corresponding to the tooth thicknesses in Table III have been compiled at the General Precision Laboratory. Because of the length and repetition of the calculations, this was only made possible through use of digital computers and the author is indebted to the members of the GPL Analysis Section. Those wishing to verify their calculations for over-pin tolerances against GPL's data can do so by contacting the author.

Pin Measurement Tables for 20-deg Spur Gears

In addition to the dimension-over-pins, D_M, the tables give the thickness factor, K_m, which is the ratio

$$K_m = \frac{\Delta D}{\Delta t} = \frac{\text{change of pin measurement}}{\text{change of circular thickness}}$$

Thus, from Table 1 for $N = 30$ teeth (external), and diametral

pitch of $P_d = 6$, $D_M = 32.4102$ in. and K_m 2.38. Since $P_d = 6$, divide D_M by 6. Hence $D_m = 5.4017$ in.

The standard circular tooth thickness is equal to $\pi/2P_d = 0.2618$ in. Assume, however, that a thickness of 0.2600 in. is desired to allow for backlash. Then $\Delta t = 0.0018$ in., and $\Delta D_M = 0.0018 \times 2.38 = 0.0043$ in. Thus, the corrected dimension-over-pins is 5.3974 in.

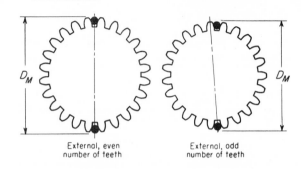

External, even number of teeth   External, odd number of teeth

Table 1.   Pin Measurements of External Standard 20-deg Involute Spur Gears
(1 Diametral Pitch; 1.728-in. Pin Diameter)

From AGMA Standard 231.52

Number of Teeth N	Dimension over Pins, in. D_M	Thickness Factor K_m	Number of Teeth N	Dimension over Pins, in. D_M	Thickness Factor K_m	Number of Teeth N	Dimension over Pins, in. D_M	Thickness Factor K_m	Number of Teeth N	Dimension over Pins, in. D_M	Thickness Factor K_m
10	12.3445	2.01	56	58.4325	2.52	101	103.4335	2.61	146	148.4513	2.65
11	13.2332	2.05	57	59.4111	2.53	102	104.4460	2.61	147	149.4430	2.65
12	14.3579	2.09	58	60.4335	2.53	103	105.4341	2.61	148	150.4515	2.65
13	15.2639	2.12	59	61.4128	2.53	104	106.4463	2.62	149	151.4433	2.65
14	16.3683	2.14	60	62.4344	2.53	105	107.4346	2.62	150	152.4516	2.65
15	17.2871	2.17									
			61	63.4144	2.54	106	108.4466	2.62	151	153.4435	2.65
16	18.3768	2.19	62	64.4352	2.54	107	109.4352	2.62	152	154.4518	2.65
17	19.3053	2.21	63	65.4159	2.54	108	110.4469	2.62	153	155.4438	2.65
18	20.3840	2.23	64	66.4361	2.55	109	111.4357	2.62	154	156.4520	2.66
19	21.3200	2.25	65	67.4173	2.55	110	112.4472	2.62	155	157.4440	2.66
20	22.3900	2.26									
			66	68.4369	2.55	111	113.4362	2.62	156	158.4521	2.66
21	23.3321	2.28	67	69.4186	2.55	112	114.4475	2.62	157	159.4443	2.66
22	24.3952	2.29	68	70.4376	2.56	113	115.4367	2.63	158	160.4523	2.66
23	25.3423	2.30	69	71.4198	2.56	114	116.4478	2.63	159	161.4445	2.66
24	26.3997	2.32	70	72.4383	2.56	115	117.4372	2.63	160	162.4524	2.66
25	27.3511	2.33									
			71	73.4210	2.56	116	118.4481	2.63	161	163.4448	2.66
26	28.4036	2.34	72	74.4390	2.57	117	119.4376	2.63	162	164.4526	2.66
27	29.3586	2.35	73	75.4221	2.57	118	120.4484	2.63	163	165.4450	2.66
28	30.4071	2.36	74	76.4396	2.57	119	121.4380	2.63	164	166.4527	2.66
29	31.3652	2.37	75	77.4232	2.57	120	122.4486	2.63	165	167.4453	2.66
30	32.4102	2.38									
			76	78.4402	2.57	121	123.4384	2.63	166	168.4528	2.66
31	33.3710	2.39	77	79.4242	2.58	122	124.4489	2.63	167	169.4455	2.66
32	34.4130	2.40	78	80.4408	2.58	123	125.4388	2.63	168	170.4529	2.66
33	35.3761	2.41	79	81.4252	2.58	124	126.4491	2.64	169	171.4457	2.66
34	36.4155	2.41	80	82.4413	2.58	125	127.4392	2.64	170	172.4531	2.66
35	37.3807	2.42									
			81	83.4262	2.58	126	128.4493	2.64	171	173.4459	2.66
36	38.4178	2.43	82	84.4418	2.58	127	129.4396	2.64	172	174.4532	2.66
37	39.3849	2.43	83	85.4271	2.59	128	130.4496	2.64	173	175.4461	2.66
38	40.4198	2.44	84	86.4423	2.59	129	131.4400	2.64	174	176.4533	2.67
39	41.3886	2.45	85	87.4279	2.59	130	132.4498	2.64	175	177.4463	2.67
40	42.4217	2.45									
			86	88.4428	2.59	131	133.4404	2.64			
41	43.3920	2.46	87	89.4287	2.59	132	134.4500	2.64			
42	44.4234	2.46	88	90.4433	2.59	133	135.4408	2.64			
43	45.3951	2.47	89	91.4295	2.60	134	136.4502	2.64			
44	46.4250	2.47	90	92.4437	2.60	135	137.4411	2.64			
45	47.3980	2.48									
			91	93.4303	2.60	136	138.4504	2.64			
46	48.4265	2.48	92	94.4441	2.60	137	139.4414	2.64			
47	49.4007	2.49	93	95.4310	2.60	138	140.4506	2.65			
48	50.4279	2.49	94	96.4445	2.60	139	141.4418	2.65			
49	51.4031	2.50	95	97.4317	2.60	140	142.4508	2.65			
50	52.4292	2.50									
			96	98.4449	2.61	141	143.4421	2.65			
51	53.4053	2.50	97	99.4323	2.61	142	144.4510	2.65			
52	54.4304	2.51	98	100.4453	2.61	143	145.4424	2.65			
53	55.4074	2.51	99	101.4329	2.61	144	146.4512	2.65			
54	56.4315	2.52	100	102.4456	2.61	145	147.4427	2.65			
55	57.4093	2.52									

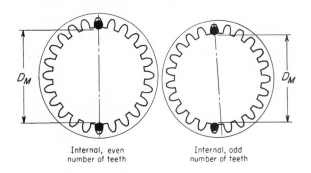

Internal, even
number of teeth

Internal, odd
number of teeth

Table 2. Pin Measurements of Internal Standard 20-deg Involute Spur Gears
(1 Diametral Pitch; 1.680-in. Pin Diameter)

Number of Teeth N	Dimension between Pins, in. D_M	Thickness Factor K_m	Number of Teeth N	Dimension between Pins, in. D_M	Thickness Factor K_m	Number of Teeth N	Dimension between Pins, in. D_M	Thickness Factor K_m	Number of Teeth N	Dimension between Pins, in. D_M	Thickness Factor K_m
30	27.6649	3.38	71	68.6856	2.95	111	108.6998	2.87	151	148.7063	2.84
31	28.6285	3.35	72	69.7032	2.95	112	109.7110	2.87	152	149.7145	2.83
32	29.6699	3.32	73	70.6867	2.94	113	110.7002	2.87	153	150.7065	2.83
33	30.6353	3.29	74	71.7038	2.94	114	111.7112	2.87	154	151.7146	2.83
34	31.6739	3.27	75	72.6878	2.94	115	112.7006	2.87	155	152.7068	2.83
35	32.6411	3.25	76	73.7044	2.94	116	113.7114	2.86	156	153.7148	2.83
36	33.6773	3.23	77	74.6888	2.93	117	114.7010	2.86	157	154.7070	2.83
37	34.6462	3.21	78	75.7049	2.93	118	115.7117	2.86	158	155.7149	2.83
38	35.6804	3.20	79	76.6897	2.93	119	116.7014	2.86	159	156.7072	2.83
39	36.6507	3.18	80	77.7054	2.92	120	117.7119	2.86	160	157.7150	2.83
40	37.6831	3.16									
			81	78.6905	2.92	121	118.7018	2.86	161	158.7074	2.83
41	38.6547	3.15	82	79.7059	2.92	122	119.7121	2.86	162	159.7151	2.83
42	39.6855	3.14	83	80.6914	2.92	123	120.7022	2.86	163	160.7076	2.83
43	40.6582	3.13	84	81.7064	2.91	124	121.7123	2.86	164	161.7152	2.83
44	41.6875	3.11	85	82.6922	2.91	125	122.7026	2.86	165	162.7078	2.83
45	42.6614	3.10	86	83.7068	2.91	126	123.7125	2.85	166	163.7153	2.83
46	43.6893	3.09	87	84.6929	2.91	127	124.7029	2.85	167	164.7080	2.83
47	44.6644	3.08	88	85.7072	2.91	128	125.7127	2.85	168	165.7154	2.83
48	45.6910	3.08	89	86.6936	2.90	129	126.7032	2.85	169	166.7082	2.83
49	46.6670	3.07	90	87.7076	2.90	130	127.7129	2.85	170	167.7156	2.82
50	47.6926	3.06									
			91	88.6943	2.90	131	128.7036	2.85	171	168.7084	2.82
51	48.6694	3.05	92	89.7080	2.90	132	129.7130	2.85	172	169.7157	2.82
52	49.6940	3.04	93	90.6950	2.90	133	130.7039	2.85	173	170.7086	2.82
53	50.6716	3.04	94	91.7084	2.89	134	131.7132	2.85	174	171.7158	2.82
54	51.6953	3.03	95	92.6956	2.89	135	132.7042	2.85	175	172.7087	2.82
55	52.6737	3.02	96	93.7087	2.89	136	133.7134	2.85			
56	53.6965	3.02	97	94.6962	2.89	137	134.7045	2.84			
57	54.6756	3.01	98	95.7090	2.89	138	135.7135	2.84			
58	55.6975	3.01	99	96.6968	2.89	139	136.7047	2.84			
59	56.6774	3.00	100	97.7093	2.88	140	137.7137	2.84			
60	57.6985	3.00									
			101	98.6974	2.88	141	138.7050	2.84			
61	58.6789	2.99	102	99.7096	2.88	142	139.7139	2.84			
62	59.6994	2.99	103	100.6979	2.88	143	140.7053	2.84			
63	60.6805	2.98	104	101.7099	2.88	144	141.7140	2.84			
64	61.7003	2.98	105	102.6984	2.88	145	142.7055	2.84			
65	62.6819	2.97	106	103.7102	2.88	146	143.7141	2.84			
66	63.7011	2.97	107	104.6989	2.88	147	144.7058	2.84			
67	64.6832	2.96	108	105.7105	2.87	148	145.7143	2.84			
68	65.7018	2.96	109	106.6994	2.87	149	146.7061	2.84			
69	66.6844	2.96	110	107.7107	2.87	150	147.7144	2.84			
70	67.7025	2.95									

Table 3. Pin Diameters for Various Pitches

Diametral Pitch P	For Standard External Gears $d = \dfrac{1.728 \text{ in.}}{P}$	For Standard Internal Gears $d = \dfrac{1.680 \text{ in.}}{P}$	Diametral Pitch P	For Standard External Gears $d = \dfrac{1.728 \text{ in.}}{P}$	For Standard Internal Gears $d = \dfrac{1.680 \text{ in.}}{P}$	Diametral Pitch P	For Standard External Gears $d = \dfrac{1.728 \text{ in.}}{P}$	For Standard Internal Gears $d = \dfrac{1.680 \text{ in.}}{P}$
1	1.7280	1.680	5	.34560	.3360	11	.15709	.15273
1½	1.1520	1.120	6	.2880	.280	12	.1440	.140
2	.864	.840	7	.24686	.240	14	.12343	.120
2½	.69120	.6720	8	.2160	.210	16	.1080	.1050
3	.5760	.560	9	.1920	.18666	18	.0960	.09333
4	.4320	.420	10	.17280	.1680			

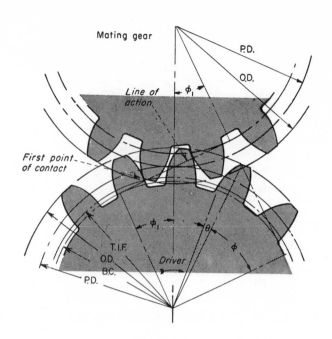

Involute geometry . . .

PD, pitch dia
OD, outside dia
BC, base circle dia
TIF, true involute form dia
ϕ, pressure angle
ϕ_1, normal pressure angle
$\theta + \phi$, degrees rolled off base circle

PROFILE DIAGRAMS . . .

a graphical way to

LeROY S. HARRIS

Manager, Research & Development
Schutte & Koerting Co., Cornwells Heights, Penn.

Troublesome gears? Here are tips on tooth-profile tolerances and how to turn them into specifications that will control errors when the gear is being made.

High accuracy can be assured, the next time you design gearing, by including a profile diagram on the gear drawing. This ties down just how much error is acceptable, and in addition makes profile checking easy for the inspection department.

Profile errors are deviations from the "true" involute of a gear tooth. They are caused by distortion in heat-treating after machining, by inaccurate gear-generating cutters, or by errors in setups or machines. The gear manufacturer can hold these errors to a minimum by selecting the proper heat-treating process, by adequate inspection methods of the machine setup, and by grinding, lapping or shaving the tooth profile after heat treating.

It is not hard to spot the errors with a commercial gear checker. The inspector can use a machine that rotates either the gear, the measuring indicator, or both about the base-circle diameter. The profile is recorded on a chart, Fig. 1. A true involute produces a straight line, and deviations can be interpreted in thousands of

TABLE I—Tooth-profile Tolerances for Spur and Helical Gears

(AGMA Standards 231.01, 232.01, and 233.01)

Class No.	Max. Speed, fpm	Diametral Pitch	Profile Error, total variation in 0.0001 in. Pitch Diameters, in.							
			¾	1½	3	6	12	25	50	100
1	—	1	—	—	—	—	50	60	70	80
		2	—	—	—	30	35	40	45	50
		4	—	—	25	30	35	40	40	40
		8	—	20	20	25	25	30	30	—
		16	15	15	15	20	20	25	—	—
2	400	1	—	—	—	—	30	35	40	40
		2	—	—	—	20	20	25	30	30
		4	—	—	10	10	11	12	12	12
		8	—	9	9	9	9	9	10	—
		16	7	7	7	7	7	7	8	—
3	2000	2	—	—	—	11	11	11	11	11
		4	—	—	8	8	8	8	8	8
		8	—	6	6	6	6	6	6	6
		16	5	5	5	5	5	5	5	—
		32	5	5	5	5	5	5	—	—
4	Over 2000	4	—	—	—	5	5	5	5	—
		8	—	—	4	4	4	4	4	4
		16	—	3	3	3	3	3	3	3
		32	3	3	3	3	3	3	—	—

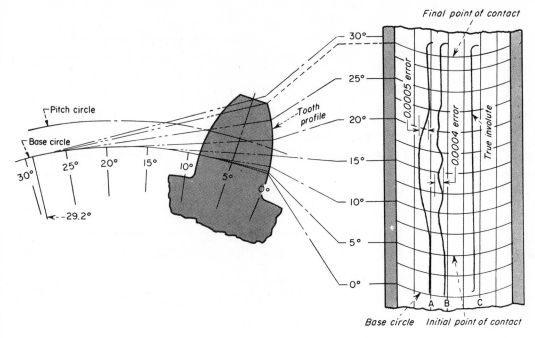

1 Profile errors . . .
are reproduced on chart by a profile checking machine. Waves in line represents deviations from "true" profile. Vertical scale is in "degrees rolled off base circle"; horizontal scale in inches. In these typical curves, tooth profile A has a higher total error than B, but is preferable for its gradual slope.

specify gear errors

an inch. (Distance between lines on the chart represent 0.0004 in.). Other methods employ optical comparators or micrometers to measure tooth thickness at various tooth heights.

Permissible deviations, Table I, have been established by AGMA (American Gear Manufacturers Association) and are based on the class of precision, diametral pitch and pitch dia of the gear. In addition, Table II gives typical tooth-profile tolerances obtainable by various gear-manufacturing methods.

WHEN A DIAGRAM IS USEFUL

No notation is needed on the drawing for ordinary commercial-grade gears where profile errors of 0.0015 to 0.003 in. or greater are permissible. But for more accurate commercial-grade gears made on gear hobbing or shaping machines, errors can be within 0.0005 to 0.001 in., and limits on the involute error should be specified.

For precision-grade gears or for critical applications where the reliability or noise of a gear is a major factor, the maximum errors should not exceed about 0.0005 in.; and the rate of change or slope also should be specified, preferably with a profile diagram. It plots the range of permissible error vs degrees rolled off the base circle, and can take the form shown in Fig. 2, which simulates an actual chart, Fig. 1, that was obtained by using an involute-checking machine. The profile diagram quickly conveys to the gear inspector or manufacturer the permissible departure of the profile from the true involute. The horizontal scale is calibrated in degrees rolled off the base circle dia, $\theta + \phi$, with

TABLE II—Typical Tooth-profile Tolerances from Various Gear-finishing Methods

Gear Finishing Method	Diametral Pitch	Pitch Dia, in.	Profile Error, in.
Grinding (Precision Gear)	6 to 12	1	0.0002
	6 to 12	7	0.0003
	12 to 24	1	0.0002
	12 to 24	7	0.0003
Shaving (Precision Gear)	6 to 12	1	0.0003
	6 to 12	7	0.0004
	6 to 12	20	0.0006
	12 to 24	1	0.0002
	12 to 24	7	0.0003
Hobbing or Shaping (Well-cut Gear)	6 to 12	1	0.0004
	6 to 12	7	0.0005
	6 to 12	20	0.0007
	12 to 24	1	0.0003
	12 to 24	7	0.0004

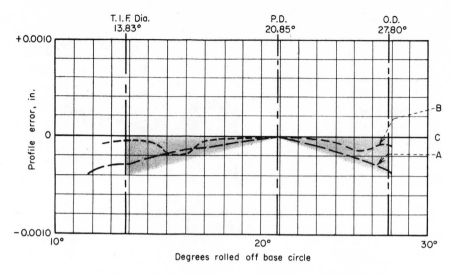

2 Profile diagram . . .
simulates typical curves in Fig. 1, in a form convenient to use on a gear drawing to tie down tooth profile errors. Area in shade represents permissible deviations from true involute. A gear represented by line B would be rejected by notation requiring error to be proportional to distance from pitch dia.

NOTE: With indicator reading zero at pitch dia. Involute profile error must be within limits shown. Variation of profile line from true involute must be proportional to distance from the pitch diameter.

values ranging from zero at beginning of the involute on the base circle, to a maximum when the involute tooth position is at the OD.

The true involute form (TIF) dia corresponds to the lowest point along the tooth at which contact occurs. This point can be determined graphically, or analytically from the intersection of the OD of the mating gear and the line of action. Actual tooth contact along the involute profile (and the range to be checked) occurs between the TIF dia and the OD, assuming there is no chamfer along the outer tip of the tooth. These limiting boundaries are also noted on the profile diagram to indicate the limits of the tooth contour included in the involute profile specifications.

The pitch dia (PD) position on the diagram is a reference or zero position on the profile from which the direction of profile error is measured or specified with a positive or negative value. For convenience, an explanatory note is added under the diagram.

A better picture of the permissible-error range can be given by shading these areas on the diagram. The colored area in Fig. 2 indicates a negative permissible profile error of 0.0004 in.; in some designs it might be desirable to permit only positive errors. Profile errors both positive and negative would be indicated by shaded areas on both sides of the zero line. For this condition the total permissible profile error would be the sum of the negative and positive values.

SLOPE OF PROFILE CURVES

The slope of the deviations in the tooth profile is not covered by AGMA specifications. Curve B in Fig.

2 shows a profile which falls within the shaded area but sharpness of the "bumps" may cause undesirable vibrations or undue wear. To control this type of error, the notation under the diagram should state limits for the slope of the profile with reference to the true involute, as in the example shown in the profile diagram above.

PROFILE MODIFICATIONS

By relieving the tooth tip to give a little clearance at the first point of contact, gears run quieter and carry more load. This or any other profile modification can be effectively specified, as in Fig. 3. The chamfer in the colored area near the tooth tip indicates that a deviation from the true involute profile is desired within the limits and dimensions shown.

SIMPLIFIED DATA SHEET

In specifying the angular position of any point on the involute profile in terms of "degrees rolled off base circle," calculations can be reduced to routine by the method shown in Table III. Calculations are shown for a typical spur gear. The data at the head of the table, such as BD = 4.229 and TIF = 4.350, are established from the basic calculations on the gear along with other pertinent data such as the addendum, dedendum, etc. The angular position of the three specific points is found to be: PD, 20.854°; OD, 27.800°; TIF dia, 13,826°. Then by selecting the desired tolerance from Tables I and II or from specific design requirements, the profile diagram is added on the gear drawing.

12 ways to LOAD-TEST GEARS

Until a set of gears is fully loaded and subjected to dynamic conditions, its performance is only guesswork. Here are a dozen methods for testing the gears with locked-in torque.

Gene Shipley
Gear Engineering
General Electric Co., Lynn, Mass.

Gear analysis and design is no more than an educated guess—until the gear goes to work. The only way to be sure that a gear will perform as well as predicted is to test it under actual conditions of load and speed. It's then that wear life, fatigue life, frictional torque and dynamic stability can be noted and measured. The same is true for purchased gearboxes and transmissions, and for testing the effects of various lubricants on gear life.

But how to apply the load—or more specifically, the torque—to a set of gears waiting in the testing laboratory or on the production floor? It's not enough to couple them to a motor on one end and a mass with a certain inertia on the other—because once the gears are running at constant speed, the torque will drop to an amount only necessary to overcome friction. Also, there are many applications where the gears operate under high horsepower requirements, so high that it becomes impractical to simulate conditions by attaching a mass to the system.

The clue is in remembering that torque is proportional to twist. So twist the shafts in the gear system the desired amount. Next "lock in" the twist, so that the twist remains constant as the gear system rotates. Now attach a motor to the system and run it at desired speed for as long as you want.

There are 12 ways of applying full torque to a gear system. All are based on the "four-square" test setup in which two gear sets are joined back to back and locked together after applying a twisting couple to one of the connecting shafts. The 12 fall into one or the other of two main categories:

1. Methods in which torque can be applied to the gear system only when the gears are stationary. These methods, shown in Figs. 1 to 3, are simple and cheap, but require the gears to start up under full load. Many systems are not designed to do this.

2. Methods in which torque can be applied while the gears are running. Figs. 4 to 12. Here, loads can be remotely controlled and varied under realistic field conditions. In many cases they can be programmed. Another advantage is time-saving—loads can be changed without shutting down the machine. Also, there is less wear on the machine because loads can be applied when lubricating conditions are most suitable.

CONTINUED ON NEXT PAGE 〉

201

LOAD-TEST GEARS (continued)

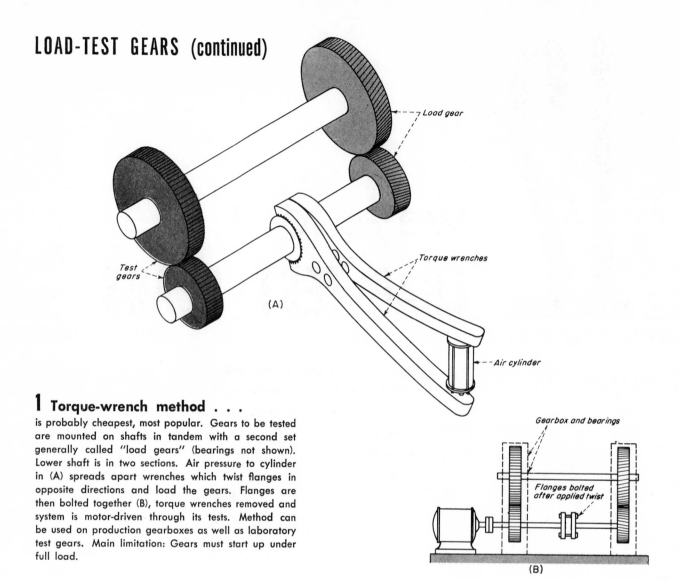

(A)

1 Torque-wrench method . . .

is probably cheapest, most popular. Gears to be tested are mounted on shafts in tandem with a second set generally called "load gears" (bearings not shown). Lower shaft is in two sections. Air pressure to cylinder in (A) spreads apart wrenches which twist flanges in opposite directions and load the gears. Flanges are then bolted together (B), torque wrenches removed and system is motor-driven through its tests. Method can be used on production gearboxes as well as laboratory test gears. Main limitation: Gears must start up under full load.

(B)

2 Screw coupling . . .

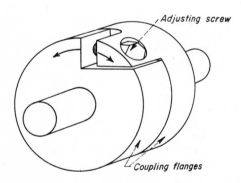

is adaptation of torque-wrench method. Turning adjusting screw displaces coupling flanges in opposite directions, applying torque to gear system. Locknut or screw with locking insert will prevent adjusting screw from loosening. System eliminates torque wrenches and air cylinder. Also, minute adjustments of torque are made easily. But gears must start up under full load.

3 Vernier coupling . . .

has two flanges with series of holes equally spaced around the outer edge. One flange has one more hole than the other. Additional hole functions as a vernier to permit fine adjustments in torque. Flanges are twisted by any method deemed best, and a pin inserted through appropriate set of holes.

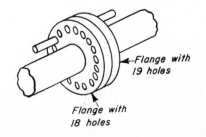

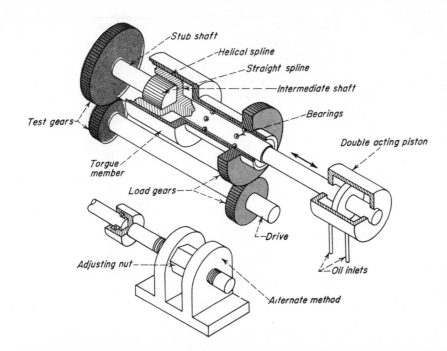

4 Spline coupling . . .

uses two spline systems to apply twist while shafts are rotating. Piston produces axial movement of the intermediate shaft, which has straight spline teeth on the ID and helical splines on the OD. This axial movement causes relative angular displacement between stub shaft and torque member. Piston pressure maintains twist as gears are driven through tests; its double action can apply torque in either direction to simulate a reversing system. Also, with a cam-driven control valve feeding pressure to the piston, the load can be controlled to simulate accelerating and decelerating conditions. A cheaper but less flexible way to apply load is to replace piston with **an** adjusting nut, as illustrated.

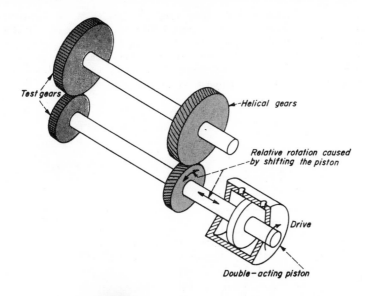

5 Helical gears . . .

mounted back-to-back with test gears provide test setup without need for additional couplings or mechanisms. When gears are stationary, helix angle allows applying a twist to the system by simply shifting the lower shaft axially either direction. The same torque will be applied if gears are running. Double-acting piston provides the shifting mechanism. Varying the pressure in piston will directly vary the torque-load on gears. Method is not practical for testing production gear-boxes. Helical-gear box and piston can be mounted on a permanent base. Wide pinion gears are used on both the helical and spur gears when possible.

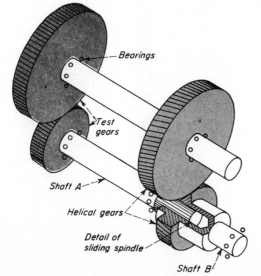

6 Sliding spline . . .

uses principle of moving a helical gear axially to produce desired torque. Spline couples two shafts A and B: Shifting helical pinion gear will not shift test gears. The lower helical gear has extra-wide face width for continual contact during shifting. Because all the mechanisms are confined to one gear-box, this system is often employed to test production gears.

(continued on next page)

LOAD-TEST GEARS (continued)

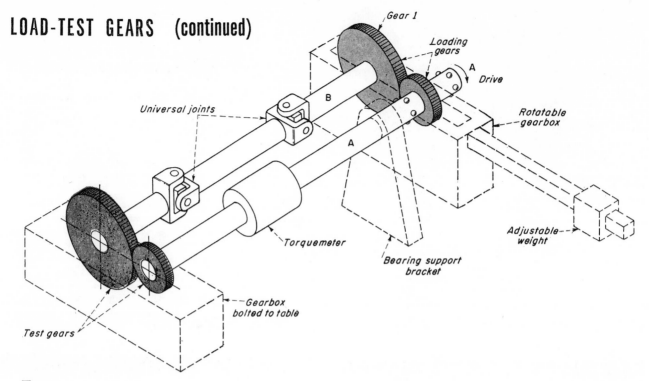

7 Rotating gearbox . . .

is free to turn about axis A-A. Test gearbox on other end is bolted to base plate or table. The boxes contain bearings for mounting the shafts. When box is rotated slightly by extending adjustable weight, gear 1 will creep over gear 2, causing angular displacement in shaft B which opposes normal rotation of the system and gives torque in the shafts. The universal joints in shaft B take up misalignment caused by relative displacement. Misalignment limits the load and speed capacity, but system is reliable, all-mechanical, easy to operate. Torque can be accurately varied by minute adjustments of weight. A torquemeter gives quick readings.

8 Bevel-gear differential . . .

applies torque simply by adding weight to the loading arm which rotates cage assembly. To assure alignment, cage is mounted in a bushing or bearing, as illustrated in section A-A.

Because shafts 1 and 2 will rotate in opposite directions, an idler is employed in the test-gear train to reverse direction—an advantage when the gearbox under test has a 3-gear spur or helical system. The full-load torque is limited by capacity of the bevel gears because it must be transmitted through the differential. Bevel gears are usually made as large as possible.

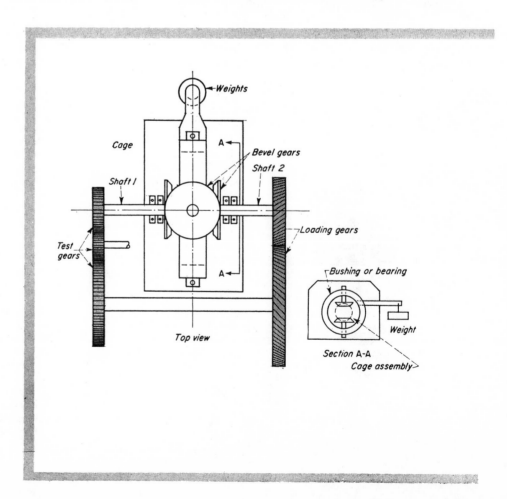

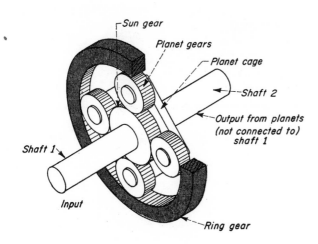

9 Planetary torque applier . . .

is similar to (8): Shaft 1 is connected to test gears and shaft 2 to loading gears. There is a constant ratio between shafts 1 and 2 as long as ring gear is stationary. Rotating it introduces relative twist between shafts. Ring gear can be accurately adjusted to give desired torque. Shaft 2 will rotate at slightly reduced speed relative to shaft 1, but in same direction.

10 Double helical sleeve . . .

operated by oil pressure which forces apart the sleeve and one of the gears. This causes a relative twist between shafts as indicated by arrows 1 and 2. Second chamber is ported to atmosphere; however, oil can be pumped in from shaft on left to convert sleeve into a double-acting piston. Method produces a wide range of torques, has given successful tests with systems up to 100,000 hp.

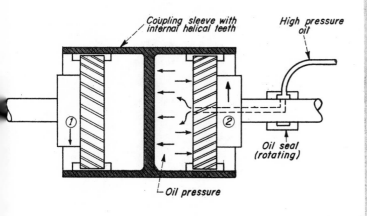

11 Multiple-shaft system . . .

applies torque to test gears by shifting shaft 4 around axis X-X. All gears marked A are same size; similarly for gears marked B. Twisting moment is controlled by adjusting weights on lever arm. System is usually limited to testing only one size of gears. All except test gears can be assembled in one loading box.

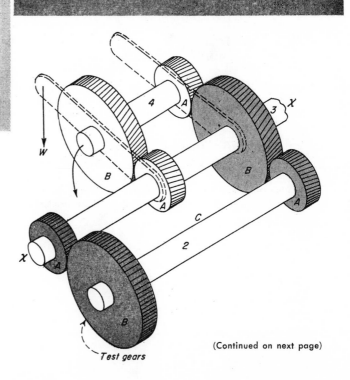

(Continued on next page)

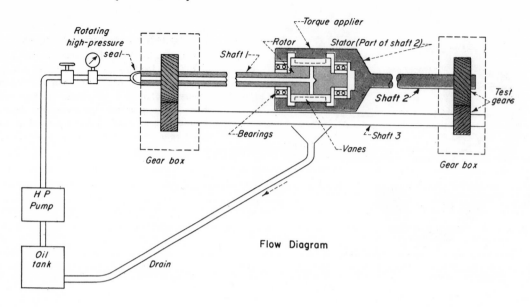

Flow Diagram

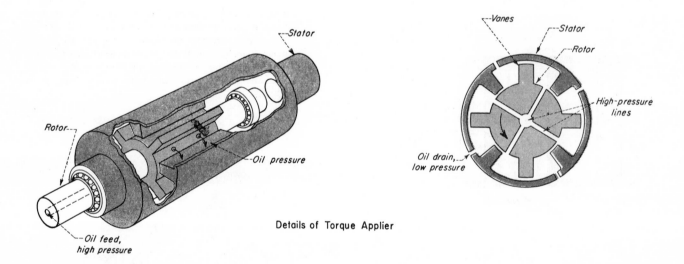

Details of Torque Applier

12 Hydraulic torque applier . . .

uses water-wheel principle to create relative twist between *shafts 1* and *2*. A high-pressure vane pushes oil through rotating shaft and out between vanes of rotor and stator. The oil forces vanes apart, causing a twist in shafts similar to a torque-wrench twist. Oil leakage to low-pressure side is bled off through holes in periphery. Oil pressure is easily adjusted to give wide range of loads. Method is ideally suited for testing production units because it works satisfactorily with a wide range of speeds and horsepowers.

Calculating Design Data from Sample Gears

Provides a means of reproducing design data for gears from a broken segment or for producing replacements for worn gears of unknown origin.

ALFRED BISHOP
Wright Aeronautical Div., Curtiss-Wright Corp.

THIS SIMPLE METHOD of calculating basic design data from a gear of unknown origin requires only two micrometer measurements, taken across consecutive numbers of teeth. It yields the base diameter of the gear plus the circular tooth and circular space thicknesses at the base diameter. From these—and the number of teeth, all other pertinent data can be calculated.

The basis of this method is that (1) a line normal to an involute curve at any point is tangent to the base circle from which the involute was produced and (2) a straight line tangent to the base circle and extended to the involute is exactly the same length as the arc of the base circle from the point of tangency of the straight line to the base of the involute. Hence, a measurement taken with the flats of a micrometer or similar instrument, tangent to the involutes of the two teeth

No. teeth between measurements	2	3	4	5	6	7	8	9
Pressure angle, deg	No. teeth in gear							
14.5	12–25	26–37	38–50	51–62	63–75	76–87	88–100	———
17	12–21	22–32	33–42	43–53	54–64	65–74	75–85	86–97
20	12–18	19–27	28–36	37–45	46–54	55–63	64–72	73–81
22.5	12–16	17–24	25–32	33–40	41–48	49–56	57–64	65–73
25	12–14	15–21	22–29	30–36	37–43	44–51	52–58	59–66

Table I—Suggested Measurements for Various Tooth Numbers

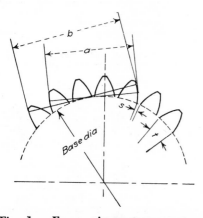

Fig. 1 — From micrometer measurements across two consecutive numbers of teeth such as a and b, the tooth thickness and space thickness at the base circle can be calculated with Eqs (3) and (4) as given in the text.

contacted is exactly equal to the length of arc subtended by the involutes from which the measurement was taken.

A limitation of the method is that pinions of a small number of teeth cannot be calculated, since it will be impossible to measure across the involutes of three teeth, the minimum required. In general, however, meas-

urements across any two consecutive numbers of teeth, excluding one as impractical, will be satisfactory. Also, if the gear or pinion is badly worn, it may be necessary to add an estimated or measured excess backlash to the measured value.

To facilitate the taking of micrometer measurements near the pitch circle of the teeth and thus avoid the possibility of including the fillets at the bottom of tooth profiles, it is suggested that the number of teeth between measurements on the sample gear be as given in Table I, where the number of teeth to be included in a measurement increases as the number of teeth in the sample gear increases.

EXAMPLE: In Fig. 1, let

a = distance across 4 teeth and 3 spaces
b = distance across 5 teeth and 4 spaces
t = arc tooth thickness at the base circle
s = arc space thickness at the base circle

Then:

$$a = 4t + 3s \quad (1)$$
$$b = 5t + 4s \quad (2)$$

Solving these equations simultaneously,

$$t = 4a - 3b \quad (3)$$

Substituting n for the number of teeth in the distance a,

$$t = na - (n - 1) b \quad (4)$$

Similarly

$$s = nb - (n + 1) a$$

The base circle pitch $= (t + s)$,

and if N is the number of teeth in the gear, the base circle diameter is

$$\frac{N(t + s)}{\pi} = \frac{(b - a) N}{\pi}$$

With the base pitch and base circle diameter, a diametral pitch can be selected that will satisfy the known factors of the gear. If the gear is standard, i.e., the tooth thickness equals the tooth space at the pitch diameter, the PD can be determined as follows: Place wires of a diameter equal to one half the base circle pitch between the gear teeth as shown in Fig. 2. The centers of these wires coincide with the pitch diameter. The PD can be evaluated by subtracting the wire diameter from the over-wire diameter.

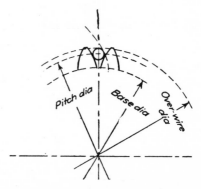

Fig. 2—A wire having a diameter equal to one half the base pitch will have its center on a radius at which the tooth thickness and space are equal.

Helicals Replace Module Spur Gears By CARL A JOHNSON

Sometimes it is necessary to replace module (metric pitch) spur gears in foreign equipment, but module gear cutting tools are not available. In such circumstances it is possible to substitute helical gears if the design of the bearing mounting will permit. The helical gears have the same transverse pitch as the spur gears but they must be cut with standard normal diametral pitch.

Because the transverse pitch is not changed, the pitch diameters of the helical gears will be the same as those for the original spur gears and the same operating center distance will apply. However the outside diameter of the helical gears must be calculated by adding addendums based on the normal diametrical pitch.

Problem:

To replace a pair of module spur gears having a module of $M = 3.25$. Here module equals the pitch diameter in millimeters divided by the number of teeth in the gear. The pinion has 25 teeth; the gear 50 teeth. Pitch diameters of the gears in inches (25.4 mm = 1 in.) are:

For pinion:

$$d = \frac{N_p \times M}{25.4} = 3.1988$$

For gear:

$$D = \frac{N_G \times M}{25.4} = 6.3976$$

Transverse diametral pitch is 7.8154. Three suitable normal diametral pitches are 8, 9 and 10. Because a low helix angle is best for the problem cited, the selection is a normal diametral pitch of 8 and a helix angle of 12°19'58".

Full-depth hobs are selected to cut the mating right and left-hand gears. Equal addendums are satisfactory. Hence, OD's are:

For pinion:

$$d_o = d + \frac{2}{P_n} = d + \frac{2}{8}$$

$$= 3.1988 + 0.25 = 3.4488$$

For gear:

$$D_o = D + \frac{2}{P_n} = D + \frac{2}{8}$$

$$= 6.3976 + 0.25 = 6.6476$$

Module	Transverse Diametral Pitch	Substitute Helical Gears Suitable Normal Diametral Pitches and Helix Angles				Module	Transverse Diametral Pitch	Substitute Helical Gears Suitable Normal Diametral Pitches and Helix Angles		
1	25.4	26 12°19'58"	28 24°53'12"	30 32° 8'57"	32 37°27'46"	6	4.2333	5 32° 8'57"		
1.25	20.32	22 22°32'11"	24 32° 8'57"	26 38°35'54"	28 43°28'18"	6.5	3.9077	4 12°19'58"	5 38°35'54"	
1.5	16.9333	17 5° 4'33"	18 19°49'25"	19 26°58'19"	20 32° 8'57"	7	3.6286	4 24°53'12"	5 43°28'18"	
1.75	14.5143	15 14°37'14"	16 24°53'12"	17 31°22'29"	18 36°15'33"	8	3.175	3.5 24°53'12"	4 37°27'46"	
2	12.7	13 12°19'58"	14 24°53'12"	15 32° 8'57"	16 37°27'46"	9	2.8222	3 19°49'25"	3.5 36°15'33"	
2.25	11.2889	12 19°49'25"	13 29°43'46"	14 36°15'33"	15 41°11'04"	10	2.54	2.75 22°32'11"	3 32° 8'57"	3.5 43°28'18"
2.5	10.16	11 22°32'11"	12 32° 8'57"	13 38°35'54"	14 43°28'18"	11	2.3091	2.5 22°32'11"	2.75 32°53'41"	3 39°40'21"
2.75	9.2364	10 22°32'11"	11 32°53'41"	12 39°40'21"	13 44°43'31"	12	2.1167	2.25 19°49'25"	2.5 32° 8'57"	2.75 39°40'21"
3	8.4667	9 19°49'25"	10 32° 8'57"	11 39°40'21"		13	1.9538	2 12°19'58"	2.25 29°43'46"	2.5 38°35'54"
3.25	7.8154	8 12°19'58"	9 29°43'46"	10 38°35'54"		14	1.8143	2 24°53'12"	2.25 36°15'33"	2.5 43°28'18"
3.5	7.2571	8 24°53'12"	9 36°15'33"	10 43°28'18"		15	1.6933	1.75 14°37'14"	2 32° 8'57"	2 25 41°11'04"
3.75	6.7733	7 14°37'14"	8 32° 8'57"	9 41°11'04"		16	1.5875	1.75 24°53'12"	2 37°27'46"	
4	6.35	7 24°53'12"	8 37°27'46"			18	1.4111	1.5 19°49'25"	1.75 36°15'33"	
4.5	5.6444	6 19°49'25"	7 36°15'33"			20	1.27	1.5 32° 8'57"	1.75 43°28'18"	
5	5.08	6 32° 8'57"	7 43°23'18"			22	1.1545	1.25 22°32'11"	1.5 39°40'21"	
5.5	4.6182	5 22°32'11"	6 39°40'21"			24	1.0583	1.25 32° 8'57"		

GEAR RATIOS AND TOOTH GEOMETRY

7

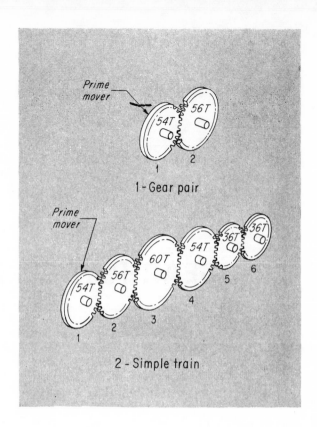

1-Gear pair

2-Simple train

How many revolutions of the input will return a complex gear train to its original alignment? The author's simplified method finds the answer by first determining the . . .

SYNCHRONIC INDEX OF GEAR TRAINS

E D KNAB, *Bell Telephone Laboratories, Whippany, NJ*

Here is a simplified way to compute the number of revolutions that the prime mover in a gear train must make before all gears return to their original alignment. It is applicable to all types of complex gear trains, and to many of the planetary-gear systems. The method also determines the number of whole-number revolutions that any one gear in the train will make relative to any other gear in the train—during the cycle of restoration.

The standard way is to work out such computations on an almost hit-or-miss basis. This is time-consuming, and for a complex gear train, the final answer cannot be easily checked for accuracy. The new method is a formal mathematical approach that will handle the various types of gear trains. It depends on a factor designated the "synchronic index" of a gear train. By definition, this is the minimum number of full revolutions of the input gear, or "prime-mover" gear, that will return all gears of the train to their initial relationship.

To describe this new term more precisely, consider two gears of unequal size, meshed together. A particular tooth on the pinion (driving gear) will make contact with a particular tooth on the driven gear at regular given intervals of rotation of the driver. That is to say, two teeth in mesh at rest will again be in mesh after the driver

has made n revolutions and, at this point, the gears will have resumed their original alignment. The number of revolutions made by the driver is the synchronic index of the train. Frequency of contact is a function of the speed of the driver and has no direct bearings on the synchronic index.

We have employed the synchronic-index method at Bell Labs for torque and angular-error analyses. For such applications it is desirable to have as low an index as possible. Where wear and life considerations are of more importance, the index should be as high as possible. Details of torque and backlash applications are given in this article.

Meshed pair

To determine the synchronic index, I_s, for two gears involves only basic arithmetic or merely intuition. But intuition can lead you astray.

For example, consider a two-gear mesh: the driver has 30 teeth; the driven gear has 60 teeth. It is fairly evident that the driver will make two revolutions before the initial alignment recurs, therefore, $I_s = 2$. However, if the driver has, say, 36 teeth, and the driven gear, 243 teeth, it would not be easily discernable that the synchronic index for this pair is 27. Hence the need for a formal approach, even in two-gear meshes.

For a two-gear mesh

$$I_s = \frac{\text{number of teeth in driven gear}}{\text{largest common factor divisible in both gears}}$$

In terms of symbols (see box below):

$$I_s = \frac{N_2}{h_{1.2}}$$

Referring to Fig 1, where $N_1 = 54$ and $N_2 = 56$, then $N_1 = 2 \times 27$ and $N_2 = 2 \times 28$. Hence

$$h_{1.2} = g \text{ of } (54, 56) = 2$$

Therefore

$$I_s = \frac{56}{2} = 28$$

Although the fact that largest common factor is equal to 2 has been determined by inspection, a table of prime factors can be consulted for more complicated cases.

Returning to the problem of the 36T:243T gear mesh, $h_{1.2} = 9$; hence $I_s = 243/9 = 27$.

The ratio $N_2/h_{1.2}$ is also called the constant of synchronization, k. For a two-gear mesh

$$I_s = k$$

Simple gear train

For a simple train of gears, that is, three or more meshed gears on independent shafts,

$$I_s = k_p$$

where k_p = lowest common multiple of all values of k related to the prime mover.

Consider a simple train comprising six gears, Fig 2, where

$N_1 = 54$	$N_4 = 54$
$N_2 = 56$	$N_5 = 36$
$N_3 = 60$	$N_6 = 36$

The constants of synchronizations are:

$$k_{1.2} = \frac{56}{2} = 28 = 2^2 \times 7$$

$$k_{1.3} = \frac{60}{6} = 10 = 2 \times 5$$

$$k_{1.4} = \frac{54}{54} = 1$$

Symbols

g = symbol signifying the operation of finding the greatest common factor for two values.

h = greatest common factor (g) for any two gears

I_s = synchronic index

k = constant of synchronization for any two gears = $N_2/h_{1.2}$

k_p = lowest common multiple of all values for k related to the prime mover.

L' = the greatest common factor (g) for k'_p and ($k_p \times r$)

N = number of teeth in a gear

r = ratio between two gears

R = number of revolutions a gear within a train makes during the synchronic cycle.

• Numerical subscripts relate various factors to gears of similar notation. Hence, $h_{1.2}$ means greatest common factor for Gears 1 and 2.

$$k_{1.5} = \frac{36}{18} = 2$$

$$k_{1.6} = \frac{36}{18} = 2$$

Thus

$$I_s = k_p = 2^2 \times 7 \times 5 = 140$$

To find the number of revolutions, R, made by any particular gear in the train during the synchronic cycle, multiply I_s by the ratio, r, of the prime mover and the gear of interest. Example:

$$R_5 = I_s \times r_{1.5} = (140)\left(\frac{54}{36}\right) = 210$$

Compound train

A compound train of gears consists of two or more simple gear trains so arranged that at least one gear of each train is fixed to a shaft common to both trains.

To compute I_s, first compute k_p for each simple train within the system. If one of the trains has only a meshed pair of gears, it too can be considered for the purpose of computation as a simple train.

Referring to Fig 3 where

$N_1 = 54$	$N_6 = 36$
$N_2 = 56$	$N_7 = 96$
$N_3 = 60$	$N_8 = 36$
$N_4 = 54$	$N_9 = 84$
$N_5 = 36$	

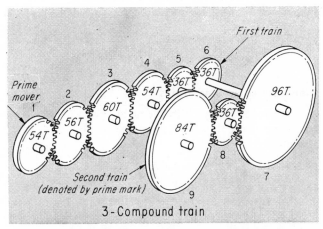

3 - Compound train

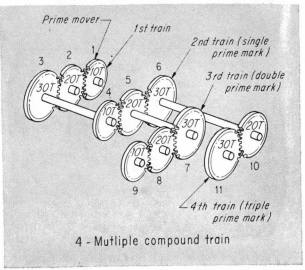

4 - Mutliple compound train

Then, as with a simple train, k values for the first gear train are:

$$k_{1,2} = 28 = 2^2 \times 7 \qquad k_{1,5} = 2$$
$$k_{1,3} = 10 = 2 \times 5 \qquad k_{1,6} = 2$$
$$k_{1,4} = 1$$

Therefore $k_p = 140$.

Assigning a prime mark to functions relating to the second gear train gives:

$$k'_{7,8} = 3; \qquad\qquad k'_{7,9} = 7$$

Therefore

$$k'_p = (3)(7) = 21$$

For the compound train of gears

$$I_s = \frac{k_p \times k'_p}{L'}$$

where L' is the greatest common factor for k'_p and $(k_p \times r)$ Thus

$$L' = g\left\{21, \left(140 \times \frac{54}{36}\right)\right\}$$
$$= g\{21, 210\} = 21$$

Therefore

$$I_s = \frac{k_p \times k'_p}{L'} = \frac{(140)(21)}{21} = 140$$

Thus, progressing from simple to compound gearing did not change the synchronic index computed for the first train. However, if in the above example N_9 was 39, then k'_p would be 39 and L' would be 3.

In this case,

$$I_s = \frac{(140)(39)}{3} = 1820$$

Multiple compound trains

The process may be extended to cover compound systems involving three or more simple trains. The general formula is

$$I_s = \frac{k_p \times k'_p \times k''_p \ldots k_p^n}{L' \times L'' \ldots\ldots L^n}$$

Example: Find I_s for the gear train in Fig 4. (For clarity, multiples of 10 are used for numbers of teeth in this problem.)

Solution—First train:

$$k_{1,2} = \frac{N_2}{h_{1,2}} = \frac{20}{10} = 2$$

$$k_{1,3} = \frac{N_3}{h_{1,3}} = \frac{30}{10} = 3$$

$$k_p = (2)(3) = 6$$

Second train:

$$k'_{4,5} = \frac{N_5}{h_{4,5}} = \frac{20}{10} = 2$$

$$k'_{4,6} = \frac{N_6}{h_{4,6}} = \frac{30}{10} = 3$$

$$k'_p = (2)(3) = 6$$

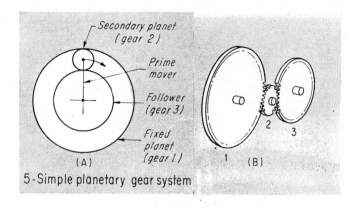

5-Simple planetary gear system

Gear N_4 is considered the prime mover for the second train. Hence

Third train:

$$k''_{7,8} = \frac{N_8}{h_{7,8}} = \frac{20}{10} = 2$$

$$k''_{7,9} = \frac{N_9}{h_{7,9}} = \frac{10}{10} = 1$$

$$k''_p = (2)(1) = 2$$

Fourth train:

$$k'''_{10,11} = \frac{N_{11}}{h_{10,11}} = \frac{30}{10} = 3$$

$$k'''_p = 3$$

The synchronic index is equal to :

$$I_s = \frac{k_p \times k'_p \times k''_p \times k'''_p}{L' \times L'' \times L'''}$$

where

$$L' = g[k'_p, (k_p \times r_{1,3})] = g[6, 2] = 2$$
$$L'' = g[k''_p, (k'_p \times r'_{4,5})] = g[2, 3] = 1$$
$$L''' = g[k'''_p, (k''_p \times r'_{4,6})] = g[3, 2] = 1$$

Substituting these values in the equation for I_s gives

$$I_s = \frac{(6)(6)(2)(3)}{(2)(1)(1)} = 108$$

The revolutions, R, for any particular gear in a compound system during the synchronic cycle is similar to that for simple gearing except that the ratio, r, is taken from the prime mover to the gears of first reduction and repeated until the gear of interest has been reached.

Example:
Referring again to Fig 4, find R for Gear 8.
Solution:

$$R_8 = I_s \times r_{1,3} \times r'_{4,5} \times r''_{7,8}$$
$$= (108)(1/3)(1/2)(3/2) = 27$$

Thus, after 27 revolutions of Gear 8, the entire gear system will have returned to its initial relationship. During that time, of course, Gear 1 would have rotated 108 times.

Planetary or epicyclic gearing—simple and compound

Planetary or epicyclic gearing involves an arrangement of gears to provide a compact transmission. The gears may be arranged in a countless number of ways, and for this reason a simple and all-inclusive formula for determin-

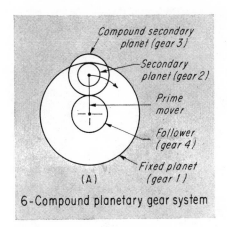

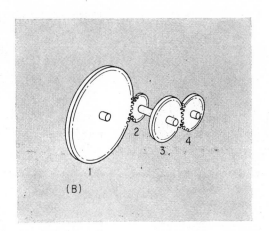

Compound secondary
planet (gear 3)
Secondary
planet (gear 2)
Prime
mover
Follower
(gear 4)
Fixed planet
(gear 1)

(A)

(B)

6-Compound planetary gear system

ing the synchronic index has not been investigated. However, some common applications of planetary gearing can be expressed as either a simple or compound spur-gear train, in which case the formulas presented here are valid with slight modifications.

For example the simple internal planetary gear train in Fig 5 (A) can be expressed as the simple gear train in Fig 5 (B). The compound planetary drive in Fig 6 (A) is expanded in Fig 6 (B) as a compound spur-gear drive. The fixed planet gear is considered the prime mover in these examples only for the purpose of computation because the planet arm (the actual prime mover) does not have gear teeth, and does not lend itself to the formulas.

Value for I_s is computed from Fig 5 (B) and 6 (B) in a way similar to that for simple or compound spur gearing. To find R, however, for a particular gear in the simple train, Fig 5 (B):

$$R_2 = I_s \times r_{1,2}$$
$$R_3 = I_s + (I_s \times r_{1,3})$$

For the compound system in Fig 6 (B)

$$R_3 = I_s \times r_{1,2}$$
$$R_4 = I_s + [I_s \times r_{1,2} \times r_{3,4}]$$

Applications at Bell Labs

One application of the synchronic-index method is the torque analysis of a precisely accurate gear reduction unit employed in a servo application. Our inspector was asked to examine the input static torque to certify the low-torque design requirements. Assuming that the gear ratio is other than an even whole number, through how many turns of input shaft rotation must torque measurements be made to ensure that all conditions of internal gear alignment have been tested? Computing the synchronic index gave the answer.

In another application, Fig 7, Gear A drives a position-sensitive electronic component, and Gear C drives a mechanical stylus which precisely records its behavior. Gear B is merely an idler that connects the two components. An eccentricity error in Gear B reflects an angular error to each of Gears A and C, giving one a negative error and the other a positive error as related to their normal direction of rotation (four positions of rotation are shown). A knowledge of the synchronic index quickly indicates the frequency at which exact coincidence could be expected between Gears A and C.

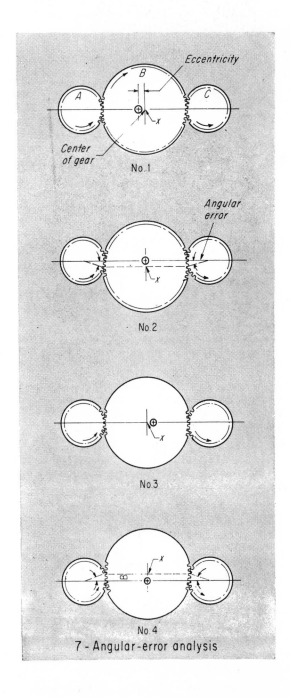

No. 1

No. 2

No. 3

No. 4

7 - Angular-error analysis

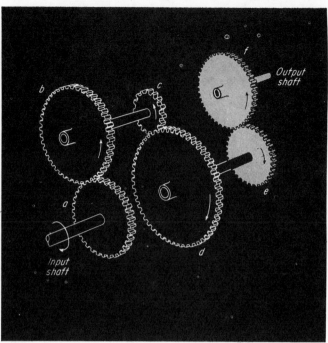

FIVE WAYS TO FIND GEAR RATIOS

For precise gear ratios the authors offer their automatic-calculator method that needs less trial and error, less manual computation . . . and consequently, gives a faster answer. Article also discusses four alternatives.

C T McCOMB and W U MATSON
US Naval Research Laboratory

The four-gear arrangement is still the most positive one for obtaining a speed ratio with five-place accuracy between a power source and its workload. And six gears, as shown in sketch, can give you seven-place accuracy. Selecting the proper change gears is fairly easy when the desired ratio is a simple fraction. But if it is an odd decimal and a high degree of accuracy is required, a systematic approach will save much time and effort.

A number of methods for calculation of change gears have preceded our new automatic-calculator method:

1. Logarithm method—utilizes a table of gear-ratio logarithms.
2. Smithson conjugate-fractions method—an arithmetical solution worked out from a table of decimals and fractions.
3. Rappaport algebraic method—it reduces non-linear equations with four unknowns to linear equations with two unknowns.
4. Gray calculator method—an earlier version of the new method.
5. McComb-Matson calculator method—needs least trial and error, and manual computation.

1 LOGARITHM METHOD (MACHINERY HANDBOOK, 16TH EDITION)

Selection of change gears can often be solved effectively by this method—particularly when your familiarity with logarithm tables lets you choose suitable gear ratios with minimum trial and error. This method calls for tables containing 6-place logs of all gear combinations between 16 and 120 teeth. Sketch above shows gears in train.

Example: Find four gears $a/b \times c/d = 2.105,399$. Log $2.105,39 = 0.323,334$. To keep the ratios, a/b and c/d, about equal, select from the tables that set of gears whose logarithm is equal to about one-half the ratio logarithm, such as log $57/37 = 0.187,673$. By subtracting this from the log $2.105,399 = 0.323,334$, the other logarithm is found to be 0.135661. From the table, log $41/30 = 0.135,663$. The result obtained is $57/37 \times 41/30 = 2337/1110 = 2.105,405$, with an error of only $0.000,006$.

2 SMITHSON'S METHOD OF CONJUGATE FRACTIONS

Sometimes a simple arithmetical method can be just as effective. In this method, two fractions, a/b and c/d, representing two pairs of change gears, are said to be conjugate if the difference of their cross-product is unity; that is, when

$$ad - bc = \pm 1$$

Also, if the numerators and denominators of such fractions are added, the resulting fraction is conjugate with either of the two original fractions. That is:

$$(a + c)b - (b + d)a = bc - ad = \pm 1$$

or

$$(a + c)d - (b + d)c = ad - bc = \pm 1$$

This means that the fraction formed by adding the numerators and denominators is the one with the smallest numerator and denominator lying between a/b and c/d. This is important in selecting change gears because it offers a fast method of finding all fractions between them which might result in suitable change gears. Two examples illustrate the effectiveness of this method—one simple; the other more difficult.

Example: Obtain a set of change gears with ratio equal to 0.528,19, and accurate within 0.000,05. From a table of equivalent fractions the following are obtained:

$$47/89 = 0.528,090 \quad \text{and} \quad 28/53 = 0.528,320$$

Adding the numerators and denominators result in:

$(47 + 28)/(89 + 53) = 75/142 = 0.528,17$, which is within the 0.000,05 limit. Resolving this fraction gives:

$75/142 = (3 \times 25)/(2 \times 71) = (60 \times 25)/(40 \times 71) = 0.528,17$. The change gears are: $a = 60$, $b = 40$, $c = 25$, $d = 71$.

Example: Obtain a set of change gears which give the ratio of 3.927,63 with a 0.000,02 tolerance. To use this method invert this ratio into a fraction, find change gears, and invert back to original ratio. Reciprocal of the ratio is:

$$1/3.927,65 = 0.254,605,2, \quad \text{and} \quad 1/3.927,61 = 0.254,607,8$$

From a table of equivalent fractions the following are the closest obtainable:

$$14/55 = 0.254,545,5, \quad \text{and} \quad 13/51 = 0.254,902,0$$

These two fractions are then conjugated in various steps until suitable change gears are obtained. These steps are summarized in the adjoining table.

Fractions in steps (a) and (b) are placed at top and bottom of the table and all the resulting conjugate fractions are placed in between, in the steps shown. The factors for resulting gear train are shown in right hand column. The reciprocal of this gear train is the required set of change gears. They are: $a = 33$, $b = 87$, $c = 49$, $d = 73$, and their ratio is $1/0.254,605,6 = 3.927,644$.

Gerald Smithson's article, "Converting Decimal Fractions to Actual Gear Ratios," appeared in PRODUCT ENGINEERING, Aug, 1952, p 158. The article gave a detailed explanation of conjugate fractions and their application. See also the use of matrix arithmetic, p 220.

3 RAPPAPORT ALGEBRAIC METHOD

Depending on the complexity of such problems and the inclination of the designer, the algebraic method is quite useful. In this method, nonlinear equations involving four unknowns, representing four change gears, are converted to linear equations of two unknowns. The basis for these equations is that the unknowns must be whole

ARITHMETICAL STEPS (a) TO (h)

Step	Formed From	Fraction	Decimal	Gears
(a)		$14/55$	0.254,545,5	
(d)	(a) and (c)	$\dfrac{69 + 14}{271 + 55} = \dfrac{83}{326}$	0.254,601,2	
(f)	(d) and (e)	$\dfrac{83 + 152}{326 + 597} = \dfrac{235}{923}$	0.254,604,6	
(g)	(f) and (e)	$\dfrac{235 + 152}{923 + 597} = \dfrac{387}{1520}$	0.254,605,2 0.254,605,3	$\dfrac{60}{27} \times \dfrac{76}{43}$
(h)	(g) and (e)	$\dfrac{387 + 152}{1520 + 597} = \dfrac{539}{2117}$	0.254,605,6	$\dfrac{87}{33} \times \dfrac{73}{49}$
(e)	(d) and (c)	$\dfrac{83 + 69}{326 + 271} = \dfrac{152}{597}$	0.254,606,4 0.254,607,8	
(c)	(a) and (b)	$\dfrac{13 + 4 \times 14}{51 + 4 \times 55} = \dfrac{69}{271}$	0.254,612,5	
(b)		$13/51$	0.254,902,0	

numbers—particularly applicable here since all change gears must represent whole numbers (of teeth).

Example: Given four change gears $a/b \times c/d = \frac{1}{2}$, and center distance between shafts = 70 teeth; then

$$a + b = c + d = 70$$

This reduces a nonlinear equation in four unknowns to a linear equation in two unknowns. Solving these equations by the algebraic method, all four change gears are represented in terms of one unknown, as follows:

$$a = w - 70; \quad b = 140 - w; \quad c = \frac{70(140)}{w} - 70; \quad d = 70\left(\frac{2 - 140}{w}\right)$$

From the original stipulation that all change gears must be whole numbers it follows that w must also be a whole number. Accordingly, it follows that 70 is less than w; w is less than 140. Also, $70(140)/w$ and $140/w$ must be whole numbers. Now the prime factors of $70(140)$ are $2 \times 2 \times 2 \times 5 \times 5 \times 7 \times 7$. Consequently, w consists of some of these prime factors, and lies between 70 and 140. Only $7 \times 7 \times 2$ and $5 \times 5 \times 2 \times 2$ fulfill these conditions; therefore $w_1 = 98$, $w_2 = 100$ are the numbers. Substituting these values above, the following two sets of change gears result in:

$$a_1 = 28, \quad b_1 = 42, \quad c_1 = 30, \quad d_1 = 40$$
$$a_2 = 30, \quad b_2 = 40, \quad c_2 = 28, \quad d_2 = 42$$

Sigmund Rappaport's article, "Whole-number Solutions of Indeterminate Equations for Gearing," appears on p 218.

Series	Key Board	Middle Dial	Upper Dial	From Table
A	R	Rb	b	ac
	0.103,872,541,1	10.906,616,815,5	105	
	Rb	Rbd	d	
1	10.906,616,82	1134.288,149,28	104	
2	"	1123.381,532,46	103	
3	"	1112.474,915,64	102	
4	"	1101.568,298,82	101	
5	"	1090.661,682	100	
6	"	1047.035,214,72	96	
7	"	817.996,261,5	75	$818 = 409 \times 2$
8	"	698.023,476,48	64	$698 = 349 \times 2$
9	"	578.050,691,40	53	$578 = 17 \times 34$
10	"	468.984,523,20	43	$469 = 7 \times 67$
Series B	R	Rb	d	
	0.103,872,541,1	10.802,744,274,4	104	
	Rb	Rbd	b	
1	10.802,744,27	1112.682,659,81	103	
2	"	821.008,564,52	76	$821 = \text{No Factor}$
3	"	766.994,843,17	71	$767 = 13 \times 59$
4	"	712.981,121,82	66	$713 = 23 \times 31$
Series C	R	Rb	b	
	0.103,872,541,1	10.698,871,733,3	103	
	Rb	Rbd	d	
1	10.698,871,73	1091.284,916,46	102	
2	"	994.995,070,89	93	$995 = 5 \times 199$
3	"	888.006,353,59	83	$888 = 24 \times 37$

4 GRAY CALCULATOR METHOD

The familiar change-gear problem, $a/b \times c/d = R$ is set up in this method as $a/b \times c/d = xR/x$, in which x is an arbitrary integer. To select the proper gears, a, b, c, d, you select an arbitrary integer, x, which can be factored into numbers suitable for gears, b and d, and whose product with the ratio, R, can be factored into numbers suitable for gears, a and c. The product, xR, is set up in a calculator by locking the ratio, R, on the keyboard of the calculator and repeatedly touching the plus bar until the product, xR, appearing on the accumulator dial, approaches an integar.

Once suitable preliminary selections of x are made, the calculator can quickly give the resulting product, xR, which may border a suitable integer. Although, this method needs considerable manual computation, it works well where all practical values of a given range of gear sizes are desired.

Henry C. Gray's article, "Converting Decimal Fractions Into Compound Gear Trains," appeared in PRODUCT ENGINEERING, June 1953, p 215.

5 McCOMB-MATSON CALCULATOR METHOD

For a designer faced with the problem of selecting an accurate set of change gears, an automatic calculator and a table of equivalent factors in gear ratios is all that is required. The calculator should have the following characteristics:

• Three registers on which the number, product, and multiplier are indicated simultaneously. They are referred to here as key board, middle dial and upper dial respectively.

• The key board must have 10 places for high-degree accuracy.

• There must be an optional restraint (or "non-shift" as on Marchant 10FA) of shifting the multiplier carriage.

The table of factors could be any table of numbers and equivalent factors which cover the range of gears available. For example, "Numbers and Equivalent Factors in 14,000 Gear Ratios," R. N. Page, Industrial Press.

Example:

A typical problem with an odd decimal may be stated as follows: Find four change gears a, b, c, d that will give a gear ratio, $R = ac/bd = 0.103,872,541,1$ with a tolerance of 0.000,002. The gears available range from 23 to 105 teeth inclusive.

The above equation can be written as: $Rbd = ac$. In this equation the factors R, b, and d are set on an automatic calculator in a series of steps so as to rapidly give a whole number, ac, whose factors are easily obtained from a standard table as mentioned above.

The table above shows in tabular form all the steps taken in the various series until the proper change gears

AUTOMATIC CALCULATOR with 10-place keyboard, three registers, and optional non-shift button is basis for new method of obtaining change gears.

are found. Start with Series A, letting $b = 105$, the largest gear available. Set R on the keyboard dial and multiply by 105 to get Rb. Now set this product in keyboard dial and clear the other dials—middle and upper. Step "A1" is obtained by multiplying $Rb \times d$. In this step, d is the next largest gear available, 104.

By restrictions set in the problem the resulting product must be an integer factorable by a and c, whose teeth are between 23 and 105. The closer Rbd is to an integer, the more accurate will be the answer. For this reason the two digits to the right of the decimal in the product, Rbd, are underlined to indicate closeness to an integer. The succeeding steps in Series A are obtained as follows:

1. Lock the traverse with indicator on the last digit in 104 in the upper dial by pressing "non-shift" button.
2. Negatively multiply by 1, by holding negative multiplier button down while successively pressing multiplier button "1."
3. When the first two digits after the decimal in the middle dial reach .99, .00, .01, .02, check the nearest resulting integer for factors a and c.

The first step, "A7," which gives a close integer for Rbd indicates that $818 = 409 \times 2$ might be suitable for a and c. However, no such gears are available, and the series is continued until the next integer shows up in the middle dial. Again, no gears are available for $698 = 349 \times 2$, the resulting factors. Series A is continued until integer, $469 = 7 \times 67$, appears. The 7-tooth gear in this case is unavailable. Also, the product, 469, is less than $23 \times 24 = 552$, the product of two smallest available gears. This indicates that a new series, "B," must be started, using $b = 104$ and proceeding again with a new series of d gears. In this series the first close integer to appear in the middle dial is 821, which is not factorable. Next integer, $767 = 13 \times 59$, contains factor, 13, for which there is no gear available. Series B is continued until the next integer, $713 = 23 \times 31$ shows up. The resulting factors here indicate available gears, so a check of the gear ratio is made:

$$\text{Actual } R = ac/bd = 713/104 \times 66 = 0.103,875,291,3$$
$$\text{Desired } R \qquad\qquad\qquad = 0.103,872,541,1$$
$$\overline{\qquad\qquad\qquad\qquad\qquad 0.000,002,750,2}$$

This actual R is slightly larger than specified, so a new series, "C" is started. Here gear $b = 103$ is selected and in step, "C1" of this series $d = 102$ is selected. Negative multiplications by 1 gear are made as in the previous series until the first integer, $995 = 5 \times 199$, appears in the middle dial. Since no gears are available for the resulting factors, the series is continued until next integer, $888 = 24 \times 37$ is reached. Since this number indicates that the resulting factors represent available gears, a check of the gear ratio is made again:

$$\text{Desired } R \qquad\qquad\qquad\qquad = 0.103.872,541,1$$
$$\text{Actual } R = (24 \times 37)/(103 \times 83) = 0.103,871,797,8$$
$$\overline{\qquad\qquad\qquad\qquad\qquad\quad 0.000,000,743,3}$$

which is less than the 0.000,002 specified. Therefore, the desired change gears are: $a = 24$, $b = 103$, $c = 37$, $d = 83$.

Once familiarity with this method is established, the solution of such problems should not require more than 5 to 30 minutes.

The method can be extended to more difficult problems involving three pairs of gears. The equation for such problems is:

$$R = ace/bdf$$

By assigning the ratio, $R_1 = e/f$ and solving above equation for a resulting ratio, the problem is put into the familiar form:

$$R_t = R/R_1 = ac/bd$$

This equation is then solved in the same manner as described above. Using R_t and its solution finds ac with trials of b and d as before. However, since four gears are now being chosen for trials, several choices of $R_1 = e/f$ will be required before suitable gears are found.

Whole Number Solutions of Indeterminate Equations for Gearing

SIGMUND RAPPAPORT
Project Supervisor
Ford Instrument Co.*

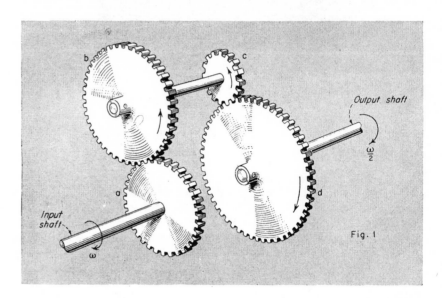

Fig. 1

INDETERMINATE EQUATIONS, those that have an infinite number of solutions, frequently arise in gearing problems, especially planetary gearing. But only a few of the many solutions, or perhaps none at all, will satisfy the practical conditions of a given problem. For example, the solutions may have to be positive integers, as where the unknowns in the equations refer to the number of gear teeth; or the solutions may have to be minimal values, or lie between given limits. In problems that are not too complicated, a trial-and-error method is often successful, although one can never be sure that the solutions found are the best ones, or even whether or not other solutions exist.

Assume the following equation were given:

$$3x + 5y - 172 = 0$$

In addition, x and y are to be positive integers, and their difference to be a minimum. The trial-and-error method might proceed like this:

To satisfy the condition that the difference between x and y is minimal, a first approach could be: $x = y$.

Substituting this condition into the given equation:

$$3x + 5x - 172 = 0,$$

Therefore

$x = 21.5$, which is not an integer.

* Division of The Sperry Corp.

A second attempt might be

$$x = 20.$$

Then,

$y = 22.4$, which is still not an integer.

After repeated trials the equation may be solved for $x = 24$, $y = 20$; but it is still uncertain as to whether or not more suitable solutions exist.

A method developed almost 2,000 years ago, but little known among engineers, solves this kind of problem. It is generally referred to as "Diophantine Equations". Applied to the preceding example, it proceeds as follows: Given,

$$3x + 5y - 172 = 0.$$

Therefore,

$$x = \frac{172 - 5y}{3}$$

$$x = 57 + 1/3 - y - \frac{2y}{3}$$

Or $\quad x = 57 - y + \frac{1 - 2y}{3} \qquad (2)$

which is obtained by first separating and then gathering the integral and the fractional terms.

It is required that x and y be integers, hence $\frac{1 - 2y}{3}$ must be an integer. Call this integer u. Then:

$$\frac{1 - 2y}{3} = u; \qquad (2a)$$

Solving equation (2a) for y and again separating integral and fractional terms:

$$y = -u - \frac{u}{2} + \frac{1}{2}$$

$$y = -u + \frac{-u + 1}{2} \qquad (3)$$

Since by definition both y and u are integers, $\frac{-u+1}{2}$ must be an integer, and can be designated by v. Then,

$$v = -\frac{u + 1}{2} \text{ or} \qquad (4)$$

$$u = 1 - 2v$$

This eliminates all fractional terms. Substituting Eq (4) into Eq (3)

$$y = 3v - 1. \qquad (5)$$

Similarly, from Eqs (2), (2a), (4) and (5)

$$x = 59 - 5v \qquad (6)$$

Therefore, x and y will be integers for all integral values of v. For example, if $v = 1$, $x = 54$, and $y = 2$.

Considering the other initial condition that the difference between x and y be as small as possible, Eq (5) is set equal to Eq (6). Hence,

$$59 - 5v = 3v - 1$$

or

$$8v = 60:$$

The values of v that satisfy this condition as closely as possible are $v = 7$ and $v = 8$. Substituting $v = 7$ in Eqs (5) and (6), $x = 24$ and $y = 20$ (the solution found previously by trial-and-error). Similarly, for $v = 8$, $x = 19$, and $y = 23$. Thus, the analyzer has his choice of two possible solutions. Example 2: Given,

$$23x - 59y - 88 = 0.$$

Find the solution giving the smallest positive integers for x and y. The procedure follows that of the first example.

$$x - 2y - \frac{13y}{23} - 3 - \frac{19}{23} = 0$$

$$x - 2y - 3 - \frac{13y + 19}{23} = 0 \qquad (7)$$

Setting the last term on the right hand side equal to integer u,

$$13y + 19 - 23u = 0, \text{ or}$$

$$y + 1 + \frac{6}{13} - u - \frac{10u}{13} = 0, \text{ and}$$

$$y + 1 - u + \frac{6 - 10u}{13} = 0$$

Setting the last term on the left hand side equal to integer v,

$$6 - 10u - 13v = 0, \text{ or}$$

$$-u - v + \frac{6 - 3v}{10} = 0 \qquad (8)$$

Again, equating the last term on the left hand side to integer w,

$$-3v + 6 - 10w = 0$$

$$v - 2 + 3w + \frac{w}{3} = 0 \qquad (9)$$

If the last term on left hand side is replaced by integer p,

$$w = 3p \qquad (10)$$

Substituting Eq (10) in Eq (9):

$$v = -10p + 2 \qquad (11)$$

Similarly,

$$u = 13p - 2 \text{ and} \qquad (12)$$

$$y = 23p - 5 \qquad (13)$$

Non-Linear Equations

Diophantine equations are applicable to linear equations only. However, by modifying the method, some classes of non-linear equations can be handled.

Suppose the gearing shown in the Fig. 1 must have an overall ratio

$$\frac{a}{b} \cdot \frac{c}{d}$$

of 1:2, where a, b, c and d designate the numbers of teeth. Input shaft and output shaft are co-linear, hence the center distance between a and b equals the center distance between c and d. Further, this center distance is given such that the sum of a and b as well as the sum of c and d is 70 teeth. Then,

$$\left(\frac{a}{b}\right)\left(\frac{c}{d}\right) = 1/2 \qquad (14)$$

$$a + b = 70 \qquad (15)$$

$$c + d = 70 \qquad (16)$$

These three equations with four unknowns can be reduced to one equation with two unknowns. This, and the stipulation that all values be whole numbers, is a characteristic Diophantine condition. Substituting Eqs (15) and (16) in Eq (14)

$$\left(\frac{a}{70 - a}\right)\left(\frac{c}{70 - c}\right) = 1/2$$

Solving for a in terms of c,

$$a = (4900 - 70c) / (70 + c), \text{ or}$$

$$a = 70 - \frac{140c}{70 + c} = 70 - u, \qquad (17)$$

where

$$\frac{140c}{70 + c}$$

has been set equal to integer u. Therefore,

$$140c = 70u + uc$$

$$c = \frac{70u}{140 - u}$$

$$c = -70 + v \qquad (18)$$

Finally, from Eqs (7), (12) and (13)

$$x = 59p - 9$$

To make x and y the smallest positive integers, p is taken as unity and the solution is $x = 50$, $y = 18$.

where integer $v = \dfrac{70(140)}{140 - u}$

Therefore,

$$70(140) = 140v - uv$$

$$uv = 140(v - 70)$$

Solving for u,

$$u = \frac{140(v - 70)}{v} = 140 - \frac{70(140)}{v}$$

Or

$$u = 140 - w \qquad (19)$$

where integer $w = \dfrac{70(140)}{v}$;

Then,

$$v = \frac{70(140)}{w} \qquad (20)$$

Substituting Eq (20) in Eq (18),

$$c = -70 + \frac{70(140)}{w} \qquad (21)$$

Similarly, from Eqs (17) and (19),

$$a = 70 - 140 + w = w - 70 \qquad (22)$$

Also, $b = 70 - a = 140 - w \qquad (23)$

Finally, $d = 70\left(2 - \dfrac{140}{w}\right) \qquad (24)$

Since a and b must be positive integers; from Eqs (22) and (23), w must satisfy the condition:

$$70 < w < 140$$

Also, it follows from Eq (20) that 70 (140) must be divisible by w. The prime factors of 70 (140) are 2 x 2 x 2 x 5 x 5 x 7 x 7. Consequently, w consists of some of these prime factors, and lies between 70 and 140. Only 7 x 7 x 2 and 5 x 5 x 2 x 2 fulfill these conditions; therefore, $w_1 = 98$, $w_2 = 100$. Substituting these values in Eqs (21), (22), (23) and (24),

$$a_1 = 28, b_1 = 42, c_1 = 30, d_1 = 40,$$

and

$$a_2 = 30, b_2 = 40, c_2 = 28, d_2 = 42$$

Thus, as it may have been expected, gears a and b can be interchanged with c and d, the other pair.

Use matrix arithmetic
To find gear ratios

ARNE BENSON, staff consultant, Sanders Associates Inc., Nashua, N. H.

FINDING suitable change gears for a specified ratio is actually two problems in one: 1) find a rational number N/D close enough to the specified ratio R; 2) factor N and D into acceptable gear-tooth numbers. The first part of this problem has received a great deal of attention while the second part still must be solved by the cut-and-try method, or by reference to factor tables.

Matrix arithmetic offers a systematic way to determine all fractions within specified limits of a given ratio. It converges rapidly to the region in which R lies; it is not subject to cumulative error; and it is easily arranged for paper-pencil, desk calculator or computer work.

The method is based upon conjugate fraction theory which states that two fractions a/b and c/d *are* conjugate if $(ad - bc) = \pm 1$. This means that between two conjugate fractions there exists no other fraction with smaller numerators or denominators. In this sense each conjugate fraction is "a best approximation" of the other. If the numbers a and b are chosen so that a/b is a fraction in lowest terms greater than the desired ratio R, and if c and d are similarly chosen so that c/d is a fraction in lowest terms less than the desired ratio R, the basic matrix can be formed:

$$B = \begin{vmatrix} a & c \\ b & d \end{vmatrix}$$

In matrix theory B is unimodular if $(ad - bc) = 1$.

A multiplier matrix is now needed and can be formed from the unit matrix:

$$X = \begin{vmatrix} 1 & 0 \\ 0 & 1 \end{vmatrix}$$

SYMBOLS

G, B, X	—2 row unimodulator matrices
B^{-1}	—is inverse of B matrix
R	—gear-train ratio
N	—numerator of R
D	—denominator of R
a, b, c, d	—elements of B matrix, positive integers
a/b and c/d	—gear ratios taken from standard tables

---

by adding adjacent row elements and interposing their sum:

First stage
$$\begin{vmatrix} 1 & 1 & 0 \\ 0 & 1 & 1 \end{vmatrix}$$

Second stage
$$\begin{vmatrix} 1 & 2 & 1 & 1 & 0 \\ 0 & 1 & 1 & 2 & 1 \end{vmatrix}$$

Third stage
$$\begin{vmatrix} 1 & 3 & 2 & 3 & 1 & 2 & 1 & 1 & 0 \\ 0 & 1 & 1 & 2 & 1 & 3 & 2 & 3 & 1 \end{vmatrix}$$

This is a new concept in matrix arithmetic which was specially developed by the author to solve gear ratios. Note that:

• The unit matrix can be expanded indefinitely;
• When regarded as a fraction each column is in lowest terms;
• The fractions follow a descending order;
• Each adjacent 2×2 sub-matrix is unimodular;
• No fraction with smaller terms exists with value between adjacent fractions.

These unique properties of the multiplier matrix can be used to find the specified ratio R which lies between the two ratios in the basic matrix by multiplying the two matrices: $BX = G$.

To show how this works in a practical way, suppose

$$R = 1/\sqrt{\pi} = 0.564,189$$

is to be approximated within $\pm$ 0.000,02. From gear ratio tables select

$$a/b = 35/62 = 0.564,516$$
$$c/d = 22/39 = 0.564,102$$

Since
$$35 \times 29 - 22 \times 62 = 1$$

the basic matrix is unimodular, and it can be multiplied by the X matrix to give:

Unit Matrix:
$$\begin{vmatrix} 35 & 22 \\ 62 & 39 \end{vmatrix} \times \begin{vmatrix} 1 & 0 \\ 0 & 1 \end{vmatrix} = \begin{vmatrix} 35 & 22 \\ 62 & 39 \end{vmatrix}$$

First stage:
$$\begin{vmatrix} 35 & 22 \\ 62 & 39 \end{vmatrix} \times \begin{vmatrix} 1 & 1 & 0 \\ 0 & 1 & 1 \end{vmatrix} = \begin{vmatrix} 35 & 57 & 22 \\ 62 & 101 & 39 \end{vmatrix}$$

A good way to start using this new mathematical tool is to apply it to an old problem

Second stage: $\begin{vmatrix} 35 & 22 \\ 62 & 39 \end{vmatrix} \times \begin{vmatrix} 1 & 2 & 1 & 1 & 0 \\ 0 & 1 & 1 & 2 & 1 \end{vmatrix} =$

$= \begin{vmatrix} 35 & 92 & 57 & 79 & 22 \\ 62 & 163 & 101 & 140 & 39 \end{vmatrix}$

Third stage: $\begin{vmatrix} 35 & 22 \\ 62 & 39 \end{vmatrix} \times \begin{vmatrix} 1 & 3 & 2 & 3 & 1 & 2 & 1 & 1 & 0 \\ 0 & 1 & 1 & 2 & 1 & 3 & 2 & 3 & 1 \end{vmatrix}$

$= \begin{vmatrix} 35 & 127 & 92 & 149 & 57 & 136 & 79 & 101 & 22 \\ 62 & 225 & 163 & 264 & 101 & 241 & 140 & 179 & 39 \end{vmatrix}$

The desired ratio lies between 101/179 and 22/39 so further effort needs to be concentrated only in this area of the G matrix. This can be done by expanding the lower portion of the X matrix to give:

$$\mathbf{X} = \begin{vmatrix} 1 & 2 & 1 & 1 & 0 \\ 3 & 7 & 4 & 5 & 1 \end{vmatrix}$$

Multiplying this with a portion of the B matrix gives:

$$BX = G$$

$\begin{vmatrix} 101 & 22 \\ 179 & 39 \end{vmatrix} \times \begin{vmatrix} 1 & 2 & 1 & 1 & 0 \\ 3 & 7 & 4 & 5 & 1 \end{vmatrix} = \begin{vmatrix} 101 & 224 & 123 & 145 & 22 \\ 179 & 397 & 218 & 257 & 39 \end{vmatrix}$

It appears now that $145/257 = 0.564{,}202$ is acceptably close to the specified R. Unfortunately, 257 is a prime number and it cannot be factored into suitable sets of change gears.

To proceed from here the lower portion of the X matrix

$$\begin{vmatrix} 1 & 1 & 0 \\ 4 & 5 & 1 \end{vmatrix}$$

can be expanded further and the resulting X matrix can then be multiplied:

$\begin{vmatrix} 123 & 22 \\ 218 & 39 \end{vmatrix} \times \begin{vmatrix} 1 & 2 & 1 & 1 & 0 \\ 4 & 9 & 5 & 6 & 1 \end{vmatrix} = \begin{vmatrix} 123 & 268 & 145 & 167 & 22 \\ 218 & 275 & 257 & 296 & 39 \end{vmatrix}$

This might yield a favorable ratio.

A second alternative is to select one or two new B matrices from the above G matrix and multiply them by an expanded unit matrix to give:

$\begin{vmatrix} 123 & 145 \\ 218 & 257 \end{vmatrix} \times \begin{vmatrix} 1 & 2 & 1 & 1 & 0 \\ 0 & 1 & 1 & 2 & 1 \end{vmatrix} = \begin{vmatrix} 123 & 391 & 268 & 413 & 145 \\ 218 & 693 & 475 & 732 & 257 \end{vmatrix}$

or

$\begin{vmatrix} 145 & 22 \\ 257 & 39 \end{vmatrix} \times \begin{vmatrix} 1 & 2 & 1 & 1 & 0 \\ 0 & 1 & 1 & 2 & 1 \end{vmatrix} = \begin{vmatrix} 145 & 312 & 167 & 189 & 22 \\ 257 & 553 & 296 & 335 & 39 \end{vmatrix}$

In these multiplications the unit matrix may be expanded to any length necessary to produce a suitable ratio.

HOW MATRIX ARITHMETIC IS USED

A Matrix is a rectangular array of numbers. In this application only 2-row matrices are used.

Matrices can be added, multiplied and raised to a power. Equations can be written with matrices as terms.

Multiplication is possible when the first matrix has as many columns as the second has rows. For example, if

$$A = \begin{vmatrix} a & c \\ b & d \end{vmatrix} \text{ and } B = \begin{vmatrix} e & g \\ f & h \end{vmatrix}$$

then their product

$$A \times B = \begin{vmatrix} (ae + cf) & (ag + ch) \\ (be + fd) & (bg + dh) \end{vmatrix}$$

Multiplication is not commutative; that is, the product $A \times B$ does not necessarily equal $B \times A$.

Determinant is a function of a square matrix, which, in the 2×2 case:

$$det: \begin{vmatrix} a & c \\ b & d \end{vmatrix} = (ad - bc)$$

Unimodular matrix has a determinant equal to ± 1. Inverse of a 2×2 unimodular matrix

$$B = \begin{vmatrix} a & c \\ b & d \end{vmatrix}$$

$$B^{-1} = \begin{vmatrix} d & -c \\ -b & a \end{vmatrix}$$

Thus the matrix equation $AX = B$ can be solved for X by finding the inverse of A, then $X = A^{-1}B$.

All rational numbers can be formed by adding adjacent numerators and denominators starting with 1/0 and 0/1. This is the easiest way to find the desirable ratio when the tolerance is wide, but for close tolerance ratios the labor of repeated long division is excessive and the chance of error increases accordingly. Also a single addition error invalidates all that follows.

With a properly formulated matrix equation you can quickly locate the region in which acceptable ratios lie, and they can be readily programmed for either calculator or computer operation. Computationally, it is less subject to error, and, by solving a matrix equation, you identify the tolerance region in one step without involving recurrence processes and continual testing of fractions. Mathematically, the properties of rational numbers is actually just a part of the larger theory of matrices.

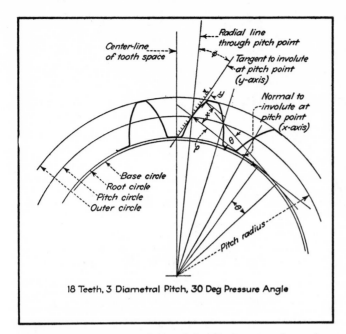

18 Teeth, 3 Diametral Pitch, 30 Deg Pressure Angle

Fig. 1—Nomenclature and legend for development of equations for calculating the coordinates of involute gear tooth profiles.

Fig. 2—Plot of tooth profile for an 18 tooth, 3 diametral pitch, 30 deg pressure angle gear. Table IV gives detailed calculations.

Involute Curve Ratio Constants
For Plotting Gear Tooth Profiles

EDWARD C. VARNUM

Mathematician, Barber-Colman Company

TEDIOUS MATHEMATICAL OPERATIONS are associated with the use of formulas in laying out involute gear tooth profiles. With tables of numerical ratios, most of such work can be avoided when calculating the coordinates of plot points. These tables can be readily compiled from the equations and constants given in "Involute Curve Calculations," PRODUCT ENGINEERING, October 1949, p. 135, by Mr. Allan Candee.

The tables of numerical ratios given here have been compiled for standard involute gear teeth of three pressure angles; $14\frac{1}{2}$, 20, and 30 degrees.

Using the nomenclature of Fig. 1,

x = plot point coordinate measured along the X-axis, that is, the tangent to the involute curve at the pitch point O, in.

y = plot point coordinate measured perpendicularly from the x-point on the X-axis to the point P on the curve, in.

r = pitch radius of gear, in.

ϕ = pressure angle at pitch point O, deg.

θ = angular displacement of the normal around the base circle, deg.

the formulas for the x and y coordinates as given by Mr. Candee are

$$x = r\ [\sin\phi\sin\theta \pm \cos\phi\ (\theta\sin\theta - \text{vers}\ \theta)] \qquad \text{(A1)}$$

$$y = r\ [\sin\phi\ \text{vers}\ \theta \pm \cos\phi\ (\sin\theta - \theta\cos\theta)] \qquad \text{(B1)}$$

with the suggestion that the expressions in brackets be called constants C_A and C_B, respectively, so that

$$C_A = x/r, \text{ and } C_B = y/r$$
$$\text{or} \quad x = rC_A, \text{ and } y = rC_B$$

In accord with this suggestion, Tables I, II, and III have been compiled for

values of C_A and C_B for three pressure angles. For each of the pressure angles, values of θ were chosen to make C_A a convenient multiplier. With the selected values of θ, corresponding values of C_B were calculated. With this tabular data, full size plot point coordinates are obtained simply by multi-

Table I—Involute Curve Constants for Plotting Above Pitch Point

$C_A = x/r$	$C_B = y/r$		
	$\phi = 14\frac{1}{2}$ deg	$\phi = 20$ deg	$\phi = 30$ deg
0.200000	0.052769	0.044476	0.034581
0.180000	0.043588	0.036569	0.028270
0.160000	0.035208	0.029380	0.022563
0.140000	0.027636	0.022916	0.017467
0.120000	0.020887	0.017189	0.012990
0.100000	0.014983	0.012218	0.009141
0.080000	0.009955	0.008028	0.005936
0.060000	0.005852	0.004653	0.003393
0.050000	0.004167	0.003285	0.002377
0.040000	0.002742	0.002140	0.001535
0.030000	0.001591	0.001228	0.000872
0.020000	0.000732	0.000557	0.000391
0.018000	0.000598	0.000453	0.000318
0.016000	0.000476	0.000360	0.000251
0.014000	0.000367	0.000277	0.000193
0.012000	0.000272	0.000204	0.000142
0.010000	0.000191	0.000143	0.000099
0.008000	0.000123	0.000092	0.000063
0.006000	0.000070	0.000052	0.000036
0.004000	0.000031	0.000023	0.000016
0.002000	0.000008	0.000006	0.000004

plying C_A and C_B by the pitch radius of the gear. As an additional convenience in calculation, the C_A values are in equal increments so that all x-values can be plotted by successive addition along the X-axis.

The data given in Tables I, II and III are for pressure angles of $14\frac{1}{2}$, 20 and 30 deg, since the majority of gears and splines fall in one of these three groups. Adjustments for pressure angles other than these can be made by finding a point on the desired involute at which the pressure angle is $14\frac{1}{2}$, 20 or 30 deg. This point can then be considered as the pitch point and its radius taken as the pitch radius for purposes of calculation and layout. Because of the popularity of the three pressure angles chosen, however, such adjustments will not often be necessary.

In Table I are given constants for points on the involute above the pitch circle. For these calculations, the plus signs preceding the term $\cos\phi$ in the brackets of Eqs (A1) and (B1) are used. Table II gives the constants with which plot points on the involute below the pitch circle can be readily calculated. Table III gives values of C_A and C_B for which the respective involute curves intersect their base circles.

For gears with 50 or more teeth, the values given in the upper part of Table I will be most useful. For 30 to 50 teeth, the values given midway in Table I will give the best results. The values in the lower part of Table I will be more suitable for pinions and splines. The data in the first column are used in every problem, whereas, only one of the other columns is used for a specific gear as determined by its pressure angle.

The last three columns of Table II are terminated at different points because the involute curve does not extend to the next step. In Table III are given the exact values of C_A and C_B at which the involute curves intersect their base circles. The values in Table III were calculated by evaluating the brackets in the following equations:

$$x = r\,[\cos\phi \text{ vers } (\tan\phi)] \qquad (A2)$$
$$y = r\,[\sin\phi - \cos\phi \sin(\tan\phi)] \qquad (B2)$$

Equations (A2) and (B2) are obtained by substituting θ equals $\tan\phi$ in Eqs (A1) and (B1) with the minus signs in effect. Radian measure was used for θ, that is, the number of radians in θ equaled the tangent of the pressure angle.

The procedure used when plotting an involute gear tooth with the tabular data given is indicated in the following example.

EXAMPLE: Plot the profile of an involute tooth for a gear of 18 teeth, 3 diametral pitch, and 30 deg. pressure angle.

SOLUTION: From the conventional gear formula

$$r = \frac{18}{3} \times \frac{1}{2} = 3 \text{ in.}$$

This value of r is multiplied by values of C_A and C_B to obtain values of x and y, respectively, as indicated in Table IV. The coordinates x and y are then plotted as shown in Fig. 2 to obtain the tooth profile. For clarity of presentation the plot in Fig. 2 is shown enlarged.

Table II—Involute Curve Constants for Plotting Below Pitch Point

$C_A = x/r$	$C_B = y/r$		
	$\phi = 14\frac{1}{2}$ deg	$\phi = 20$ deg	$\phi = 30$ deg
0.002000	0.000008	0.000006	0.000005
0.004000	0.000033	0.000024	0.000016
0.006000	0.000074	0.000054	0.000036
0.008000	0.000134	0.000096	0.000065
0.010000	0.000212	0.000150	0.000101
0.012000	0.000309	0.000218	0.000146
0.014000	0.000426	0.000302	0.000198
0.016000	0.000565	0.000392	0.000261
0.018000	0.000726	0.000500	0.000331
0.020000	0.000912	0.000621	0.000410
0.030000	0.002306	0.001451	0.000935
0.040000	....	0.002696	0.001686
0.050000	....	0.004457	0.002675
0.060000	....	0.006977	0.003917
0.080000	....	....	0.007231
0.100000	....	....	0.011828
0.120000	....	....	0.018081
0.140000	....	....	0.027087

Table III—Involute Curve Constants for Base Circle Intersection

$C_A = x/r$	$C_B = y/r$		
	$\phi = 14\frac{1}{2}$ deg	$\phi = 20$ deg	$\phi = 30$ deg
0.032197	0.002782	....	....
0.061559	....	0.007502	....
0.140372	....	....	0.027319

Table IV—Coordinates of Tooth Profile for an 18 Tooth, 3 Diametral Pitch, 30 Degree Pressure Angle Gear

	C_A	$x = rC_A = 3C_A$, in.	C_B	$y = rC_B = 3C_B$, in
	From Table I		From Table I	
Points above Pitch Line	0.020000	0.060000	0.000391	0.001173
	0.040000	0.120000	0.001535	0.004605
	0.060000	0.180000	0.003393	0.010179
	0.080000	0.240000	0.005936	0.017808
	0.100000	0.300000	0.009141	0.027423
	0.120000	0.360000	0.012990	0.038970
	From Table II		From Table II	
Points below Pitch Line	0.020000	0.060000	0.000410	0.001230
	0.040000	0.120000	0.001686	0.005058
	0.060000	0.180000	0.003917	0.011751
	0.080000	0.240000	0.007231	0.021693
	0.100000	0.300000	0.011828	0.035484
	0.120000	0.360000	0.018081	0.054243
	0.140000	0.420000	0.027087	0.081261
	From Table III		From Table III	
Base Circle Intersection	0.140372	0.421116	0.027319	0.081957

Contact Ratio Charts for Designing Spur Gears

WAYNE H. BOOKMILLER
Gear Engineering Division, General Electric Company

CONTACT RATIOS of various sets of gears are frequently calculated to help determine which set will give best service in a particular installation. Usually each gear set is considered as an individual case and a special calculation or layout made. These charts have been prepared to minimize calculation and avoid repetition.

A series of tooth form layouts were made to cover a wide range of tooth numbers for 20 deg pressure angle gears. Layouts were made for standard and long-short addenda gears. The data compiled from these layouts are presented in chart form.

As an example of the use of the charts, consider this pinion and gear that have standard addenda:

	Pinion	Gear
Number of teeth............	28	85
Diametral pitch	20	20
Pressure angle, deg..........	20	20
Pitch diameter, in..........	1.400	4.250
Addendum, in.	0.050	0.050
Tooth thickness, in..........	0.077	0.077

The contact ratio, M_P, for this gear set is 1.735, as read on the graph of Fig. 1.

The graph of Fig. 2 gives the addenda for equal Y

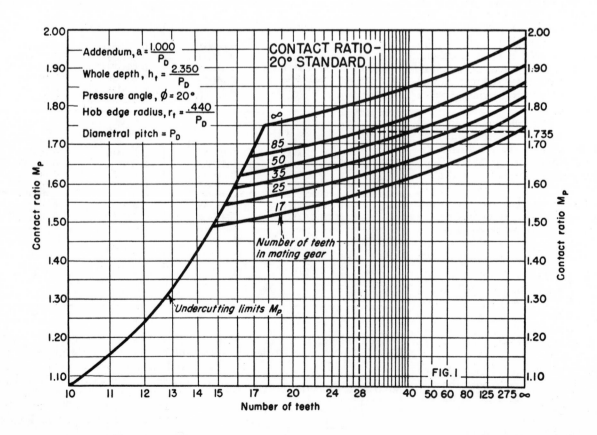

FIG. 1

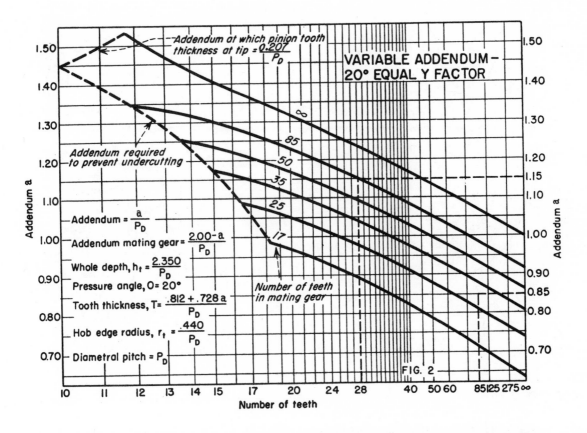

Addendum = $\dfrac{a}{P_D}$

Addendum mating gear = $\dfrac{2.00 - a}{P_D}$

Whole depth, $h_t = \dfrac{2.350}{P_D}$

Pressure angle, $O = 20°$

Tooth thickness, $T = \dfrac{.812 + .728a}{P_D}$

Hob edge radius, $r_t = \dfrac{.440}{P_D}$

Diametral pitch = P_D

VARIABLE ADDENDUM –
20° EQUAL Y FACTOR

FIG. 2

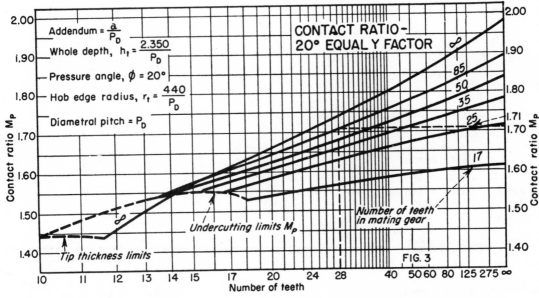

Addendum = $\dfrac{a}{P_D}$

Whole depth, $h_t = \dfrac{2.350}{P_D}$

Pressure angle, $\phi = 20°$

Hob edge radius, $r_t = \dfrac{440}{P_D}$

Diametral pitch = P_D

CONTACT RATIO –
20° EQUAL Y FACTOR

FIG. 3

factor gears and indicates the formula for calculating tooth thickness.

As an example: Design a long-short addenda gear set using the initial data of the previous problem. On Fig. 2, read the indicated addenda: 0.85, gear; 1.15, pinion. By dividing the diametral pitch into these values, the actual addenda are found: 0.0425, gear; 0.0575, pinion. The tooth thicknesses are then calculated to be: 0.0715, gear; 0.0825, pinion. The contact ratio for this gear set is read on Fig. 3 as 1.710.

It should be noted that strengthening the pinion, by using a long addendum pinion and a short addendum gear, reduces the number of teeth in the arc of action.

When it was found, Fig. 3, that the line for a rack as a mating gear crossed over other mating gear lines, additional layouts were made and carefully checked. The fact that the lines did cross was confirmed, and the crossover point was accurately located.

Basic Gear Geometry and Tooth Proportions

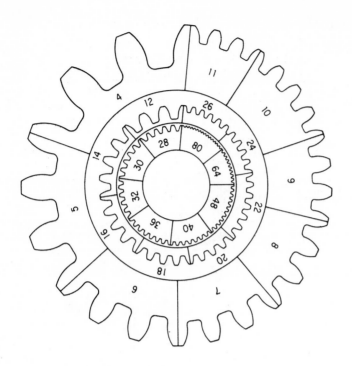

This figure courtesy of Barber-Colman Company, Rockford, Ill.

SIZES OF GEAR TEETH OF DIFFERENT DIAMETRAL PITCHES (TO SCALE)

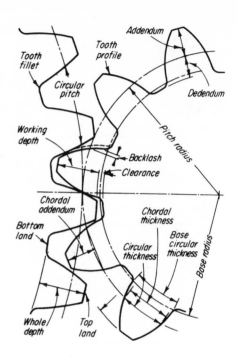

GEAR NOMENCLATURE

(From Dudley, "Gear Handbook", p.4-29, McGrow-Hill Book Company, New York, 1962.)

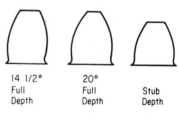

| 14 1/2° Full Depth | 20° Full Depth | Stub Depth |

COMPARISON OF TOOTH PROFILES

Spur - gear Formulas

To find	Having	Formula
Diametral pitch	Number of teeth and pitch diameter	$P = \dfrac{N}{D}$
Diametral pitch	Circular pitch	$P = \dfrac{3.1416}{p}$
Pitch diameter	Number of teeth and diametral pitch	$D = \dfrac{N}{P}$
Outside diameter	Pitch diameter and addendum	$D_o = D + 2a$
Root diameter	Outside diameter and whole depth	$D_R = D_o - 2h_t$
Root diameter	Pitch diameter and dedendum	$D_R = D - 2b$
Number of teeth	Pitch diameter and diametral pitch	$N = D \times P$
Base-circle diameter	Pitch diameter and pressure angle	$D_b = D \times \cos\phi$
Circular pitch	Pitch diameter and number of teeth	$p = \dfrac{3.1416\,D}{N}$
Circular pitch	Diametral pitch	$p = \dfrac{3.1416}{P}$
Center distance	Number of gear teeth and number of pinion teeth and diametral pitch	$C = \dfrac{N_G + N_P}{2P} = \dfrac{D_G + D_P}{2}$
Tooth thickness	Diametral pitch	$t_t = \dfrac{1.5708}{P}$

System for Spur and Helical Gears

To find	Having	Formula
Addendum	Diametral pitch	$a = \dfrac{1.000}{P}$
Addendum	Circular pitch	$a = 0.3183p$
Dedendum	Diametral pitch	$b = \dfrac{1.200}{P} + 0.002$
Dedendum	Circular pitch	$b = 0.3820p + 0.002$
Working depth	Diametral pitch	$h_k = \dfrac{2.000}{P}$
Working depth	Circular pitch	$h_k = 0.6366p$
Whole depth	Diametral pitch	$h_t = \dfrac{2.200}{P} + 0.002$
Whole depth	Circular pitch	$h_t = 0.7003p + 0.002$
Clearance	Diametral pitch	$c = \dfrac{0.200}{P} + 0.002$
Clearance	Circular pitch	$c = 0.0637p + 0.002$

8
ANGULAR ERRORS
AND BACKLASH CONTROL

latest equations for
ANGULAR ERRORS IN GEARS

There's no equipment available that can measure these errors quickly and accurately. Instead the author presents equations—some for the first time—which predict the effects of the seven key factors that influence angular accuracy. He also develops statistical-mean formulas that simplify estimates of probable total error.

T C NIELSEN, *associate engineer*
IBM, Federal Systems Division
Owego, NY

SYMBOLS

a — Arc length of angle $(\sigma + \theta)$, in.
B — Linear backlash, in.
ΔB — Deviation from nominal tooth thickness, in.
ΔC — Center distance tolerance or change, in.
D — Pitch diameter, in.
e — Eccentricity, in.
F — Face width of gear, in.
N — Number of teeth
p — Circular pitch, in.
Δp — Pitch error
R — Pitch radius, in.
R_b — Base-circle radius, in.
ΔR — Increase in radius due to misalignment, in.
S — Radial separation, in.
V — Pitch-line velocity of a perfect gear, in./sec.
t — Nominal tooth thickness, in.
t_e — Effective tooth thickness, in.
α — Misalignment angle, deg.
β — Angular backlash, radians
γ — Angular indexing error, radians
θ — Involute function of angle σ, radians
σ — Angle between radius vector and radial line to point of tangency of generating line with base circle, radians
λ — Tilt angle
λ_e — Lead-angle error, deg
δ — Angle of rotation radians
ϕ — Basic pressure angle, degrees or radians
ω — Angular velocity

The prime function of many gears these days is to accurately transmit angular displacement. The power they transmit is negligible. One problem with this type of application is that there are no commercially available machines for measuring angular errors in gears.

Some gear manufacturers do have specially designed test equipment, but they report that it often takes several days to check one set of gears. Also, because the phase relationship among the various factors that contribute to angular error is unpredictable, the actual total angular error is indeterminate.

To overcome these limitations we conducted a study to determine the following:

• Equations relating backlash and index error to variations in gear and mounting dimensions. Some of the equations presented here are well known; others are derived and are published for the first time.

• Maximum errors that can result from these variations for a set of Precision III gears (which are the highest quality gears listed in AGMA's standard 236.04).
The errors are based on the maximum manufacturing deviations permissible for this class of gears in accordance with the standard.

• Statistical average of the total of such errors.

With these equations and Table 1, you can quickly calculate the maximum or probable angular error for a given quality class. If the error is too high, more precise or specially matched gears must be specified.

GEARING ERRORS

In general, all gearing errors resolve themselves into one of two types: backlash, and indexing error (sometimes

TABLE I—FACTORS CONTRIBUTING TO ANGULAR ERRORS
(Based on Precision III Tolerances)

Factors	Maximum Permissible Tolerance, in.	Maximum Backlash Error		Maximum Indexing Error	
		Minutes of Arc	Percent of Total	Minutes Of Arc	Percent of Total
Center-distance Tolerance	+0.0005	2.500/D	18.0	0	0
Running Clearance	+0.0003	1.500/D	10.8	0	0
Tooth Thickness	− 0.0002	1.375/D	9.9	0.687/D	7.5
Pitch Error	±0.0001	1.375/D	9.9	1.375/D	15.1
Profile Error	±0.0002	2.935/D	21.2	2.935/D	32.1
Pitch-diameter Runout	0.00025	1.250/D	9.0	1.718/D	18.8
Lead-angle Error (F = 0.1)	0.0005/in.	0.344/D	2.5	0.344/D	3.8
Lateral Runout (F = 0.1)	0.0005/in. rad.	0.344/D	2.5	0.344/D	3.8
Bearing Tolerances (ABEC-7)	0.0002	1.000/D	7.2	1.375/D	15.1
Shaft & Bearing Clearance	− 0.0002 +0.0001	0.250/D	1.8	0.344/D	3.8
Housing & Bearing Clearance	− 0.0000 +0.0004	1.000/D	7.2	0	0
Totals		13.873/D	100.0	9.121/D	100.0

called angular-displacement error). In the design of precision gearing, both of these malfunctions seriously affect the proper functioning of the over-all system.

Backlash is the error which results from the tooth space of one gear being larger than the tooth thickness of the mating gear at the operating pitch circle. The true value of backlash is sometimes indeterminate because the engaging tooth may come to rest at any position within the tooth space. The term angular backlash refers to the amount of angular motion a meshed gear has when its mating gear is held stationary.

Indexing error is the deviation from the theoretical linear relationship between input and output angles of rotation of a gear mesh.

Although backlash and indexing error are independent of each other, they often are a result of the same error source. In unidirectional systems, in which the driving and driven surfaces always remain in intimate contact, backlash does not contribute to the angular inaccuracy of the system. Indexing error, on the other hand, is a cyclic factor and affects every revolution. In the case of bidirectional systems, the total angular error of a gear system is dependent upon the combined effect of both malfunctions.

CAUSES OF ANGULAR ERROR

The key factors which contribute to angular errors are:

(1) center distance
(2) tooth thickness
(3) pitch error
(4) involute-profile error
(5) pitch-diameter eccentricity
(6) lead angle
(7) wobble

From the data obtained from this study, it was calculated that a pair of one-inch-pitch-dia, Precision III, 20° pressure-angle gears can introduce an indexing error (independent of backlash) amounting to $2 \times 9.12 = 18.24$ minutes of arc, as shown in the table above. These same individual errors will cause backlash in the gear mesh amounting to 13.87 minutes of arc.

EFFECT OF CENTER-DISTANCE VARIATION

An increase in the nominal center distance between mounting holes in a gear case will cause a radial separation of meshed gears (see diagram). The effect of a radial

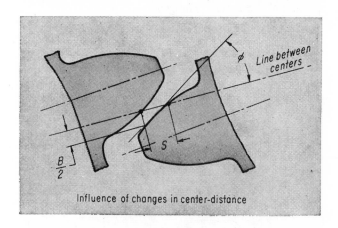

Influence of changes in center-distance

separation, (usually caused by a change in center distance) on the linear backlash, B, is determined by the equation:

$$B = 2S \tan \phi$$

where B is in inches. The angular backlash will then be equivalent to:

$$\beta_1 = \frac{2S \tan \phi}{R}$$

Due to a characteristic of mating involute curves, a variation from the theoretical center distance will not affect indexing error. Hence, output will be constant (for a constant input) even if the gears are mounted farther apart than usual.

TOOTH-THICKNESS VARIATION

A deviation in tooth thickness will affect both the backlash and index accuracy of a gear mesh. However, if the tooth-thickness error is consistent throughout the gear, it will only affect backlash of the mesh. The indexing error is introduced only when tooth thickness varies.

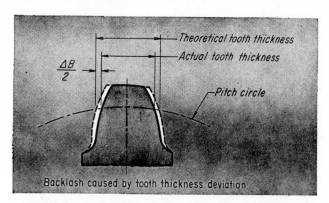

Backlash caused by tooth thickness deviation

The angular backlash, β_2, resulting from a deviation from the nominal tooth thickness, ΔB, may be expressed as:

$$\beta_2 = \Delta B / R$$

where ΔB is equivalent to the linear backlash found by taking the difference between the theoretical tooth thickness (no backlash) and the actual tooth thickness (see diagram).

For a variation in the thickness of the gear teeth, the indexing error, γ_2, is equal to:

$$\gamma_2 = \frac{\Delta B_0 - \Delta B_1}{2R} = \frac{\Delta B_0 - \Delta B_1}{D} \text{ (radians)}$$

where ΔB_0 and ΔB_1 are the linear backlash values of the respective teeth under consideration. Only half of the linear backlash values contribute to the error because only one side of the tooth is engaged.

ERRORS CAUSED BY PITCH DEVIATIONS

Pitch error is the amount by which the circular arc between corresponding sides of adjacent teeth, measured on a given circle, differs from the value obtained by dividing

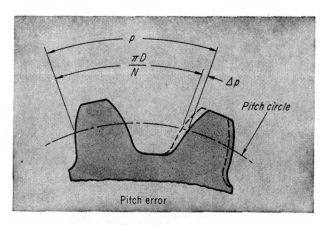

Pitch error

the circumference of the circle by the number of teeth. Referring to the diagram above:

$$\Delta p = p - \frac{\pi D}{N}$$

and

$$\beta_3 = 2\Delta p / D$$

A variable tooth spacing (pitch variation) will result in a tooth-indexing error. This angular displacement between any two teeth may be determined from:

$$\gamma_3 = 2(\Delta p_1 - \Delta p_2)/D$$

where Δp_1 and Δp_2 are the pitch errors of the respective teeth under consideration.

INVOLUTE-PROFILE ERROR

Involute-profile error is the deviation of the actual tooth form from the theoretical or true involute. An error in tooth profile produces the same effect as tooth-thickness error, with the exception that the profile error may occur anywhere along the active profile of the tooth. A variation in profile error from tooth to tooth will produce an indexing error.

Two positions of an involute gear-tooth profile generated

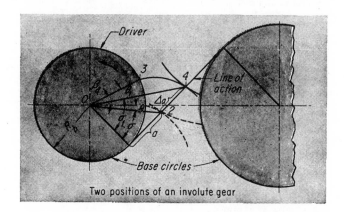

Two positions of an involute gear

from a base circle having a radius R_b are illustrated. As the driver is rotated through an angle β_4, the base or origin of the involute, point 1, moves to point 3. At the original point of contact, point 2, the involute function of the angle σ is equal to angle θ.

After rotation, these angles have increased in magnitude to σ_1 and θ_1, respectively. As the diagram shows, the angle or rotation is equivalent to the difference between the sums of the two sets of angles:

$$\beta_4 = (\theta_1 + \sigma_1) - (\theta + \sigma) = (\theta_1 - \theta) + (\sigma_1 - \sigma)$$

or:

$$\beta_4 = \Delta(\theta + \sigma)$$

By definition of an involute, the arc subtended by angle $(\sigma + \theta)$ is equal to distance a. Because a is measured along the line of action, a profile error Δa, measured perpendicular to the tooth surface may be substituted, and the equation for β_4 becames:

$$\beta_4 = \Delta a / R_b$$

This equation gives the backlash resulting from a profile error of Δa. The indexing error caused by a difference in the profile errors Δa_1 and Δa_2, of two arbitrary teeth, is:

$$\gamma_4 = \frac{\Delta a_1 - \Delta a_2}{R_b}$$

PITCH-DIAMETER ECCENTRICITY

To determine the error of angular transmission caused by the eccentricity of a gear, assume that the gear ratio is one-to-one and that the driver is a "perfect" gear without eccentricity. The mating gear has an eccentricity of e inches. The initial point of reference is when the center of the eccentric gear lies on the line joining the centers of rotation and between these centers. This results in the position of tightest mesh or minimum backlash.

The angle of error, γ_5, is the difference between the number of degrees which each gear has turned as a

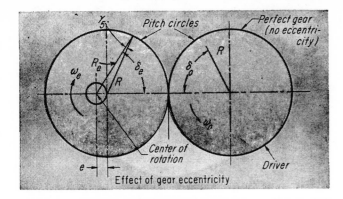

Effect of gear eccentricity

result of a given input (diagram above):

$$\gamma_5 = \delta_e - \delta_o$$

Because the driven gear is eccentrically mounted on its shaft, the angular velocity, ω_e, of the driven shaft will vary because of the variation in the effective pitch radius R_e.

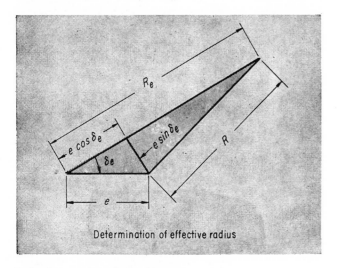

Determination of effective radius

Referring to the above diagram:

$$R_e = e \cos \delta_e + (R^2 - e^2 \sin^2 \delta_e)^{1/2}$$

Because e is small compared to R, the last term within the parentheses may be neglected; thus:

$$R_e = e \cos \delta_e + R$$

By differentiating the equation for γ_5 and substituting the equivalent valves for $d\sigma_e$ and $d\sigma_o$:

$$d\delta_e = \omega_e \, dt = \frac{V}{R_e} \, dt$$

$$d\delta_o = \omega_o \, dt = \frac{V}{R} \, dt$$

$$d\gamma = d\delta_e - \frac{R_e}{R} \, d\delta_e$$

where ω_o = constant input velocity. Substituting for R_e in the equation for γ_5, and integrating, gives

$$\gamma_5 = -\frac{e}{R} \sin \delta_e$$

This is an expression of the indexing error caused by eccentricity. Backlash resulting from the varying distance between the theoretical gear centers also should be considered. Because of eccentricity, the amount of radial separation will be the difference between the maximum pitch $(R + e)$ and the effective pitch radius (R_e) at any position of the eccentric gear:

$$S = (R + e) - R_e$$

Substituting in the equation for β_1 gives:

$$\beta_5 = 2 \frac{e}{R} (1 - \cos \delta_e) \tan \phi$$

This equation represents the instantaneous backlash caused by eccentricity for a given angle of rotation. If both gears are eccentric and the velocity ratio is one-to-one, the total effect of backlash and angular error is dependent upon the phase relationship of the two gears. Where velocity ratios other than one-to-one are used, the number of combinations is unlimited and the total effect depends upon the phasing of the gears, the velocity ratios, and the amount of eccentricity.

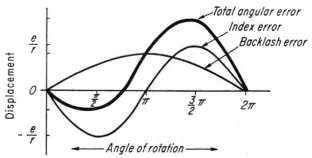

Variations in gear-eccentricity errors

The graph above illustrates the relationship between the indexing error, backlash error, and total error which results from eccentrically mounted gears.

LEAD-ANGLE ERROR

A constant lead-angle error increases the effective tooth thickness of the gear. This means the operating-center distance must be increased to avoid binding, which results in excessive wear and additional angular error.

Angular backlash caused by a lead-angle error, λ_e, is:

$$\beta_5 = \frac{F}{R} \tan \lambda_e$$

Max angular error caused by a variable lead angle is:

$$\gamma_6 = \frac{F}{R} \tan \lambda_e$$

ERRORS CAUSED BY WOBBLE

Assume that a perfect gear drives a similar gear that has been mounted so that the central axis of the gear teeth does not coincide with the axis of its shaft. This causes the gear to wobble like an elliptical gear with a varying lead angle.

An exact analysis of wobble error becomes rather cumbersome if all parameters are considered. Therefore, a simplified approach is taken by analyzing only two positions of the wobble gear. Each revolution has two distinct positions considered as "pure shaft misalignment," and two positions considered "pure tilt"—diagrams (A), (B), next page.

POSITION I—PURE SHAFT MISALIGNMENT

Assumptions in this analysis are:

• Both gears are perfect except for misalignment (nonparallel condition between the axes of rotation).

• Axes of rotation lie within the same place.

• Axis of rotation of the misaligned gear is coincident with the geometric axis of the gear teeth.

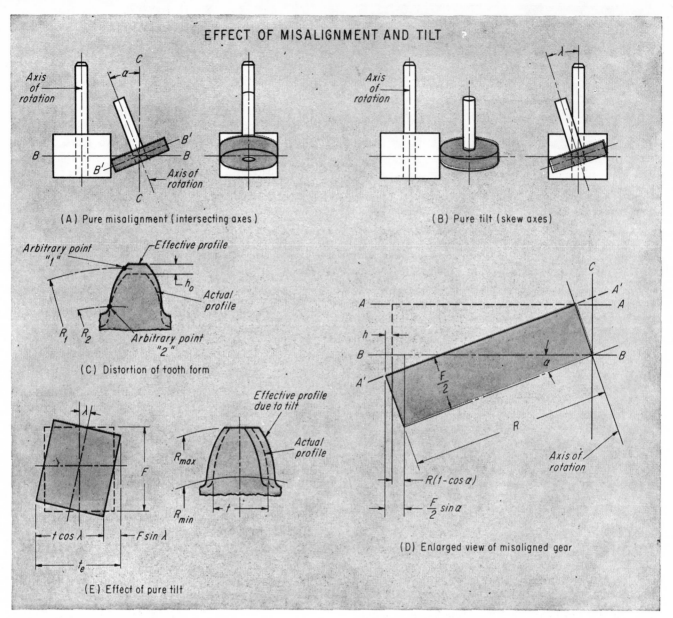

(A) Pure misalignment (intersecting axes)

(B) Pure tilt (skew axes)

(C) Distortion of tooth form

(D) Enlarged view of misaligned gear

(E) Effect of pure tilt

• Gear ratio is one-to-one, and only one tooth is in active engagement at any time.

The effective tooth profile is distorted from the true involute form diagram C). Point contact will occur in the plan A'A', perpendicular to the axis of rotation of the misaligned gear (diagram D).

Difference in the radial position of any point on the involute curve can be found from:

$$h = \frac{F}{2} \sin \alpha - R(1 - \cos \alpha)$$

To avoid binding, the center distance must be increased by an amount equal to h. This increases linear backlash by $\Delta B = 2h \tan \phi_1$.

The resultant angular backlash is:

$$\beta_7 = \frac{\Delta B}{R} = \frac{[F \sin \alpha - 2R_1(1 - \cos \alpha)] \tan \phi_1}{R}$$

where ϕ_1 is the pressure angle and R_1 is the radius at arbitrary point 1 (diagram C).

Because α is usually small, $\cos \alpha$ approaches 1, and the above equation can be simplified to:

$$\beta_7 = \frac{F}{R} \sin \alpha \tan \phi_1$$

This equation indicates that the angular backlash varies as a function of the instantaneous pressure angle. Hence, an angular error will occur for each tooth in the misaligned gear. This displacement error is equivalent to:

$$\gamma_7 = \frac{F}{R} \sin \alpha(\tan \phi_2 - \tan \phi_1)$$

where ϕ_2 is the pressure angle at arbitrary point 2.

POSITION II—PURE TILT

Assumptions in this analysis are:

• Geometric axis of the wobble gear and the axis of its shaft must lie within the same plane, and this plane must be parallel to the axis of rotation of the mating gear.

• Both gears are perfect except for tilt angle.

• Center of the tilted gear coincides with the intersection between the two axes as seen in a plane perpendicular to the plane in which the two axes of rotation are parallel.

At any point along the tooth profile (diagram E)

$$t_e = t \cos \lambda + F \sin \lambda$$

The angular backlash caused by tilt is:

TABLE II—SUMMARY OF EQUATIONS
(converted from radians to minutes of arc)

Error Source	Angular Backlash, Minutes of arc	Index Error, Minutes of arc
1. Center—distance tol. ΔC, or running clearance $\circlearrowleft$	$1375\dfrac{\Delta C}{D}\tan\varphi$	O
2. Tooth—thickness variation ΔB	$6875\dfrac{\Delta B}{D}$	$3438\dfrac{\Delta B_0-\Delta B_1}{D}$
3. Pitch error Δp	$6875\dfrac{\Delta p}{D}$	$6875\dfrac{\Delta p_0-\Delta p_1}{D}$
4. Involute—profile error Δa	$6875\dfrac{\Delta a}{D\cos\varphi}$	$6875\dfrac{\Delta a_1-\Delta a_2}{D\cos\varphi}$
5. Pitch—diameter eccentricity $2e$	$6875\dfrac{e}{R}(1-\cos\theta_e)\tan\varphi$	$\pm\,3438\dfrac{e}{R}\sin\theta_e$
6. Lead—angle error λ_e	$6875\dfrac{F}{D}\tan\lambda_e$	$6875\dfrac{F}{D}\tan\lambda_e$
7. Lateral Runout, λ	$3438\dfrac{F}{R}\sin\lambda$	$3438\dfrac{F}{R}\sin\lambda$

Maximum and probable totals of angular errors

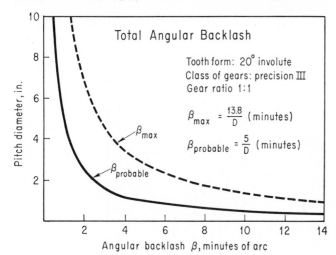

Total Angular Backlash

Tooth form: 20° involute
Class of gears: precision III
Gear ratio 1:1

$\beta_{max} = \dfrac{13.8}{D}$ (minutes)

$\beta_{probable} = \dfrac{5}{D}$ (minutes)

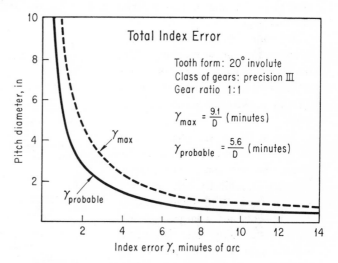

Total Index Error

Tooth form: 20° involute
Class of gears: precision III
Gear ratio 1:1

$\gamma_{max} = \dfrac{9.1}{D}$ (minutes)

$\gamma_{probable} = \dfrac{5.6}{D}$ (minutes)

$$\beta_8 = \frac{t_e - t}{R}$$

Combining equations

$$\beta_8 = \frac{t(\cos\lambda - 1) + F\sin\lambda}{R}$$

Because angle λ is usually very small, its cosine is nearly equivalent to 1. Therefore, a close approximation is:

$$\beta_8 = (F\sin\lambda)/R$$

Maximum angular error caused by wobble occurs in the vicinity of the pure tilt position and is equivalent to:

$$\gamma_8 = F\sin\lambda\left[\frac{1}{R_{max}} - \frac{1}{R_{min}}\right]$$

where R_{max} and R_{min} define radii of the active profiles.

These equations are summarized in Table II.

STATISTICAL ANALYSIS OF ANGULAR ERROR

The statistical value of the total angular backlash between two gears is found by taking square root of the sum of the squares of the individual factors:

$$\beta_8 = \sqrt{\beta_1{}^2 + \beta_2{}^2 + \beta_3{}^2 + \cdots \beta_n{}^2}$$

Substituting the values as found in Table 1 gives the approximate general equation for Precision III gears:

$$\beta_8 = 5.02/D \text{ (minutes)}$$

Maximum total backlash is

$$\beta_{max} = 13.8/D \text{ (minutes)}$$

The statistical value of index error between two gears is:

$$\gamma_8 = 4.02/D \text{ (minutes)}$$

Also

$$\gamma_{max} = 9.1/D$$

This figure, however, is based upon the assumption that one gear is "perfect." If two like gears are meshed, then

$$\gamma_8 = \sqrt{2\left(\frac{4.02}{D}\right)^2} = 5.6/D$$

and

$$\gamma_{max} = 18.2/D$$

A quick approximation of the inaccuracies can be found from the above graphs.

As an example of the magnitudes involved, if two ½-in.-pitch-dia, Precision III, 20° involute gears are meshed, there will be a probable backlash of 10.04 minutes of arc, and a probable index error of 11.2 minutes of arc.

Effect of Mounting Tolerances

Problem: The characteristics of a gear system (see diagram) are: Fit of mounting plate to bearing OD = 0.0004, bearing run-out = 0.0002, fit of bearing ID to shaft = 0.0004, fit of shaft to hub = 0.0002, fit of mounting plate to motor = 0.0004, run-out of motor shaft = 0.0004, diametral pitch = 96, number of teeth on gear X = 90, number of teeth on gear Y = 17. If the center-to-center distance tolerance is plus 0.0008 and minus nothing, find: 1) gear classes that will give gear X a backlash of not more than five degrees of arc when gear Y is locked, 2) the center-to-center distance.

Step 1. Convert seconds of arc of backlash to radians by multiplying by 4.848×10^{-6}, ie, $5 \times 60 \times 60 \times 4.848 \times 10^{-6} = 8.72 \times 10^{-2}$ radians.

Step 2. For either gear compute the backlash from the formula: $B = (\widehat{B} \times N \cos \phi)/(2DP)$ in. Now compute the backlash for gear X: $B = (8.72 \times 10^{-2} \times 90 \times 0.9397)/(2 \times 96) = 3.84 \times 10^{-2}$ in. Note: This is for a 20° pressure angle and 96 diametral pitch; this same procedure should be done for each pressure angle and diametral pitch contemplated.

Step 3. Divide the backlash between the two gears. The proportionality of the division of the backlash between the two gears is arbitary and any ratio can be used. With small backlash requirements and a large gear ratio a proportionality that gives the smaller gear a larger part of the backlash will yield a lower class for that gear.

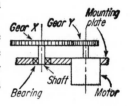

We choose to divide the backlash between the two gears evenly. The backlash for each gear is $B = 0.5 \times 3.84 \times 10^{-2} = 1.92 \times 10^{-2}$ in. Do all subsequent steps for each gear.

Step 4. Divide the plus tolerance of the center-to-center distance between the two gears. The proportionality is arbitrary and any ratio can be used. With small backlash requirements and a large gear ratio a proportionality that gives the smaller gear a smaller part of the plus tolerance of the center-to-center distance will yield a lower class for that gear.

Add all the eccentricities and run-outs of the mounting hardware for that gear and that portion of the tolerance of the center-to-center distance allocated to that gear.

We choose to divide the center-to-center distance plus tolerance evenly between the two gears. This is, $0.5 \times 0.0008 = 0.0004$ in. Addition for gear X is 0.0016. Addition for gear Y is 0.0012.

Step 5. Compute the pitch diameter from the formula: $P = N/DP$ in. The pitch diameters are, for gear X, $P = 90/96 = 0.938$ in. and for gear Y, $P = 17/96 = 0.177$ in.

Step 6. On chart 1 or 2 draw a line at 45° at the intersection of the

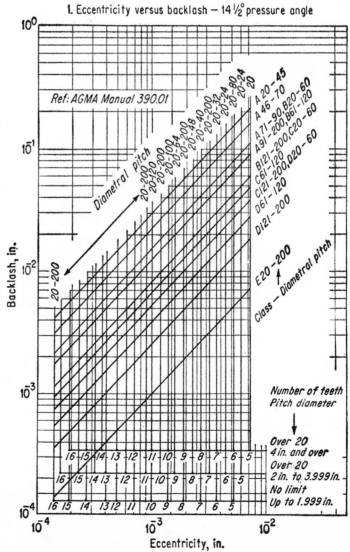

1. Eccentricity versus backlash — 14 ½° pressure angle

Ref: AGMA Manual 390.01

Number of teeth
Pitch diameter

Over 20
4 in. and over
Over 20
2 in. to 3.999 in.
No limit
Up to 1.999 in.

on Backlash

Brian Powell, Sr Engineer of Litton Systems Inc, Cal.

backlash and the lowest quality-number line that qualifies for both pitch diameter and number of teeth. Diametral pitch is of no consequence at this point. Subtract the eccentricity value obtained in step 4 from the value of the eccentricity of the quality number previously found. At the point where the drawn 45° line intersects this value draw a horizontal line to the left and a vertical line down. Any class designation and quality number that qualify for the diametral pitch, pitch diameter, and number of teeth are acceptable.

The lowest quality number that contains 90 teeth and a pitch diameter of 0.938 is 5 which has an eccentricity of 5.2×10^{-3} in. This number is also satisfactory for 17 teeth.

Subtracting the value of step 4, for gear X, we have $5.2 \times 10^{-3} - 1.6 \times 10^{-3} = 3.6 \times 10^{-3}$ in. For gear Y, $5.2 \times 10^{-2} - 1.2 \times 10^{-2} = 4.0 \times 10^{-2}$ in. From chart 2 the lowest gear class for gears X and Y is class C and quality number 8.

Step 7. The minimum center-to-center distance is the sum of the pitch radii and the eccentricities of the two gears. The pitch radii are one half of the pitch diameters found in step 5. From chart 2 the eccentricity of the chosen gear quality number is 1.9×10^{-3} for each gear. The center-to-center distance is $0.5 \times (0.938 + 0.177) + 2 \times 1.9 \times 10^{-3} = 0.469 + 0.0885 + 0.0038 = 0.5613$ in.

Step 8. The plus tolerance on the center-to-center distance is the sum of the differences between the acceptable and actual eccentricities of the gears. The acceptable eccentricities of the gears are, for gear X, 3.6×10^{-3}; for gear Y, 4.0×10^{-3}. The actual eccentricities of the gear are, for gears X and Y, 1.9×10^{-3} in. The plus tolerance on the center-to-center distance is, $3.6 \times 10^{-3} - 1.9 \times 10^{-3} + 4.0 \times 10^{-3} - 1.9 \times 10^{-3} = 1.7 \times 10^{-3} + 2.1 \times 10^{-3} = 3.8 \times 10^{-3}$ in.

And that's it. Actually, the above procedure is reversible, thus making it possible to analyze an existing design. When doing this the plus tolerance on the center-to-center distance must be added to the eccentricity of the gear before going to the charts. You can divide the plus tolerance between the two gears however suits you best.

2. Eccentricity versus backlash—20° pressure angle

NOTATION

B	= linear backlash
$\widehat{B}$	= angular backlash
R	= gear ratio
DP	= diametral pitch
P	= pitch diameter
N	= number of teeth
ϕ	= pressure angle

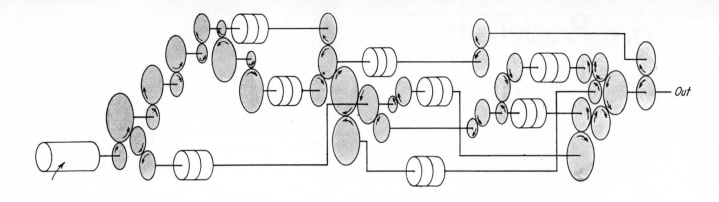

Lost Motion in Gear Trains

This interesting method provides a simple yet accurate technique of determining the lost motion or angular error in a gear train (Fig 1), which results from backlash at each of its gear meshes. The total lost motion, θ, is of utmost importance in instrument gearing. It can be computed by means of the following expressions (presented by Paul Dean, Jr of General Electric who heads the AGMA fine-pitch committee):

$$\theta = Sc \sum \frac{B}{\frac{1}{2}D \times S} + \frac{B'}{\frac{1}{2}D' \times S'} + \frac{B''}{\frac{1}{2}D'' \times S''} + \cdots + \cdots$$

Where

B, B', B'', = Backlash, in. at the pitch line between
etc G and its mating gear
D, D', D'' = Pitch diameters, in., of the successive gears in the train
S, S', S'' = Relative speeds of the shafts on which gears D, D'' are mounted
c = Constant due to the units desired for θ

When θ is in: radians, $c = 1.0$; in degrees, $c = 57.296$; in minutes, $c = 3438$.

Speed, S, refers to the rotational speed of any one shaft, relative to any other specific shaft. Usually, the slowest shaft is assigned 1 speed. Thus in Fig 4, the output can be assigned unit speed. Shaft *2* will turn at

$$\frac{144}{24} \times 1 = 6 \text{ speed}$$

Shaft *4* will turn at

$$\frac{80}{20} \times \frac{144}{24} \times 1 = 24 \text{ speed}$$

Shaft *6* will turn at

$$\frac{120}{15} \times \frac{80}{20} \times \frac{144}{24} \times 1 = 192 \text{ speed}$$

Assume that the maximum backlash or play between the mating teeth of the meshes are shown in Table I. (These values can be obtained from a table of permissible backlash ranges for various backlash classes, as given in the new manual, AGMA 370.01, and in the classification manual previously mentioned AGMA 390.01.)

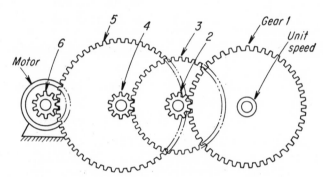

1. Backlash at gear meshes results in lost motion.

Table I — Example of Lost Motion Calculation

Gear number	Pitch dia, D	Backlash of mesh, B	Gear ratio	Shaft speed, S	$\frac{B}{D/2 \times S}$, rad
1	3.00	0.0015	6:1	1 S	0.0010
3	1.25	0.0030	4:1	6 S	0.0008
5	1.25	0.0030	8:1	24 S	0.0002
					0.0020 Total

Thus, to find θ in minutes:

$$\theta = (6 \times 1)(4 \times 1)(1) \sum \frac{0.0015}{(3/2)(1)} + \cdots$$

Or, as shown in Table I:

$\Sigma = 0.0010 + 0.0008 + 0.0002 = 0.002$ radians
$\theta = 3438(0.002) = 7.14$ minutes

This method shows which meshes make the greatest or least contribution, and indicates how much remedial action might be taken. Thus, if the backlash in mesh 5-6 was reduced 50% to 0.015 in., the total angular error would be 6.78 min, a reduction of only 5%. If, however, the backlash in mesh 1-2 was reduced 50%, to 0.00075 in., the angular error would be 5.36 min, a reduction of 25%.

WHAT IS BACKLASH?

It is simply the clearance between a tooth and its mating gap. Backlash can be measured several ways: for example, with a feeler gage slipped between the two mating teeth, or by holding one gear stationary and measuring the angle through which its mate rotates. The latter method measures angular backlash and it could be related approximately to the feeler-gage measurement (linear backlash) by the formula

$$B = \frac{\theta}{360}\, \pi D$$

where B = linear backlash, in.; θ = angular backlash, deg; D = pitch dia of gear being rocked.

Many factors contribute to backlash—radial play in bearings, shaft eccentricity, pitch dia runout, and incorrect center-to-center distance are the most common culprits. However, some of these same factors can cause the gear train to jam. Two methods are recommended to prevent jamming:

• Increase the center distance C by an amount Δ C equal to 70% to 100% of the total composite error (TCE) permissible in the gears.

• Reduce the tooth thickness as specified in AGMA standard 236.04 (Inspection of Fine-Pitch Gears). The standards contain tabulated values for the reduction of tooth thickness for all gear quality classes

Thinning of the gear teeth reduces the gear pitch diameters below nominal, resulting in a predetermined range of backlash values listed in Table 10 of the AGMA standard 236.04.

Backlash is a cyclically fluctuating value. It can approach zero at point of tightest engagement of a pair of gears. The gears may be so phased, however, that this theoretical minimum is never attained. Generally speaking, backlash fluctuates between min and max values, which may be somewhat different even for two gear trains of identical design.

It is not always necessary to eliminate backlash throughout an entire gear reduction train. Backlash of any one gear mesh reflected at the output is divided by the gear ratio of that particular mesh and the output. A 1° backlash at the first mesh of a 60:1 speed reducer is only 1-min. at the output. Thus, only the last pass or two may need attention.

18 WAYS TO CONTROL BACKLASH IN GEARING

No matter how accurately a pair of gears are made and meshed together, backlash is bound to creep in. Solution: an antibacklash device. Here are 18, including some of the latest, that the author rounded up when backlash became a serious problem in a servo-positioning device he was designing for ITT Labs.

FREDRICK T GUTMANN, *development engineer*
ITT Laboratories, Nutley, N J

It is quite impossible to eliminate backlash completely from a gear train—and still have a train that will run smoothly without binding. Another difficulty with backlash is inability to predict it quantitatively—a set of precision gears mounted on loose bearings may have more backlash than a set of commercial gears. But over the years quite a few effective devices have been developed that practically eliminate backlash. These devices may tap some of the power available but, on the other hand, they allow relaxing some of the accuracy requirements of gears. Here are 18 of the latest available. *(CONTINUED NEXT PAGE)*

Adjustable-center Types

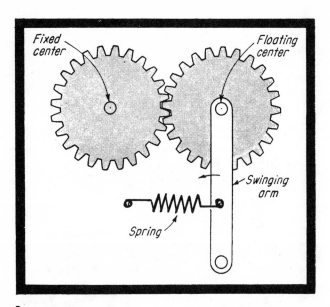

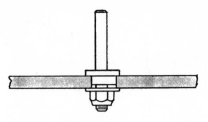

1 Floating center . . .
One gear center is stationary; the other, allowed to float or pivot. Spring force holds the gears in intimate contact, thereby completely eliminating backlash. Applications: gear-rolling inspection fixtures, worm-wheel drives.

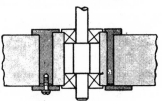

2 Single eccentric . . .
is one of the two well-known adjustable-center types. It can do the job when there are only two gears in mesh.

3 Double eccentric . . .
Is the second type, employed when a gear runs between two other gears with fixed centers. In a gear train, adjustable centers are usually alternated with fixed centers.

4 Adjustable tooth width . . .
Controls backlash with a split gear that adjusts the effective tooth width. The two halves are rotated with respect to one another, at assembly of the gear train, until there is no appreciable backlash at point of tightest engagement. They are then fixed in their relative positions by screws. This method, as with the adjustable-center method, eliminates backlash only at point of tightest gear mesh and proportionately reduces backlash at the other points.

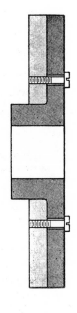

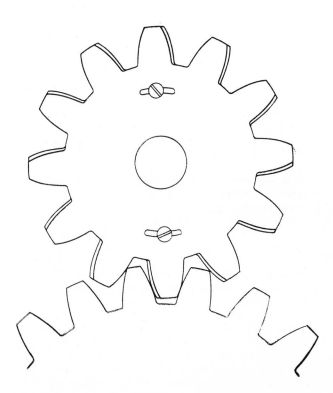

Spring-loaded Spur Gears

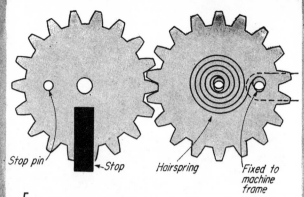

Stop pin **Stop** **Hairspring** **Fixed to machine frame**

5 Hairspring type . . .

This device is useful for limited rotation only. One of the gears — usually the output — is loaded by attaching one end of a hairspring to a shaft, other end to a fixed pin.

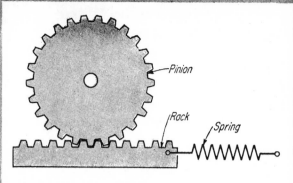

Pinion **Rack** **Spring**

6 For rack and gears . . .

The loading spring can be a coil spring, or preferably a spring of constant tension. For some designs, counterweights may be substituted, but they add mass and may require undue accelerating and breaking forces.

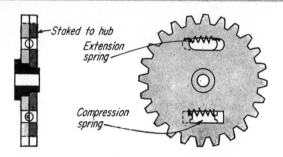

Staked to hub **Extension spring** **Compression spring**

7 Unlimited rotation . . .

Can be obtained with split spring-loaded gears. One gear is staked to the common hub; the other is free to rotate, but loaded by springs in one direction. The mating gear is solid and should be slightly wider than the split gear. The spring force can be varied by winding the split gear to vary amount of scissors action on mating gear.

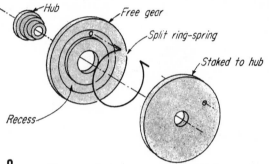

Hub **Free gear** **Split ring-spring** **Staked to hub** **Recess**

8 Another variation . . .

Uses a ring-spring. This requires an annular groove which may be simpler to machine than milled slots for extension springs. Spring loading, of course, adds to the tooth load and wear, and requires a higher driving torque as compared with unloaded gearing.

Spring-loaded Loops

Wide gear **Split gear** **Split gear, spring-loaded**

9 Eliminating backlash . . .

between two or more components, such as synchros mounted side by side, is a handy application for spring-loaded loops. The intermediate, dual gear in this 3-shaft loop contains one gear staked to the shaft, and one free to rotate, but has no loading springs.

Thin gear **Split gear, spring-loaded** **Wide gear** **Thin gear**

10 For 4-shaft arrangements . . .

The intermediate gears must be on different levels because each mates with one portion of the spring-loaded gear.

(CONTINUED NEXT PAGE)

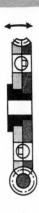

Spring-loaded Worm and Bevel Gears

11 In worm wheels . . .
Backlash can be eliminated by employing a split-type wheel. But, because of needed clearance in the bore, the free portion of the split wheel tends to rock in the direction shown by the double-headed arrow as rotation of the worm is reversed.

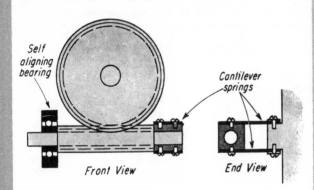

Self aligning bearing

Cantilever springs

Front View

End View

12 Second method . . .
Uses a floating mesh. One end of the worm is supported by a self-aligning bearing; the other end floats and is spring-loaded into the worm wheel by a force acting in direction of the arrow. This force can be supplied by mounting the floating bearing on two cantilever springs which act as a parallelogram.

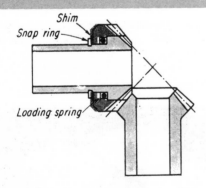

Shim

Snap ring

Loading spring

13 Spring-loaded bevel gears . . .
are feasible if the blanks of the two parts are nested together and cut simultaneously. A design of such gears, available from stock from PIC Design Corp., is shown above.
It is also possible to load helical gears in the same manner as spur gears, but this is rarely, if ever, done.

Special Gear Types

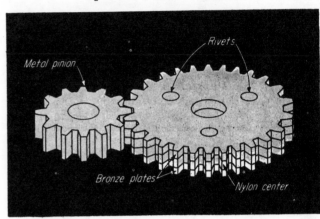

Metal pinion

Rivets

Bronze plates

Nylon center

14 Sandwich gears . . .
having nylon lamination sandwiched between two phosphor-bronze outer plates are employed successfully, shown above, in a drilling machine manufactured by Industrial Technics Ltd., Southampton, England. The three layers are machined to the same dimensions, but the nylon lamination swells slightly and is thus larger than the metal sections. This permits tighter meshing with standard pinions because of nylon's high elasticity combined with good abrasive resistance. This design is said to be applicable to both spur and worm gear meshes, but has been only used in slow-speed drives, and no information is available on performance at high speeds and under heavy loads.

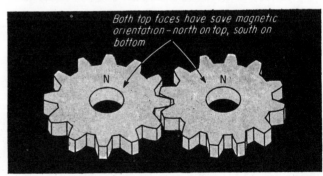

Both top faces have save magnetic orientation - north on top, south on bottom

N

N

15 Magnetic gears . . .
Another type reported from England depends on the repelling forces between magnetized gears. Gear faces of the same magnetic orientation face the same way.
This design is useful only for very small torques, in order of 0.01 to 0.02 in.-oz.

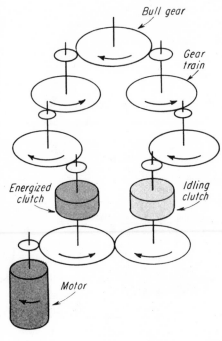

Dual Gear Trains

16 Mirror-image method . . .

Runs two identical gear trains against each other. There is no backlash at the output gear. The arrangement (left) has eliminated backlash in radar antennas. A motor drives two counter-rotating gears placed on input shafts of two clutches. Each clutch is associated with one of the gear trains, and both trains terminate at the output "bull" gear. If one clutch is energized, the associated gear train rotates the bull gear, say, clockwise. At the same time, the bull gear drives gear train associated with the second clutch backwards.

When the second clutch is engaged, the reverse happens. In this manner, both gear trains are always loaded toward the bull gear: the active gear train by the clutch, and the driven one by friction in the gears and bearings, which is reflected at the bull gear as a fairly high torque.

Springs or other loading devices are not needed in this setup, but added torque from the input motor is called for to overcome friction in the inactive but rotating gear train. Also, high tooth loads result at the pinions meshing with bull gear. Pinions driving the bull gear should be approximately, but not exactly, 180° apart. This will decrease side thrust on the bull gear, but also put the two gear trains slightly out of phase to make the output motion free of excess toothiness.

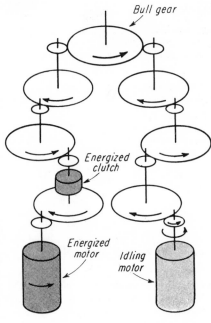

17 Dual-motor drive . . .

is a variation of mirror-image method. The two motors are controlled individually, and if one motor drives, say, counterclockwise, the clockwise motor will be de-energized. However, it is necessary to insure that the de-energized motor, when driven through the gear train at high speed, does not act as generator and overload the entire arrangement. Whether to use the two motors and one clutch rather than one motor and two clutches depends on components' cost.

18 Spring-loaded dual trains . . .

A clutchless dual gear train is also possible, but needs spring-loaded gears for proper operation. Shown here is a speed reducer of this type, made by Metron Inc., and available in several reduction ratios. Each gear mesh in the reducer is loaded by means of torsion springs. The designer can thus choose a ready-made backlash-free device to suit his needs.

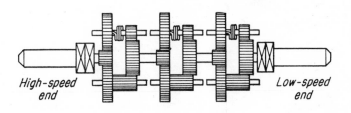

High-speed end   Low-speed end

MORE WAYS TO CONTROL BACKLASH

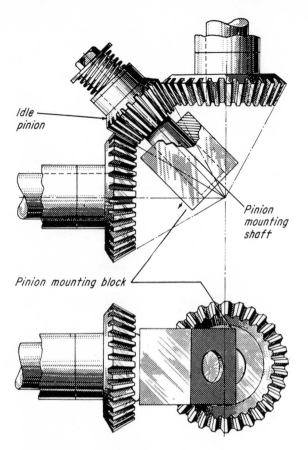

SPRING-LOADED PINION is mounted on a shaft located so that the spring forces pinion teeth into gear teeth to take up lost motion or backlash.

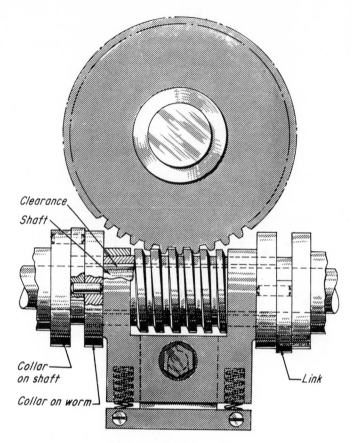

HOLLOW WORM has clearance for shaft, which drives worm through pinned collars and links. As wear occurs, springs move worm into teeth.

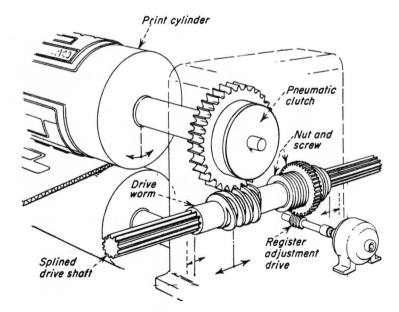

A drive that avoided backlash build-up . . .

inherent in long gear trains was required for the four-section, split-construction design because accumulated backlash would seriously affect register. Solution was a splined drive shaft running full length of the machine, with an independent worm-gear power-takeoff at each feeding, printing and slotting station. The worm drives absorb, as thrust, much of the cyclic load patterns of the kicker and slotter mechanisms. Running register of the print cylinders, and the upper slotting shaft is adjusted by shifting the drive worm axially. Main drive motor and slotting station are in the fixed section. Each worm drive includes a remotely controlled pneumatic clutch. Stations can be driven in any position.

9
GEAR EFFICIENCY, LIFE, AND VIBRATION

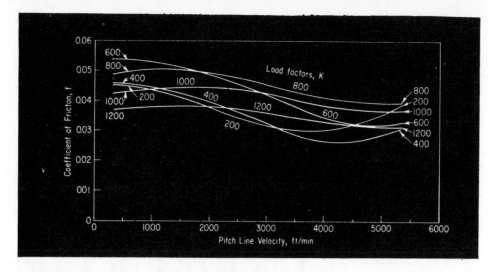

1 Coefficient of friction in . . . mating gears varies with load and speed. Curves are for 20° pressure angle, case-carburized and ground spur gears, with average lubrication fed in at 120-F inlet temperature.

How to predict
EFFICIENCY of GEAR TRAINS

**The sliding action of involute gear teeth causes a power loss.
It can be computed from new formulas and test data given here.**

EUGENE E. SHIPLEY
Advance Gear Engineering, General Electric Co., Lynn, Mass.

New formulas and test data now make it possible to acurately predict power losses in gear meshes. Knowing such losses will help select oil reservoirs, pumps and coolers of proper capacity. Also, gear-mesh losses may make a difference when deciding on size of the power source.

The losses are important enough to warrant attention even though gears give high power-transmitting efficiency—in the upper 90% range. Almost all the energy loss is converted into heat. In high-power applications, this affects size of the lubricating system which removes much of the heat.

This Equation Does It

Gear-tooth surfaces move across each other with a combination of rolling and sliding motion. Therefore, the loss is a function of the average sliding motion and coefficient of friction. In terms of percent power loss for a simple set of gears, the power formula is

$$P = \frac{50f}{\cos \theta} \left(\frac{H_s^2 + H_t^2}{H_s + H_t} \right)$$

where

$$H_s = \left(\frac{N + 1}{N} \right) \left(\sqrt{\frac{r_o}{r} - \cos^2 \theta} \ - \sin \theta \right)$$

and

$$H_t = (N + 1) \left(\sqrt{\frac{R_o}{R} - \cos^2 \theta} \ - \sin \theta \right)$$

The percent efficiency is then:

$$E = 100 - P$$

LIST OF SYMBOLS

D — pitch dia of pinion, in.
E — efficiency of gear mesh, %
f — coefficient of friction, dimensionless
F — face width of thinner gear, in.
H — specific sliding velocity of teeth (ratio of sliding velocity to pitch-line velocity); subscript s denotes value during start of approach, subscript t during recess action
K — load factor or index to tooth loading
n — pinion speed, rpm
N — gear ratio where $N = 1$ or larger
P — power loss of gear mesh, %
r — pitch radius of pinion, in.
r_b — base radius of pinion, in.
r_o — outside radius of pinion, in.
R — pitch radius of larger gear, in.
R_o — outside radius of larger gear, in.
s — path of approach, in.
t — path of recess, in.
V — pitch-line velocity, ft/min
W — driving load at pitch radius, lb; W_t = tangential component; W_n = normal component
Z — friction force; $Z = f W_n$
θ — pressure angle, deg
ω — angular velocity of pinion, rad/sec

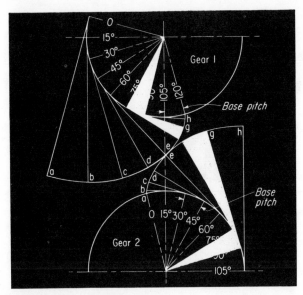

2 Involute profiles . . .

of mating teeth have unequal segments, causing a sliding as well as rolling action.

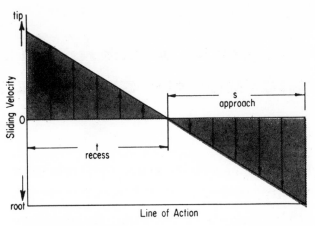

3 Sliding velocity . . .

between mating teeth varies from max values at tip and root of tooth, to zero at pitch line.

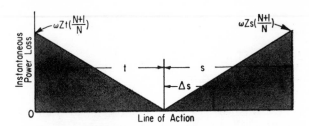

4 Instantaneous power loss . . .

is proportional to sliding velocity. Shaded area represents average power loss.

All factors above are easily determined from geometry of the gear system, with exception of the coefficient of friction f. A common value for lubricated case-hardened and ground gears has been $f = 0.06$ but lengthy tests now show it to be on the conservative side. More accurate values, given in chart, Fig. 1, are based on various load factors K and pitch-line velocities V, which in turn are computed for a particular gear set from these two equations:

$$V = 0.2618 Dn$$

$$K = \frac{W_t}{FD}\left(\frac{N+1}{N}\right)$$

Nature of f Curve

The chart shows that for any given load factor, f starts out high, dips to a minimum, then climbs again. These curves follow hydrodynamic theory in that high friction values occur during boundary lubrication conditions and then dip and climb with continued increase in speed.

Additional tests are underway by the author to study the effects of higher speeds, improved surface finishes, viscosity changes and temperature variations on the coefficient of friction.

The Sliding Motion

In normal operation there is always some sliding motion between gear teeth—an inherent characteristic of involute gearing. The involute profiles of the mating teeth can be developed by unwinding a string from a base circle while keeping it taut, Fig. 2. For two involute gears to transmit a smooth and constant angular motion, the base pitches of both must be equal. However, the equivalent segments on the involute profile are not constant—being very small at beginning of development (segment gh for gear 1 and ab for gear 2), and growing larger toward the tooth tips (segment ab for gear 1 and gh for gear 2).

Thus, for both gears to have equal angular displacements during operation, segment gh on gear 1 must correspond to segment gh on gear 2, and sliding motion (in addition to rolling motion) occurs. Sliding velocity is a maximum at the first point of contact, reduces to zero at the pitch dia, changes direction and increases again to a maximum at the last point of contact, as shown in Fig. 3. For a 1:1 gear ratio, max values at both ends are equal, and length of recess action equals length of approach action.

Deriving Power Formula

The instantaneous power loss is proportional to the sliding velocity. Total power loss for a single-tooth engagement is represented by the shaded area in Fig. 4. Average power loss P_L is

$$P_L = \frac{\omega Z}{(s+t)}\left(\frac{N+1}{N}\right)\left(\int_0^s s\,ds + \int_0^t t\,dt\right)$$

which gives

$$P_L = \frac{\omega Z}{2}\left(\frac{N+1}{N}\right)\left(\frac{s^2+t^2}{s+t}\right)\frac{\text{ft-lb}}{\text{time}}$$

Average power transmitted is

$$P_T = \omega W_n r_b$$

The percent power loss P is then the ratio of P_L to P_T, multiplied by 100. This ratio, when substituting values for s and t results in the power formula given on the previous page.

how various operating conditions affect . . .
FRICTION OF GEAR TEETH

Important new tests show that lubricant viscosity, load and sliding velocity are related to gear-tooth friction. Results have led to a simple formula that can be a valuable aid in gear design.

G H BENEDICT
B W KELLEY
Caterpillar Tractor Co, Peoria, Ill

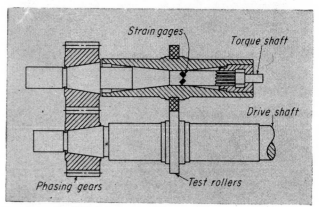

1..GEAR-ROLLER TEST MACHINE simulates gear-tooth action

In the analysis of gear-tooth surfaces, it has been fairly well established that when scoring occurs it is a result of welding, which in turn is caused by failure of the lubricant. This failure is believed to occur at a critical temperature, which is reached by frictional heating of the surfaces.

Coefficient of friction is usually assumed to be constant, and some value is picked based on experience. When such a value is used to calculate the correlation of scoring and temperature-flash the value is reliable for gears of a given size operating in a rather limited speed range. When the speed range is extended, however, the constant coefficient is no longer good enough, and more accurate information must be used to predict scoring failure.

Tests were carried out with a gear-roller test machine (Fig 1) which simulates the rolling and sliding action of gear teeth. Results show that, in general, friction varies with lubricant, viscosity, load and velocity.

Gear-roller analogy

A common approximation for the profile of gear teeth is to assume that they can be simulated by circular arcs, Fig 2, having the same radius of curvature as the gear teeth at the point of contact. All contact-stress calculations are made using this approximation. Also, the gear tooth surfaces have a tangential velocity which is constant for a given point on the profile at a constant gear speed. The velocities of the mating surfaces are in the same direction, and for a given pair of gears, always have the same velocity ratio at a point on the line of action; this is similar to the roller test with a fixed phasing gear ratio. By properly choosing speed, phasing gears, and roller diameters, any condition on the line of action of any gear pair can be duplicated exactly under steady state conditions. The essential difference between the rollers and gear teeth is that the tooth profile curvature is continually changing, and hence its velocity is changing, while in roller tests, the radii and velocities do not change. How important these

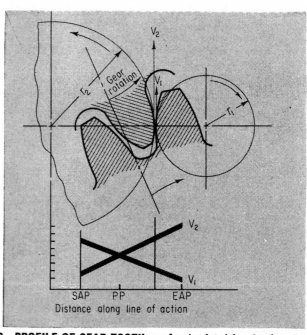

2..PROFILE OF GEAR TOOTH can be simulated by circular arcs.

differences are is hard to guess. Differences arising from these effects, however, are considered to be negligible.

Temperature measurements

The thermocouple probe was lightly suspended in the oil film on the entering side of contact, Fig. 3, as close to the contact area as possible. The oil lubricant supply was on the outlet side to extract the maximum amount of heat from the rollers.

Loads and speeds

Each of four loads was applied at each speed setting. After a load was applied, the data was gathered by recording the friction torque and probe temperature several times as they rose to their maximum values. This procedure

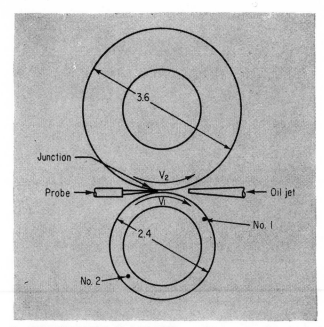

3..THERMOCOUPLE IS LOCATED as close as possible to test rollers. It was found that probe temperature was very sensitive to the heat-dissipation conditions on the roller surfaces.

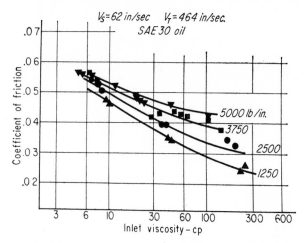

4..VISCOSITY, as indicated by oil temperatures, affects friction coefficient.

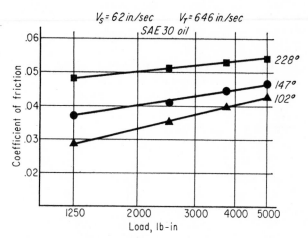

5..EFFECT OF LOAD on friction.

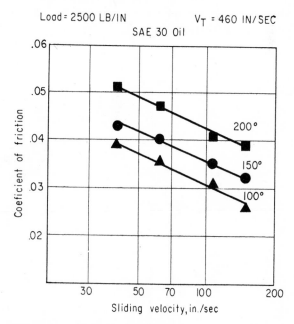

6..EFFECT OF SLIDING VELOCITY on friction.

was repeated at about five driving speeds between 300 and 8000 rpm for five phasing gear-sets which gave velocity ratios from 0.333 to 0.815.

TEST RESULTS

Figs 4 to 7 reveal how the friction coefficient is affected by various factors.

Velocities are combined as sliding velocity (V_s, inch/sec $= V_2 - V_1$) and sum velocity (V_t, inch/sec $= V_2 + V_1$). The value of V_s was always positive because V_2 was greater than V_1.

ANALYSIS

The type of lubrication and its effect on friction are illustrated in the typical hydrodynamic-lubricated curve. Fig 8. The parameter $\mu N/P$ is called the hydrodynamic parameter, and it evolved directly from Reynolds hydrodynamic equations. It partially characterizes the state of lubrication in a journal bearing and is included in the Sommerfeld number—a more familiar parameter.

Region B is a mixed lubrication region where part of the load is borne by hydrodynamic action and part by boundary lubrication. Most gear applications fall in region B, but there are a few exceptions. Lightly loaded high-speed gears might reach full fluid-film lubrication and heavily loaded low-speed gears show signs of being in the boundary region. These are the extremes, however, and in the results here and in gears normally used, mixed lubrication prevails. This concept of the transition from hydrodynamic to boundary lubrication or part metallic contact can be used to explain friction results.

An equation can represent these test results.

$$ f = 0.0127 \log_{10} \left[\frac{3.17 \times 10_8}{\mu_0 \, V_s \, V_t^2/W} \right] $$

A line calculated from the equation is drawn in Fig 9.

Lubricant viscosity grades

In this series of tests, a significant point is that there was no great difference in the friction of the different

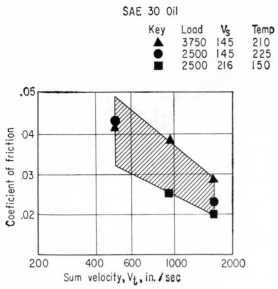

SAE 30 Oil

Key	Load	V_S	Temp
▲	3750	145	210
●	2500	145	225
■	2500	216	150

7..EFFECT OF SUM VELOCITY on friction.

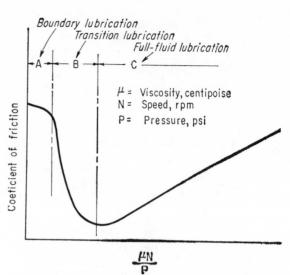

μ = Viscosity, centipoise
N = Speed, rpm
P = Pressure, psi

8..TYPICAL HYDRODYNAMIC-LUBRICATION CURVE shows section (B) of curve where most gear applications operate.

viscosity grades if coefficient is plotted against the inlet viscosity. It is concluded that the viscosity is the controlling factor for like oils, and not viscosity grade.

Conclusions

In a gear contact, as simulated on a roller test machine, it has been shown that the instantaneous coefficient of friction follows the concept of transition from boundary to hydrodynamic lubrication. The coefficient has been found to increase with increasing load, and to decrease with increasing sum velocity, sliding velocity and oil viscosity, where the viscosity is determined by the temperature of the oil entering contact and the viscosity-temperature characteristics of the lubricant. The results have been combined in a formula which closely represents the data.

When this formula is used in scoring calculations, the same type of U-shape load-speed curve is obtained as has been found on several gear test rigs.

Speculation on the significance of this work questions whether, for instance, a state of lubrication should be considered a cause of gear failure. If the lower coefficients obtained in these tests do represent a fair amount of hydrodynamic support, and tests with known gear-blank temperature verify the critical temperature hypothesis, then the hypothesis will prove to be still valid for marginal hydrodynamic conditions.

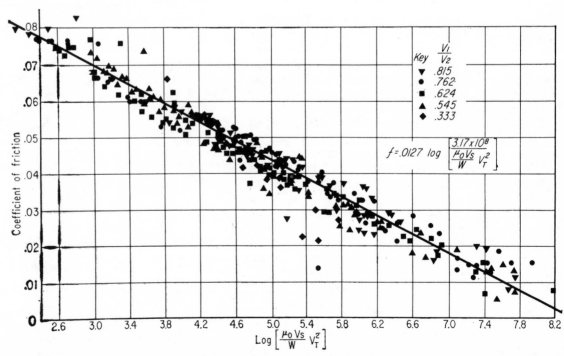

Key	$\frac{V_1}{V_2}$
▼	.815
●	.762
■	.624
▲	.545
◆	.333

$$f = .0127 \log \left[\frac{3.17 \times 10^8}{\frac{\mu_0 V_S}{W} V_T^2} \right]$$

9..EFFECT of $\log_{10} (\mu_0 V_s V_t^2 / W)$ on friction for all data.

New formulas and charts help predict
Operating life of gears

If you're worried about reliability of your gear systems, try the method that has produced fatigue-free reduction units for Raytheon.

CLARENCE A. UNDERWOOD, *Gear Consultant, Raytheon Co., Wayland, Mass.*

HERE is a simplified way to predict operating life of gears. This method offers new equations for computing the inertia effects of gears and integral shafts and provides new charts for relating gear loads to gear life. The charts cover an assortment of gear materials, including ferrous, nonferrous, and laminated-phenolic materials.

The method has been applied successfully to the design of many gear systems where reliability has been the primary goal—under a wide variety of speed and load conditions—and there has been no evidence of fatigue or strength failures during the predicted life of the gears.

The predicted life is the number of cycles, or hours, that gears will operate before detrimental wear occurs. Other methods and design charts for predicting this factor have been published, but they are based on somewhat vague load terms, such as "light," "moderate" and "shock," and are not applicable to light-load or no-load operations. Also those methods usually pertain to hardened steel gears only.

The method presented here acknowledges the fact that in many modern applications the masses of gears and their integral shafts—or, in the case of rotating ring gears, their integral housings—can contribute to loading on gear teeth at least as much as the loads to be transmitted. Hence, formulas have been developed for quickly determining the moments of inertia of the gear systems—including the effect of every change in contour. The method also employs life charts based on actual tests of various material combinations. Before the charts can be used, however, a careful estimate of the surface loading on the gear teeth must be first made (after the inertia analysis). Buckingham's equations are drawn upon for this stage of the analysis.

In brief, the method determines sequentially these factors in a gear train:

1. Mass moment of inertia of each geared shaft
2. Average effective mass per mesh
3. Average force for accelerating the gear masses
4. Load to deform teeth
5. Average accelerated load
6. Total load, applied plus dynamic
7. Predicted life from charts

Two examples are included based on actual gear systems to illustrate the method. It will be shown that in one gear mesh, induction-hardened gears resulted in a predicted life of 960,000 hours of operation, while a lower hardness drastically reduced life to 485 hours.

Inertia equations

The mass effects of the gears and rotating components can be computed from the following formulas.

The polar moment of inertia, I, of a disk is

$$I = Mk^2$$

where k = radius of gyration (see Symbols, p. 68). Therefore

$$k^2 = \frac{r^2}{2} = \frac{D^2}{8}$$

The mass, M, of a disk is equal to

$$M = \frac{W}{g} = \frac{\pi (D/2)^2 Fw}{g}$$

Hence

$$I = \frac{\pi D^4 Fw}{32\,(32.2)}$$

or
$$I = 0.003,05\ wFD^4$$
$$I = CFD^4$$

where $C = 0.003,05\ w$.

Values of C for eight common materials are given in the following table:

Material	Density, w lb/in.^{3}	Values for C, $\times 10^{-5}$
Steel	0.2816	86
Bronze	0.315	96
Brass	0.306	93
Aluminum bronze	0.270	82
Aluminum	0.092	27
Magnesium	0.065	20
Nylon	0.036	11
Laminated phenolics	0.042	13

The polar moments of inertia for other likely configurations have been similarly derived and are given

1..Typical configurations for inertia analysis

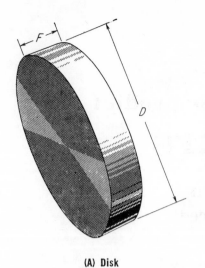

(A) Disk

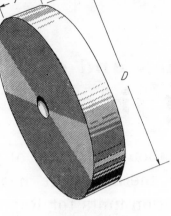

(B) Disk with small bore

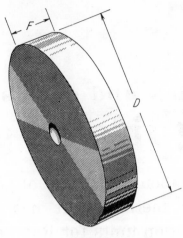

(C) Gear with pitch dia, D

below. The effect of small bores on the final results are negligible (as will be shown later) and may be omitted.

For a gear or disk, with or without a small bore (illustrations A, B, and C in Fig. 1):

$$I = CFD^4 \qquad (1)$$

For a gear or disk with hub or integral shaft, and with or without a small bore (illustrations D and E in Fig. 1):

$$I = C[F_1 D_1^4 + F_2 D_2^4] \qquad (2)$$

For a gear or disk with web, hub, and large bore (illustration F in Fig. 1):

$$I = C[F_1(D_1^4 - D_2^4) + F_2(D_2^4 - D_3^4) + F_3(D_3^4 - D_4^4)] \qquad (3)$$

Additional terms may be added inside the brackets of Eq. 2 or Eq. 3, or the equations may be combined to be applicable to more complex configurations.

Comparison with tabular methods

Let us compare the use of the above equations with tabular methods for computing inertia, such as the one described in "The Significance of Wk^2 and How to Calculate It" (*PE—June* 27, '60, *p.* 39). The comparison will show that bores of sizes typically employed in rotating compounds can be safely considered as having negligible effects on inertia computations.

The total Wk^2 values for the flywheel analyzed in the above article and illustrated in Fig. 2 amounts to

$$Wk^2 = 360.78 \text{ lb-ft}^2$$

To convert units, multiply by 144 in.² and divide by the acceleration due to gravity, 32.2 ft/sec². Hence,

$$I = 360.78 \left(\frac{144}{32.2} \right) = 1615 \text{ lb-in}^2\text{-sec}^2/\text{ft}$$

2..Inertia calculations for a flywheel by use of tables

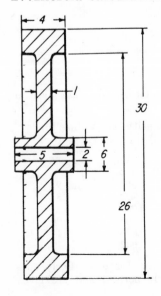

RIM	30 in. disk	155.92 per in. thickness (from table, right)
	less 26 in. disk	87.96
		67.96 × 4 in. thick = 271.84
WEB	26 in. disk	87.96
	less 6 in. disk	0.25
		87.71 × 1 in. thick = 87.71
HUB	6 in. disk	0.25
	less 2 in. disk	0.003
		0.247 × 5 in. thick = 1.23
		Total 360.78

$$I = 360.78 \left(\frac{144}{32.2} \right) = 1615$$

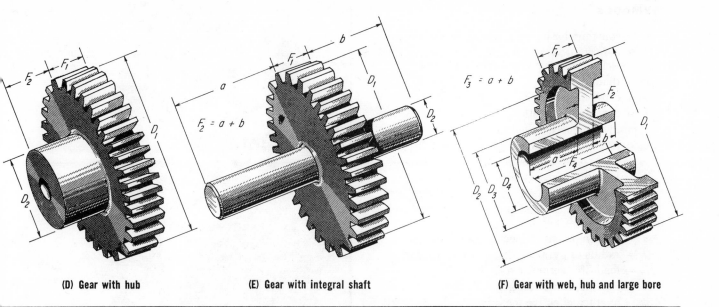

(D) Gear with hub   (E) Gear with integral shaft   (F) Gear with web, hub and large bore

Actually, if the effect of the 2-in. bore is dropped from the computations, it would not change the above value for I unless the computations were carried out to three decimal places.

For comparison purposes, the same problem is solved with the use of Eq. 3 and a value of C = 0.000,86:

$$I = 0.000,86 \left[4\,(30^4 - 26^4) + (26^4 - 6^4) + 5\,(6^4 - 2^4) \right]$$
$$I = 1612 \text{ lb-in}^2\text{-sec}^2/\text{ft}$$

The K factors

On the basis of tests conducted by G. Talbourdet and W. C. Cram of United Shoe Machinery Corp. over a number of years, a series of K factors have been determined for various materials and hardness combinations, and for various cycles of operation. The test fixture, a schematic of which is shown in Fig. 3, employs rolls of various materials. The rolls are driven by gears to provide a rolling action between test rolls (gear ratio is then 1 to 1) or a combined rolling and sliding action (gears are then of unequal size).

Load is applied gradually by means of a calibrated spring until the desired load is fully exerted on the rolls. Tests are stopped when either roll begins to show appreciable wear. The number of revolutions or stress cycles is noted and a load-life factor is calculated from the following equation:

$$K_r = P \left[\frac{1}{r_1} + \frac{1}{r_2} \right] \tag{4}$$

where K_r is the load-life factor for rolls, psi; P is the load per inch of contact length, lb/in.; and r_1 and r_2 are the radius of the test rolls.

continued, next page

INERTIA OF SOLID STEEL CYLINDERS, 1 IN. LONG

Dia, in.	Wk^2 Lb-Ft^2	Dia, in.	Wk^2 Lb-Ft^2	Dia, in.	Wk^2 Lb-Ft^2
1	0.000192	16	12.61	31	177.77
2	0.00308	17	16.07	32	201.8
3	0.01559	18	20.21	33	228.2
4	0.049278	19	25.08	34	257.2
5	0.12030	20	30.79	35	288.8
6	0.2494	21	37.43	36	323.2
7	0.46217	22	45.09	37	360.7
8	0.78814	23	53.87	38	401.3
9	1.262	24	63.86	39	445.3
10	1.924	25	75.19	40	492.78
11	2.818	26	87.96	41	543.9
12	3.991	27	102.30	42	598.8
13	5.497	28	118.31	43	658.1
14	7.395	29	136.14	44	721.4
15	9.745	30	155.92	45	789.3

SYMBOLS

C = constant based on material, lb-sec²/in³-ft

C_e = deformation factor, lb. Represents load to deform teeth per assumed error in action.

D = diameter, in.

e = error-in-action, in.

E = modulus of elasticity, lb/in.²

f_a = average accelerated load, lb

f_1 = force for accelerating gear masses, lb

f_2 = load to deform teeth, lb

F = length, or face width of gear blank, in.

g = acceleration due to gravity = 32.2 ft/sec²

H = combined effect of pressure angle and radius of curvature, per in.

I = mass moment of inertia, lb-in.²-sec²/ft

k = radius of gyration, in.

K_r = load-life factor for rolls, lb/in.²

K = load-life factor for gears, lb/in.²

m = average effective mass per mesh, lb-sec²/ft

M = mass of a rotating body, lb-sec²/ft

n = gear reduction ratio

P = load per inch of contact length of test rolls, lb/in.

P_a = Load to cause an assumed deflection, lb

r = radius of a test roll, in.

R = radius to pitch line of gear teeth, in.

V = pitch line velocity, ft/min

w = weight of material, lb/in.³

W = weight of a disk, lb

W_a = applied load, lb

W_d = total load (applied plus dynamic), lb

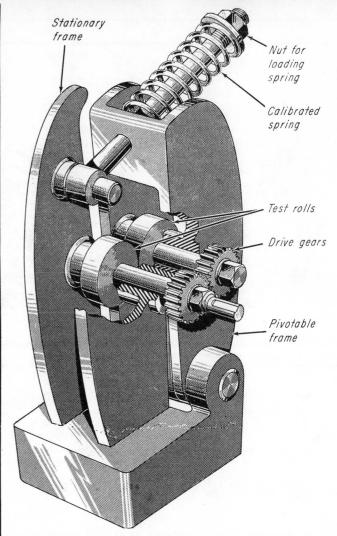

3..Testing machine for load-life factors

Load-life charts

Tabulated values of K_r have been published in "How to Predict Wear Life of Rolling Surfaces" (*PE—May 9, '60, pp. 44-47*), and in "Surface Endurance Limits of Engineering Materials" (ASME paper 54-LUB-14). For use with gears, however, the values must be multiplied by one-fourth the sine of the gear pressure angle:

$$K = \frac{K_r \sin \theta}{4} \qquad (5)$$

where K is the load-life factor for gears, psi.

The five charts, Figs. 4 to 8, are based on published and unpublished values, converted by means of Eq. 5 to be applicable for 20° pressure angle. They are arranged for various hardness conditions and number of cycles. The combinations include:

Fig. 4—Steel vs steel of various Brinell hardness ratings, including induction- and case-hardened gears

Fig. 5—Cast iron vs cast iron, including Meehanite

Fig. 6—Manganese- and phosphor-bronze vs steel

Fig. 7—Aluminum and magnesium both with oxide hardcoat, also cast aluminum vs cast iron

Fig. 8—Laminated graphitized phenolic vs steel

Design problem I

We have applied the inertia equations and life-load charts to the design of gear-reduction units for radar an-

tenna drives. The type analyzed in this problem has three speed-reduction stages, Fig. 9. Input from a hydraulic motor passes through the coupling to the first gear mesh. The gear of the first mesh is mounted integrally with the pinion of the second mesh. In the third mesh, the pinion drives a ring gear with internal teeth whose housing is fastened to the outer housing which, in turn, rotates the radar antenna.

Details and dimensions of the four geared units are given in Figs. 10 to 13. The mass moment for each must be computed separately.

For the shaft in Fig. 10, employing Eq. 2:
$$I_1 = 0.000,86 \left[\left(\tfrac{1}{2}\right)\left(\tfrac{9}{16}\right)^4 + \left(1\tfrac{3}{4}\right)\left(\tfrac{1}{2}\right)^4 \right] = 0.000,14$$

For the shaft in Fig. 11, employing a combination of Eqs. 2 and 3:
$$I_2 = C \left[F_1 D_1{}^4 + F_2 D_2{}^4 + F_3 \left(D_3{}^4 - D_4{}^4\right) + F_5 D_5 \right]$$
$$= 0.000,86 \left[\left(\tfrac{7}{16} + \tfrac{7}{16}\right)(0.472)^4 + \left(\tfrac{5}{8}\right)\left(\tfrac{3}{4}\right)^4 \right.$$
$$\left. + \left(\tfrac{1}{2}\right)(2.344^4 - 2^4) + \left(\tfrac{5}{8}\right)(1)^4 \right]$$
$$I_2 = 0.006,85$$

For the shaft in Fig. 12, employing the same equation as above:
$$I_3 = 0.000,86 \left[\left(\tfrac{7}{16} + \tfrac{7}{16}\right)\left(\tfrac{11}{16}\right)^4 + \left(\tfrac{3}{4}\right)\left(1\tfrac{1}{2}\right)^4 \right.$$
$$\left. + \left(\tfrac{1}{2}\right)(2\tfrac{14}{32}^4 - 2\tfrac{14}{32}^4) + \left(1\tfrac{7}{16} + \tfrac{1}{8}\right)\left(\tfrac{3}{4}\right)^4 \right]$$
$$I_3 = 0.012$$

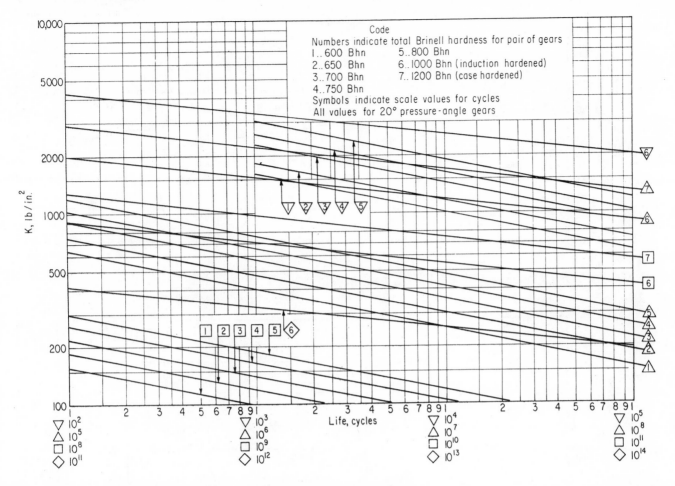

4..Steel vs steel (SAE 1020), various hardness ratings

For the ring-gear housing in Fig. 13:

$$I_4 = 0.000{,}86 \left[(\tfrac{11}{16})(12\tfrac{3}{4} - 12^4) + \tfrac{1}{4}(14\tfrac{1}{4} - 12\tfrac{3}{4}) \right.$$
$$+ (1\tfrac{1}{4})(15\tfrac{3}{4} - 14\tfrac{3}{4}) + (\tfrac{3}{16})(17\tfrac{1}{4} - 15\tfrac{3}{4})$$
$$\left. + (\tfrac{3}{4})(16\tfrac{1}{8} - 15\tfrac{1}{4}) \right]$$

$$I_4 = 34.89$$

The effective mass for the first mesh and driven radii can now be obtained:

$$m_1 = \frac{I_1}{R_1{}^2} = \frac{0.000{,}14}{(9/32)^2} = 0.0018$$

$$m_2 = \frac{I_2}{R_2{}^2} + \left(\frac{R_3{}^2}{R_2{}^2}\right)\left(\frac{I_3}{R_4{}^2}\right) + \left(\frac{R_3{}^2}{R_4{}^2}\right)\left(\frac{I_4}{R_6{}^2}\right)$$

$$m_2 = \frac{0.006{,}85}{1.172^2} + \left(\frac{0.5^2}{1.172^2}\right)\left(\frac{0.012}{1.25^2}\right)$$

$$+ \left(\frac{0.75^2}{1.25^2}\right)\left(\frac{34.89}{6^2}\right)$$

$$m_2 = 0.356$$

The average effective mass is

$$m = \frac{m_1\, m_2}{m_1 + m_2} = \left[\frac{(0.0018)(0.356)}{0.0018 + 0.356} \right]$$

$$m = 0.0018$$

The next step is to compute the total load on the teeth. Buckingham's method is followed (*Analytical Mechanics of Gears* by E. Buckingham, McGraw-Hill). Given data for the drive are:

> Input pinion speed = 3300 rpm
> Input torque = 20.8 in.-lb
> Reduction ratio of the first stage = 4:1

The resulting applied load, W_a, on teeth of first pinion is

$$W_a = 20.8/(\tfrac{9}{32}) = 74 \text{ lb}$$

With a pinion speed of 3300 rpm, the pitch line velocity at the first mesh is

$$V = \pi (9/16)(3300)/12 = 486 \text{ ft/min.}$$

The combined effect, H, of pressure angle and radii of curvature at the meshing point is equal to:

$$H = 0.0012 \left(\frac{1}{R_1} + \frac{1}{R_2} \right)$$

where factor 0.0012 is for 20° pressure angle gears. Hence

$$H = 0.0012 \left(\frac{1}{0.281} + \frac{1}{1.172} \right) \quad 0.0053$$

continued, next page

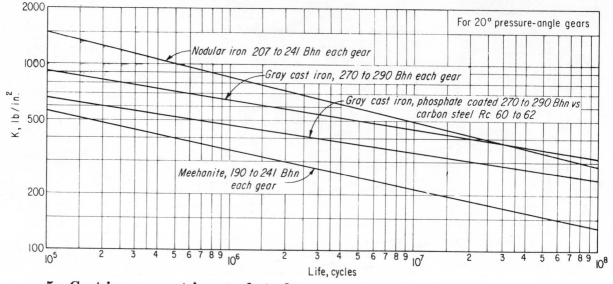

5..Cast iron vs cast iron and steel

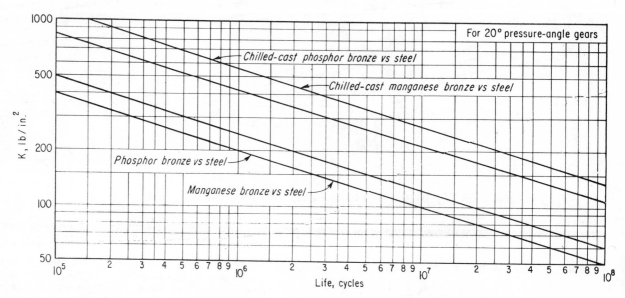

6..Manganese bronze and phosphor bronze vs steel of Rc 60 to 65

The accelerating force is

$$f_1 = HmV^2$$
$$= (0.0053)\ (0.0018)\ (486)^2$$
$$f_1 = 2.25$$

The load to deform teeth is

$$f_2 = FC_e + W_a$$

where C_e is the tooth loading caused by errors in tooth profile

$$C_e = \frac{0.111\ e}{\left(\dfrac{1}{E_1} + \dfrac{1}{E_2}\right)}$$

For steel, and for a 0.001 in. error-in-action, $C_e = 1660$ lb. Thus

$$f_2 = (0.5)\ (1660) + 74 = 904$$

The average accumulated load is

$$f_a = \frac{f_1 f_2}{f_1 + f_2} = \frac{(2.25)\ (904)}{906} = 2.24$$

The total load on the teeth (including applied and dynamic loads) is

$$W_d = W_a + \sqrt{f_a\ (2f_2 - f_a)}$$
$$= 74 + \sqrt{2.24\ (1808 - 2.24)}$$
$$W_d = 137\ \text{lb}$$

The relationship between total load and the load-life factor K is

$$W_d = KDFQ$$

where

$$Q = \frac{2n}{n + 1} = \frac{2\ (4)}{(4) + 1} = 1.6$$

Hence, $$K = \frac{137}{(0.562)\ (0.5)\ (1.61)} = 303\ \text{lb/in.}^2$$

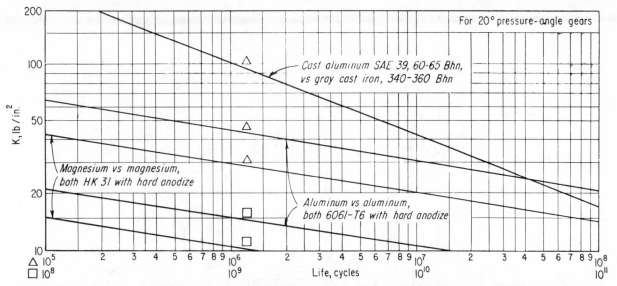

7..Aluminum vs aluminum, magnesium vs magnesium, aluminum vs cast iron

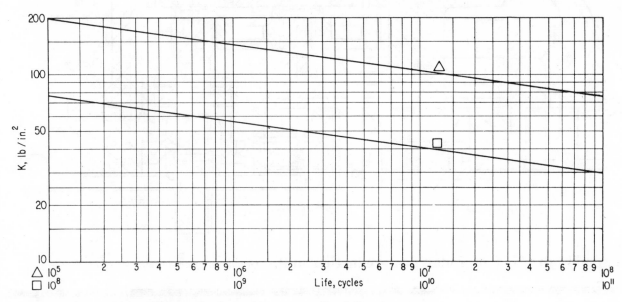

8..Laminated graphitized phenolic vs steel of Rc 60 to 62

According to the chart for induction-hardened steel, Fig. 4, for $K = 303$ lb/in.^{2} the predicted life is 1.9×10^{11} cycles. To use this chart, determine the total Brinell hardness desired for the gear pair. The pinion is usually made harder than the gear (a rough rule of thumb is that the ratio of Brinell hardness between pinion and gear should be inversely proportional to the ratio of their diameters).

Now check the hardness code on the chart. This indicates which of the lines numbered 1 through 6 should be employed. The symbols—triangle, diamond, etc.—indicate which of the scales to employ. Thus, when the horizontal line for 303 intersects the line for the diamond-6 line, read down to the 10^{11} scale to obtain 1.9×10^{11}.

The number of hours will be

$$\text{life} = \frac{\text{cycles}}{\text{rpm} \times 60} = \frac{(1.9)(10^{11})}{(3300)(60)} = 960{,}000 \text{ hr}$$

If a pair of steel gears with a lower total hardness, say Bhn = 800, is chosen, the predicted life will be drastically reduced. The number of cycles will be 9.6×10^7, and the predicted life will be only:

$$\text{life} = \frac{(9.6)(10^7)}{(1.98)(10^5)} = 485 \text{ hr}$$

The life for the second mesh is now checked by following the same procedure as for the first mesh (but with an input pinion speed of 794 rpm):

$$m_1 = \frac{I_2}{R_2{}^2} + \left[\frac{R_2{}^2}{R_3{}^2} \right] \left[\frac{I_1}{R_1{}^2} \right]$$

$$= \frac{0.006{,}85}{0.5^2} + \frac{(1.172)^2 (0.000{,}14)}{(0.281)^2}$$

$$m_1 = 0.037$$

continued, next page

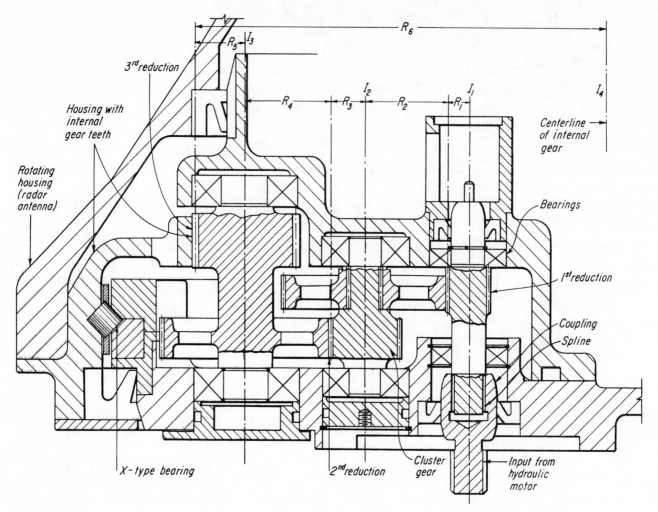

9.. Assembly drawing of radar gear-reduction unit. *Details of geared shafts at right*

$$m_2 = \frac{I_3}{R_4{}^2} + \left[\frac{R_5{}^2}{R_4{}^2}\right]\left[\frac{I_1}{R_6{}^2}\right]$$

$$= \frac{0.012}{1.25^2} + \left[\frac{0.75^2}{1.25^2}\right]\left[\frac{39.8}{6^2}\right]$$

$m_2 = 0.357$

$m = 0.034$

$H = 0.0012\,(1/0.5 + 1/1.25) = 0.003{,}38$

$V = (0.262)\,(1)\,(794) = 208$

$f_1 = (0.003{,}38)\,(0.034)\,(4.34)\,(10^4) = 5.0$

$f_2 = (0.5)\,(1660) + 180 = 1010$

$f_a = 5.0$

$W_d = \sqrt{180 + 5.0\,(2020 - 5.0)} = 280$

The ratio for this mesh is 2.5:1.

$$Q = \frac{2\,(2.5)}{2.5 + 1} + 1.43$$

$$K = \frac{280}{(1)\,(0.5)\,(1.43)} = 396$$

For induction-hardened gears (the diamond-6 line), the predicted number of cycles $= 1.8 \times 10^{11}$.

$$\text{life} = \frac{(1.8)\,(10^{11})}{(794)\,(60)} = 3{,}780{,}000 \text{ hr}$$

For the third mesh (with a pinion speed of 320 rpm)

$$m_1 = \frac{I_3}{R_5{}^2} + \left[\frac{R_4{}^2}{R_3{}^2}\right]\left[\frac{I_2}{R_3{}^2}\right] + \left[\frac{R_2{}^2}{R_3{}^2}\right]\left[\frac{I_1}{R_1{}^2}\right]$$

$$= \frac{0.012}{0.75^2} + \left[\frac{1.25^2}{0.75^2}\right]\left[\frac{0.006{,}85}{0.5^2}\right]$$

$$+ \left[\frac{1.172^2}{0.5^2}\right]\left[\frac{0.000{,}19}{0.281^2}\right]$$

$m_1 = 0.107$

$$m_2 = \frac{I_4}{R_6{}^2} = \frac{34.89}{36} = 0.97$$

$$II = 0.0012\,(1/0.75 - 1/16) = 0.0014$$

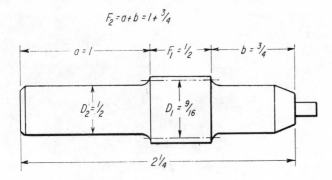

$F_2 = a+b = 1\frac{3}{4}$

$a = 1$ | $F_1 = \frac{1}{2}$ | $b = \frac{3}{4}$

$D_2 = \frac{1}{2}$ $D_1 = \frac{9}{16}$

$2\frac{1}{4}$

10..Pinion shaft

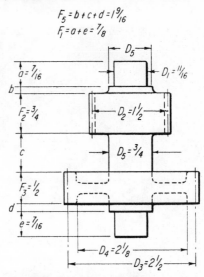

$F_5 = b+c+d = 1\frac{9}{16}$
$F_1 = a+e = \frac{7}{8}$

D_5 $D_1 = \frac{11}{16}$

$a = \frac{7}{16}$
b
$F_2 = \frac{3}{4}$ $D_2 = 1\frac{1}{2}$
c $D_5 = \frac{3}{4}$
$F_3 = \frac{1}{2}$
d
$e = \frac{7}{16}$

$D_4 = 2\frac{1}{8}$
$D_3 = 2\frac{1}{2}$

12..Second cluster shaft

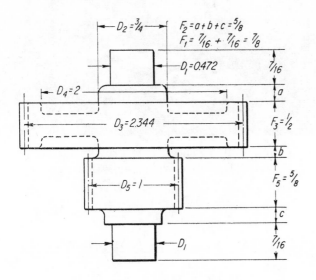

$D_2 = \frac{3}{4}$ $F_2 = a+b+c = \frac{5}{8}$
$F_1 = \frac{7}{16} + \frac{7}{16} = \frac{7}{8}$

$D_1 = 0.472$ $\frac{7}{16}$
a
$D_4 = 2$
$D_3 = 2.344$ $F_3 = \frac{1}{2}$
b
$D_5 = 1$ $F_5 = \frac{5}{8}$
c
D_1 $\frac{7}{16}$

11..First cluster shaft

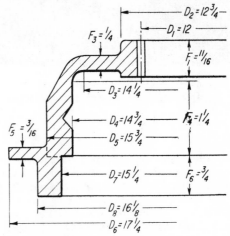

$D_2 = 12\frac{3}{4}$
$F_3 = \frac{1}{4}$ $D_1 = 12$
$F_1 = \frac{11}{16}$
$D_3 = 14\frac{1}{4}$
$F_5 = \frac{3}{16}$ $D_4 = 14\frac{3}{4}$
$D_5 = 15\frac{3}{4}$ $F_4 = 1\frac{1}{4}$
$D_7 = 15\frac{1}{4}$ $F_6 = \frac{3}{4}$
$D_8 = 16\frac{1}{8}$
$D_6 = 17\frac{1}{4}$

13..Ring-gear housing

The negative sign occurs because of the internal drive.

$$V = (0.262)(1.5)(320 \text{ rpm}) = 126$$
$$f_1 = (0.0014)(0.96)(15810) = 21.4$$
$$f_2 = (0.687)(1660) + 300 = 1441$$
$$f_a = 21.1$$
$$W_d = 300 + \sqrt{21.1(2882 - 21.1)} = 545$$

The ratio for this mesh is 8:1. Hence

$$Q = \frac{(2)(8)}{8-1} = 2.18$$

$$K = \frac{545}{(1.5)(0.687)(2.28)} = 232$$

For induction-hardened gears, the predicated number of cycles, from Fig. 4, is 2×10^{13}.

Design problem II

We have also applied the method to the redesign of a pinion that meshes with two other gears. In the original design, Fig. 14, the pinion was on one end of a 9-in.

PRODUCTION UNIT OF ASSEMBLY SHOWN ABOVE.

14..Pinion redesign eliminated excessive wear on teeth (left) by reducing its effective mass.

New pinion (right) has provided more than five years of service with negligible wear.

long solid shaft. Field reports indicated scoring early in the wear life of the gear. After study it was decided that wear life could be increased by reducing the mass of the pinion and shaft. Hence, the shaft was made hollow, and the gear flanged, as shown in Fig. 14. The life of the redesigned gear was predicted as follows:

Given: W_a = 43-lb tangential applied load

R = 0.625 radius to pitch line of gear, in.

F = 0.344 gear blank face width, in.

Speed = 9000 rpm

V = 2940 ft/min

C_e = 1660 lb (for steel for 0.001-in. error-in-action)

D_1 = 0.625, D_2 = 0.50

G = 12 × 10⁶ torsion modulus (steel)

L = length of hollow shaft = 1.625 in.

P_a = load to cause an assumed deflection

From mechanics:

$$R\theta = \frac{32 P_a R L}{\pi (D_1^4 - D_2^4) G}$$

Assuming a 0.001-in. deflection for $R\theta$ (consistent with the assumed 0.001-in. error-in-action):

$$P_a = \frac{\pi (0.001) (D_1^4 - D_2^4) G}{R^2 L g}$$

$$P_a = 168$$

$$f_2 = W_a + \frac{F P_a C_e}{F P_a + C_e}$$

$$= 43 + \frac{(0.344) (1660) (168)}{(0.344) (1660) + 168}$$

$$f_2 = 173$$

Assume also that $m = 0.001$; hence

$$f_1 = (0.0012) (0.001) (8.66 \times 10^6) \left(\frac{1}{0.625} + \frac{1}{1.312} \right)$$

$$= 24.6$$

$$f_a = \frac{(25) (173)}{25 + 173} = 21$$

$$W_d = 43 + \sqrt{21 (346 - 25)} = 126$$

$$Q = \frac{2 (84/40)}{2 + (84/40)} = 1.35$$

$$K = \frac{125}{(1.25) (0.312) (1.35)} = 238$$

Predicted life for induction-hardened gears, from the diamond-6 line in Fig. 4, is 1.8×10^{13}, or

$$\text{life} = \frac{1.8 \times 10^{13}}{9000 \times 60} = 3.34 \times 10^7 \text{ hr}$$

The redesigned pinions have been employed in several hundred radar units for five years with no report of significant wear.

From Mechanical World

Cushioned Gear Drives

When flexibility is required in a gear drive it may range from very slight movement to complete slippage. The devices surveyed cover the full power range from the very light to the very heavy, and take account of unusual working conditions

By R. WARING-BROWN

A MODERN trend in all classes of machine construction is replacement of belt drives by gearing. Belt drives when properly designed with correct belting materials for the job are highly efficient. A notable instance is the new U.S.A., positive drive flat belt of neoprene-nylon construction, for which a 99% efficiency is claimed. Further, belts present one very important advantage in their capacity to absorb load reactions, shock and vibration. Obviously then, it is the space that a belt takes up and the fact that it is generally exposed that has led to a return to gearing in some applications.

The gear drive while normally more intricate and expensive lends itself to total enclosure, and enables the designer to produce a compact arrangement especially suitable to the unit construction now the vogue for all classes of machinery and mechanism. At the same time it will be appreciated that a solid gear drive may be subjected to wear and possible damage resulting from shock, vibration and alternate stressing. These objections have been overcome by the introduction of resiliency in the drive, usually by adopting some spring element in the gears themselves, or by some form of flexible coupling mounted between the gear and the driven element of the mechanism. This resiliency may range from something very small to quite a considerable amount of springing, depending on the requirements of the drive.

In practice many components require couplings giving complete rotational slip, hence a great divergence in design and construction of cushioning devices will be observed. As an example, the fuel injection pump of a compression ignition engine is driven by a gear from the engine. This is illustrated in Fig. 1, where it will be noted that a gear and shaft A driven from the engine has splined to it a coupling disc B, which has two tongues formed upon its outer rim. Similarly keyed to the camshaft E of the fuel injection pump is another coupling disc C, also constructed with two tongues. Between these two discs are placed the coupling rings D, having four

slots machined in it to accommodate the tongues of the two coupling discs B and C.

The camshaft of a fuel injection pump is a very hard working component, designed to withstand the very heavy operating pressures and both torsional and bending effects. Experience has shown that the type of drive illustrated is highly suitable for the particular purpose, the coupling ring being of woven cotton or nylon impregnated with plastic or synthetic rubber heated and compressed to approximately a third of its original thickness. The degree of flexibility obtained is ideal for fuel pump transmission and the coupling stands up well to fatigue caused by pulsation of the pump while also providing for the minute angular deflexion so essential and allowing for the small but desirable transverse misalignment between the connected shafts of up to 0·05 in.

Where longitudinal shaft deviation is not envisaged, infinitesimal angular deflexion can often be efficiently dealt with in gear drives by installing a gear wheel made of similar materials to those enumerated previously, but if a more enhanced flexibility is necessary this may be obtained by instituting an arrangement similar to that illustrated in Fig. 2. Here it will be seen that the gear wheel is formed into two components, a gear ring A and the hub B, with the interposition of a rubber jointing as a cushioning element. This basic principle allows of several constructional methods, but referring to the example shown, which is used for light gear drives, there is disposed between the gear ring A and the hub B a composite unit, consisting of inner and outer sleeves D and F respectively, between which is interposed the rubber cushion E. The hub of the gear wheel is keyed to its shaft C and the composite rubber/steel unit is pressed into position with heavy pressure. Alternatively, it is possible and practical to insert the rubber to make a joint directly between the gear ring and the hub, thereby eliminating the necessity for two inner sleeves.

The method of assembly will perhaps be better appreciated when it is considered that the longitudinal fibres

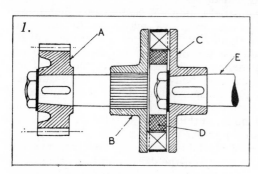
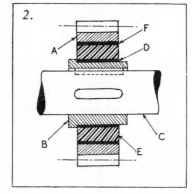
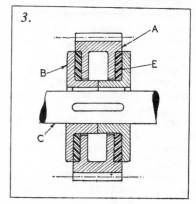

Flexible gear drives for: fuel pump (Fig. 1 above); power press (Fig. 2 centre); textile machines (Fig. 3 right)

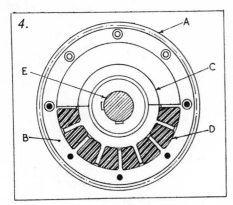
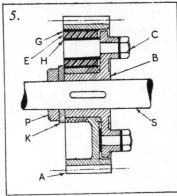
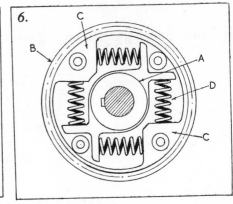

Flexible gear drives for: rock crusher (Fig. 4 left); line shafts (Fig. 5 centre); heavy oil engine (Fig. 6 right)

of the elastic ring are stretched, and when in position between the inner and outer surfaces, the effort made by the ring to re-establish its original shape exerts a radial compressive force of such intensity that no slip can take place between the two components.

The design of the drive ultimately depends not only on the torque to be transmitted but also upon the degree of flexure determined from actual operation, and frequently the atmospheric conditions. Constructions such as those described have been successfully applied to power presses, absorbing the shocks and cushioning the intermittent loads to which this class of machine is subjected: not only so, but the durability of the driving gear has shown improvement of as much as 400%.

The special advantage of rubber cushioning is the ability to deal with torsional and longitudinal deflection singly or in combination. Where somewhat heavier duty power transmission is involved, and where it is desirable to incorporate the advantages of rubber as the resilient medium, more robust constructions become essential. One such is illustrated in Fig. 3. Here a toothed gear ring has two flanges A, and two hubs B, keyed to the gearshaft C. Synthetic rubber rings E are welded to the metal components in a pre-load condition.

Variations of flexure are easily attainable in either natural or synthetic rubbers, but the latter have advantages in that they are impervious to humidity, oils, spirits, acids, and high and low temperatures within wide limits. Such gear drives have in practice proved highly satisfactory on certain textile machinery where a continuously efficient performance at high rotational speed and rates of flexure are involved. The example shown permits any reasonable misalignment and eccentricity, while cushioning the machinery against overloads, vibration and shock. It is equally satisfactory in either direction of rotation, while its extreme simplicity assures great durability.

Coming now to gear drives where duties are heavy, and again wishing to retain the favourable conditions appertaining to the embodiment of rubber, the assembly illustrated in Fig. 4 is of interest. Here a resilient gear wheel is based on the employment of rubber blocks in compression between the blades of a driving and driven member. The construction comprises five distinct parts, and in the figure it will be noted that a large gear A has a series of internal blades B, while a hub member C formed with external blades D is keyed to the shaft E. Successively between the two sets of blades, rubber blocks F, which are initially in compression during assembly, are inserted. This initial compression is to preclude complete unloading under the most severe

torque conditions, while ensuring that full resilience is available at all times. An important feature of this particular construction is the adoption of a maximum volume of rubber, which allows of very high shock absorption and damping so desirable in colliery work, rolling mills and rock crushers.

There are of course many other types of resilient gear drive based on the inclusion of rubber as the cushioning medium, and in some instances, while flexibility is desirable, an essential feature is that the whole may be quickly and easily demountable. In Fig. 5 is illustrated a construction of this kind occasionally used for line shafting and special machine assemblies. It will be evident that it fulfils the requirements for it is in effect a pin type coupling and gear wheel combined. The gear wheel A is a solid element formed with a central bearing machined and fitted to be capable of part rotation around the hub of the driven flanged member B. This latter is mounted by keys on the shaft S and carries a series of pins C, the larger portions of which are located in pre-loaded rubber bearing components having inner and outer sleeves G and H. These rubber bearings are assembled into bores machined in the gear wheel A and are maintained in position with the flange member by a collar K and a nut P screwed on to the driven shaft S.

It will be evident that while torsional flexibility is catered for, no longitudinal deviation has been considered in this particular design, nevertheless this construction has proved very satisfactory in practice, being noiseless in action and requiring a minimum of maintenance.

Flexibility, load carrying capacity, vibration and damping are all calculable so that practically any desired result is attainable. Hence by reason of general simplicity, it is obvious that rubber type flexible bearings are likely to play a greater part in the future development of cushioned gear drives.

Certain industrial conditions such as very high power gear transmission, excessive maximum or minimum temperatures, humidity, salt water and certain alkalis may preclude the use of rubber or plastic cushioning devices, and definitely render necessary the use of all-metallic gear drives embodying flexibility. The absorption of shock in the transmission of power to reciprocating pumps, air compressors, shearing machines, planing and shaping machines, turbine gearing, etc., is often most important for durability.

The degree of flexibility to be embodied will more or less decide the design of the drive, and in machinery or plant where divergencies in loading are strictly limited it is often possible to obtain the required resilience merely

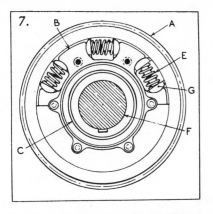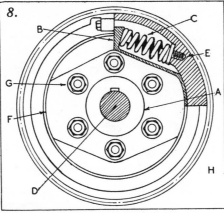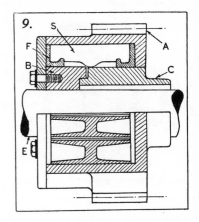

Flexible gear drives for: locomotives (Fig. 7 above);
reciprocating pump (Fig. 8 centre); heavy machinery (Fig. 9 right)

by equipping the installation with a driving pinion of rawhide or compressed paper, fabric, or phenolic resin, all of which will institute some degree of cushioning.

To ensure definite slip it is sometimes advisable to introduce a clutch in the drive, using friction, hydraulic or electro-magnetic devices, but this is largely influenced by special requirements.

In general it is highly desirable to embody the cushioning elements as far as possible within the dimensions of the driving pinion or gear wheel. In Fig. 6 is illustrated such an assembly comprising an element A keyed to the driving shaft in the usual manner, which has mounted upon it a gear wheel B. Lugs C are formed integrally with the internal diameter of the gear wheel and also on the hub member, while disposed between them is a series of helical springs D through which the load must always be transmitted in either direction of rotation. Completing the assembly is a cover plate to render it dust proof. This is a very substantial construction used in various industrial gear drives and is obviously ideally suited to dealing with heavy and fluctuating loads.

Another gear drive which differs considerably from the preceding example, except that the cushioning effect is again taken through coil springs, has proved very satisfactory over years of service in divergent applications. It is illustrated in Fig. 7 and it will be seen that it consists of a gear ring A on the internal surface of which are formed four short arms B, the gear ring being mounted to run freely on the steel hub C, which is also furnished with four projecting arms. These latter are provided with sufficient clearance to accommodate the projecting arms B, although the only connection between these sets of arms is through the heavy helical springs E by which means power is transmitted from the gear to the hub fixed to the shaft F. Ball headed pins G are fitted at the ends of the springs to ensure correct positioning on the arms. Cover plates bolted to the hub afford protection to the interior against foreign matter, the whole forming a compact assembly.

Coming now to the flexibility of exceedingly heavy gear drives, it is difficult to construct cushioning devices capable of being completely contained within the normal width of the gear, hence the general design will more or less be dictated by the space available. In Fig. 8 is illustrated a cushioned drive assembly for a very heavy duty reciprocating pump, and in this particular instance the resilient gear is of the split rim type of construction. This drive might well be adapted to air compressors, metal planers and shapers, and punching and shearing machinery, though it is not used to any great extent for

such purposes, no doubt because of the additional expense.

In the figure it will be noted that helical compression springs C, arranged for shock absorption, are disposed side by side, that is in duplicate, on one central hub member A keyed to the driven shaft D. The hub has a series of abutments B, while a similar number of stud abutments E are fitted on the interior of the gear wheel H, the compression springs C being located between them. In such gear drives the springs are generally designed for loads of five times the normal gear load. End covers F, held by studs G, complete the assembly.

It will be apparent that torsional flexibility can be adequately provided, but that transverse deviation is not so easily dealt with, and in heavy duty assemblies is more or less ignored. For this reason many industrial gear drives are designed in conjunction with flexible shaft couplings, but such assemblies are largely dependent of space. Fig. 9 illustrates such a design, wherein a toothed gear A is fitted with a component carrier B in which are located a series of special plate springs S. The hub C is keyed to the driven shaft E, and is also provided with a number of slots to take the ends of the plate springs. There is no connection between the driving and driven members A and C except through the plate springs, these being further supported by the end rings F. The locating arrangement is such that the end rings float in the interior of the gear and are capable of rotary movement about the hub axis.

Important advantages accrue to this construction in that a wide variation in the degree of flexibility is obtainable by changing the number and gauge of the plate springs, which enables the torsional stiffness to be varied without in any way impairing the required ultimate strength of the drive. In the design of such components for heavy intermittent and irregular loading the spring force is usually based on two or three times the actual gear tooth load.

As mentioned, it is often desirable that the cushioning of a gear drive may be effected by spring controlled friction to deal with shock or overload. Such devices can be comparatively simple, but if subject to shaft misalignment it is essential to make special arrangement, often culminating in the adoption of a flexible shaft coupling. The writer has often overcome this problem of alignment by fitting some form of self-aligning bearing between the hub and the gear.

Where gear drives are employed for looms with fixed reeds, some flexibility *via* slip is very necessary. In Fig. 10 is illustrated a construction which has been found very satisfactory for the purpose over a number of years. As

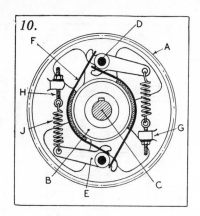
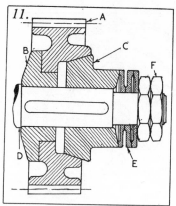
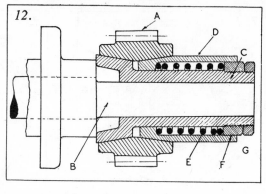

Flexible gear drives for: looms (Fig. 10 left); jig crane (Fig. 11 centre); machine tools (Fig. 12 above)

will be seen it is incorporated in the gear wheel which has two parts, an outer toothed ring A which is free on the hub of an inner brake drum B keyed to the shaft C. The flange of the gear wheel carries two pins D on which are mounted two levers E having ears which carry the ends of brake straps F lined with friction material to operate on the brake drum B. The gear wheel flange also carries two lugs G to which are fastened screwed hooks H, which latter are connected to tension springs J mounted on the ends of levers E.

The tension on the brake bands may be regulated through the springs and is so adjusted that the maximum starting torque is transmitted without slip, but if the loom to which the shaft is connected is suddenly stopped, slip between the drum and the brake band occurs, thus reducing the shock on the gear teeth.

When it is necessary to protect an electric motor a spring will act as a safety device. This is particularly the case on motor-driven jig cranes, and one such device as fitted by a well-known manufacturer is shown in Fig. 11. Here a gear A meshing directly with the motor pinion is mounted upon the flanged hub B and a bronze cone C, both of which are keyed to the driven shaft D. By means of the three tempered steel Belleville type springs E and the two adjusting nuts F, the desired axial force may be transmitted to the clutch members B and C. The combination A, B, C amounts to a conical slip clutch. The angle that the element of the cone makes with the axis 50°. The assembly gives a desired torsional flexibility for starting a motor drive and has proved very efficient.

A final construction on similar lines adopted for machine tool drives is illustrated in Fig. 12. As utilized on a table feed mechanism its function is to permit the pinion A to slip when the load on the cutter becomes excessive. The driving shaft B has keyed to it a bronze sleeve C on which is slidably mounted another sleeve D. It will be seen that parts of the sleeves are made conical to fit the conical bore of the pinion A. The frictional force required to operate the table is obtained by the pressure of the spring E inside the sleeve D. By means of the adjusting nuts F and G the spring pressure may be varied to suit any particular operating conditions.

It will be apparent that in this and similar devices where due regard has been given to torsional flexibility, little attention is paid to misalignment, but it will be appreciated that this may be infinitely small and be covered by tooth clearance in the gears.

In conclusion, the attention of designers is drawn to the variations in type and style of cushioned gear drives as outlined herein, and to the ever-increasing demand for new ideas in this field. It would appear that rubber will play an important part in the future where drives not only require flexibility, but also are to negate various forms of misalignment. It is also possible that a combination of rubber and metal constructions may prove most effective where extreme temperature conditions do not prevail, while being capable of meeting heavy loading together with silent operation, durability and minimum maintenance.

Noise Standards for Gears

The AGMA has made the first tentative step toward control. Specification 295.01 suggests noise limits for high-speed helical and herringbone gear sets. Their ceilings: 107 db between 20 and 75 cps; 99 db between 75 and 150 cps; 94 db between 150 and 300 cps; 92 db up to 10,000 cps.

Longest of the papers in a 38-page report (299.02), "Noise Control in Automobile Helical Gears," is that devoted to design considerations, by Evan Jones of Chrysler and W. D. Route of Chevrolet. They point out that gear noise is predominantly tooth contact frequency and it is presumed that noise sources are the events at beginning and end of contact. Design parameters suspected to mitigate gear noise are:

1) Fine pitch for overlap and tooth action close to the pitch point.

2) Low pressure angle for low tooth stiffness.

3) Helix and face width for helical overlap.

4) Recess action for reduction of normal tooth loads resulting from frictional effects.

5) In planetary arrangements, there should be:

a) Phasing of tooth contact events on the pinions to accomplish wave interference.

b) "Floating" elements to avoid tight mesh forces produced by deflection of supports.

c) Carrier structural rigidity.

6) In countershaft arrangements, the highest degree of stiffness permitted by the size and weight limitations of the transmission must be provided.

Other areas of control are important in noise reduction. These include involute and lead modification, dimensional specifications, inspection, and noise measurement.

Critical Speeds of Geared Shafts

Here are curves that shorten calculations for two-bearing machines with overhung weight

REYTON F WOJNOWSKI

Applied science representative, IBM
Rochester, NY

THOMAS R FAUCETT

Prof of Mechanical Engineering
Univ of Missouri, Rolla, Mo

Critical speeds of rotating machinery are speeds of maximum vibration, and therefore are critical for life also. One common configuration is an overhung weight—gear, grinding wheel, propeller—on a two-bearing shaft. The designer needs to know critical speed so that he can either avoid it or shift it.

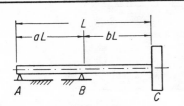

SYMBOLS

a = Fraction of shaft length between supports
b = Fraction of shaft length overhung
E = Modulus of elasticity, psi
g = Gravitational constant, in./sec²
I = Moment of inertia, in.⁴
L = Length of shaft, in.
N = Ratio W/W_b
N_c = Critical speed, rpm
p = $(W\omega^2/gEI)^{1/4}$
W = Weight of end weight, lb
W_b = Weight of shaft, lb
y = Shaft deflection, in.
ω = Circular frequency of vibration, rad/sec

The method shown below finds the first three critical speeds of a two-bearing machine with overhung weight on a shaft with uniform cross-section. It has two advantages: It is short because of the charted parameters; it covers a wide range of cases, because the parameters are dimensionless.

Critical speeds

Assuming that critical speed and natural frequency are almost the same, critical speed

$$N_c = \frac{60}{2\pi} \left[\frac{(pL)^4 gEI}{wL^4} \right]^{1/2}$$

Computer solutions of the complicated equation which gives values for pL are shown in the graphs. These theoretical values check closely with

measurements from a mechanical model for the first critical speed. At the second critical speed, errors show up and increase as the overhung weight becomes heavier and deflections larger. Although no comparisons have been made, the error is probably larger for the third critical speed.

The reason for the errors is the disregard of the angular inertia of the end weight as it rotates about an axis perpendicular to the centerline of the shaft. Stated more simply, the disk is assumed to be rotating in a vertical plane, while it is really tilted. Since the common formulas for critical speeds rarely include rotary inertia, accuracy is up to usual standards.

When the ratio of shaft weight to

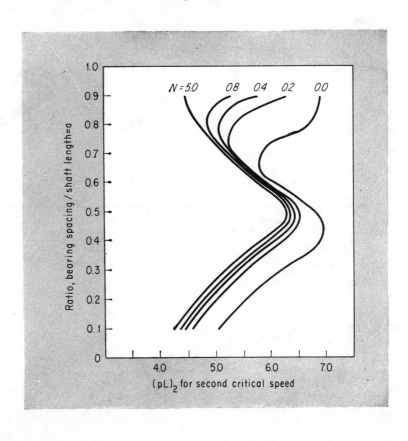

$(pL)_2$ for second critical speed

end weight exceeds five $(N > 5)$, a simpler analysis can be made with small error by neglecting the end weight.

Displacement

Between the bearings deflection $y_1 = C_1 \sin px + C_2 \sinh px$.

Along the overhang deflection $y_2 = D_1 \sin px + D_2 \cos px + D_3 \sinh px + D_4 \cosh px$.

In terms of D_4 the constants are:

$$C_1 = -D_4/(\sin apL)$$
$$C_3 = D_4/(\sinh apL)$$

$$D_1 = [D_4 \cos bpL + \cosh bpL + \\ \sinh bpL \,(\text{ctnh } apL - \text{ctn } apL)]/ \\ (\sinh bpL + \sin bpL)$$

$$D_2 = -D_4$$

$$D_3 = [-D_4 \cos bpL + \cosh bpL + \\ \sin bpL \,(\text{ctn } apL - \text{ctnh } apL)]/ \\ (\sinh bpL + \sin bpL)$$

EXAMPLE: What are the first three critical speeds and their corresponding displacement curves for a 3-in.-dia steel shaft if $L = 10$, $W = 100$ and $a = 0.3$?

Solution: Steel weighs 0.283 lb/in.³, so $W_b = 266$ lb. For $a = 0.3$ and $N = 100/266 = 0.375$, the graphs give $(pL)_1 = 1.80$, $(pL)_2 = 5.40$, and $(pL)_3 = 9.43$ which drop into the equation for critical speed.

$$N_{c1} = \frac{60}{2\pi} \frac{(pL)^4 gEI}{wL^4}$$

$$= \frac{60}{2\pi} \left[\frac{\begin{array}{c}(1.80)^4 \times 32.2 \times 12 \times \\ 30 \times 10^6 \times 3.97\end{array}}{266 \times 120^3} \right]^{1/2}$$

$$= 310 \text{ rpm}$$

$N_{c2} = 2790$ rpm
$N_{c3} = 8460$ rpm

Data necessary for calculating normal elastic curves for the three critical speeds and the curves plotted from them are shown below.

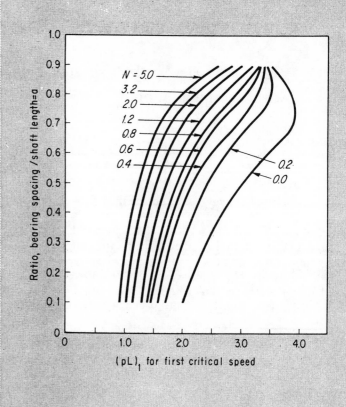

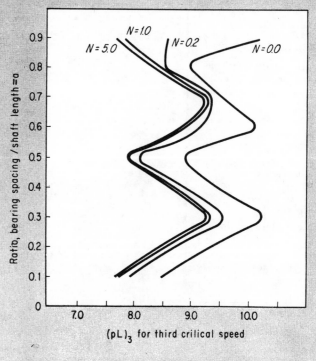

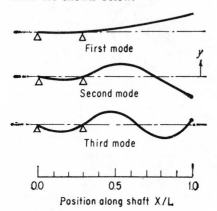

pL	C_1	C_3	D_1	D_2	D_3	D_4
1.80	$-1.94D_4$	$1.77D_4$	$1.08D_4$	$-D_4$	$-0.73D_4$	$+D_4$
5.40	$-1.00D_4$	$0.411D_4$	$2.14D_4$	$-D_4$	$-1.02D_4$	$+D_4$
9.43	$-3.24D_4$	$0.118D_4$	$5.09D_4$	$-D_4$	$-1.0D_4$	$+D_4$

Bearing Loads on Geared Shafts

ZBIGNIEW JANIA

Project Engineer, Ford Motor Company

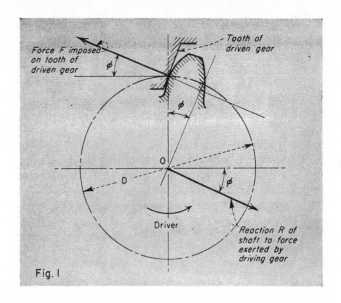

Fig. 1

PROBLEMS THAT INVOLVE bearing loads imposed on a shaft, on which two or more driven or driving gears are fastened, usually require that both the magnitude and the directions of the bearing loads or reactions be found. Such problems can be solved graphically when the magnitudes and directions of the forces upon the gear teeth are known.

For evaluating bearing loads, most techniques resolve the tooth forces acting on the gears into tangential and radial components and then solve for conditions of equilibrium. A graphical method, which is less complicated, is here described wherein the tooth forces

and reaction forces are not required to be resolved into any system of directional components.

In Fig. 1 is shown a tooth of involute profile on a driving gear and the tooth load or force that is imposed on the tooth of the mating driven gear. In the plane of the gears, the line of the reaction to the force exerted by the driving gear is opposite in direction and passes through the axis of its shaft. Where

T = torque transmitted by the gear, lb in.
D = pitch diameter of the gear, in.
F = tooth load or force applied in the line of tooth action, lb
φ = pressure angle of gears, deg

the gear tooth load is calculated from

$$F = (2T)/(D \cos\varphi)$$

Since a parallel displacement of a single force through a given distance can be effected, without disturbing the equilibrium of the system, by adding to the system a moment equal to the product of the force and the distance it is moved, a tooth load F can be referred to a plane that is perpendicular to the axis of the shaft and that passes through a supporting bearing by adding the appropriate moment to the system.

For example, in Fig. 2 the force F_b of the tooth load imposed on gear B is

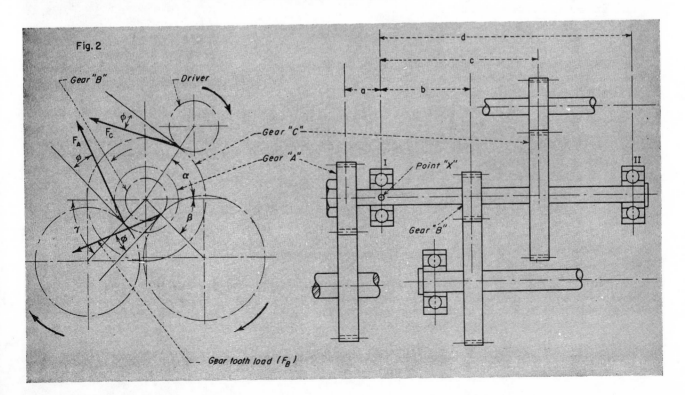

Fig. 2

Gear tooth load (F_B)

equivalent to an equal parallel force applied in a plane perpendicular to the axis of the shaft passing through point X plus a moment equal to the product of the tooth load F_b and the arm b. In establishing a moment, a sign convention must be adopted to distinguish between clockwise and counter-clockwise directions.

For a shaft supported in two bearings and having two or more gears fastened to it, to maintain equilibrium, the resultant of the moments introduced by referring their tooth forces to the plane of one bearing must be equal and opposite to the moment of the reaction of the second bearing about the first bearing.

The moment of the reaction of the second bearing as referred to the first bearing is readily found by means of a moment vector diagram. With this moment known, the reaction load at the second bearing is the moment divided by the distance between the bearings.

Since the forces acting on the gear teeth plus the bearing reactions on the shaft must also be in equilibrium, the reaction of the first bearing on the shaft can be found by closing a force vector diagram of the forces acting on the gear teeth including the reaction of the second bearing on the shaft as found from the moment vector diagram.

This procedure, which is set forth in the following example, of calculating loads on bearings can be applied to shafts carrying any number of gears. The method, however, is limited to cases that are statistically determinate.

EXAMPLE. Find the bearing loads acting on the central shaft of the gear train shown in Fig. 2; for which the given data are:

Pressure angle ϕ of all gears is 20 degrees.

Pitch diameter of gears, in.:

$A = 2.00$; $B = 1.50$; $C = 4.00$; Driver = 1.75

Moment arms, in.:
  $a = 1.50$; $b = 3.50$; $c = 5.00$; $d = 7.00$

Angles, deg; $\alpha = 55$; $\beta = 48$; $\gamma = 45$

Torque on driver = 100 lb in.

Torque delivered by A = 40 per cent of torque on center shaft

Torque delivered by B = 60 per cent of torque on center shaft

SOLUTION: Before drawing the vector diagrams, calculate these data:

Table I—Summarized Data for Constructing Vector Diagrams

Gear or Bearing	Dist. from Ref. Point X	Force, lb	Moment, lb, in.	Angular Position, deg
A	-1.50	97	-145.5	$180 + \gamma - 90 - \varphi = 115$
I	0	P_I	0	Reference plane
B	3.50	195	621	$360 - \beta - 90 - \varphi = 202$
C	5.00	121.5	608	$\alpha + 90 + \varphi = 165$
II	7.00	P_{II}	$7P_{II}$	δ, unknown

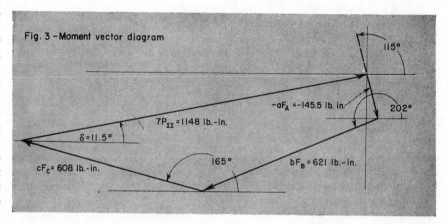

Fig. 3 – Moment vector diagram

$115°$

$-aF_A = -145.5$ lb. in.

$202°$

$7P_{II} = 1148$ lb.-in.

$\delta = 11.5°$

$cF_C = 608$ lb.-in.

$165°$

$bF_B = 621$ lb.-in.

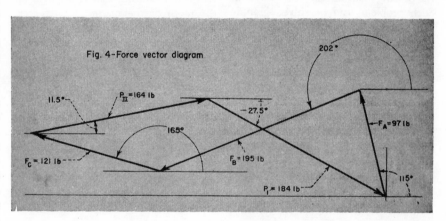

Fig. 4 – Force vector diagram

$202°$

$11.5°$

$P_{II} = 164$ lb

$-27.5°$

$-F_A = 97$ lb

$165°$

$F_C = 121$ lb

$F_B = 195$ lb

$P_I = 184$ lb

$115°$

Torque on center shaft =

$$\frac{100 \times 2 \times 4.00}{1.75 \times 2} = 228 \text{ lb in.}$$

Gear tooth loads:

$F_A = (0.4 \times 2 \times 228) / (2.00 \cos 20 \deg) = 97$ lb

$F_B = (0.6 \times 2 \times 228)/(1.50 \cos 20 \deg) = 195$ lb

$F_C = (2 \times 228)/(4.00 \cos 20 \deg) = 121.5$ lb

After performing these calculations, the data needed to construct the vector diagrams are assembled in a summary as in Table I for ready reference.

Draw the moment vector diagram, Fig. 3, to determine the direction and magnitude of the vector $7 P_{II}$. From this diagram, it is found that this vector measures 1,148 lb in., and has an angular position δ of 11.5 degrees.

The reaction of bearing II then is

$P_{II} = 1,148/7 = 164$ lb at 11.5 deg

With the magnitude and direction of P_{II} known, the force vector diagram, Fig. 4, can be drawn to determine the magnitude and direction of the reaction of bearing I. On completing the force vector diagram, the load on bearing I is found by measurement to be

$P_I = 184$ lb at -27.5 deg

After the magnitude and direction of all forces acting on the shaft have been found, bending moment and shear diagrams for any axial plane of the shaft can be drawn using the components of the forces as resolved in the axial plane under consideration.

10
GEAR MATERIALS
AND LUBRICATION

Materials for Gears

by Norman E. Woldman, *Consulting Engineer*

The satisfactory performance of a gear depends to a large degree upon the materials of which it is made. A large number of gear materials are available.

from MATERIALS IN DESIGN ENGINEERING

Westinghouse Electric Corp.

Huge marine reduction gear *typifies problem of selecting right material for specific end application. This gear is built up of plain carbon steel hub and webs to which are welded gear rims of carbon or molybdenum-vanadium steel.*

Gear Materials—
A Quick Summary of Their Properties

Carburizing Steels	Combine maximum surface hardness and wear resistance with interior toughness and shock resistance. Best suited for heavy duty service. Offer high resistance to wear, pitting and fatigue. Low carbon content provides maximum ductility; higher carbon content provides maximum core strength. Core and case properties depend largely on type of heat treatment.
Nitriding Steels	Case properties generally same as carburizing steels. Useful for large gears with thin sections. Nitrided steels retain hardness at temperatures up to 80) F.
Through Hardening Steels	Possess greater core strength and provide quieter operation than carburized steels. Less expensive than carburizing steels because of simpler heat treatment required. Not as ductile or as resistant to surface compression stresses as case hardened steels. Relatively shallow hardening types suitable for gears requiring only moderate strength and impact resistance. Medium to deep hardening materials suitable for gears requiring medium to high wear resistance and high load carrying capacity.
Special Gear Steels	
Leaded Steels............	Reduce machining time and costs. Permit faster metal removal through use of deeper cuts and faster cutting speeds.
High Manganese Steels....	High resistance to heat, wear and bending stresses. Surfaces become work hardened and polished after use.
Boron Steels............	Used for highly stressed gears.
Stainless Steels	High corrosion resistance combined with excellent mechanical properties.
Ultra High Strength Steels.	Combine maximum ductility with high tensile strengths (230,000 psi and over).
Pre-heat Treated Bar Steels	Used for gears that cannot be heat treated after machining. Use where distortion cannot be tolerated, tolerances are small, or where grinding to close tolerances is impractical.
Gray Cast Iron	Combine low cost with good damping capacity and good wear resistance when properly lubricated. Relatively weak compared with rolled, forged or cast steel. Quality depends to large degree on foundry practice.
Nodular or Ductile Iron	Good casting and machining properties. Wear resistant. Higher strength than gray cast iron. Respond well to heat treatment and make dependable power transmission elements
Bronzes	
Tin Bronze..............	Good hardness, strength and resilience. Recommended for general purpose worm wheels.
Silicon Bronze...........	Usually used for worm wheels that mate with case hardened worms. Recommended for medium loads and medium to high speeds.
Phosphor Bronze........	Used for worm wheels that mate with worms of high hardness and fine accuracy. Generally recommended for medium loads and medium to high speeds. Some grades can be used under heavy pressures and severe working conditions.
Leaded Bronze...........	Recommended for worm gears to mate with soft steels under low loads and low to medium speeds.
Manganese Bronze	High manganese bronze is highly wear resistant. Recommended for high loads and speeds.
Aluminum Bronze	Combines high strength and hardness with good resistance to wear and fatigue. Particularly suited for severe service where long life is required.
Nickel Bronze...........	Good yield strength, toughness and wear resistance.
Metal Powders	Low cost when produced in large quantities. Widely used for low loads and comparatively low speeds. Properties can be considerably improved by carburizing and impregnation.
Nonmetallic Materials	
Molded Plastics	Quiet operation, good resiliency, vibration and damping properties, combined with low cost in large quantities.
Reinforced Laminates.....	Properties generally same as molded materials.
Rawhide................	Quiet operation combined with high shock resistance, comformability and resiliency

Ferrous Metals

Steel is the most widely used ferrous material for gears since it can be manufactured and processed to a great many different specifications, each of which has a definite use. In general, there are two types of gear steels: surface hardening and through hardening. The surface hardened steels are hardened to a relatively thin case depth and include carburizing and nitriding steels in low carbon, plain or alloy types. Through hardening steels may be comparatively shallow hardening or deep hardening, depending on their chemical composition and method of hardening.

Carburizing steels

General properties — Carburizing or nitriding grades of steel are usually specified where maximum wear resistance is required for bearing surfaces. Carburized, case hardened gears are best suited for heavy duty service, e.g., transmission gears, and offer high resistance to wear, pitting and fatigue. Surfaces must be sufficiently hard to resist wear and of sufficient depth to prevent crushing. A rough rule for case depth is that it shall not exceed one-sixth of the base thickness of the tooth.

A case hardened gear provides maximum surface hardness and wear resistance, and at the same time provides interior toughness to resist shock. In general, case hardened gears can withstand higher loads than through hardened gears, although the latter are quieter and are less expensive because of the simpler heat treatment required.

Selection factors—A number of factors must be considered when selecting a case hardened gear:

1. High tooth pressures will crack a thin case.

2. Too soft a core will not provide proper backing for a hard case.

3. Compressive stresses in the case improve fatigue durability, and a high case hardness increases wear resistance.

4. If the ratio of case depth to core thickness is too small, excessive stresses in subsurface layers can produce poor fatigue life.

5. Residual tensile stresses are highest with low core hardness and increase with increasing case depth. These stresses can be relieved by tempering.

Grain size variations have an important effect on core properties. These variations are influenced by the type of steel and the method of heat treatment used subsequent to carburizing. Section thickness also influences core properties.

A tough tooth core may not be required in applications where a gear will not be subjected to impact loading. In these applications core properties are relatively un-important, provided the core is sufficiently hard to support the case. Considering the case alone, it is important that the surface resist wear and fatigue bending, since bending stresses vary from a maximum at the surface to zero near the tooth center.

Effects of composition—The carbon content of carburized gears is usually within the range of 0.10 to 0.25%. (A carbon content of 0.35% has also been occasionally used.) A low carbon content is usually used to obtain maximum ductility, and a high carbon content is used to obtain maximum core strength. The recent trend is toward high carbon gears with comparatively shallow cases produced by gas carburizing or activated carburizing baths.

In all typical case hardening carbon and alloy steels the desired carbon content is obtained by specifying the proper AISI (or SAE) designation. The last two digits of the steel designation indicate the mean carbon content. The 1100 series carbon steels are the free-cutting grades. Some representative AISI carburizing steels used for gears are: 1015, 1020, 1022, 1025, 1117, 1118, 2317, 2515 and 3120. Also, 3310, 4020, 4118, 4320, 4620, 4820, 5120, 6120, 8620, 8720 and 9310. Many other standard and special carburizing steels of low and high carbon and alloy content are also available.

The nickel carburizing steels are used chiefly where exceptional core toughness combined with the highest degree of wear resistance

TABLE 1—PROPERTIES OF TYPICAL CARBURIZED GEAR MATERIALS•

Steel ↓	Carbon Content	Heat Treatment (see key)	Brinell Hardness	Tensile Strength, psi	Yield Point, psi	Elongation in 2 in., %	Reduction of Area, %	Izod Impact Strength, ft-lb
1-In. Rods								
C1015	0.15	A1	149	73,000	46,000	32	71	91
C1020	0.20	A1	156	75,000	48,000	31	71	93
C1022	0.22	A1	163	82,000	47,000	27	66	81
C1117	0.17	A2	192	96,500	59,500	23	53	33
C1118	0.18	A2	229	113,000	76,500	17	45	16
½-In. Rods								
2317	0.17	B11	195	95,000	60,000	35	65	85
		B12	210	100,000	65,000	30	60	70
2515	0.17	B13	277	130,000	90,000	24	60	65
		B14	352	170,000	135,000	14	50	40
3115	0.17	B15	212	100,000	70,000	25	55	55
3310	0.10	B1	375	181,500	149,000	15	57	40
		B2	363	180,000	145,500	14	57	55
		B3	352	177,000	143,500	15	58	47
		C1	375	181,000	153,000	15	58	40
		C2	363	180,000	149,500	14	58	57
		C3	341	175,500	146,500	15	59	50
4320	0.20	B1	429	217,000	159,500	13	50	33
		B2	429	218,000	178,000	14	48	28
		B4	302	152,000	97,000	20	49	49
		C1	415	215,000	159,000	13	49	26

KEY TO HEAT TREATMENT

A—Single quench and temper

1—Carburized at 1675 F for 8 hr, pot cooled, reheated to 1425 F, water quenched, tempered at 350 F.
2—Carburized at 1700 F for 8 hr, pot cooled, reheated to 1450 F, water quenched, tempered at 350 F.

B—Recommended practice for maximum case hardness

1—Direct quench from pot: quenched in oil, tempered at 300 F.
2—Single quench and temper for good case and core properties: pot cooled, reheated to 1500 F, quenched in oil, tempered at 300 F.
3—Double quench and temper for maximum grain refinement of case and core: pot cooled, reheated to 1500 F, quenched in oil, reheated to 1450 F, quenched in oil, tempered at 300 F.
4—Same as B3 except final quench from *1425 F.*
5—Same as B3 except final quench from *1475 F.*
6—Same as B2 except final quench from *1475 F.*
7—Same as B2 except final quench from *1550 F.*
8—Same as B3 except pot cooled, reheated to *1550 F,* quenched in oil, reheated to *1475 F,* quenched in oil, tempered at 300 F.
9—Same as B2 except reheated to *1450 F.*
10—Same as B3 except pot cooled, reheated to *1475 F,* quenched in oil, reheated to *1425 F,* quenched in oil, tempered at 300 F.
11—Same as B3 except reheated to *1525 F,* quenched in oil, reheated to *1375 F,* quenched in oil, tempered at 300 F.

and greatest surface compressive strength is required. These steels are the nickel steels (2000 series), nickel-chromium steels (3000 series), nickel-molybdenum steels (4600 and 4800 series), and nickel—chromium—molybdenum steels (4300, 8600, 8700 and 9300 series). The carbon-molybdenum steels (4000 series) are used where exceptional toughness and good resistance to temper brittleness are required.

Another advantage of the more highly alloyed steels is the ability of heavy sections to harden more completely. This greater hardenability promotes better core strength properties than can be achieved with shallow hardening steels quenched in the same size section.

Effects of heat treatment—
Table 1 lists the properties that can be expected of typical case hardened carbon and alloy steels. As shown, three methods of quenching and tempering can be used to obtain varying ranges of case and core properties. Method A consists of a single quench and provides a good compromise of case and core properties. Method B is recommended for maximum case hardness and Method C is recommended for maximum core toughness.

In the past, double quenching was recognized as the best method for obtaining optimum case and core properties. However, to avoid the danger of distortion inherent in double quenching treatments, most carburized gears are cur-

rently given a single quenching. The development of a single quench treatment has been made possible by steels with controlled grain size that maintain a fine grained structure after heat treatment.

Nitriding steels

Nitriding steels can be used in many gear applications where a hard, wear resistant case, good fatigue strength, low notch sensitivity and some degree of corrosion resistance are desired. Since the hardness of nitrided steel surfaces does not change at temperatures as high as 800 F, these steels are quite useful in applications where hardness must be maintained at temperatures that would destroy a carburized surface. In addition, nitriding

TABLE 1—PROPERTIES OF TYPICAL CARBURIZED GEAR MATERIALS[a]—continued

Steel	Carbon Content	Heat Treatment (see key)	Brinell Hardness	Tensile Strength, psi	Yield Point, psi	Elongation in 2 in., %	Reduction of Area, %	Izod Impact Strength, ft-lb
½-In. Rods—continued								
		C2	415	212,000	173,000	13	51	29
		C4	293	146,000	94,500	22	56	49
4620	0.17	B1	311	148,000	116,500	17	56	47
		B2	277	119,000	83,500	20	59	52
		B5	248	122,000	77,000	22	56	64
		C1	302	147,500	116,000	17	58	43
		C2	248	115,500	81,000	21	64	69
		C5	235	115,000	77,000	23	62	78
4820	0.21	B1	415	205,000	165,500	13	53	33
		B6	415	207,500	167,000	14	52	44
		B3	415	204,500	165,500	14	52	31
		C1	401	200,500	170,000	13	53	30
		C6	415	205,000	184,500	13	53	47
		C3	401	196,500	171,500	13	53	29
8620	0.23	B1	388	192,000	150,000	13	50	28
		B7	388	188,500	150,000	12	52	26
		B8	269	133,000	83,000	20	57	56
		C1	352	181,000	134,000	13	51	34
		C7	341	168,000	121,000	14	53	30
		C8	262	130,000	77,000	23	52	66
9310	0.11	B1	375	179,500	144,000	15	59	57
		B9	363	173,000	135,000	16	60	61
		B10	363	174,500	139,000	15	62	54
		C1	363	178,000	146,500	15	60	46
		C9	341	168,000	137,500	16	60	39
		C10	352	169,500	138,000	15	62	63

12—Same as B11 except final quench in water.
13—Same as B3 except reheated to 1500 F, quenched in oil, reheated to 1350 F, quenched in oil, tempered at 300 F.
14—Same as B3 except reheated to 1525 F, quenched in oil, reheated to 1400 F, quenched in oil, tempered at 300 F.
15—Same as B3 except reheated to 1550 F, quenched in oil, reheated to 1400 F, quenched in oil, tempered at 300 F.
C—Recommended practice for maximum core toughness
 1—Direct quench from pot: quenched in oil, tempered at 450 F.
 2—Single quench and temper for good case and core properties. pot cooled, reheated to 1500 F, quenched in oil, tempered at 450 F.

 3—Double quench and temper for maximum grain refinement of case and core: pot cooled, reheated to 1500 F. quenched in oil, reheated to 1450 F, quenched in oil, tempered at 450 F.
 4—Same as C3 except final quench from 1425 F.
 5—Same as C3 except final quench from 1475 F.
 6—Same as C2 except reheated to 1475 F.
 7—Same as C2 except reheated to 1550 F.
 8—Same as C3 except pot cooled, reheated to 1550 F, quenched in oil, reheated to 1475 F, quenched in oil, tempered at 450 F.
 9—Same as C2 except reheated to 1450 F.
10—Same as C3 except pot cooled, reheated to 1475 F, quenched in oil, reheated to 1425 F, quenched in oil, tempered at 450 F.

steels make it possible to surface harden the teeth of large gears having thin sections that might be impractical to carburize and quench.

Nitrided gears are relatively free from wear up to the load at which surface failure occurs, but at this load they become badly crushed and pitted. Thus, nitrided gears are generally not suitable for applications where overloads are liable to be encountered.

Hardness of the case is a function of the amount and nature of the nitride forming elements present, carbon content and nitriding temperature. Carbon tends to form carbides with the nitride forming elements and removes them from solid solution so that they are not available for hardening during nitriding. Nitriding steels containing no aluminum have lower surface hardness, but their case is considerably tougher and they can be peened without causing any spalling.

Several grades of Nitralloy steels are available for surface hardening by the nitriding process. These include: Nitralloy 115, 125, 135, 135 Modified, 225, 230 and N.

Nitralloy N, a nickel nitriding steel, is a precipitation hardening alloy which attains a core strength and hardness after nitriding that are considerably in excess of its original properties. Both Nitralloy N and 135 Modified are outstanding for heavy duty gears that are highly stressed (see Table 2 for typical core properties). Little change in tensile strength of nitriding steels occurs if the tempering temperature used for treating the core is at or above the nitriding temperature. However, because of the increased hardness of the case, the elongation, ductility and impact strength of both alloys are considerably reduced after tempering, though not to the

---

Alloy steel gears: a wide range of properties

Philadelphia Gear Works, Inc.

SAE 4320 *steel carburized and hardened to 75 to 85 Shore is used for this pair of spiral bevel miter gears.*

Ajax Electric Co., Inc.

SAE 1340 *automobile transmission gears are case hardened to a depth of 5 mils by liquid cyaniding.*

SAE 4140 *clutch spur gear is induction hardened to 520 Brinell.*

Philadelphia Gear Works, Inc.

SAE 4640 *cast steel helical gear is flame hardened and quenched to hardness of 65 to 75 Shore.*

Philadelphia Gear Works, Inc.

same extent; Nitralloy N develops a tougher and softer surface and a stronger core than Nitralloy 135 Modified.

Any of the AISI steels that contain nitride forming elements, such as chromium, vanadium or molybdenum, can also be nitrided. The steels most commonly nitrided are 4140, 4340, 6140 and 8740. In some applications the 0.50% carbon grades are also used.

In some applications it may be desirable to nitride the gear teeth only and leave the rest of the gear soft. Such selective hardening can be done with a stop-off coating such as tin plate. The entire gear is tin plated to 0.3 to 0.5 mil and the tin plate removed from the teeth before nitriding.

Through hardening steels

By virtue of their higher carbon content, through hardening steel gears possess greater core strength than carburized gears. They are not, however, as ductile or as resistant to surface compressive stresses and wear as case hardened gears. Hardness of gear surfaces may vary all the way from 300 to 575 Brinell.

Typical of the relatively shallow hardening, carbon steel gear materials are types 1035, 1040, 1045, 1050, 1137, 1141, 1144 and 1340. These steels are water hardening, but not deep hardening, types and are suitable for gears requiring only a moderate degree of strength and impact resistance.

In general, the more highly alloyed through hardening steels harden more completely when quenched in heavy sections. This greater hardenability provides greater strength than can be attained with shallow hardening steels quenched in the same size section.

Typical of the low alloy, medium to deep hardening gear materials are (in order of increasing hardenability): 4042, 5140, 8640, 3140, 4140, 8740, 6145, 9840 and 4340. These steels, as well as many other alloy steels with the proper hardenability characteristics and a carbon content of 0.35 to 0.50%, are suitable for gears requiring medium to high wear resistance and high load carrying capacity.

The core properties of representative through hardening carbon and alloy gear materials are

listed in Table 3. Other standard and special through hardening steels are available. In selecting a through hardening steel it should be borne in mind that a higher carbon and alloy content is accompanied by greater strength and hardness (but lower ductility) of the surface and core.

Fully hardened and tempered

TABLE 2—PROPERTIES OF TYPICAL NITRIDING STEELS·

Alloy ➡	Nitralloy 135 Modified		Nitralloy N	
	Before Nitriding	After Nitriding	Before Nitriding	After Nitriding
Tensile Strength, psi.............	138,000	138,000	132,000	190,000
Yield Point, psi...................	120,000	110,000	114,000	180,000
Elongation in 2 in., %............	26	4	22	6–15
Reduction of Area, %............	60	17	59	43
Brinell Hardness.................	320	310	277	415
Charpy Impact Strength, ft-lb......	44	21	—	—

aNitrided for 90 hr at 900 F, quenched in oil and tempered at 1200 F.

TABLE 3—CORE PROPERTIES OF THROUGH HARDENED CARBON AND ALLOY STEEL GEAR MATERIALS

Type of Steel ↓	Tensile Strength, psi	Yield Point, psi	Elongation in 2 in., %	Reduction of Area, %	Izod Impact Strength, ft-lb	Brinell Hardness
WATER QUENCHED CARBON STEELS (1-IN. BARS)						
C1030.........	122,000	92,000	16	47	9	475
C1040.........	130,000	95,000	17	50	10	475
C1050.........	160,000	116,000	12	34	9	475
C1137.........	216,000	168,000	4	17	7	415
C1144.........	290,000	188,000	5	11	2	555
OIL QUENCHED CARBON STEELS (1-IN. BARS)						
C1137.........	156,000	136,000	6	22	10	352
C1144.........	127,000	92,000	16	36	7	277
WATER QUENCHED ALLOY STEELS (½-IN. BARS)						
2330.........	250,000	205,000	12	50	30	460
4130.........	232,000	196,000	12	43	32	460
8630.........	228,000	210,000	11	52	26	460
OIL QUENCHED ALLOY STEELS (½-IN. BARS)						
1340.........	235,000	205,000	10	42	7	475
3140.........	242,000	218,000	12	50	22	475
4140.........	240,000	220,000	12	47	9	475
4150.........	237,000	220,000	10	39	12	475
4340.........	248,000	216,000	13	49	13	475
4640.........	243,000	227,000	12	47	13	475
5150.........	245,000	223,000	10	39	7	475
6150.........	235,000	216,000	10	40	11	475
8740.........	240,000	220,000	12	47	15	475
8750.........	237,000	218,000	12	42	14	475
9440.........	236,000	213,000	12	50	11	475
9255.........	233,000	215,000	8	23	6	475
OIL QUENCHED ALLOY STEELS (1-IN. BARS)						
2345.........	240,000	220,000	13	40	9	475
3145.........	245,000	218,000	8	28	5	475
3250.........	243,000	214,000	9	37	8	475
4340.........	242,000	222,000	12	45	14	475
4640.........	238,000	218,000	11	43	14	475
8645.........	235,000	210,000	11	40	12	475

Properties of special gear steels

TABLE 4—PROPERTIES OF MARTENSITIC STAINLESS STEEL GEAR MATERIALS

AISI Type ➡	403	410	414	416	420	431	440A
Tensile Strength, psi	180,000	180,000	190,000	180,000	230,000	195,000	260,000
Yield Strength, psi	140,000	140,000	145,000	140,000	195,000	150,000	240,000
Elongation in 2 in., %	15	15	15	13	8	15	5
Izod Impact Strength, ft-lb	35	35	45	20	10	45	4
Brinell Hardness	375	375	400	375	500	400	510

TABLE 5—PROPERTIES OF TYPICAL ULTRA HIGH STRENGTH STEELS[a]

Steel ➡	Hy-Tuf	Super Hy-Tuf	Tricent
Tensile Strength, psi	235,000	294,000	297,000
Yield Strength, psi	191,000	241,000	242,000
Elongation in 2 in., %	14	10	8
Reduction of Area, %	50	35	23
Rockwell Hardness	C47	C54	—
Impact Strength, ft-lb	33 (Izod)	—	18 (Charpy)

[a] 1-in. dia; oil quenched and tempered at 500 F.

TABLE 6—PROPERTIES OF TYPICAL PREHEAT TREATED BAR STEELS

Steel ➡	Stressproof	Fatigue-proof	Ryax	Rycrome	Elastuf A-2
Tensile Strength, psi	135,000	145,000	114,000	150,000	135,000
Yield Strength, psi	107,000	135,000	75,000	128,000	125,000
Elongation in 2 in., %	13	7	23	18	20
Brinell Hardness	285	300	225	311	285

medium carbon alloy steels possess an excellent combination of strength and toughness at room temperature and at lower temperatures. However, toughness can be substantially decreased by temper brittleness, a form of embrittlement developed in some alloy steels by slow cooling through the temperature range of 850 to 1000 F, or by holding or tempering in this range. Because of their good hardenability and immunity to temper brittleness, molybdenum steels have been widely used for gears requiring good toughness at room and low temperatures.

Special gear steels

Leaded steels—The principal advantage of leaded steels is that they reduce machining time and cost. These steels permit faster metal removal through the use of deeper cuts and higher cutting speeds, and require less driving power when being machined. Also, burring of teeth after cutting is often completely eliminated or at least greatly minimized.

Both the 1000 series of carbon steels and low alloy grades have been leaded with little change in tensile, bending and impact properties. In general, lead causes a 20% reduction in the fatigue strength of steels with a tensile strength of 265,000 psi, an 8% reduction in steels of 150,000 psi, and a negligible reduction in steels of 120,000 psi.

High manganese steels — High manganese-nickel alloy steels are noted for their high resistance to heat, wear and bending stresses. The teeth of gears made from 13% manganese steel, with or without nickel, bend under excessive loads but do not usually break. The tooth faces also become work hardened and polished after continuous use.

Inasmuch as the 13% manganese steel cannot be economically machined, the gears are usually cast to shape and finished by grinding. An addition of 4% nickel enables the alloy to resist changes in grain structure resulting from overheating.

Boron steels—The presence of boron in lean alloy steels improves their heat treatment properties and enables them to be used for highly stressed gears. These steels have a minimum content of 0.0005% boron which is added as an intensifier to insure adequate hardening during quenching. Typical of the carburizing boron steels are: 46B12H, 94B15 and 94B17H. Typical through hardening types are: 50B46H, 81B45H and 86B45H.

Stainless steels—Stainless steels of the martensitic type are principally used for gears requiring a high degree of corrosion resistance. As shown in Table 4, these materials also possess excellent mechanical properties.

High strength steels — Ultra high strength steels with increased silicon and manganese content hold considerable promise for gears where maximum ductility is required at tensile strengths of 230,000 psi and over. These steels still retain their ductility when heat treated to as high as 285,000 psi. They may be considered as modified 4300 series steels with or without other alloying elements such as boron, vanadium and aluminum. Typical properties are listed in Table 5.

Preheat treated bar steels—A

Properties and selection of cast irons for gears

TABLE 7—TYPICAL MECHANICAL PROPERTIES OF STANDARD GRAY IRONS

ASTM Class ↓	Tensile Strength, psi	Compressive Strength, psi	Torsional Shear Strength, psi	Modulus of Elasticity, 10^6 psi		Reversed Bending Fatigue Limit, psi	Brinell Hardness
				Tension	Torsion		
20..........	22,000	83,000	26,000	9.6–14.0	3.9–5.6	10,000	156
25..........	26,000	97,000	32,000	11.5–14.8	4.6–6.0	11,500	174
30..........	31,000	109 000	40,000	13.0–16.4	5.2–6.6	14,000	201
35..........	36.500	124,000	48,500	14.5–17.2	5.8–6.9	16,000	212
40..........	42,500	140,000	57,000	16.0–20.0	6.4–7.8	18,500	217
50..........	52,500	164,000	73,000	18.8–22.8	7.2–8.0	21,500	228
60.........	62,500	187.500	88,500	20.4–23.5	7.8–8.5	24,500	252

TABLE 8—RECOMMENDED OPERATING CONDITIONS FOR NICKEL CAST IRON GEARS

Properties Needed ↓	Composition, %						Average Brinell Hardness [b]
	Total Carbon	Nickel	Silicon [a]	Manganese	Chromium	Molybdenum	
Moderate Strength and Wear Resistance	3.2-3.4	1.0–1.2	2.1–2.3	0.60–0.90	0.30–0.40	—	215
	3.2-3.4	1.0–1.2	1.7–1.9	0.60–0.90	—	—	215
	3.2-3.4	1.4–1.6	2.0–2.2	0.60–0.90	0.20–0.30	0.3–0.5	220
	3.2-3.4	1.4–1.6	1.7–1.9	0.60–0.90	—	0.3–0.5	220
Excellent Wear Resistance	3.5 min	1.8–2.0	1.3–1.5	0.65–0.85	0.20–0.40	0.4–0.6	220
	3.3 min	1.8–2.0	1.4–1.6	0.65–0.85	0.20–0.40	0.4–0.6	235
Good Wear Resistance and High Strength	3.0–3.2	1.8–2.0	1.8–2.0	0.65–0.85	0.20–0.30	0.6–0.8	235
	3.0–3.2	2.0–2.2	1.4–1.6	0.70–0.90	—	0.6–0.7	235
	3.0–3.2	1.2–1.4	1.8–2.0	0.70–0.90	0.20–0.30	0.2–0.3	227
Very High Strength	2.4–2.8	1.0–1.2	2.4–2.8	0.80–1.00	0.10–0.20	1.0–1.2	300
	2.6–2.8	1.2–1.4	1.9–2.3	0.80–1.00	0.15–0.25	0.4–0.6	300

aSilicon contents are for ¾ and 1¼-in. sections. For lighter sections, higher silicon contents are advisable; for heavier sections, lower silicon contents are advisable.
bMean of values taken over range of section thickness.

TABLE 9—PROPERTIES OF AS-CAST MEEHANITE

Type →	GA	GC	GE
Tensile Strength, psi..	50,000	40,000	30,000
Transverse Strength, lb	3,400	3,100	2,300
Shear Strength, psi...	48,000	40,000	30,000
Brinell Hardness......	207	192	174

TABLE 10—PROPERTIES OF DUCTILE IRON GEAR MATERIALS

Type →	60–45–10	80–60–03	100–70–03	120–90–02
Tensile Strength, psi....	60–80,000	80–100,000	100–120,000	120–150,000
Yield Strength, psi......	45–60,000	60–75,000	70–90,000	90–125,000
Elongation in 2 in., %...	10–25	3–10	3–10	2–7
Brinell Hardness........	140–190	200–270	240–300	270–350
Charpy Impact Strength (unnotched), ft-lb.....	60–115	15–65	35–50	25–40

number of proprietary bar steels are available for gears that cannot be heat treated subsequent to machining. These steels (see Table 6 for properties) are used where distortion cannot be tolerated, where tolerances are small, or where grinding to close tolerances is impractical. They are obtainable in hardnesses up to 400 Brinell and can be machined to very close o.d. dimensions.

Cast irons

Gray and alloy irons—Cast iron provides three principal advantages when used as a gear material. These are: 1) low cost, 2) good damping capacity, and 3) good wear resistance when properly lubricated. The material is used in many applications where strength is not too important but good wear resistance is required. In such applications the high carbon irons are usually specified because of the presence of large amounts of graphite, which acts

American Foundry Equipment Co.

Shot peening *is used to relieve tensile stresses in gear surfaces and improve fatigue resistance.*

as a lubricant and improves wear resistance.

Cast iron is relatively weak and brittle when compared with rolled, forged or cast steel and has a low modulus of elasticity (10-20 x 10⁶ psi). Because of shrinkage and porosity tendencies, the quality of cast iron gears is also dependent to a large degree on foundry practice.

Cast iron gears requiring maximum dimensional stability should be specified with a stress relieving treatment consisting of: 1) heating to a minimum temperature of 900 F for 1 hr per in. of section, and 2) furnace cooling at a rate not exceeding 50 deg per hr to a temperature of 200 F or less. Where improved hardness, wear resistance and strength are required, gears of simple shape and section can be oil quenched from 1550-1600 F and tempered at 700 to 1000 F. Depending on composition and drawing temperature,

hardness is approximately doubled and tensile strength increased by 20 to 30% after heat treatment.

Typical mechanical properties of the ordinary grades of cast iron are listed in Table 7. The alloy cast irons, such as nickel, nickel-chromium, nickel-chromium-molybdenum, are usually used where greater strength, hardness and wear resistance are required. Compared to the ordinary gray irons, these materials provide a higher ratio of endurance limit to tensile strength, relatively low notch sensitivity, and a finer-grained, denser structure.

The various grades of nickel cast iron gears recommended for specific operating conditions are listed in Table 8. In general, the plain nickel and nickel-molybdenum gray irons are most satisfactory where gears are subjected to impact loading. The nickel-chromium and nickel-chromium-molybdenum types are best suited for

constant load applications requiring high wear resistance.

A type of cast iron known commercially as Meehanite is also supplied for gear applications. Typical properties of three different grades of this material are listed in Table 9.

Corrosion resistant austenitic cast iron is used in many special applications such as pump gears for handling corrosive liquids. Typical of this grade of cast iron is Ni-Resist, an alloy containing 2% chromium, 6 copper, 15 nickel, 3 carbon, 2 silicon, 1% manganese, and balance iron.

Ductile iron—Ductile iron has also proved quite successful as a gear material because of its desirable combination of properties. Its combination of good casting and machining characteristics and high strength provides important cost and strength advantages. Ductile iron gears respond well to heat treatment, are wear resistant, and are dependable power transmission elements.

Typical mechanical properties of four types of ductile iron for gears are given in Table 10. Modulus of elasticity of the materials is about 25 x 10⁶ psi, as compared to 30 x 10⁶ for steel, and is substantially higher than that of ordinary gray iron.

Bronzes

Various types of bronzes are used for gears. Gear bronzes are typed according to their major alloying element such as tin, silicon, phosphor, lead, manganese, aluminum and nickel bronzes, and beryllium copper. Properties vary with the kind and amount of alloying elements and with the type of heat treatment. Only the aluminum bronzes and beryllium copper respond to heat treatment for improved mechanical properties.

Tin bronze—Because of their good combination of properties,

including good hardness, strength and resilience, the tin bronzes are recommended for general purpose worm wheels. These materials have a tensile strength of about 40,000 psi with 20% elongation.

Since an increase in tin content over 10% causes a decrease in tensile strength, ductility and shock resistance, the tin content of gear bronzes is limited to a maximum of 12%. Phosphorus content is limited to that amount that will insure good fluidity. Phosphorus is a powerful hardener

in that it forms a hard load bearing constituent which, however, is quite brittle.

A lead content over 0.5% lowers the yield point in compression and tends to promote pitting. Nickel raises the yield point in compression and thereby helps to resist pitting. It also increases tensile and impact strength.

Silicon bronzes—These materials are usually used for worm wheels that mate with case hardened worms, and are recommended for medium loads, medium-to-high speeds and severe service. Their average tensile strength is about 50,000 psi and their elongation is 15%.

Phosphor bronzes — Phosphor bronzes are also recommended for worm wheels to mate with hardened steel worms of high hardness and fine accuracy. They can be used under conditions of severe service, medium loads, and medium to high speeds. Phosphor bronzes of 86-89 copper, 9-11 tin, 1-3 zinc, 0.2 lead and 0.02% phosphorus are recommended for severe working conditions and heavy pressures.

Leaded bronzes—Leaded bronzes containing 78-91 copper, 9-11 lead, 0.5-1.0 zinc and 0.25% phosphorus are recommended for worm gears to mate with soft steel worms under low loads and low to medium speeds. These materials have a tensile strength of 25,000 psi and an elongation of 80%.

Manganese bronzes—Manganese bronze with a content of 59 copper, 39 zinc, 1 tin, and 1% iron is recommended for spur and bevel gears used at low speeds and low tooth pressures. A high manganese bronze containing 3-4 manganese, 2-4 aluminum and 1-1.5%

iron is recommended for severe service, high loads and high speeds. This bronze is highly wear resistant and has a tensile strength of 90,000 psi with a Brinell hardness of 120.

Aluminum bronzes—These materials (see Table 11) combine high strength and hardness with good resistance to wear and fatigue. Aluminum bronze gears are particularly suited for severe service where long life is required. Because of their high hardness they are especially recommended where gears are subject to grit, dust and scale. Also, because of

Bronze gears have good wear resistance

American Smelting and Refining Co.

Continuous cast bronze *stock (left) is used for pump rotors. Only one rough cut and one fine cut are made on teeth (right) before use.*

American Smelting and Refining Co.

Bearing bronze *alloy gears drive armature of floor polisher motor.*

SAE 65 bronze *is used for lower gear and induction hardened 4140 steel for upper gear of this spiral gear set. Combination of two materials reduces friction and eliminates possibility of seizing.*

Philadelphia Gear Works, Inc.

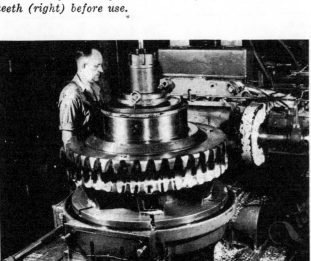

Brad Foote Gear Works

SAE 62 bronze *worm gear is used in large steel mill drive.*

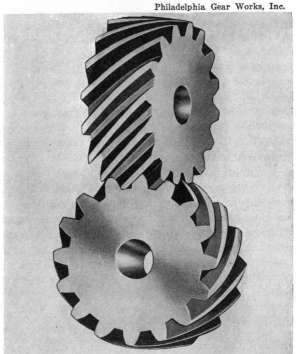

their high strength, they are suitable for gears that are subject to shock and overloads.

Aluminum bronzes and aluminum-nickel bronzes are used for highly stressed worm wheels designed to mate with hardened steel worms. They are also used for spur and bevel gears designed to mate with hardened steel pinions under severe service and heavy loads.

Aluminum bronzes with high aluminum content and iron additions respond favorably to a strengthening heat treatment. Bronzes with a low aluminum content cannot be heat treated satisfactorily. In general, the low-aluminum bronzes are satisfactory for slow moving gears, but are not recommended for worm wheels subject to high friction loads.

Nickel bronzes—Nickel bronzes, such as the Ni-Vee bronzes with 5% nickel, possess a number of properties that recommend their use for gears that are subject to heavy and irregular loading. As shown in Table 12, they possess good yield strength and toughness and are also quite wear resistant. Their superior as-cast properties can be elevated still further through simple heat treatments which control a nickel-tin-copper precipitation reaction.

Properties of gear bronzes

TABLE 11—TYPICAL PROPERTIES OF ALUMINUM BRONZE GEARS

Alloy	Condition	Tensile Strength, psi	Yield Strength, psi	Elongation in 2 in., %	Brinell Hardness (3000-kg load)
89.5 Cu–9.5 Al–1.0 Fe	Sand Cast......	73,000	30,000	35	120
	Heat Treated....	85,000	44,000	26	150
88.5 Cu–10.5 Al–1.0 Fe	Sand Cast......	75,000	42,000	16	136
	Heat Treated....	88,000	46,000	14	172
88.5 Cu–8.6 Al–2.9 Fe	Sand Cast......	65,000	24,000	25	116
	Hot Rolled......	100,000	60,000	20	180
88.5 Cu–10.2 Al–3.3 Fe	Sand Cast......	75,000	34,000	20	145
	Centrifugal Cast.	82,000	37,500	18	150
85.0 Cu–11.0 Al–3.7 Fe	Heat Treated....	95,000	50,000	7	215
81.3 Cu–10.7 Al–4.0 Fe–4.0 Ni	Heat Treated....	100,000	55,000	12	220

TABLE 12—TYPICAL PROPERTIES OF NICKEL BRONZE GEAR MATERIALS

Alloy →	Ni-Vee A[a]			Ni-Vee B[b]	
	As Cast	Tempered	Heat Treated	As Cast	Tempered
Tensile Strength, psi....	50,000	65,000	85,000	45,000	60,000
Yield Strength, psi......	22,000	40,000	55,000	20,000	30,000
Elongation in 2 in., %....	40	10	10	30	8
Reduction of Area, %....	50	—	26	—	—
Brinell Hardness........	85	130	180	80	120
Izod Impact Strength (unnotched), ft-lb.....	85	80	110	—	—

[a] 88 Cu-5 Ni-5 Sn-2% Zn.
[b] 87 Cu-5 Ni-5 Sn-2 Zn-1% Pb.

Nonmetallics

Nonmetallic materials are selected for gears principally because of their silence of operation, resiliency, vibration damping properties and low cost in large quantities.

Molded plastics

A number of cotton and glass fiber-reinforced phenolic molding materials are used for gears. Cotton-filled phenolic gears are especially noted for their quiet operation, but their mechanical properties are not as good as those of glass-filled gears. Glass-filled gears possess a number of desirable properties including: 1) resistance to high and low temperatures, 2) high tensile and flexural strength, 3) high impact resistance, 4) good dimensional stability, and 5) good resistance to deformation under load.

Molded polystyrene materials are also available for gears. These materials have good dimensional stability over a wide humidity range.

Nylon—One of the most popular plastics molding materials available for gears is nylon. Because of their so-called self-lubricating properties, nylon gears do not require any lubrication and require little or no maintenance. The gears also have a low coefficient of dynamic friction, are quiet and exceptionally resilient, and possess good tensile, flexural and impact strengths. Nylon gears operate quite well against steel pinions, as well as against other nylon gears provided that they are not highly stressed or operated at elevated temperatures.

Because of their compliability, i.e., their ability to yield under pressure and return to their original state when pressure is released, nylon gears are able to absorb shock and to deform slightly under impact loads. This property also helps to compensate for any dimensional inaccuracies that crop up during molding.

When used within a safe working stress of 4500 to 4900 psi and

Nonmetallic gears provide quiet operation

Bakelite Co.

Phenolic-glass *molding is specified for engine timing gear because of its high strength and impact resistance.*

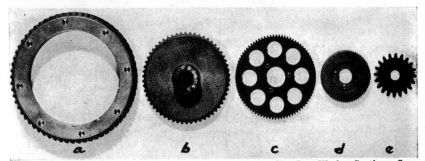

Quaker City Gear Works; Synthane Corp.

Cotton-base phenolic *laminate gear (A) with structural brass plates is used in voice recording instrument. Canvas-base gear (B) is used in noiseless elevating equipment. Fabric-base gear (C) is used in electronic instruments. Nylon fabric laminate gear (D) is part of motion picture sound projector. Similar gear (E) is used for television set control.*

below 150 F, nylon can often outperform steel. The material has excellent service life when used within its strength and heat range. However, because of their low heat conductivity (1.7 Btu/sq ft/hr/°F/in.), nylon gears may be subject to deformation and undue wear because they are not able to dissipate heat fast enough. For this reason it may be necessary to provide some means of heat dissipation.

Some geometric corrections are usually required on tooth profiles of nylon gears to compensate for mold shrinkage. It is a well-known fact that nylon absorbs moisture. For this reason sufficient back lash should be allowed in the teeth to compensate for swelling under high humidity conditions.

It is possible to compensate for swelling by immersing the material in water (preferably at elevated temperatures) until its weight increases by 2½%. Since gears treated in this manner have already been subjected to moisture swelling, their dimensions are less likely to change during actual operation at high humidities.

Reinforced laminates

In general, the properties of reinforced plastics laminate gears are the same as those of reinforced molded gears. Reinforced laminates are produced by impregnating a reinforcing material with a thermosetting resin, laminating the material into multiple layers, and curing with heat and pressure to form a dense, hard solid with good mechanical strength.

A cotton or linen fabric (or occasionally paper) impregnated with a phenolic resin is probably the most popular combination of reinforcing material and resin used in laminated gears. Materials of this type possess a number of significant properties including: 1) good shock resistance, 2) high resiliency, 3) low dynamic coefficient of friction on hardened steel, 4) good wear resistance, 5) good dimensional stability, 6) high resistance to corrosive atmospheres and liquids, 7) low weight and low moment of inertia, and 8) good machinability.

The power transmission capacity is approximately the same as that of cast iron.

In addition to the above phenolic laminates, vulcanized fibre is also used for gears. This material is made by laminating chemically gelled paper, leaching out the chemical and calendering. The resulting material is tough, easy to machine, resistant to wear and abrasion and makes an extremely quiet gear.

Rawhide

For many years rawhide has been a popular material for gears and pinions designed to mate with metallic gears in applications where quiet operation is required only to use one rawhide gear in a pair of gears. Because of the relatively high cost of rawhide it is usually specified for the smaller gear. A cast iron gear is usually used for the mating gear because of its low cost and because it will transmit about the same load that a rawhide pinion of like pitch will transmit. The strength of a rawhide pinion is about 65% of that of a cast iron spur pinion of the same dimensions.

The blanks from which rawhide gears are made are built up of layers of prepared steer hides. The laminations are coated with a special adhesive compound and subjected to pressure until they adhere firmly. The blanks are then assembled with rivets and brass or steel flanges which serve to protect and support the material. After assembly, blanks are turned to size on a lathe and then milled.

Despite their resilience and elasticity, rawhide gears should not be used in applications where they are subject to severe reciprocating or intermittent motion. They should not be allowed to become wet, as water will swell the rawhide and destroy the bonding cement. The material also has a tendency to shrink and lose its shape after drying. The best lubricants for rawhide gears are hard grease and graphite. Mineral oils should not be used as they tend to soften the material.

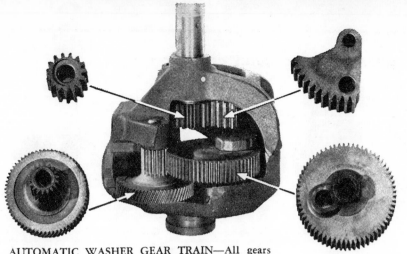

HELICAL MIXER GEAR—Gear is pressed in sintered iron. Rotating dies are used to form dense teeth and to avoid stripping them when compact is removed.

AUTOMATIC WASHER GEAR TRAIN—All gears are high-carbon, copper-impregnated iron (C 1%, Cu 15-20, bal Fe). Originally made of pearlitic malleable cast iron (Z metal) except for a steel pinion. Standard sintered iron (low carbon, Cu 7%, bal Fe) was tried, but the higher-strength heat-treated iron was found necessary for wear and strength. Helical teeth on the dual gear are cut, because of withdrawal problems. Stresses on these gear teeth are calculated as 70,000 psi. Over 50 percent savings by shifting from cut malleable iron to sintered impregnated gears.

CONTROL CLUTCH FOR AUTOMATIC WASHER—Complicated combinations of shapes can be obtained in a single molded piece. In the case shown control cams and gear teeth were molded, but the C-ring undercut below the teeth was machined. Reentrant surfaces are not possible in single-pressure-axis powder metallurgy.

Why and When to Specify Gears

Five processes, in addition to cutting (of cast, forged, or fabricated blanks) and stamping, are used to produce precision gears. Each of the five solves one or more production problems—cost, unmachinable material, complex shape, or the like.

Gears formed by the compression and sintering of powdered metals can be made cheaply in large quantities. Complex shapes are simple to manufacture and can be used for low and medium power transmission. Non-symmetrical powdered gears can usually be fabricated at one-half to one-third the cost of a cast or cut gear.

Die-cast gears of zinc, aluminum and brass are inherently accurate, and are used for high quantity applications where speed and load are low. Investment cast gears can transmit more power than die-cast gears, but are only economical for small quantities where the design is intricate or the material unmachinable with too high a melting point for die casting. This is one of the most accurate common casting methods. The shell molding process offers a method of precision casting at conventional sand mold costs. It has not been used extensively for gears to date, but indications are that this process will yield accurate cast gears which require little finish machining.

Extruded gears are used for high quantity, low duty applications. The extruded stock has a higher tensile strength, smoother surface, sharper edges and denser grain structure than sand castings of the same material. Types of material which can be extruded include copper, brass, bronze, aluminum, and some nickel-silver and phosphor-bronze alloys.

Cold-drawn gear rod can be formed at high production rates, is accurate enough to require no finish machining and has a higher durability and load carrying capacity than cut gears of the same material. This forming method is particularly useful in that cam-shaped and non-circular gears can be made as well as the involute and cycloidal sections.

The injection molding of plastic gears yields a product that is cheap, silent and often self-lubricating. In some cases plastic gears stand up better than their counterparts in steel or brass.

In general the advantages and limitations of these processes apply to products other than gears.

DUAL GEAR FOR AUTOMATIC WASHER—At bottom center is the original gear machined from pearlitic malleable iron, at top left the sintered blank, and at top right the finished sintered gear. A combination of helical and spur gears cannot be molded, so one must be cut. The 18 spur teeth are molded to 20 DP, 20 deg PA, and 0.9 in. PD. The 65 helical cut teeth have 25 deg RH helix, 26 DP, 20 deg PA and 2.835 in. OD.

NINE SINTERED GEARS FOR APPLIANCES—Largest is the 3.25 in. PD crank gear at top left, with about 7 sq in. area normal to the pressure axis. It requires at least a 250-ton press. By contrast, the little fan gear at bottom center has only 1 sq in. area normal to the pressure axis, and can be produced with ease on a high-speed 50-ton press. Largest tooth is on the 9-pitch wringer-head pinion at center, the smallest on the 40-DP fan spin gear at bottom center.

Formed by

1. *Powder Metallurgy*
2. *Casting*
3. *Extrusion*
4. *Cold Drawing*
5. *Plastic Molding*

1. Powder-Metal Gears

POWDER-METAL GEARS are used for low and medium duty applications. With this method of forming, complicated gear shapes can be produced by a simple, high-production process. Proper process and quality control in manufacturing results in good performance characteristics and low cost.

To form powder-metal gears, fine (325 mesh) metal powders are compacted under pressures of 20 to 60 tons per sq in. in suitable dies. The resulting briquette is then sintered. This increases the gear strength by bonding the compressed powder. With the standard (93% Fe, 7% Cu) powdered iron (ASTM-B222-47T) a tensile strength of 35,000 psi is obtained, with low elongation and impact strength.

By cold working and resintering, or by molten-metal impregnation the properties of the gear may be improved. Although cold working and resintering can be repeated with the part getting denser and stronger after each treatment, the process of molten-metal impregnation has proved the most promising method to date.

In a typical molten-metal impregnation process, a slug of lower-melting alloy, such as copper, is placed in contact with the sintered iron part and heated in a reducing atmosphere until the slug melts and is absorbed by the sintered part. By this method the tensile strength of 93% Fe, 7% Cu can be more than doubled and the impact strength quadrupled.

Process Limitations

Powder-metal gears are subject to certain dimensional manufacturing errors as listed in Table II. The first three are common to both conventional gears and sintered gears, while the last four are unique in powder-metal gears.

SHAPE—The molding tools limit the variety of shapes possible in powder-metal gears. Subsequent machining or forming operations are possible, but generally, powder-metal gears are made in *single-draw* tools only (single pressure axis—no reentrant surfaces). Typical shapes are blanks with rims or hubs molded integrally. Face notches or interruptions can be molded in if they are not major masses. The maximum advantage occurs when a gear has a non-symmetrical outline. Here the cost is usually $\frac{1}{2}$ or $\frac{1}{3}$ that of a cast or cut gear.

Tooth shape depends on the method of tool production. Usually a set of broaches is prepared by form-grinding the teeth in a *milling machine* setup so undercut is not possible. These broaches cut the die barrel. A *stripper* similar to a broach is ground for forming the ejecting floor of the cavity.

The tooth root is made a pure radius by some tool vendors because of the fragility of the wheel and general difficulty when grinding a conventional 3-radius root. Otherwise, the teeth (flanks and OD) are supplied in conventional form.

SIZE—The range of manufacturable sizes depends on the strength of the punches in the small sizes and the capacity of the presses in the large sizes.

Table 1—Limiting Tolerances for Various Gear-Manufacturing Processes

Process	Assumed size of Part	Degree* of Care	Tooth-to-Tooth Spacing	Concentricity Full Indicator	Profile Inv. Var.	Helix	Composite Check Total Tooth Error	Tooth Thickness	Tooth Surface Finish
Sintering	0 to 1 1/2 in. dia 0 to 3/4 in. face width 10 to 32 DP	Commercial	0.0007	0.0025	0.0005	0.0007	0.003	0.001	45
Sintering	1 1/2 to 4 in. dia 0 to 1 in. face width 10 to 20 DP	Commercial	0.001	0.004	0.001	0.0007	0.004	0.002	45
Stamping	0 to 1 in. dia 0 to 1/16 in. face width 20 to 128 DP	Commercial	0.001	0.003	0.001	0.0005	0.004	0.001	128
Extruding	0 to 1/2 in. dia 0 to 1/4 in. face width 20 to 128 DP	Commercial	0.001	0.005	0.002	0.001	0.007	0.001	64
Hobbing	0 to 6 in. dia 0 to 1 in. face width 20 to 128 DP	Commercial	0.0005	0.003	0.001	0.001	0.0035	0.002	63
		Precision	0.0003	0.002	0.0006	0.0005	0.0025	0.001	32
		Very Best	0.0002	0.001	0.0004	0.0003	0.0012	0.0005	16
Grinding	0 to 8 in. dia 0 to 1 in. face width 8 to 64 DP	Commercial	0.0005	0.002	0.001	0.0005	0.0025	0.0015	40
		Precision	0.0003	0.0012	0.0004	0.0003	0.0015	0.001	25
		Very Best	0.0002	0.0006	0.0003	0.0001	0.001	0.0003	10

* Commercial—Quality obtainable in good job shops.
  Precision—Quality obtainable with new or first-class machine tools and precision tooling.
  Very Best—Quality at 25 to 50 percent extra cost. Best tools and personnel.

High-strength gears (85,000 psi) from $\frac{3}{16}$ in. dia up to $3\frac{1}{2}$ in. dia have been made. Web perforations would allow the same area to be enlarged to a greater pitch diameter. The thickness of the blank is unimportant as far as molding goes; it matters only when warpage may distort the tooth flank and prevent adequate uniformity of contract. In addition, thin blanks (less than $\frac{3}{32}$ in.) are difficult to strip from the mold without injury to teeth, and excessively heavy sections may have low density because of briquet-

ting pressure loss from excessive wall friction. The diametral pitches have varied from 40 to 10.

Teeth can be molded in the hub of a dual gear and optimum strength obtained by blending the flanks and root with the hob for minimum stress concentration. Simple gears can be made with helical teeth molded in.

Splines, keyways, and double-D bores can be molded in when a concentricity tolerance of about 0.005 in. TIR can be accepted. Round bores are frequently bored or ground in respect

to the pitchline for optimum concentricity after sintering.

STRENGTH—The usual base materials are brass or iron. In the original first sintered condition, the gears are only useful for low-duty gearing such as timing devices, toys, or small appliances. Table III shows the relatively low tensile strength and percent elongation of four alloys in their original sintered condition.

By suitable extra processes such as cold working and resintering, and molten-metal impregnation the mechanical properties can be improved. Greater percent elongation is obtained through the use of more expensive iron powders. Table IV shows the improvement in the mechanical properties, of three of the four alloys mentioned above, by copper impregnation.

QUANTITY—The minimum economic quantity varies with the cost of other methods, performance advantages, consistency of quality, and availability. For a typical 1 in. dia gear, the tools might cost $2,000. Thus, for a production run of 20,000, the tool component of cost would be ten cents per gear. For a complicated 3 in. dia high-strength gear whose tool cost (including broaches) might be $10,000, a minimum run of 50,000 would

Table II—Dimensional errors which can occur in the manufacture of powder-metal gears, with the usual cause of each error.

ERROR	USUAL CAUSES
1. PD not concentric	Play in molding tools due to wear
2. PD too large	Die barrel worn
3. Tooth chordal distance too large	Die barrel worn
4. Tooth face taper	Die barrel wear always greater at the top than at the bottom
5. PD not round (pear-shaped), etc.	Uneven shrinkage because of unsymmetrical blank
6. Tooth-tooth spacing uneven.....	Uneven shrinkage, different powder densities, correct powder feed into mold
7. Tooth face not parallel to bore	All reasons given above

PROJECTOR GEARS—Cellulose-acetate-butyrate (black, at left) worm-driven helical gears have replaced nylon (right) in a projector at reduced cost, although the nylon gears ran continuously under test friction load of 2 in.-lb at 100 rpm. For light loads, gears can be made thin, but a heavier cross-sectional area is more effective in conducting heat away from tooth surfaces.

FOOD MIXERS – The GE assembly at upper right has a 24t, 32p, 20 deg PA, 33 deg + LH helix gear on a steel shaft. The 12t, 32p, 30 deg PA, 24 deg + LH helix pinion drives three beaters through a phenolic and two sintered-iron gears and is strong enough to withstand loading to stall the motor. Now a complete plastic pinion and gear are being considered. The second mixer assembly (left) is now under field test. The clustergear (56t, 43p, 30 deg PA, 20 deg + RH helix) may have to be redesigned to reduce excessive wear from the right-angle drive. The three 42t, 32p, 30 deg PA gears in the lower housing are working very well. Gear at center (29t, 1.18 pd, 20 deg PA, 19½ deg RH helix) is molded for Westinghouse of Canada. The gear cavity is cast in beryllium copper. At lower right are a pair of gears from an early food mixer. Mold-cavity inserts were hubbed and incorporated a draft angle. Original PA was 14½ deg, but this was changed to 18 deg, 45 min in a recently redesigned version to allow for shrinkage. This maintained the dimensional accuracy of the finished gear.

make the cost of the dies not more than 20 cents per gear. Heavy-duty gears in the above sizes, in 100,000 lots, have been manufactured to sell at low prices, about 12 cents for the 1 in. size and 60 cents for the 3 in. size. Light duty gears run as low as 2.5 cents for a 1 in. size.

Process Advantages

Costs—The main advantage of powder-metal gears is the reduction in labor and time of tooth formation. In most applications, the gear is passed through a press just once, and at a rate of perhaps 2 to 15 sec per gear. Sintering is done in a conveyorized furnace, as are other treatments such as further metal impregnation or steam or oil treatments. There is almost no scrap loss in this process, so that even the relatively high cost of most metal powders does not appear as a high piece price.

Selective properties—By powder-metal techniques, a variety of results can be achieved. Selective impregnation, teeth only, and lamination of different powders in the same mold (i.e., adding a thin bronze clutch face integrally to one side) have been tried. Differences in porosity can be created for special lubrication needs. Carbon can be added uniformly or selectively for improvements of properties. Brazing may be done either during the sintering operation, or later.

Selective hardening of powder-metal

Table III—Typical commercial sintered metal mechanical properties:

	Tensile Strength psi	Elongation, %/1 in.
Alloy A (Fe + 7 1/2% Cu)	35,000	0.8
Alloy B (Fe + 1% C)	35,000	0.8
Alloy C (Fe)	20,000	15.0
Alloy D (80–20 Brass)	30,000	20.0

	Tensile Strength psi	Elongation, %/1 in.
Alloy A (above) Cu impregnated...............	70,000	1
Alloy B (above) Cu impregnated...............	85,000	1
Alloy C (above) Cu impregnated...............	60,000	20

gears can be accomplished by localized induction hardening of high-carbon (1%) pieces. Strength and abrasion resistance are improved, although the exact mechanism and metallurgy are not fully understood.

The molding process inherently produces a good tooth surface finish. Tools are lapped to a mirror finish to minimize wall friction. The resulting gear teeth can be made smooth to a few micro-inches. No other single machining process can approach it. Combinations of hobbing and shaving, or grinding and lapping would otherwise be required.

Quality Control

Quality Control of powdered metals is a problem. Since the structure is quite heterogenous and inherently porous, hardness is not a good criterion of mechanical properties. Only the finished part is available, so that the best method for testing quality is a destructive test of the part itself. In the case of gears, the method commonly used is to measure the shear strength of teeth in a tensile machine.

The method of testing sintered iron structural parts is specified in ASTM Spec B222-47T. Samples are tested in a fixture which loads a tooth evenly at the pitch diameter. The minimum allowable tooth strength taken in this manner varies from 1,000 lb for a automatic washer crank gear tooth to 2,700 lb for a standard washer rack pinion. In actual practice the strengths vary from close to the minimum to almost twice that value. This is particularly true of copper-impregnated gears. On unimpregnated gears of similar dimensions the value would be lower but more consistent. The gears are also inspected for dimensional accuracy. The gear tolerances shown in Table I under *sintering* are representative of the accuracy which is now reached commercially, causing a scrap loss of only 1% or so in quantities exceeding 100,000 per year.

2. Cast Metal Gears

UNTIL RECENTLY cast gears have been manufactured by three processes: sand casting, die casting and investment casting. The advent of shell molding and other new casting methods may radically change the outlook for cast gears, but until now little is known or admitted about these new processes.

Sand-cast steel gears are only used today for large diameter parts. The teeth are cut since modern tolerances are usually too close to permit the use of the rough casting.

Die-cast gears of zinc, aluminum or brass are used for light duty applications where large quantities are required, such as in appliances.

Gears are made by investment casting where the material required is difficult to machine, complex in shape or has too high a melting point for die casting. Compared to other casting methods, the process is very accurate, but the gears are not in the class of those formed by precision machining.

Reports on shell molding and other new casting methods indicate that they show good prospects, with claims that tolerances within a thousandth or less can be obtained in quantity production. These methods are still under development and little information is forthcoming beyond the fact that they are claimed to be more precise and considerably cheaper than investment casting. Costs for the shell molding process are estimated to be about the same as for conventional sand casting.

Die Casting

Zinc, aluminum, and brass die castings are being made in spur, bevel, and other gear forms, particularly for instruments, cameras, business machines, washing-machine gear pumps, small speed reducers, and lawn-mowers. Die-cast gears are specified when large quantities enable high die costs to be amortized. Such gears are not used at high speeds or heavy tooth loading.

Good concentricity, tooth spacing, and surface finish for the above applications are obtained because the accuaracy is built into the tools. When face width approaches or exceeds $\frac{1}{4}$ in., some draft or taper is usually required to allow for removal of the piece from the die. In such cases, it is essential that the flash trimming die also broach or shave the teeth.

Spur gears, as made by one shop, normally range from 4 to 6 in. pitch diameter, but smaller sizes are produced. Bevels are made smaller than the range cited but are heavier. If the gears must be run at higher speeds than normal for die-cast products a steel or bronze insert is used for the bore.

Brass provides the strongest die-cast gears. But if the application demands both accuracy and strength, it is best to cut the teeth.

Investment Casting

Limited numbers of gears have been made by investment casting, where the material required was difficult to machine, complex in shape, or had too high a melting point for die casting. The process is very accurate compared to more familiar casting methods, but does not compare with precision ma-

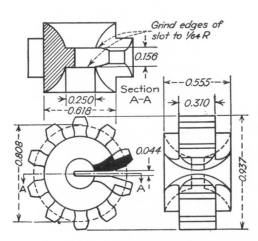

Fig. 1—Wire-twister gear which mates with a rack, investment-cast on a production basis by Haynes Stellite; the only one thus made. Gears for pickling tanks have been similarly cast. Complicated shapes make it economical to produce gears of this type by the investment-casting process.

chining. Usual accuracies specified are ±0.005 in. per linear inch on large pieces up to about 5 in. wide, ±0.003 in. on pieces under ¾ in., with angular tolerances to ±0.5 deg. Gears have been cast, however, with precision steel molds for the patterns, to ±0.001 in. on tooth dimensions and ±0.25 deg on angularity.

Wax, plastic (generally more accurate), and frozen mercury (larger pieces) have all been used as pattern materials, and some types of gears are being made successfully, one being pictured here in Fig. 1. However, the process seems most applicable in gear making where one or a few gears, particularly of intricate design or unmachinable material, are involved. If finish-machining is to be done, usual allowances are from 0.010 in. on small gears to 0.040 on large ones. As-cast surface finishes are normally 70 to 80 mu-in., rms, but with plastic patterns this has been cut to 20 mu-in., rms.

Strengths of investment-cast gears usually range midway between the longitudinal and transverse values for rolled bars of the same material, 10 percent or so above the properties of sand castings. Even tool steel and nitriding steels can be cast, but must be annealed for machining. AISI 3140, heat-treated, will give a tensile up to 220,500 psi, heat-treated 420 stainless to 120,000 psi, Monel H or S to 120,000 psi and a hardness of Rc30. Aluminum-bronze gives about 85,000 psi and can be heat-treated to Rb90. Beryllium copper, heat-treated, about 170,000 psi. These materials can be investment cast, but this method is only used if no other process is suitable since production costs are high.

Table V—Typical Extruded Sections

	Number Teeth	Outside Diameter	Pitch Diameter	Root Diameter	Pitch	Tooth Thickness
	18	1.241	——	0.972	16	0.098
	7	1.487	1.062	0.775	6	0.261
	8	0.985	0.800	0.544	10	0.157
	7	1.430	——	0.712	6	0.261
	15	2.020	——	1.588	8	0.1963
	9	1.325	——	0.920	14	0.196
	16	1.285	——	0.985	15	0.112
	8	1.150	——	0.750	Tolerance = 0.010	
	12	1.8	1.5		8	CP = 0.3927
	11	1.8	——	1.265 hole drawn to 1.215		
	6	Segment for Easy washer, body 0.720, radii 1/32 in.				
	12[1]	5.25	4.438	3.625	Tolerance = ± 0.034	

[1] Aluminium

3. Extruded Gear Stock

SPECIAL CROSS-SECTIONS, such as pinions and cams, can be extruded in copper, brass, bronze, and some nickel-silver and phosphor-bronze alloys for cutting to length and machining of hubs, collars, and the like in screw machines and turret lathes. Aluminum alloys can be extruded in similar shapes, and undoubtedly will be as the use of stamped aluminum-alloy gears increases. Such sections are commercially constant in cross-section, thereby eliminating considerable machining. These rods have a higher tensile strength, smoother surface free of pits and porosity, sharp and clean edges, and a denser grain structure than sand castings. They are readily machined, have a smooth surface, and can be easily polished or plated.

Some extruded bars are cold-drawn afterwards for harder surfaces and more exact cross-section. Smaller diameters and a wider range of alloys are also available. Either drawn or extruded stock may be gripped in special collets, or when pitch is small can use standard round collets. Special collets are also required for segment shapes or pinions with eccentric machining or holes.

Most commonly extruded materials are Naval brass (Cu 60, Zn 39.25, Sb 0.75) and extruded architectural bronze (Cu 56, Zn 41.5, Pb 2.5). Desirable tooth form is involute rather than cycloidal; some typical tooth forms are shown in Table V, plus commoner sizes. While extruded shapes below 1 in. dia were formerly made, presently available ones are between 1- and 2-in. diameters. Smaller sizes are made by cold-drawing rod.

Cost of tooling (hardened steel) for extrusion is generally estimated to be amortized at roughly a cent a pound. Press capacities limit section weight to about 180 lb in brass. The aluminum gear sketched was not made, but is included to show that section characteristics are like those of brass. The area would be a 15.455 sq in. and weight per foot about 18.55 lb, depending on the alloy.

A new process of hot extrusion, in which glass is used as a lubricant, will allow the extrusion of solid shapes such as gears from the more refractory stainless steels and highly alloyed and high melting point metals. Substitution of glass for the carbonaceous materials previously used as lubricants extends die life and improves surface quality.

4. Cold-Drawn Pinions and Small Gears

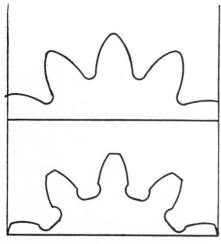

Fig. 2—Cold-drawn steel pinion has much higher hardness near the surface; cutting action gives only a slight increase in hardness over the center.

COLD-DRAWING, like injection molding, produces a dense, cold-worked surface with high durability, accurate enough so no machining of the drawn shape is required. The process is adaptable to any drawable material, which includes not only aluminum alloys, brasses, and bronzes, but also most steels (including high-carbon and some stainless alloys) silver, beryllium copper, nickel and its alloys, and copper and its alloys. With certain limitations in tooth form, cold-drawn rod can be finished at high production rates in screw machines, producing pinions and small gears with collars or shanks, drilled holes, and the like. Thus cold-drawn gears are used in watches, clocks, business machines, motion-picture projectors, carburetors, magnetos, small motors, switches, timers, telephone and scale mechanisms, cameras, appliances, toys, and similar units.

Accuracies of cold-drawn rod are claimed to be as high as those of cut teeth, with higher durability and load-carrying capacity if the material has good cold-working properties. This is because cold drawing produces compressive stresses which increase working strength of the teeth and give a smooth, hardened surface. Tooth form and dimensional accuracy are also claimed greater than for extruded rod.

The amount of cold working to which a given material is subjected determines not only the skin hardness of the pinion rod, but also the depth of the case, the hardness at different points in the cross-section, and the density of the grain structure.

The curves in Fig. 2 show the comparative hardness of small pinions that were cold-drawn and similar pinions that were cut. Both pinions were 0.55 in., 12-tooth gears made from AISI 1112 steel. The cold drawn pinion shows a Vickers hardness number of 270 at the tooth tip, dropping to 180 at the core. The cut gear shows a hardness of 182 at the tooth dip dropping to 168 at the core. In a similar test of drawn and cut phosphor bronze pinions, the cold-drawn pinion was 235 at the tooth tip, 165 at the core, as compared with 175 and 150 for the cut pinion.

Based upon the hardness numbers, the tensile strength of the steel pinion at its outer surface was approximately 126,500 psi, which is quite high for AISI 1112 steel. The hardness of the cut pinion near its outer surface indicated a tensile strength of 85,000 psi, which is normal for this type of steel. From this, it is evident that the strength of the cold drawn steel pinion is greatly increased by the cold working of the material.

In a similar manner, cold-drawn pinions are more durable than cut pinions of the same material. Table VI gives the results of tests on six types of cold-drawn and cut pinions. In each test, the number of stress cycles required to produce 0.001 in. of wear on tooth thickness was determined. As indicated the durability gain of the cold drawn pinion over

Fig. 3—For cold-drawing, the 11-tooth pinion below is enlarged by the AGMA-ASA standard to the form above, avoiding undercut and providing radiused rather than sharp corners.

Table VI—Comparative Durability of Cold Drawn and Cut Pinions

Material	Diametral Pitch	No. of Teeth	Pitch Dia., in.	Tooth to Tooth Composite Error, in.	Number of stress cycles*		Gain in Durability of Cold-drawn Pinion, percent
					Cut Pinion	Drawn Pinion	
Hard Phosphor Bronze	100	15	0.15	0.0005	2,100,000	4,500,000	214
	60	21	0.35	0.0005	2,960,000	6,200,000	210
Brass	32	16	0.50	0.0003	4,200,000	9,100,000	217
AISI 1113 steel	100	15	0.15	0.0003	4,100,000	8,310,000	203
	80	20	0.25	0.0003	4,300,000	9,080,000	214
	32	18	0.56	0.0003	4,010,000	12,850,000	320

* Stress cycles required to produce an average of 0.001 in. wear on tooth surfaces; based on 10 tests.

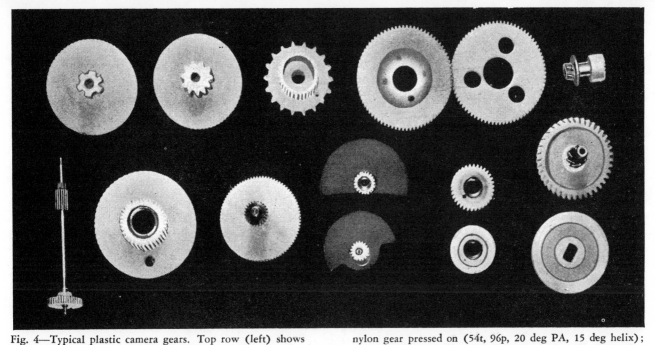

Fig. 4—Typical plastic camera gears. Top row (left) shows front and rear of a combination face and pinion gear with splined hub. It replaced cut or shaved gears held to ± 0.001 TIR. Next to it is a one-piece sprocket and gear, and at right a 78t, 48p set of gears molded at one-third previous cost for the RCA 16-mm projector. At 1,440 rpm, dry, they outwear steel three times.

Lower-row gears are machined from nylon blanks to ± 0.002 TIR by Revere Camera. At left is a shutter shaft with nylon gear pressed on (54t, 96p, 20 deg PA, 15 deg helix); a 1-piece cluster gear with bronze bearing (pinion: 36t, 48p, 14½ deg PA; gear: 58t, 48p, 14½ deg PA, 45 deg helix); a steel pinion and nylon gear, shutter drive gear (18t, 64p, 20 deg PA); and shutter pinion (21t, 64p, 14½ deg PA); molded blank with steel insert and engaging cam gear (34t, 48p, 14½ deg PA), in which the steel insert takes load of a locknut and drives a phenolic and steel gear at 960 rpm; blank on a steel shaft, and a face gear (24t, 64p, 14½ deg PA).

the cut pinion was more than 200 percent in each case.

In addition to the greater skin hardness of the cold-drawn pinion, the quality of the surface finish has considerable effect on the life. The average surface finish on cold-drawn rod will range between 10 and 30 mu-in. rms. This is better than the finish on the average cut pinion.

This method is versatile in that not only involute and cycloidal gear forms can be produced, but also ratchets, sectors, cam-shaped and non-circular gears. Tooling is moderate, and the use of such stock presupposes quantities that permit low writeoff.

Tooth form must be modified for best results in cold-drawing. The ASA 20 deg involute Fine-Pitch System (B6-7-1950) can be used, except that the teeth are made with full radius at tip and root to avoid tearing. The modification of an 11-tooth pinion for drawing is shown in Fig. 3. Excessively thin, deep teeth are not recommended, nor are undercut involute pinions, particularly in materials with poor drawing properties.

Cold-drawn pinion rod has shown savings of as high as ten to one over cut pinions, and averages three to five times cheaper. Durability ranges from five to seven times greater.

5. Molded Plastic Gears

PLASTIC GEARS formed by injection molding are inexpensive, silent and often sulf-lubricating. For high quantity applications the latest production methods show considerable cost saving by injection molding with fast cycles in permanent steel molds.

Recent work by material manufacturers has centered on the development of plastics having a higher heat resistance. Commercial polystyrene materials have shown much improvement along these lines. The polystyrenes are lowest in cost @ 35¢ per lb. Dimensional stability over a wide humidity range not only makes these materials outstanding among thermo-polastics, but also permits them to compete with the phenolics in spite of a slight material cost disadvantage.

Heat-resistant cellulose acetate (180F) materials in the high acetyl form or low-plasticizer, hard-flow types have been recommended for small precision parts. A life test on a worm drive and helical gear, run dry and without dissimilar materials, gave unexpectedly favorable results. Investigation of this tough thermoplastic for small gears in a lower-temperature zone (below 180 F) would seem to justify further study because of low manufacturing costs.

A majority of thermoplastic gear applications today appear to favor the use of polyamide materials, more commonly called nylon. Tests have shown that nylon thread has a higher strength than steel wire of the same diameter.

Table VII—Nylon Safe Working Stresses

(Static stress Factor = 6,000 psi)

Pitch-line Velocity, fpm	Nylon FM-10001	Pitch-line Velocity, fpm	Nylon FM-10001
100	4,500	1,000	2,250
150	4,071	1,500	2,029
200	3,750	2,000	1,909
250	3,500	2,500	1,833
400	3,000	3,000	1,781
500	2,786	4,000	1,715
800	2,400		

Based on MIT formula: Velocity factor $= \dfrac{150}{200 + PLV} + 0.25$. As the equivalent for metal gears is $\dfrac{600}{600 + PLV}$. velocity factor for non-metallic gears is higher at speeds above 1,000 fpm. Nylon tensile strength is 10,530 psi at room temperature, according to duPont (ASTM D638–44T). Flexural fatigue tests (ASTM D671–49T) show 6,000 psi for nylon FM–10001. These values are used in the Lewis formula:

$$Hp = \frac{9.5 \times 10^{-5} \times SWS \times FW \times PLV}{DP}$$

While this strength relationship has not been proven for heavier sections, the fact remains that with proper orientation by injection molding or by post treatments, strength values can be improved on molded parts.

Nylon gears are at present in food mixers, projectors, cameras, business machines, fishing reels, lawn sprink-

Table VIII—Variations from shot to shot, for 15 consecutive shots in a production run of injection molded gears.

Cycles	OD	Over-pin Dia.	Root Dia	Hole
1	0.711	0.7655	0.533	0.3051
2	0.710	0.7644	0.5315	0.3051
3	0.7105	0.7643	0.5325	0.305
4	0.710	0.7645	0.532	0.305
5	0.7115	0.7655	0.533	0.3051
6	0.710	0.7655	0.532	0.305
7	0.710	0.7654	0.532	0.305
8	0.7103	0.7655	0.532	0.3051
9	0.710	0.766	0.532	0.3051
10	0.710	0.765	0.532	0.3051
11	0.7103	0.7654	0.533	0.3052
12	0.710	0.7658	0.533	0.305
13	0.710	0.7657	0.532	0.3051
14	0.710	0.7656	0.533	0.3051
15	0.7104	0.7658	0.533	0.3051

lers, hydraulic pressure gages, automotive speedometer gears and electric windshield wiper gears, washing machines, and home temperature indicators. First application of nylon at Eastman Kodak was in the smallest bevel gear ever made. Nylon has the highest heat resistance among the lower-priced thermoplastic materials and thereby dominates in these applications. Even-higher-heat-resistant materials are available, among these trifluoro-chloro-ethelyne (Kel-F)

Design Considerations

In the less-rigid thermoplastic gears, such as nylon, resiliency absorbs the inevitable inaccuracies of cut gears. The self-lubricating characteristic offsets friction of mating gear-tooth surfaces. Nylon possesses the property which Louis Martin of Eastman Kodak has dubbed *conformability*—the ability to deform slightly to improve mating and absorb shock when run with a steel pinion.

Theoretically, plastic and steel gears deflect similar amounts when subjected to loads proportional to their moduli of elasticity—a tooth in a plastic gear under 1 psi load will deflect as much as a similar steel tooth under 75 psi. Horsepower loads can be determined based on the strength of a single tooth from the Lewis formula, with the safe working stress factors given in Table VII.

Rigid metal gears with a slight inaccuracies of tooth form and spacing can create high dynamic loads which have to be carried by a single tooth. The resiliency of nylon plastic, however, may allow contact of two or more teeth under load conditions. Consequently, computations from the Lewis formula will be conservative.

Tooth wear on plastic gears is effected by relative sliding of gear-tooth surfaces, excessive compressive stress at progressive points of contact, and cutting the flank of the driving tooth by the tip of the entering tooth. A nylon driving pinion mated with a steel gear is relieved of this undue wear by using the all-addendum form on the pinion, thus reducing the angle of approach and eliminating the trochoidal path of the tip of the entering tooth. Tooth strength is increased 50 to 75 percent. This procedure is more desirable than the modified tip of the steel-driven gear tooth which would reduce tooth-contact duration and would not remove the tendency to interfere. When plastic-driven gears are considered, these modifications are necessary only in the sense of eliminating excessive undercut. In all cases of a steel versus nylon wear surface, the best condition is with a highly polished, hardened-steel pinion.

Nylon gears have the low heat conductivity of 1.7 Btu/hr/sq ft/degF/in. Experimental gears have been made with copper shims to facilitate conduction of the heat from the gear-tooth area to prevent undue wear and deformation from the heat generated at this point. In more critical cases of gear design where the favorable silent wear characteristics of nylon may be desired, a gear blank or insert can be used in the mold with a thin cross-section of plastic molded over the surface of the insert. The problem of heat dissipation can then be minimized while increasing the beam strength with a metal insert.

In cameras, projectors, and fishing reels, nylon gear applications have been extensive for either one or both of two reasons: One, silent operating, self-lubricating characteristics; Two, cost savings by injection molding a complete cluster or original assembly of separate parts. In a Shakespeare fishing reel, the spool pinion (20t,

53.2509p, 24° PA, 12° + RH helix) and drive gear (77t, 53.2509p, 24° PA, 12° + LH helix) have been cut from injection-molded blanks for several years. Here the requirement for silent operation with high pitch-line velocity is essential.

Production Methods

The greatest cost saving factor in the production of thermoplastic gears is that the complete gear is produced in one operation in a permanent mold.

Most machines today have automatic controls for time, temperature, pressure and rate of injection. These machines range from 1 to 300 oz per shot. Most of the small-part production, however, is being confined to 8 oz and smaller machines that can be operated on automatic cycles. Smaller machines with small molds represent lower investment in capital equipment to achieve high production.

The time cycle in injection molding is fast, being on the average 100 cycles per hr on 8 oz machines to as rapid as 600 per hr on the small machines. Cycle speeds are controlled by the quality required of the molded part.

Mold temperatures range from the refrigerated types to approximately 200 F and should be controlled within 5 F over the entire mold surface. This will maintain a definite shrinkage factor for the material, which is required to maintain uniform production. For example, shrinkage in tooth length would be greater than in width, so the mold cannot be of true involute form. Further nylon absorbs moisture or oil, and expands as a result. This is frequently a highly desirable property in nylon gears because it tends to make them self-lubricating.

Consistent production and close temperature control are mandatory in producing accurate molded parts. Table VIII is a record of a production run of 15 consecutive shots, showing slight variations from shot to shot.

The camera gear set, Fig. 4, shows a case where corrective measures were required to true up a pd where off-center core pins caused slight variations in the finished product. It should be noted, however, that these gears were injection molded within approximately 0.003 limits on the pd. These limits were held within ±0.001 by a second shaving operation.

which INSTRUMENT-GEAR MATERIALS wear least?

As precision requirements go up in the instruments field, gear wear must come down. To help you choose among the increasing number of gear materials, including plastics, here are results of wear tests. The gears were meshed in various combinations and operated under different conditions of lubrication and temperature.

R J BENSON, *Bell Telephone Laboratories, Whippnay, NJ*

New developments in metals, plastics, platings and surface treatments have made available to the designer a variety of gear materials with low rates of wear. The wide choice has come mainly in answer to demands for higher precision in instrument, servo-mechanism and data-transmission applications. Gear wear in such applications adds to backlash and angular errors.

For wear comparison, we performed a series of life tests on spur gears hobbed from a number of different materials. Sixty-eight pairs (meshes) of gears have been tested, and their wear determined by noting the increase in backlash with reference to elapsed running time. Backlash changes are indicative of wear, and from plotted backlash-versus-time curves we constructed a table giving the relative qualities of the materials tested.

Each pair or test mesh consisted of gears with a 20° pressure angle and 48 diametral pitch. The drive pinion had 48 teeth, and the driven gear 97 teeth. Each test setup accommodated four gear meshes. All gears were hobbed to full depth involute form, had a 0.250-in. face width, and were AGMA Precision Class 1. This class of precision indicates that the gear pairs have a maximum total-composite-error of 0.001 in. and a maximum tooth-to-tooth composite error of 0.0004 in.

The gear materials

Various combinations of gear materials and surface treatments tested during the program include:
- **Aluminum**—untreated, anodized, electrolized and chrome-plated.
- **Steel**—(stainless) untreated and electrolized.
- **Rulon B**
- **Nylon 3001**
- **Nylasint**—six compositions: Nylasint 64, 66, 64-6G, 66-6G, 64-MS, 66-MS
- **Sinite D-10S**

See the table on the next page for specific mesh combinations that were tested. The gears were operated with and without externally applied lubrication. Some materials have built-in lubrication; these are listed with the unlubricated groups because no external lubrication was applied.

Aluminum gears were machined from Type 2024-T3 alloy. This material has good strength and machining qualities.

Steel gears were of stainless steel Type 416, heat-treated to Rockwell C 25-30 before machining.

Rulon B is a plastic-bearing material. Its manufacturers have compounded or modified Teflon to produce a new product. The gears were hobbed from extruded rod.

Nylon gears were hobbed from nylon Type 3001 (similar to Zytel 31).

Sinite D-10S is a self-lubricating bearing material comprising 50% dry lubricative pigments and 50% bronze and tin powders. The dry lubricants are graphite and molybdenum disulphide, which are fused with the metallic materials. Sinite has mechanical properties and appearance similar to Oilite, a well-known bearing material. After hobbing, the Sinite gears were impregnated in a bath of Dow Corning 510 silicon oil (recommended by the manufacturer) for 30 minutes at 200 F.

Nylasint is a tradename referring to a group of sintered nylon products. This material is manufactured by

RELATIVE WEAR OF GEAR-MATERIAL COMBINATIONS

(Bold face denotes top performers; tests were scheduled for either 2000 hr, or for 480 hr at a higher operating speed—other figures indicate mesh was removed because of excessive wear.)

Unlubricated Gear-meshes

48-TOOTH PINION	97-TOOTH GEAR	HOURS RUNNING	INCREASE IN BACKLASH, IN. (WEAR)	RELATIVE WEAR, FROM 1 (BEST RANK) TO 68
Aluminum	Aluminum	421	0.0172	66
		421	0.0107	53
Anodized aluminum	Aluminum	2000	0.0181	67
		2000	0.0244	68
	Anodized aluminum	1156	0.0129	65
		1156	0.0124	64
Electrolized aluminum	Aluminum	300	0.0084	48
		700	0.0113	58
	Anodized aluminum	600	0.0112	57
		600	0.0109	54
	Electrolized aluminum	96*	0.0120	61
		138*	0.0076	43
	Stainless steel	2000	0.0031	37
		1000	0.0122	62
Chrominum-plated aluminum	Chromium-plated aluminum	2000	0.0010	22
		2000	0.0011	23
Stainless steel	Aluminum	2000	0.0071	42
		2000	0.0092	49
	Anodized aluminum	2000	0.0011	24
		2000	0.0011	25
	Chromium-plated aluminum	2000	0.0098	51
		2000	0.0080	45
Electrolized Stainless steel	Aluminum	1000	0.0079	44
		1000	0.0080	46
	Anodized aluminum	2000	0.0123	63
		2000	0.0112	56
	Electrolized aluminum	480*	0.0022	34
		480*	0.0009	20
	Stainless steel	**2000**	**0.0008**	**17**
		2000	0.0082	47
Rulon B	Anodized aluminum	700	0.0106	52
		700	0.0093	50
Nylon 3001	**Anodized aluminum**	**2000**	**0.0008**	**13**
		2000	**0.0008**	**14**
Nylasint 64	**Anodized aluminum**	**2000**	**0.0012**	**26**
		2000	**0.0005**	**6**
66	**Anodized aluminum**	**2000**	**0.0009**	**18**
		2000	**0.0007**	**10**
64-6G	**Anodized aluminum**	**2000**	**0.0017**	**30**
		2000	**0.0010**	**21**
66-6G	Anodized aluminum	2000	0.0022	33
		2000	0.0024	36
64-MS	**Anodized aluminum**	**2000**	**0.0009**	**19**
		2000	**0.0008**	**15**
66-MS	Anodized aluminum	2000	0.0012	27
		2000	0.0017	31
Sinite D-10S	**Anodized aluminum**	**2000**	**None****	**1**
		2000	0.0006	8
	Stainless steel	2000	0.0007	9
		2000	0.0002	2

Lubricated Gear-meshes				
48-TOOTH PINION	97-TOOTH GEAR	HOURS RUNNING	INCREASE IN BACKLASH, IN. (WEAR)	RELATIVE WEAR, FROM 1 (BEST RANK) TO 68
Aluminum	Aluminum	2000	0.0007	11
		2000	0.0003	4
Anodized aluminum	Aluminum	2000	0.0020	32
		2000	0.0013	29
	Anodized aluminum	2000	0.0008	16
		2000	0.0012	28
Stainless steel	Aluminum	2000	0.0007	12
		2000	0.0004	5
	Anodized aluminum	2000	0.0002	3
		2000	0.0005	7
Gears Subjected to Temperature Variations, —65 F to 200 F				
Aluminum (lubricated)	Aluminum	480*	0.0068	41
Anodized aluminum (lubricated)	Anodized aluminum	383*	0.0117	59
Stainless steel (unlubricated)	Aluminum	238*	0.0120	60
	Anodized aluminum	238*	0.0111	55
Stainless steel (lubricated)	Aluminum	480*	0.0031	38
	Anodized aluminum	480*	0.0022	35
Nylon 3001 (unlubricated)	Aluminum	480*	0.0048	40
	Anodized aluminum	480*	0.0045	39

*Pitch-line velocity 471 ft/min.; all others 113 ft/min.   ** No measurable backlash

cold-pressing nylon powders and sintering below the melting point. The sintering medium used on the test-gear material was Meprolene, a well-known heat-transfer oil.

The Nylasint 66 materials have the same chemical structure as nylon FM10001, while the Nylasint 64 materials are combinations of nylons, with FM10001 as the major constituent. Types 66-6G and 64-6G have graphite added; the 66-MS and 64-MS have molybdenum disulphide added to the basic material to impart certain properties. Because Meprolene is a good heat-transfer medium but a poor lubricant, the gears made from Type 66 and 64 materials were immersed in acetone after hobbing—to leach out the Meprolene oil. These gears were then impregnated in a bath of lubricating oil (SAE 30) at 200 F for 30 minutes. The Nylasint gears containing graphite and molybdenum disulphide received no such treatment.

Surface treatments

Anodizing indicates that the gears were subjected to a chromic process after machining.

Electrolizing applies an additive finish to the outer surface of the base material. The finish as deposited is approximately 90% chromium and has a Rockwell hardness value of C 70-72.

Chromium-plated aluminum gears were plated by the Hardalume process, a procedure developed by Tiarco Corp., Clark, NJ. With this process, the chromium layer is applied directly to the aluminum without any other intervening metal or anodic film. This process avoids cumulative errors in plating thickness which could become excessive and destroy the initial gear accuracy. Also, it avoids preplating, which may produce an undesirable electrolytic couple.

After machining, all gears to be anodized, chromiumplated, or electrolized were measured over pins. They were then surface-treated as required, and measured again.

Anodizing decreased the measurements over pins by an average of 0.0001 in., with a consequent reduction in tooth thickness.

Electrolizing, on the other hand, caused an average increase in over-pin measurements of approximately 0.0002 in. Plotted curves of the measurements showed that the alloy was not deposited uniformly and in some cases exceeded the maximum limit established for this dimension.

Chromium plating also increased the measurement over pins for all gears. Plating was uniform in most cases.

Lubrication of teeth

A Mil G-3278 grease (Beacon 325) was chosen wherever lubrication was indicated for the test meshes. This general-purpose lubricant has a temperature range of —65 to +250 F. It was applied to the teeth of both pinion and gear before the start of a test; the meshes were then manually rotated and excess lubricant removed. This was the only application during the test.

Operating speeds

Most gears were operated with an average pitch-line velocity of 113 ft/min.—others (noted by an asterisk in the table) were operated at 471 ft/min. When operated at 113 ft/min., the drive gears were rotated at 3450 rpm for one hour, then rotated at 300 rpm for 23 hours. The rotational direction was reversed every 4½ minutes with a 30-sec delay period between reversals.

To obtain 471 ft/min., the drive gears were rotated at a constant speed by a 1800-rpm synchronous motor. The rotational direction was reversed every 5 minutes with a 20-sec delay period between reversals.

Backlash measurements

Backlash measurements were taken with a 24X microscope mounted on a micrometer slide (photo next page). A

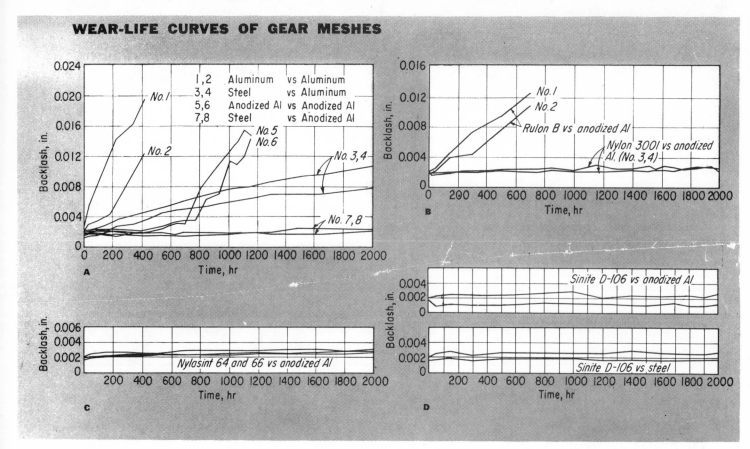

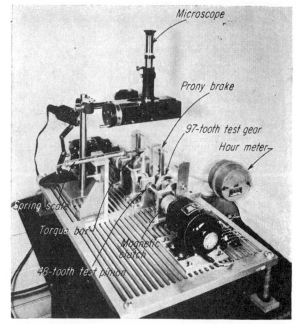

TEST SETUP for loading and measuring gear wear.

fixture with a 4-in.-long bar was attached to the drive-shaft; the driven gear was held fast to prevent rotation. Deflection at end of the bar, when a torque of 200 gm-in. was applied, was converted to backlash in inches at the pitch line of the gears.

The average of three such readings taken equidistant along the pitch circle of the 97-tooth gear constitute the backlash reading for the test mesh.

Tests with temperature variations

One group of gears were subjected, while operating, to high and low temperatures inside an environmental test chamber. The motor was mounted outside the test chamber and the gears driven by a driveshaft coupled from motor with a universal joint and flexible coupling.

Temperature extremes to which the gears were subjected were −65 F and +200 F—one temperature cycle per 24 hours. Temperature of the gears was lowered to approximately −65 F and held for one hour, then increased to +200 F and held there for one hour, and finally, lowered to room temperature of about +75F. Temperature data were taken by means of a copper-constantan thermocouple attached to the test setup adjacent to the gears. Measurements to obtain backlash were taken at room temperature.

Gear-tooth loading

To simulate conditions of loading, the 97-tooth gear of each test mesh was loaded by means of the prony brake with a torque of 200 gm-in. This was maintained on all meshes throughout the running time of the life test.

The torque loading of a test mesh is converted to a value of compressive stress by use of Dudley's equations (*Practical Gear Design* by Darle W. Dudley, pp 48-50) which in turn are derived from studies made by Hertz.

Some computed compressive-stress values were: all-stainless-steel mesh = 9150 psi; all-aluminum mesh = 5670 psi; stainless steel versus aluminum mesh = 7868 psi. The equations do not take into consideration such factors as surface treatment, surface finish and lubrication.

Relative wear of gear materials

Backlash measurements were taken at appropriate intervals of time during the test and data plotted. The table gives the relative wear of the gear meshes, both for nonlubricated and lubricated conditions. This can be used to estimate the wear life of a gear mesh involving these materials. The "increase in backlash" shown in the table is directly proportional to wear and is between initial and final backlash measurements. Key conclusions from these tests are:

Sinite D-10S gears showed superior wear properties—these gears were meshed with anodized aluminum or stainless steel gears. One mesh gave no measurable increase in backlash—this was the best performing mesh. The low rate of wear of Sinite meshes is undoubtedly caused by the inherent lubrication (oil, graphite, molybdenum disulphide) rather than basic material characteristics.

Anodizing the surface of aluminum gears greatly reduced wear—this is interesting to note because aluminum gears are frequently used where inertia is a problem and lightweight materials are desired. Without anodizing, the aluminum gears gave generally unsatisfactory wear performance.

Lubrication of gears also reduced wear—although applied only once during the test. Meshes that fail under ordinary conditions, such as a mesh of unanodized aluminum gears, give good wear life when lubricated.

Unplated stainless-steel pinions meshed with anodized aluminum gears gave excellent results.

Plastic gears showed excellent wear life—these were the Nylon 3001, Nylasint 64, 66 and 64-MS gears. They had approximately the same rates of wear.

Rulon B gears gave unsatisfactory performance—but the wear was primarily on the anodized-aluminum gears rather than the Rulon pinions.

Electrolized-aluminum gear pairs gave poor wear results—unsatisfactory performances were also obtained with electrolized-aluminum gears meshed with anodized aluminum or unanodized gears: such pairs showed rapid and large increases in backlash.

Electrolized stainless-steel pinions performed satisfactorily with electrolized-aluminum gears—but poorly with anodized-aluminum gears. The tests showed that electrolizing the steel pinion of a steel versus aluminum mesh does not decrease the wear rate—in fact, wear is greater.

Electrolized aluminum gears meshed with stainless steel gears gave a wide variance in results. A similar situation exists with the combination of electrolized steel pinions meshed with steel gears. On the basis of these test results neither of the two material combinations should be used.

Chrome-plated aluminum gears meshed together have excellent wear properties. On the basis of previous life tests this combination exhibited far better wear properties

than any other all-aluminum mesh, regardless of finish. Using an all-aluminum mesh offers distinct advantages where minimum weight is desired.

Chrome-plated aluminum gears meshed with stainless gears gave poor results. This combination is not satisfactory for gearing.

Temperature cycling tests showed that the lubricated mesh of a stainless-steel pinion and an anodized aluminum gear wore the least. The anodized gears again showed less wear than unanodized ones.

Backlash-versus-time curves

For an evaluation of the wear rates of some of the significant meshes, see charts A to D:

Aluminum, steel, anodized aluminum—Chart A. Only four meshes completed the 2000-hr test (Mesh 3, 4, 7, and 8). These were combinations of stainless-steel pinions versus anodized and unanodized aluminum gears. The other four meshes were discontinued during the test because of excessive wear. Mesh 5 and 6 did not show a definite increase in backlash until the anodized finish began to break off (at 500 hr). The finish wore entirely off the face of the gears at 750 hr. From this point on, the rate of wear was about the same as for Mesh 1 and 2. Mesh 7 and 8 produced only about 0.001-in. backlash throughout the 2000 hours.

Rulon, nylon, anodized aluminum—Chart B. Mesh 1 and 2 began to wear rapidly from the beginning and were removed after 700 hours because of extreme tooth wear. Wear products in the form of a very fine black powder were deposited profusely. Mesh 3 and 4 continued without interruption for the required 2000 hours, with an increase in backlash of less than 0.0001 in. No wear products were seen.

After testing, all gears were cleaned and examined. Most wear on Mesh 1 and 2 occurred on the anodized aluminum gears rather than the Rulon pinions. The Rulon pinions were loaded with aluminum oxide particles, which undoubtedly caused wear on their mating gears.

Nylasint, anodized aluminum—Chart C. All meshes of various Nylasint-material pinions meshed with anodized-aluminum gears completed the 2000-hr life test without significant increase in backlash.

A slight amount of wear products was noticed under the meshes (not shown) whose pinions were hobbed from the Nylasint material having the graphite filler.

Sinite, anodized aluminum, steel—Chart D. These curves show clearly the slight wear suffered by the two material combinations composed of Sinite D-108, anodized aluminum and stainless steel. There was no measurable increase in backlash on one test involving anodized aluminum. In fact, the final backlash figure was less than the initial backlash reading. No wear products were in evidence beneath the test meshes.

DESIGN OF Nylon Resin Spur Gears

Method of calculating the load-carrying capacity of gears of Zytel nylon resin. Design recommendations are given and design charts to establish torque capacity or gear size to transmit a given torque.

K. W. HALL, Professor
H. H. ALVORD, Associate Professor
University of Michigan

A thermoplastic material, Zytel is supplied in the form of granulated powder from which many articles can be molded by conventional molding techniques. Molded material has relatively light weight, is rigid and tough, has low coefficient of friction, and good resistance to abrasion. Zytel is made in several nylon molding powder compositions, of which Zytel 101 is recommended for mechanical parts.

Physical properties are affected by temperatures; yield strength decreases with increase in temperature, while impact strength increases. Moisture content of the molded material also affects properties. Material used is conditioned to have 2½% moisture by weight, equivalent to 50% relative humidity and 73 F

Besides the load-carrying ability of gear teeth, the design recommendations of Table I are based on test in-formation and investigations with gears of Zytel nylon resin.

GEARS TESTED

Gear testing machines ran with two pair of gears loaded against each other. In this way the driving motor supplied only the power to overcome friction and windage.

Tooth load is applied by twisting a torsion bar which connects the gear shaft. Lead shot in buckets is used as weight, thus permitting any desired amount of twist by changing the amount of shot in the buckets. Test machine is driven by a constant-speed motor through a speed variator, giving a 600 to 5000 rpm speed range. Oil mist lubrication is used, which is shut off for dry gear runs.

Gears were machined from molded disks of Zytel. Teeth were cut with hobs after the disks had been ma-chined to size. Pitch diameters ranged from 1.375 to 3.75 in., with diametral pitches of 16, 20, 32 and 48.

In addition to the cut gears, 2½ in. dia, 20 pitch and 32 pitch molded teeth gears were tested. They have a 20° pressure angle, full-depth teeth and ½ in. face width. A ⅛ in. web joins the hub and rim.

TESTS AND RESULTS

An extensive life testing program was first conducted with the machined gears having cut teeth, so that variations in gear diameter, diametral pitch, face width, pressure angle, and others, could be more easily accomplished. Pitch line velocities ranged from 785 to 3730 fpm, both with and without lubrication of the teeth. Tests were made with pairs of identical gears running together, and with small pinions in contact with larger

---

Table I—Some Design Recommendations for Zytel Gears

Size of teeth, as with metal gears, should be the smallest possible which are capable of transmitting the necessary power. Limitations of the load-carrying calculations should be considered when determining teeth size.
Pressure angle and tooth form best suited for the nylon resin are full-depth teeth with a 20° pressure angle. Full-depth teeth may be strengthened by making the blank oversize, but care must be exercised as strength gained by the thicker base may be offset by the weakening effect of higher temperature caused by increased normal force and sliding velocity.

The 16-pitch, 27-tooth, 20° pressure angle gears made with ⅛ in. oversize blank diameters had greater strength than the same gear with standard full-depth teeth. Similar gears with 20° and 30° stub teeth had less life than the standard full-depth teeth.
Backlash must be provided, but performance does not seem to be affected by reasonable variations in backlash. Recommended values measured at room temperature: 16 diametral pitch, 0.004 to 0.006 in.; 20 DP, 0.003 to 0.005 in.; 32 and finer, 0.002 to 0.004 in. For high speed or very heavy load operation which may heat the teeth, backlash should be increased to compensate.
Gear proportions, rims, webs, and hubs should be generously proportioned to give ample support for the teeth. This is important for heavily loaded teeth.

In general, the rim thickness beneath the root of the teeth should be at least 3 times the thickness of the teeth at the pitch circle; web thickness should be about equal to the rim thickness; hub diameter at least 1½ times the shaft diameter; and the hub length at least equal to the shaft diameter.

If the gear is large and the teeth heavily loaded, the Zytel gear should be mounted on a metal flange rather than key or spline the gear directly to the shaft.

Molded gears should be designed to reduce or eliminate the formation of residual stresses, and the mold should have a ring gate to maintain concentricity.
Treatment after molding includes a full anneal if there is any evidence of residual stresses being present. It is better to design the gear to prevent formation of these stresses, rather than to rely on annealing, as annealing may affect the teeth accuracy if the residual stresses are not uniformly distributed.

294

gears. Ambient temperature and humidity were not controlled. Temperature varied from 68 to 92 F; relative humidity from 29 to 85%.

Results obtained with cut teeth gears are shown in Figs. 1 and 2. Similar but less extensive data were obtained for 20 and 40-pitch gears.

Results obtained with molded teeth are shown in Figs. 3 and 4. Some had been tested as molded; some had been moisture conditioned; and some were annealed. Fig. 4 indicates that the life of 32 pitch molded gears is greater at any given load than of gears with cut teeth. Fig. 3, however, does not show this same increase in life.

One reason advanced for the low values of the 20-pitch molded teeth, particularly in the higher stresses, was the presence of residual stresses in the rim of most of the molded gears, and the thin web which did not give enough support for the rim and teeth at high loads. The line of Fig. 3 is drawn to represent the test results obtained with annealed gears. Annealed gears will fail at the root in a typical fatigue failure.

At higher loads there was evidence of web distortion in the 20-pitch gear, and this probably adversely affected the life of the molded gears. Neverthe-

Table II—Tooth Form Factor Y for Load Near Pitch Point

Number of Teeth	20° Full-Depth Form	20° Stub-Depth Form
15	.....	0.556
16	.....	0.578
17	.....	0.587
18	.....	0.603
19	.....	0.616
20	0.544	0.628
22	0.559	0.648
24	0.572	0.664
26	0.588	0.678
28	0.597	0.688
30	0.606	0.698
34	0.628	0.714
38	0.651	0.729
43	0.672	0.739
50	0.694	0.758
60	0.713	0.774
75	0.735	0.792
100	0.757	0.808
150	0.779	0.830
300	0.801	0.855
Rack	0.823	0.881

less, molded teeth are an improvement over cut teeth in the sizes tested and probably in 16 and 48-pitch molded gears as well. Thus, Fig. 5, which establishes the allowable stress curves includes the 16 and 48-pitch.

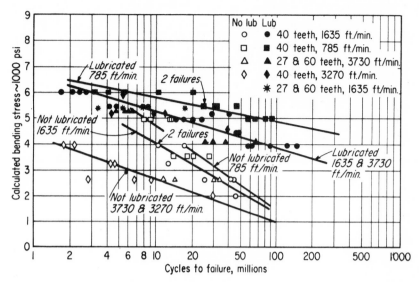

Fig. 1—Fatigue life of 16-pitch cut gear teeth.

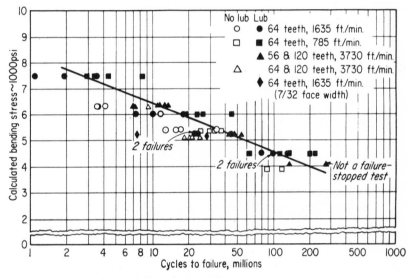

Fig. 2—Fatigue life of 32-pitch cut gear teeth.

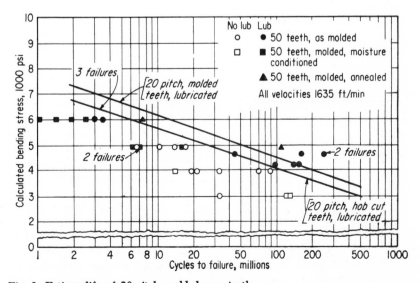

Fig. 3—Fatigue life of 20-pitch molded gear teeth.

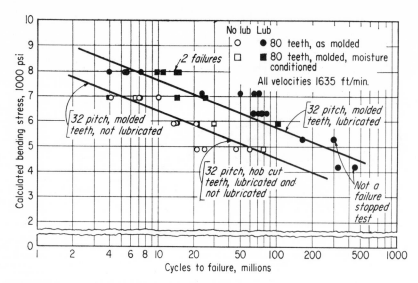

Fig. 4—Fatigue life of 32-pitch molded gear teeth.

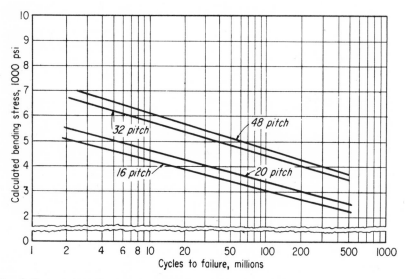

Fig. 5—Recommended maximum allowable stresses for molded Zytel gear teeth, 20° pressure angle, full-depth, oil lubricated for the indicated pitch gears.

Table III—Values of Design Factor K for Use With Lewis Equation and Design Charts

Teeth	Lubri-cation	Velocity ft/min	Pitch	Factor K
molded	yes	below 4000	16-48	1.00
molded	yes	above 4000	16-48	0.85
molded	no	below 1635	16-20	0.70
molded	no	above 1635	16-20	0.50
molded	no	below 4000	32-48	0.80
cut	yes	below 4000	16-48	0.85
cut	yes	above 4000	16-48	0.72
cut	no	below 1635	16-20	0.60
cut	no	above 1635	16-20	0.42
cut	no	below 4000	32-48	0.70

smaller teeth, which has the effect of weakening the larger teeth.

Curves of Fig. 5 show the maximum stresses which should be used when the load is reasonably steady with virtually no shock or impact loading. These should be used with judgment, and reduced to compensate for overloading, impact or shock loading, or other conditions of operation that may be present in the design.

Design factor, K, Table III, compensates for designs using cut rather than molded teeth, for lack of lubrication, and for velocities in excess of 4000 rpm. Other than this no correction need be made for velocity.

It is possible to approximate allowable stresses for pitches other than shown in Fig. 5. However, approximation of stresses for gears with teeth much larger than tested should not be attempted. The combination of large loads and high sliding velocities may exceed some limiting value not reached with the present test. For example, the teeth may fail by surface deterioration rather than by bending fatigue. Here, load-carrying capacity would be limited by resistance to wear rather than by bending fatigue.

Teeth having pitches finer than 48 should not be approximated using Fig. 5, unless the dimensions of the teeth are very carefully controlled. Errors could reduce the carrying capacity of such small teeth.

To facilitate the design of gear teeth of Zytel, the equations and the stresses of Fig. 5 have been combined to form the design charts of Fig. 6 (A), (B), (C) and (D). Charts can be used to establish the size of gear required to transmit a given torque. A design factor K, Table III, is used with the charts to increase the torque which must be transmitted, or to decrease the torque which the gear is capable of transmitting.

LOAD-CARRYING CAPACITY OF GEARS OF ZYTEL

The Lewis equation which was used to calculate the stresses in the gear teeth can also be used to calculate the load-carrying capacity of the teeth by using the test data to establish the allowable stresses. However, this equation is more useful when rewritten in either of the following forms:

$$T = SDFYK/2P$$

where
T = torque gear can transmit, lb-in.
S = allowable stress, psi, from Fig. 5
D = pitch diameter of pinion, in.
F = face width, in.
Y = form factor for pinion, Table II
K = design factor, Table III
P = diametral pitch

or

$$HP = SDFYNK/126{,}000\,P$$

where
HP = horsepower gear can transmit
N = pinion speed, rpm

What are considered to be the maximum recommended stresses for molded and lubricated teeth are shown in Fig. 5. To give a margin of safety, the lines have been reduced 25% from the represented failure of the gears on test. Fig. 5 indicates higher allowable stresses for small teeth than for large teeth. Reason: load is distributed among more teeth when the teeth are small, even though the method of calculation considers the entire load to be carried by one tooth. Also, larger teeth tend to run at a higher temperature than the

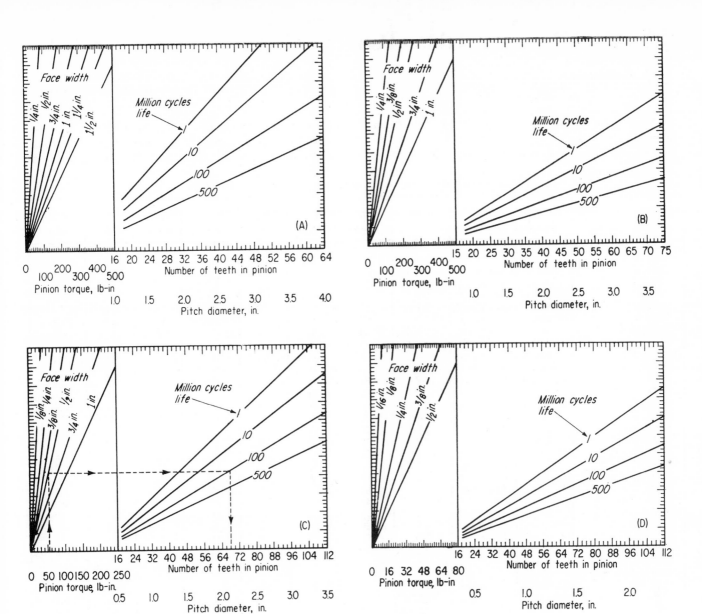

Fig. 6—Design charts for full depth molded Zytel gear teeth, 20° pressure angle, oil lubricated. (A) 16-pitch gears. (B) 20-pitch gears. (C) 32-pitch gears. (D) 48-pitch gears.

EXAMPLES USING DESIGN CHARTS

Example 1—A 32 pitch pinion having a ½ in. face width is to transmit a torque of 54 lb-in. at 1760 rpm. Teeth will be molded and lubricated. A life of 100 million revolutions is desired. What should be the pitch diameter of this pinion?

Design factor K, Table III, is 1.0 for molded and lubricated teeth with a pitch-line velocity below 4000 rpm; hence the design torque equals the actual torque of 54 lb-in. Solution is shown by dotted lines of Fig. 6 (C), and finally downward to find number of teeth, 68, and the pitch diameter, 2.125 in.

Comparisons among Nylon, Delrin, and Lexan Gears

Properties—Units	Delrin 500	Delrin 100	Nylon (Zytel 101) 2% Moisture	Nylon (Zytel 101) 2.5% Moisture	Lexan
Yield strength, psi	10,000		11,400	8,500	9–10.5
Shear strength, psi	9,510		9,600	—	—
Impact strength (Izod)	1.4	2.3	0.9	2.00	12–16
Elongation, %	15.0	75.0	60.0	300.00	60–100
Modulus of elasticity, psi	410,000		410,000	175,000	—
Hardness, Rockwell	M94 R120		M79 R118	M59 R108	—
Coefficient of linear thermal expansion, in./in. °F	5.5 x 10⁻⁵		5.5 x 10⁻⁵	5.5 x 10⁻⁵	3.9 x 10⁻⁵
Water absorption 24 hrs, %	0.25		1.5	—	0.3
Saturation, %	0.9		8.0	—	—
Specific gravity	1.425		1.14	1.14	—

Powder Metals for Gears

MARVIN FEIR
Dixon Sintaloy, Inc.

From Materials in Design Engineering

As the table shows, data are included for most common metals, including irons, steel, stainless steel, and copper and nickel and their alloys. For heat treatable ferrous metals, properties are given for both the as-sintered and heat treated conditions. For nonferrous metals, properties are given only for the as-sintered.

MECHANICAL PROPERTIES OF FERROUS AND NONFERROUS P/M PARTS

Material	PMPMA[a] Designation	Condition	Density, gm/cc	Ult Ten Str, 1000 psi	Yld Str, 1000 psi	Elong, %	Transv. Fiber Str, 1000 psi	Shear Str, 1000 psi	Impact Str, ft–lb	Hardness, Rockwell
Steels										
99Fe–1C	F–0010–P	As–sintered	6.1–6.5	35	27	1.0	89	22	1	50R$_B$
99Fe–1C	F–0010–P	Sintered, h.t.	6.1–6.5	47.7	—	0.5	—	—	4.5	90R$_B$
99Fe–1C	F–0010–S	As–sintered	7.0	60	—	3.0	120	—	2	—
99Fe–1C	F–0010–S	Sintered, h.t.	7.0	65	—	0.5	120	—	5.0	100R$_B$
99Fe–1C	F–0010–T	As–sintered	7.3	68	—	3.0	140	—	3.0	—
99Fe–1C	F–0010–T	Sintered, h.t.	7.3	127	—	2.5	235	100	6.0	105R$_B$
SAE 1080	—	Wrought, ann.	7.8	90	54	24	—	75	4.5	15R$_C$
90Fe–10Cu	FC–1000–N	As–sintered	5.8–6.2	30	25	0.5	75	17	—	—
90Fe–10Cu	FC–1000–N	Sintered, h.t.	5.8–6.2	54	—	1.0	103	—	3.5	30R$_C$
92Fe–7Cu–1C	FC–0710–N	As–sintered	5.8–6.2	50	40	0.5	115	35	3.0	70R$_B$
92Fe–7Cu–1C	FC–0710–N	Sintered, h.t.	5.8–6.2	85	—	1.5	180	—	6.0	30R$_C$
92Fe–7Cu–1C	FC–0710–S	As–sintered	6.8	83	63	1.0	131	57	4.0	73R$_B$
92Fe–7Cu–1C	FC–0710–S	Sintered, h.t.	6.8	110	—	1.5	209.5	—	7.0	40R$_C$
80Fe–20Cu	FX–2000–T	As–sintered	7.1 Min	70	70	1.0	140	55	14.0	75R$_B$
80Fe–20Cu	FX–2000–T	Sintered, h.t.	7.1 Min	128	—	0.5	209.5	65	11.0	35R$_C$
79Fe–20Cu–1C	FX–2010–T	As–sintered	7.1 Min	110	90	1.0	190	66	11.0	95R$_B$
79Fe–20Cu–1C	FX–2010–T	Sintered, h.t.	7.1 Min	152	—	1.0	—	110	—	40R$_C$
1.5Ni–0.5Mo–0.6C	—	As–sintered	6.8	70	58	2.5	150	43.5	7.1	80R$_B$
1.5Ni–0.5Mo–0.6C	—	Sintered, h.t.	6.8	90	80	0.5	150	—	—	25R$_C$
1.5Ni–0.5Mo–0.6C	—	As–sintered	7.2	90	72	2.5	180	47	9.2	95R$_B$
1.5Ni–0.5Mo–0.6C	—	Sintered, h.t.	7.2	140	120	0.5	207	—	4.3	35R$_C$
7Ni–2Cu–1C	—	As–sintered	6.8	70	50	2.5	140	50	5	70R$_B$
7Ni–2Cu–1C	—	Sintered, h.t.	6.8	135	—	1.5	262	—	6.5	42R$_C$
7Ni–2Cu–1C	—	As–sintered	7.2	92	75	3.5	180	60	11	85R$_B$
7Ni–2Cu–1C	—	Sintered, h.t.	7.2	157	—	2.0	285	—	8.6	44R$_C$
Stainless steels										
18Cr–8Ni	SS–303L–P	As–sintered	6.0	35	32	2.0	—	—	—	52R$_B$
18Cr–8Ni	SS–303L–R	As–sintered	6.6	52	47	7.0	—	20	4.5	55R$_B$
SAE 303	—	Wrought, ann.	7.9	90	35	50	—	—	80	80R$_B$
18Cr–12Ni–2Mo	SS–316L–P	As–sintered	6.16	38.5	35	2.0	95	—	2.0	55R$_B$
18Cr–12Ni–2Mo	SS–316L–R	As–sintered	6.65	58	51	8.0	135	20	4.5	65R$_B$
SAE 316L	—	Wrought, ann.	7.9	78	32	50	—	—	110	79R$_B$
12.5Cr–0.15C	SS–410–N	As–sintered	5.9	42	41	1	90	—	2.0	85R$_B$
12.5Cr–0.15C	SS–410–N	Sintered, h.t.	5.9	85	—	—	—	—	—	15R$_C$
12.5Cr–0.15C	SS–410–P	As–sintered	6.4	55	54	1	130	—	4.5	95R$_B$
12.5Cr–0.15C	SS–410–P	Sintered, h.t.	6.4	110	—	—	—	—	—	29R$_C$
Copper and alloys										
90Cu–10Sn	BT–0010–N	As–sintered	5.8–6.2	8	7	1	30	—	—	11R$_B$
90Cu–10Sn	BT–0010–R	As–sintered	6.4–6.8	14	11	1	36	9	—	30R$_F$
90Cu–10Sn	BT–0010–S	As–sintered	6.8–7.2	20	20	2–3	42	—	—	43R$_F$
90Cu–10Sn	BT–0010–W	As–sintered	8.0	45	30	11–15	90	—	4.5	80R$_F$
90Cu–10Sn	—	Wrought, ann.	8.8	85	65	68	—	56	—	—
80Cu–20Zn(+Pb)	BZ–0218–T	As–sintered	7.2 Min	20	15	10	31	15	3.0	37R$_F$
80Cu–20Zn(+Pb)	BZ–0218–U	As–sintered	7.7 Min	23	18	12	65	25	—	42R$_F$
80Cu–20Zn(+Pb)	BZ–0218–W	As–sintered	8.0 Min	37	28	21	80	—	—	50R$_F$
80Cu–20Zn(+Pb)	—	Wrought, ann.	8.8	38	12	53	—	32	25	61R$_F$
64Cu–18Ni–18Zn	—	As–sintered	7.2	25	16	15	55	—	—	65R$_F$
64Cu–18Ni–18Zn	—	As–sintered	7.9	42	26	14	85	—	—	90R$_F$
64Cu–18Ni–18Zn	—	As–sintered	8.3	45	30	30	94	—	—	32R$_B$
64Cu–18Ni–18Zn	—	Wrought, ann.	8.73	58	25	40	—	—	—	50R$_B$
Nickels and alloys										
67Ni–30Cu–3Fe	—	As–sintered	7.0–7.4	30	18	8	110	—	—	34R$_B$
67Ni–30Cu–3Fe	—	As–sintered	8.0	52	22	19.5	119	—	—	50R$_B$
67Ni–30Cu–3Fe	—	Wrought, ann.	8.84	90	55	35	—	—	—	88R$_B$

[a]Powder Metallurgy Parts Manufacturers Association. [b]Brinell.

Stronger Gears with Silicon Bronzes

From Machinery

Silicon bronze alloys make it possible to cast stronger gears. These corrosion-resistant alloys can improve the strength and appearance of many other types of castings, too.

New, high-strength silicon bronze alloys are making it possible to get more power out of gears and gear boxes without any changes in basic designs. And these alloys have excellent resistance to corrosion as well as high strength-to-weight ratios. Silicon bronze alloys are eminently ductile, malleable and castable too, and their machinability is good.

Nominal composition of a typical silicon bronze alloy, Herculoy (American Smelting and Refining Co., New York) is 91.5 percent copper, 0.50 percent tin, 4.0 percent zinc, and 4.0 percent silicon. The addition of silicon improves the alloy's foundry characteristics—it increases fluidity and decreases tendencies of the casting to hot-crack and shrink during solidification.

Silicon bronze is replacing gun metal bronze, Navy G metal, phosphor gear bronzes and similar alloys for gears. The tensile strength, yield strength and elongation of a typical silicon bronze are up to 60 percent higher than for phosphor bronze.

One company that has tested silicon bronze gear alloys extensively is Eimco Corp., Salt Lake City. Eimco, a manufacturer of heavy industrial processing equipment, consumes 40 tons of silicon bronze per year in the production of heavy duty, worm-driven gears. It has tried out both high-tin and silicon bronzes for this purpose. Wear resistance and ability to withstand severe torque of silicon bronze proved to be far superior to those of high-tin bronzes.

Pouring temperature of the silicon bronze used for gears is 2050-2250 F for complex, detailed castings and 1900-2050 F for heavy castings. Silicon bronze pours freely and sets with no more shrinkage than conventional brasses and bronzes. There is no need to add deoxidizers or hardeners to the metal during casting, and no smoke or noxious fumes are produced during melting.

Alloys such as Herculoy can be remelted without the use of flux. Melting should take place in a slightly oxidizing atmosphere. The metal is brought to temperature as quickly as possible for best results. Risers are used extensively to prevent shrinkage. Chills help to maintain conditions for optimum casting.

Properties of the cast metal allow it to be easily brazed, silver-soldered, and gas or electric-welded.

Applications of silicon bronze alloys are not limited to gears. Corrosion-resistant valves, nonsparking car wheels, ship fittings, "hot line" tools for the electrical industry, and electrical wire and cable connectors are just a few of the many products now being cast in these modern alloys. ▲▲

SILICON BRONZE ALLOYS were used to cast these heavy duty drive gears, which range from 8 inches to 5 feet in diam. Some of these drive gears must withstand up to 1.5 million ft-lb of torque in industrial service.

Shot Peening and Nitriding of Gears

Shot peening can increase fatigue strength which permits a change in tooth design to increase scoring resistance.

Shot peening

Tests have long shown that shot peening (see Fig 1) increases the beam (bending) strength of gear teeth. There is also strong evidence that shot peening increases the fatigue strength and may improve scoring and pitting resistance in gears.

The extensive use of peened gears—thousands are being peened yearly for the automobile industry—is increasing, says J. C. Straub of The Wheelabrator Corp, Mishawaka, Ind. This is in spite of the fact that in a good many cases the full benefits of shot peening are not being realized. The size of shot should be controlled. Straub found that if the shot striking the work is not uniform in size, the gain in fatigue strength is diminished. Also, he recommends use of small shot sizes at higher velocities.

Tests with automotive type gears, with a life requirement of 1 million cycles, have shown that shot peening has increased their fatigue strength by more than 25%—or for the same stress value, the gears averaged a life span 10 times longer than the nonpeened gears.

Nitriding of gears

The overwhelming majority of gears being manufactured today can be nitrided successfully to produce longer wear qualities at lower cost per operating cycle—so claims Horace Kneer of Metlab Co in Philadelphia, a firm that has been nitriding gears since the early 1940s.

Although nitriding is not a new method, ways are continually being found to improve the technique from the manufacturing standpoint so that gears can be nitrided to a given case depth in quicker time. Gears usually need to be placed in the nitriding furnace for 72 hr and up—but this can be done in great quantities. To speed up the nitriding process, Kneer has made the surprising discovery that the alloy steel 4340, which has been found so useful in the manufacture of large marine gears for the Navy, gives up to 50% deeper case for the same time as special nitriding materials such as Nitralloy (see Fig 2). Kneer also made some interesting comparisons with other hardening techniques for gears.

The nitriding process causes the steel surface to absorb nascent atomic nitrogen by the dissociation of dry ammonia gas at about 975 to 1025 F. The absorbed nitrogen combines with other elements in the steel, producing complex nitrides in submicroscopic particles. These particles resist permanent deformation in the crystalline grains of the metal and therefore cause hardness.

A furnace may contain several tons of work, all uniformly treated. Once loaded it requires little attention, except automatic control, and no labor. The elimination of the individual quenching and tempering operations required for carburized gears, is in itself a saving of time and labor. Also, Kneer reports that a 4140 or 4340 nitrided case can be struck repeated blows with a machinist hammer and will merely peen over, while the case of a carburized-steel or Nitralloy gear would chip.

Kneer recommends 4340 for heavy gears, 4140 for light gears (its case is the same as for 4340), and Nitralloy 135 Modified where extreme hardness (with a reduction in case toughness) is called for.

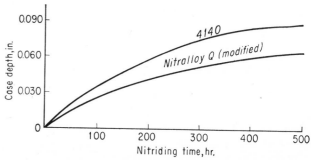

1. Peening of gears is fast becoming a popular production technique to increase bending and fatigue strength of gear teeth.

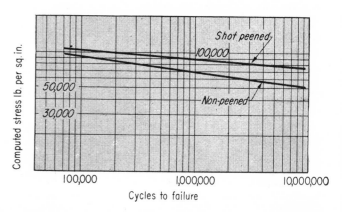

2. Surprising results show that the alloy steel 4140 gives up to 50% deeper case from nitriding than do special steels.

Comparison of Gear Heat-treatment Methods

How long a life?

At times, predicting the life of a gear seems almost as chance-like as predicting the number that will come up on a roulette wheel. Basically, gear durability depends on the fatigue strength of the material. And here's one big uncertainty. On a series of material flexural tests points of failure will be widely scattered. One bar may fail after a million cycles, while a seemingly similar bar may successfully withstand 10 million cycles.

Test, test, test. The only solution is to subject gears of differing designs and materials to various loads, then compile this mass of data to obtain fatigue limits that will give reasonable reliability. This is what General Electric has done since 1946 to some 3000 gears. The test results were reported by John Seabrook, gear metallurgist and Darle Dudley, manager, advanced gear engineering.

A summary of these results with fatigue limits for various materials and heat treatments is given in the table. GE, as you will notice, has adopted the concept of 10% and 50% failure "guarantees" borrowed from the familiar B-10 life system for rolling element bearings. According to this concept, 90% of the gears will withstand the unit load given in the 10% column and only 50% will withstand the load in the 50% failure column.

As you will also notice, GE has published the test results in terms of unit loads rather than in terms of stress, because, especially in foreign countries, misunderstandings frequently arise when stress values are given. The unit load is equivalent to the load on a tooth of one pitch in a normal plane having a 1-in. face width. The units are lb/in.² but are not stress. For a given load, the maximum tangential force (or pinion torque per pinion pitch radius) can be obtained from these equations:

For spur gears

$$W_t = U_L \frac{F}{P_d}$$

For helical gears

$$W_t = U_L \frac{F \cos \psi}{P_d}$$

Where

W_t = tangential driving force lb
U_L = unit load, lb/in.²
F = contacting face width, in.
P_d = diametral pitch
ψ = helix angle

Material and treatment	Fatigue limit (1000 psi unit load)	
	10% failure	50% failure
Carburized 9310	51	55
Carburized and shot peened 9310	58	60
Nitrided Nitralloy 135 Mod	42	
Nitrided Nitralloy N	44	46
Nitrided 5% Ni 2% Al	56	58
Nitrided H12	50	
Nitrided 4340	34	
Induction hardened 4340	38	44
Induction hardened and shot peened 4340	42	45
Induction hardened 4140	36	
Induction hardened and shot peened 4140	40	
Through hardened 4340 at 35 R_c	30	
Through hardened and shot peened 4340 at 35 R_c	36	
Through hardened modified 4340 at 42 R_c	41	

The conclusions? From the welter of data collected over the years in the tests, a number of general conclusions have emerged. Usually, Seabrook and Dudley say, case and core configurations have relatively high fatigue strengths. Within this class, nitrided and carburized teeth have the greatest strength; induction hardened are next, followed by furnace hardened. Chemical composition of base material in the groups apparently is unimportant except where it affects case or core hardness during the processing.

Shot peening, the test shows, is beneficial, except for "a nitrided nitriding steel surface." On nitrided low-alloy steels, shot peening gives moderate improvement. On hardened pinions, shot peening yields as much as 30% improvement in fatigue limit. And, a fact not often recognized, shot peening of induction hardened teeth produces considerable benefit—tests, for example, have indicated a 33% improvement in induction-hardened 4140. Such values bring fatigue limit near what is expected, based on hardness of the teeth. Presumably, peening overcomes residual tensions inherent in selective heating. One note of caution: Improperly controlled processing can undo the good done by peening, and, when scatter is taken into account, end up giving lower strengths than found in a nonshot-peened pinion.

Best with care. When properly and carefully carried out, nitriding gave the best strength values. And among

nitrided steels, best performance was obtained with H12, H11, Nitralloy N and 5% Nickel, and 2% aluminum nitriding steel.

Induction hardening of 10-pitch teeth, the tests have shown, produces tooth strength just a shade inferior to nitriding and carburizing. But again, let there be no variations in the processing, otherwise a scatter of strength will be evident.

Tempering above 350 F, after induction hardening, gives markedly poorer strength. With increasing temperatures, strength continues down to low at a tempering temperature of 700 F. Then, as higher temperatures are reached, strength increases, reaching the value expected of furnace hardened pinions.

Hard to control. Carburizing has its good and bad aspects. With such treatment some of the highest strength yields were reached, but many variables affect the process; so more can go wrong than with, say, nitriding. For good results control must be exercised over carburizing potential, and surface conditions must be maintained in the root. Often, too, complex residual stresses are present. But despite pitfalls, a carburized tooth combined with shot peening to give good surface quality will make a strong tooth.

Decarburization, whether at the outer skin of a carburized part or in a through-hardened part, is also a pitfall to beware. Decarburization saps as much as 40% of the strength.

Comparison of Surface Coatings for Gears

Surface coatings evaluated

New data on the effect of surface coatings on wear and load-carrying capacity of steel surfaces under sliding and rolling conditions were reported by G. F. Wolfe of General Electric, West Lynn, Mass.

These results may lead to improved reliability of bearings and gears operating under poor lubrication. Coatings evaluated included a new G.E. ion-nitriding treatment. Surfaces tested were lubricated with silicone, diester, or petroleum oils.

Wolfe pointed out that many of today's applications call for gears and bearings to operate at high loads and high speeds in difficult environments ranging from high temperatures (with marginal lubricants) to cryogenic fluids. These conditions are placing many designs on the verge of failure. Failures usually are scoring or scuffing and wear.

Wolfe employed a Falex machine, a roll tester, and a gear stand. His tests showed that whenever the combination of steel sliding on steel is broken up, load-carrying capacity and wear life are improved. Surface treatments and coatings evaluated included:

Electrodeposited: Steel specimens were plated with the following metals: silver, nickel, chromium, cadmium, rhodium over silver, platinum, gold, copper, indium, lead, and zinc. The coatings were applied directly to the metal specimen, with no undercoating. Thickness on the Falex specimens averaged 0.001 in., coatings on the rolls and gears 0.0002 to 0.0003 in.

Phosphate coatings: Both manganese phosphate and zinc phosphate coatings were tested. Thickness was 0.0005 in.

Sulfurized: Parts are immersed in a molten salt bath containing sulfur compounds at 1050 F. Sulfur is diffused into the metal, forming a layer of iron sulfide.

Electroless nickel: This process deposits a coating, approximately 92% nickel and 8% phosphorus, by chemical reduction of a nickel salt with hypophosphite in aqueous solution.

Diffusion coatings: This is a new metalliding process consisting of electroplating out a metal from a bath of molten alkali fluoride salts. Elements such as silicon, boron, and beryllium are diffused into the surface of metals and alloys. The surface developed is an alloy of the deposited element, rather than a coating, and is metallurgically bonded. Cases several mils thick can be obtained. The coatings are smooth, very hard, uniform, and extremely adherent. Borided specimens were evaluated in this investigation because there is essentially no surface build-up, and specimen dimensions can be closely held.

Soft-nitride: This new method of nitriding by means of a salt bath was developed in Germany and is now available in this country. The salt bath consists principally of potassium cyanide and potassium cynate, operating at 1000 to 1050 F. Many nitrogen atoms enter the surface and nitriding results. The resulting case is approximately half as hard as the case produced by conventional nitriding and is not brittle. Case depths of 10 to 18 mils are possible.

Bonded solid lubricants: Specimens were coated with several molybdenum disulfide formulations and then baked for 1 hr at 300 F. Total thickness was 0.0005 in. Specimens were phosphate-treated prior to application of the solid lubricant.

PTFE coatings: These polytetra fluoroethylene coatings were baked at 750 F for 3 hr. Thickness was 0.0005 in.

Flame-sprayed: Specimens were flame-sprayed with tungsten carbide and aluminum oxide, and finish-ground after spraying. The final thickness was 0.002 in.

Spark-discharge: This is a recently developed metal-surface-hardening process, accomplished by sparking a low-alloy steel with a tool steel or a hard carbide. Although the transfer of this metal is negligible, a steel surface can also be sparked with metals which are easily transferred in large amounts. In this investigation surfaces were sparked with silver, copper, and a cobalt alloy. The thickness averaged 0.0002 in. and the surfaces were quite rough in appearance. The surface can be improved by burnishing.

Ion-nitriding: A new G.E. glow-discharge process produces hard-case layers on steel. An electrical discharge is induced between conductors, one of which is the workpiece, separated by a nitrogen-hydrogen gas mixture (90% N, 10% H) at reduced pressure. Charged gas particles are attracted to the workpiece (cathode). Upon striking the surface, the kinetic energy is converted to thermal energy and the surface temperature rapidly rises to a high level. The combination of high surface temperatures and activated gas particles permits rapid diffusion into the metal, and chemical reactions occur. Hardness and case depths similar to conventional nitriding are obtained in much shorter times, and formation of the white nitrided layer associated with ammonia nitriding is minimized.

Test results: Table I lists various electrodeposited metals in order of increasing load-carrying ability in silicone oils as determined by the seizure-load test. Also included are results with diester oils. Table II lists results with solid lubricants, flame-sprayed coatings, and spark-deposition coatings. Table III evaluates metal platings for improving gear performance —a high scoring index is desirable. Table IV gives results obtained with rolls running against each other in an arrangement that produces both a sliding and a rolling motion.

On the basis of these tests, Wolfe has drawn the following conclusions:
1. Soft metal plates and metals with low melting points improve the load-carrying ability, which may be independent of the lubricant.
2. Silver plate is very effective in all oils—a flash coating of rhodium over silver provides the best results.
3. Coating thicknesses of 0.0002 to 0.0003 in. provide best anti-wear properties.
4. An electroless nickel coating provided excellent anti-wear properties in diester and petroleum oils—but not in silicone oils.
5. Bonded coats of molybdenum disulfide improve the seizure loads but do not significantly improve the anti-wear properties. A phosphate pre-treatment is necessary for best results.
6. PTFE coatings improve the load-carrying capacity of steel but wear off rapidly at low loads.
7. Silver deposited by a spark-discharge process outperforms a plated silver surface of the same thickness.
8. Diffusion processes which produce a definite case layer on steel outperform conventional carburizing and nitriding processes.

TABLE I—Wear and load capacity of plated surfaces

ELECTRO-DEPOSITED SURFACE	METHYLPHENYL SILICONE		METHYL CHLOROPHENYL SILICONE		DIESTER MIL-L-7808D	
	Seizure load, lb	Wt. loss (mg) 10 min at 250 lb	Seizure load, lb	Wt. loss (mg) 10 min at 250 lb	Seizure load, lb	Wt. loss (mg) 10 min at 250 lb
Uncoated steel	100	Large	450	560	1500	7
Platinum	200	Large	600	503	3000	7
Electroless nickel	250	Large	850	208	3000	1
Nickel	250	Large	1100	28	1750	1
Chromium	750	554	1500	2	1100	44
Cadmium	2000	10	2500	3	2250	4
Zinc	2000	27	3250	17	—	—
Indium	2250	71	3400	66	3400	—
Lead	3000	100	3500	87	3400	—
Rhodium	3600	10	3300	44	4500	47
Copper	3900	96	3250	5	4000	122
Gold	4000	144	3000	142	2750	75
Silver	4250	76	4500	76	3850	18

TABLE II—Surface treatments on steel

SURFACE (in diester oil)	SEIZURE LOAD, LB	LOAD TO PRODUCE 1 MIL WEAR, LB
Uncoated steel	1500	750
Aluminum oxide	1800	—
Tungsten carbide	2000	—
Molybdenum disulfide (silicone resin)	2000	750
Case-carburized	2300	1500
Phosphate	2500	750
Molybdenum disulfide (bonded over phosphate	3000	1000
Molybdenum disulfide (metal bonded)	3400	1000
Nitrided nitralloy N	3500	1500
PTFE	3750	250
Sparked copper	3750	750
Sulfurized	4000	2500
Soft-nitrided	4000	3000
Sparked Co-Cr-W,6	4100	1500
Sparked silver	4500	1500
Borided	4500	3250
Ion-nitrided	4500	4500

TABLE III—Platings on case-carburized gears

GEAR SURFACE	SCORING INDEX
Case-carburized	320
Chrome plate	330
Sparked silver	345
Rhodium over silver	360
Electroless nickel	370
Nitralloy N	385
Ion-nitrided	420

TABLE IV—Scoring with plated rolls

ROLL MATERIAL	SURFACE	SCORING INDEX
Carburized 9310	None	450
Carburized 9310	Moly disulfide (resin bonded)	575
Nitralloy N	None	525
Nitralloy N	Ion nitrided	570
Nitralloy N	Sparked silver	665
M-2 tool steel	None	550
M-2 tool steel	Sulfurized	575

Typical Methods of Providing

Below are shown various lubricating systems that can serve as guides

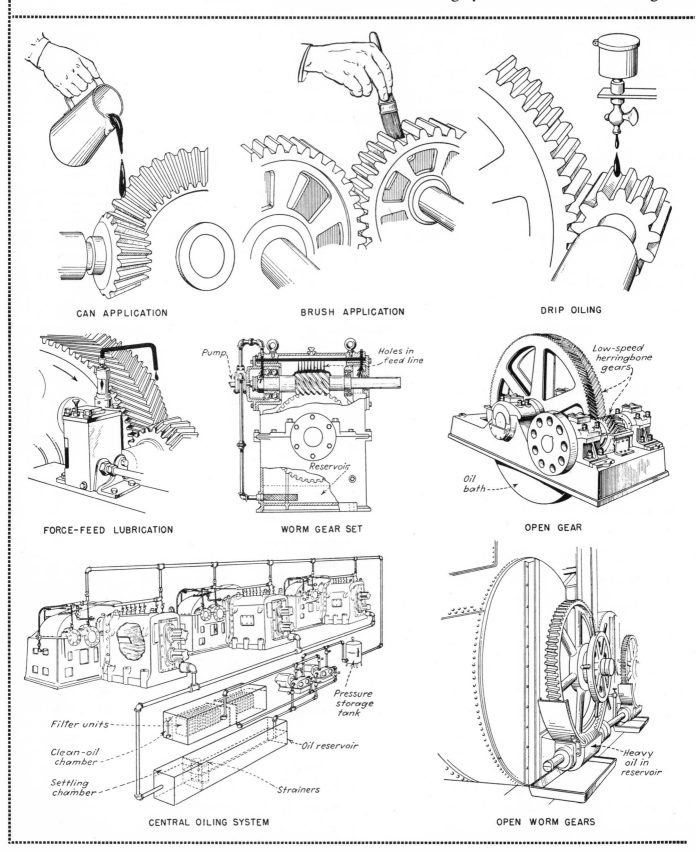

CAN APPLICATION

BRUSH APPLICATION

DRIP OILING

FORCE-FEED LUBRICATION

WORM GEAR SET

OPEN GEAR

CENTRAL OILING SYSTEM

OPEN WORM GEARS

Lubrication for Gear Systems

When designing for successful, efficient gear systems.

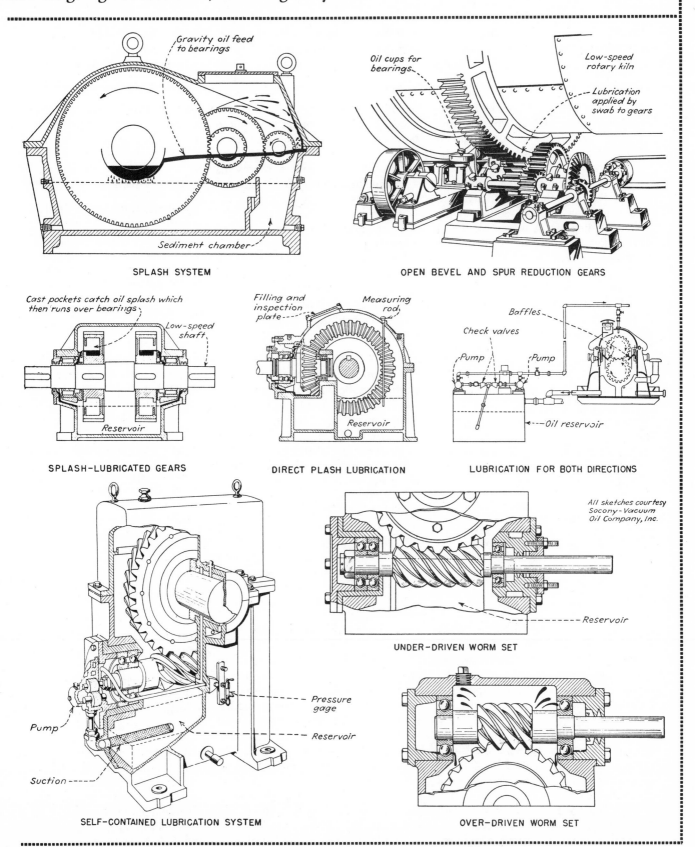

SPLASH SYSTEM

Gravity oil feed to bearings

Sediment chamber

OPEN BEVEL AND SPUR REDUCTION GEARS

Oil cups for bearings

Low-speed rotary kiln

Lubrication applied by swab to gears

SPLASH-LUBRICATED GEARS

Cast pockets catch oil splash which then runs over bearings

Low-speed shaft

Reservoir

DIRECT PLASH LUBRICATION

Filling and inspection plate

Measuring rod

Reservoir

LUBRICATION FOR BOTH DIRECTIONS

Baffles

Check valves

Pump

Pump

Oil reservoir

SELF-CONTAINED LUBRICATION SYSTEM

Pressure gage

Reservoir

Pump

Suction

UNDER-DRIVEN WORM SET

Reservoir

All sketches courtesy Socony-Vacuum Oil Company, Inc.

OVER-DRIVEN WORM SET

305

Pressurized Oil Systems for Gear Drives

D. W. BOTSTIBER AND LEO KINGSTON

Most high capacity gear drives, including marine, naval, aircraft and industrial reduction and step-up units require pressure lubrication.

Some of the heat generated by friction at the tooth flanks will be dissipated by radiation and convection. The remainder will be conducted deeper into the gear blanks and adjoining parts and dissipated from there. For a given system, the amount of heat rejected is a function of the temperature difference and volume of convecting medium. The more compact the unit, the less effective will be radiation and convection to the surrounding air. A major portion of the friction losses must be carried off by cooling oil.

Industrial gear drives of small power capacity were formerly often designed without external cooling. With increasing loads, designers resorted to improved cooling by mounting fans or impellers on the high speed shaft, and sometimes adding fins or shrouds to the housing. By properly shaping the oil sump around the gears, the oil may be guided to continuously flow back towards the gear teeth. The amount of such cooling, by direct convection, however, is small. It can be calculated as follows:

$$Q = \frac{A \times T}{C} \qquad (1)$$

where Q = dissipated heat, BTU/min
A = exposed area of gear case, ft^2
T = temperature difference between surface of gear case and ambient air
C = 30 for still air
C = 22 for free circulation
C = 10 for forced air flow

Using this formula on high capacity drives, Q will usually be equal to 2–5 per cent of the transmitted power on aircraft gear cases, and 5–15 per cent on marine and general purpose reduction gears.

System Geometry

A pressure oil system consists of a reservoir, pump, heat exchanger, and orifices which direct streams of oil to the gear teeth. The oil is heated by contact with the hot surface and drains back into the reservoir. Assuming sump is desired, the lower part of the gear case is used for this purpose. The pump is located near the bottom. It takes the oil in through a strainer, discharges it to the cooler. A pressure relief valve is added in the pressure line when pressure peaks may be frequent and of substantial magnitude. Also, an oil filter may be included. If the pump capacity is substantially greater than the required flow rate, a bypass valve may be used. This should be done only on test installations, since it represents a hazard. In the event of clogging of cooler or jets, the greater part of the flow may go through the bypass, giving only a small pressure rise and starving gears and bearings.

From the cooler, the oil goes to the jets which direct it to the gears. The spray should strike the unmeshing side of the teeth favoring the pinion. Small pinions should be struck by the oil stream radially.

Selection of Components

Oil Pump. Because of the need for continuous flow of oil under pressure, the pump should be a positive displacement type. Its drive should be mechanically positive. Shear pins or other torque limiting elements, designed to protect the pump from overload, should be avoided. They would save the pump at the expense of the gears. The pump should be near the bottom of the sump, with as little suction required for the intake as possible. Although most positive displacement pumps will develop a vacuum of 27 to 29 in. (mercury), this may be insufficient at low temperatures with ordinary oils, a clogged strainer, or with leaks in the suction line.

A good operating pressure range is 45 to 75 psi, but the pump should be capable of withstanding peaks of 250 psi. These may occur at sudden starts or at low temperatures with high oil viscosity. Pumps using the gear or sliding vane principle are well suited for this operating range. The gears may be of the external mesh type with extra deep teeth or of the internal mesh type with special tooth forms. Vane type pumps should have positive vane motion.

Intake Strainer. To protect the pump from solid particles, a strainer or screen should be installed at the intake. The size of its openings is a compromise between the filtering requirements and allowable pressure drop. Mesh 20 x 20 with 0.016 in. wire, with total free area of about five times suction line area as proved adequate for most circuits. Area of the intake line should be twice that of the discharge line to allow for the increased volume of entrained air caused by suction.

Discharge Lines. Pipe size is a compromise between economy (size, weight, and cost), which calls for smaller lines, and pressure loss which makes it necessary to maintain adequate passages. Oil jets must have positive pressure under the most unfavorable conditions. Generally, 15 psi is the permissible minimum. Losses in the lines should be in the range of 10–40 per cent of the pump pressure. For an average circuit, a flow velocity of about 6 fps is a good initial approximation. Actual losses in the lines may be calculated, based on laminar flow, using the conventional formulas of fluid dynamics.

Thermal Bypass. A bypass is normally closed; at low temperatures it opens and passes oil directly to the jets, bypassing the cooler. As the oil temperature increases, it must close gradually so the oil will apply pressure to the cooler and gradually clear its passages. Units of this type are used mainly in aircraft circuits.

Oil Filter. New filters generally have losses of 1.5 psi; their bypass valves open when the pressure across the filter rises to 5 psi. A strainer of 40 x 40 mesh with 0.012 wire will be a suitable compromise if a screen in the pressure line is desired and pressure drop must be low.

Oil Cooler. For a given flow rate and temperature of oil, the heat rejection capacity of a cooler varies with the temperature and flow rate of the cooling medium (air or water). This gives a number of variables, which are determined by tests and plotted on diagrams as systems of curves. These data are used as the basis for selecting a cooler for a given application. Usually, the order of magnitude of cooling medium flow by weight is comparable to that of oil flow. That means that if the oil flow is 10 lb/min, an air flow of about 10 lb/min may be expected.

Oil and water should flow in opposite directions since this will keep the temperature difference between oil and water uniform through the cooler.

Oil Jet. Cooled oil is sprayed on the gears through a jet which should produce a solid stream. Velocity of oil leaving the jet must be adequate to reach the gear teeth through the whirling air and penetrate the space between teeth to hit the flanks. Pressures at the jet inlet of 15 to 50 psi are generally satisfactory. The jet orifice must be adequate to pass the required flow. An empirical formula to determine the diameter for round orifices is:

$$F \text{ (gpm)} = 22 \ d^2 \ p \qquad (2)$$

where d = jet orifice dia, in.
p = pressure at jet inlet, psi

In general, one jet should be used for every 2 in. of gear face. Using rectangular orifices does not seem to improve cooling or lubrication.

Minimum permissible diameter for a jet is determined by the tendency of clogging from foreign matter. Jets of less than 0.060 in. dia are troublesome. For antifriction bearings, smaller orifices are sometimes used.

Physical Data and Average Design Values for High Capacity Gear Lube Systems

Oil Sump Temperature	70–100 C (160–212 F)
Specific Heat of Oil	0.42 BTU/lb/F
Specific Weight of Oil	7.2 lb/gal
Heat Rejection by Outside Convection (% of total)	2–15 per cent
Friction Losses	See Table I
Temperature Drop through Cooler	10–20 C, 18–36 F
Air Entrainment in Oil Circuit	10–50 per cent
Sump Capacity	1/5 to 5 × Flow Rate/min
Max Oil Temperature	230 C, 450 F
Min Cooling Water Temperature	1 C, 34 F
Min Air Temperature (Arctic Operation)	54 C, 122 F
Max Ambient Air Temperature (Climatic)	50 C, 122 F
Spec Weight of Air at Sea Level and 100 F	0.070 lb/cu ft
Spec Weight of Air at Sea Level and 59 F	0.765 lb/cu ft
Spec Weight of Air at 10,000 ft with 59 F at S/L	0.0565 lb/cu ft
Temp at 10,000 ft when 59 F at Sea Level	24 F
Flash Point of Oil per MIL–L–6082B	216 C, 420 F
Congealing Point of Oil per MIL–L–6082B	–18 C, 0 F
Congealing Point of Oil per MIL–L–7808B	Below –54 C, –65 F
Operating Oil Pressure at Jet Inlet	15–50 psi
Min Jet Orifice Dia for Gears	0.060 in.

Table I — Heat Losses in Typical Gear Drives

Type of Unit	Loss per Gear Mesh — Per Cent of Power Transmitted
General machine quality cut gears, sleeve bearings	1.5–1.8
High quality shaved or ground gears, sleeve bearings	0.8–1.2
High quality shaved or ground gears, antifriction bearings, gears in perfect condition	0.5–0.7

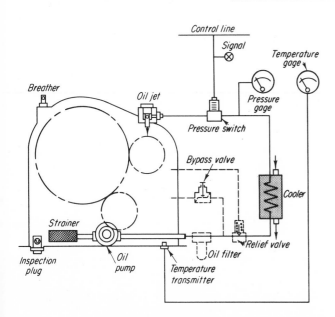

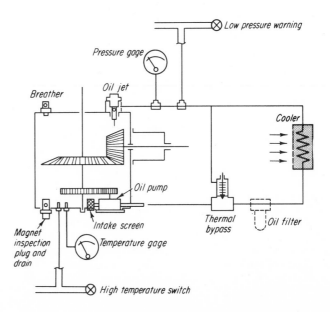

Powder lubrication

A new and remarkably simple lubrication technique has passed with honors some recent feasibility studies for high-temperature applications. The technique employs finely ground particles which are drawn into the carrier gas, usually air. The suspended particles may be of various materials, but graphite-cadmium oxide particles have produced the most promising results because of their ability to function successfully in air environments.

The technique was first tried a few years ago by a group of experimenters at Stratos Division of the Fairchild-Hiller Corp in Bay Shore, LI. They were looking for a suitable lubricant for bearings operating at very high temperatures (over 1000 F). Today, says Stanley Wallerstein, senior project engineer at Stratos, the applications are much broader. Powder lubrication can be applied to:

- **Any rotating components,** such as bearings and gears, to supplement other forced-feed lubrication systems
- **Extreme temperature applications to 1200 F.** The use of powders completely eliminates viscosity problems.
- **Space applications,** such as the lubrication of joints, doors, and mechanisms on the outside of space vehicles. Silicones tend to vaporize in such vacuum conditions
- **Radiation-resistance applications.** Powder lubrication is not affected by radiation, unlike the petroleum-based oil and gear lubricants.

Wallerstein predicts that the technique will also prove economical for many ordinary applications, such as the lubrication of automobile engines in which the suspended particles are burned up during the combustion cycle after performing their duty of lubricating the cylinder walls.

How the method works

A lubricator stores the dry-powder particles and agitates them to prevent caking. It also meters and dispenses the particles by means of the adjustable-speed drive which feeds the particles into the gas at a specific ratio.

The particles need a carrier (forced air or another gas) which is then pumped to the bearing (gears or other rotating parts) to coat the contacting surfaces. The process is a continual buildup and wearing-away of the powder on the surfaces. The excess powder is drawn away through the exhaust system. At present the powder is not recirculated, but Stratos is in the process of perfecting such a closed circuit system.

Of the lubricant combinations tested, Stratos found that there are three promising powder compositions:

1. **Molybdenum disulfide (MoS_2) in a nitrogen gas carrier.** Air could not be used because of the tendency of the molydisulfide to oxidize at about 820 F. Also, Stratos felt that the friction characteristics of the MoS_2/N_2 combination was not satisfactory at the higher temperatures (near 1000 F).

2. **Molybdenum disulfide (MoS_2) plus metal-free phtalocyanine (PCH_2).** This combination also must be used in a nitrogen gas carrier. The PCH_2 improved the high-temperature friction characteristics by increasing the adhesiveness of the molydisulfide—but the nitrogen gas carrier is a handicap in average applications.

3. **Graphite plus cadmium oxide (CdO) in an air carrier.** This combination provided low friction characteristics in an air environment and over a wide temperature range (0 to 1000 F). (The oxide improves the adhesion characteristics of graphite at high temperatures.) Hence, most of the applications employ the graphite/cadmium oxide mixture—specifically, micronized Acheson No. 38 graphite. Particle size is about 5 microns (somewhat larger than talcum powder). Stratos has been using a 5:1 ratio by weight of graphite to cadmium oxide in ratio of 20 parts air to 1 part powder.

Which gear materials for high-temp?

In high-temperature gear applications, the new lubrication technique was successful in keeping a proper film thickness between mating teeth. But Stratos found that most gear materials did not withstand the high temperatures too well. An experimental program was carried out in cooperation with Battelle Memorial Institute to find the best high-temperature gear materials from 14 contenders. Table I. The gears were required to operate at 1000 F and 15,000 rpm under conditions simulating gear operation using a solid powder lubricant.

The best materials according to the tests are Rene 41, Haynes 151, and Stellite 6B. Of these three the Stellite alloy has the greatest potential for high-temperature gears on the basis of friction, deformation and resistance to surface scuffing.

Table I — Comparison of modern high-temp gear materials

Material	Temp, F (RT=room temp)	Wear and galling resistance
Super Alloys		
Rene 41	RT 1400 1600	Very good
Hastelloy C	RT 1200 1500	Good Good
Haynes 25	RT 1500	Good
Haynes 151	RT 1500	Very good
Tool Alloy		
98M2	RT 1500	Poor in air Good in reactive gases
Haynes Stellite 6	RT 1500	Probably same as 98m2
Star J	RT 1500	Fair
Stellite 3	RT 1500	Good
Haynes Stellite 6B	RT 1500	Excellent
Cermets		
K161B	RT 1600	Low against itself
K164B	RT 1600	Poor
K162B	RT	Good

11
SPLINES, SPROCKETS, AND FRICTION DRIVES

A guide to selection of type . . .

- Fixed or flexible
- Size and face width
- How many teeth
- Pressure angle and tooth height
- Industrial practices and standards

how to design INVOLUTE SPLINES

DARLE W. DUDLEY

Trend in splines today is toward an involute profile similar to gear teeth. Main functions of a spline are to transfer torque from a shaft to gears, pulleys, fans, etc., or to couple two shafts together. Some components are loosely fitted, others solidly shrunk on and removable only with difficulty.

A spline coupling with involute teeth will transmit more torque for its size than any other type, with almost no speed limitations. For example, turbine shafts at 10,000 rpm or more, carrying several thousand horsepower, can be readily connected to a driven machine by a gear-tooth coupling; few other types of couplings can do it.

But perhaps the most important reasons for widespread use of involute splines is that they can be cut and measured on the same machines that cut and measure gear teeth, and most shops have gear-cutting facilities.

Splines differ from gears, however. The spline has no rolling action, and all teeth contact at once. Splines seldom pit or break at the root, although they do fail by fretting corrosion and by fatigue.

FIXED OR FLEXIBLE SPLINES

Fixed splines permit no relative or rocking motion between internal and external teeth, and can be either shrink fitted or loosely fitted together. A loose fit may be subject to backlash between gear and shaft when not under torque; but once under torque, the spline will center itself, take a fixed axial position, and axes of the mating splines will become coaxial.

Flexible splines permit some rocking motion, and under torque the teeth slip axially to accommodate axial expansion or runout. The axes of the mating splines usually intersect instead of being coincident. Although the angle between the axes is small, normally much less than a degree as the coupling rotates, angular misalignment will cause some rocking every revolution. Misalignment plus axial movement may cause continuous relative movement at the spline tooth.

Application, rather than tooth design, determines whether a spline is fixed or flexible. Tooth form will

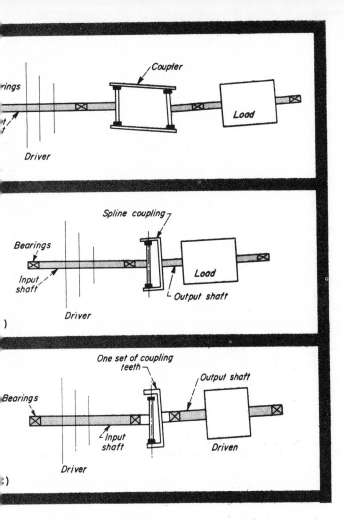

Controlling misalignment . . .

when using splined couplings. Double-spline system (A) accommodates both angular misalignment and offset. Although shown here with internal teeth, coupler usually is designed with internal teeth. Single-spline system (B) accepts only mis-alignment but uses fewer parts (note three bearings instead of four). Four-bearing, single-spline system (C) may function as an internal gear system if not aligned correctly.

not stop a flexible coupling from attempting to flex, and if the application does not want to flex, loose-fitting teeth probably will cause not flexing. However, for good design, teeth for a flexible coupling should be designed to permit flexing, and a fixed coupling should be designed to discourage flexing.

Fixed splines usually are mounted by shrinking the gear on the spline or fitting a ring at each end. The rings center the gear and allow the spline to transmit torque only.

Axial and angular misalignment when a motor is coupled to a load can be compensated for by using a double-spline system, (A) in the accompanying diagram, with the teeth of the coupling subject only to angular misalignment. Shaft offset is converted by the coupler, or "distance piece," into an angular misalignment. Increasing the coupler length will decrease misalignment.

The three-bearing system (B) accepts only angular misalignment—the two shafts must intersect at the

coupling for it to operate correctly. Also, because the output shaft has only one bearing, the coupling will carry a heavier load.

Trouble will arise if four bearings are mounted with only one spline, as in (C). With only a slight amount of misalignment the spline functions as an epicyclic gear system. The load is concentrated on only a few teeth, overloading the tooth surfaces as much as 10 to 20 times. In addition, the reaction places a serious load on the bearings. However, this system can function correctly if the shafts are limber and bend to intersect at the coupling center.

HOW TO ESTIMATE SPLINE SIZE

First step in designing a spline is to make a preliminary estimate of the size required, using chart, opposite page, relating spline capacity to pitch dia for seven types of service. The chart is based on a study of spline performance and indicates the normal range of torque for flexible couplings and for fixed splines.

Average commercial practice is represented at (A) with involute teeth of 20° or 25° pressure angle, and tooth depth about 75% of full-depth gear teeth. Teeth are of low-hardness steel and finished by cutting. Face width is $\frac{1}{4}$ to $\frac{1}{3}$ of the pitch dia. Misalignment is generally less than 0.003 in. per in.

High-capacity couplings (B) have tooth design similar to (A), but the face width may be wider, from $\frac{1}{4}$ to $\frac{2}{3}$ of the pitch dia. Misalignment may be held as low as 0.001 in. per in. Material is usually alloy steel hardened to R_c 55.

Aircraft flexible couplings (C) have a 30° pressure angle and hardnesses ratings in the range of R_c 55 to 65. Tooth proportions range from the 50%-depth proportions of the ASA B5:15-1950 spline standard to as high as 67%.

Aircraft splines with 1.250 in. and smaller pitch dia generally have face width equal to or greater than the pitch dia. On larger dia, flexible splines face width is narrower, $\frac{1}{4}$ to $\frac{2}{3}$ the pitch dia. It is debatable whether the wide faces of splines below 1 in. dia actually increase load capacity.

By coincidence, this rating is typical of a single key used as a fixed coupling. When used for single keys, the length is 1 to 1$\frac{1}{4}$ times the shaft dia. Stress in the shaft is about 7,500 psi, neglecting stress concentration at the keyway.

High-capacity single keys (D) set up 9,500-psi stress in the shaft away from the keyway. Length and size of key is the same as for (C). Shafts and keys are made of heat-treated steel of moderately high hardness. Design is typical of commercial flexible couplings keyed to motor or generator shaft extensions.

Multiple-key fixed spline (E) has involute teeth with pressure angles from 14$\frac{1}{2}$° to 45°; tooth depth is usually 50% or less. Length of spline generally ranges from 75% to 125% of the pitch dia. Shaft hardness is BHN 200 to 300.

High-capacity aircraft splines (F) will have spline length from 50% to 100% of pitch dia. Shafts are generally hollow to save weight, and hardness ranges up to R_c 58 for case-hardened parts.

Boundary line of spline design practice (G) is for a solid shaft with 65,000-psi shear stress. For a hollow

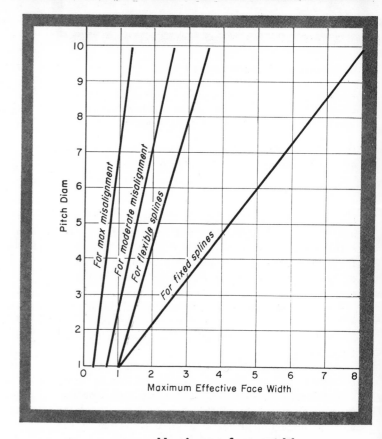

Quick estimate of size . . .

can be made from diameter-torque relationships. Example shows that a 1.6-in.-dia spline is required to transmit 800 in.-lb torque with a commercial spline.

A Commercial flexible
B High-capacity flexible
C Aircraft flexible or single-key commercial
D Single-key, high-capacity
E High-capacity fixed
F Aircraft fixed
G Limit of spline design (65,000-psi solid shaft)

Torque values based on application factor of 1 (turbine driving a generator)

Range of capacity for flexible splines
Range of capacity for fixed splines

Pitch dia. of splines or OD of keyed shaft, in.

Torque, lb – in.

shaft with a bore 75% of the outside dia, shear stress would be 95,000 psi. Such splines can be built to work. Several years ago, the author designed a fixed aircraft spline 1.860 in. dia and tested several models to destruction. The spline took an appreciable permanent set at 100,000 lb-in. of torque, and ruptured at about 160,000 lb-in. The steel was 35 R_C and a helix correction was used. From the chart the torque rating is 75,000 lb-in. at 1.860 in.

HOW WIDE THE FACE

For fixed splines with width ⅓ the pitch dia, the teeth have same shear strength as the shaft—providing they are spaced accurately so that all are loaded uniformly. Since there is always some error in tooth spacing, the face width should be at last ⅔ of the pitch dia, which will balance tooth and shaft strength with only half the teeth working. If weight is not too important, it is good design to keep the face width equal to the pitch dia.

In flexible splines, a wide face width often contributes little or nothing to load-carrying capacity even though conventional design formulas indicate otherwise. Unfortunately, there is neither good theory nor test data on just how much face width is useful. Flexible splines are generally misaligned since the accomodation of misalignment is one of the main functions. The spline teeth are so stiff that they have practically no deflection, and most of the load will fall on the ends of misaligned spline teeth. A wide face width will not relieve the intensity of end loading but will extend the time required for the wear in the middle of the spline to catch up with wear at the ends.

Maximum effective width can be estimated from

For max misalignment
For moderate misalignment
For flexible splines
For fixed splines

Pitch Diam

Maximum Effective Face Width

Maximum face width . . .

for fixed and flexible splines. Wide face widths in splines often contribute little to load-carrying capacity.

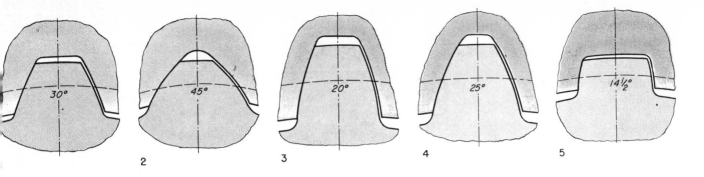

Five common spline tooth forms

TABLE I—SPLINE TOOTH PROPORTIONS
for 1 Diametral Pitch

Tooth Form No.		1*	2*	3	4	5
ASA Standard No.		B5:15–1950	B5:26–1950	. . . . .	. . . . .	. . . . .
Pressure Angle		30°	45°	20°	25°	14¼°
Nominal Depth (Compared to Gear Teeth)		50%	40%	75%	70%	30%
Addendum of External, a_e	Side-bearing fit	0.409	0.500	0.750	0.700	0.340
	Major-dia fit	0.500	. . . . .	0.950	0.900	0.430
Dedendum of External, b_e	Normal design (flat root)	0.600	. . . . .	1.000	0.950	0.390
	Full dedendum (fillet root)	0.900	0.500	1.250	1.050	. . . . .
Addendum of Internal, a_i		0.500	0.300	0.750	0.700	0.250
Dedendum of Internal, b_i	Side-bearing fit (flat root)	0.500	. . . . .	0.950	0.900	0.430
	Major-dia fit (flat root)	0.500	. . . . .	0.950	0.900	0.430
	Full dedendum (fillet root)	0.900	0.700	. . . . .	. . . . .	. . . . .
Chamfer on External Tip for Major-dia fit	Minimum height chamfer	0.091	. . . . .	0.200	0.200	0.090
	Slope of chamfer, β	55°	. . . . .	50°	50°	45°
Fillet on Internal for Major-dia Fit	Maximum radial height	0.085	. . . . .	0.200	0.200	0.090
	Approximate radius, R_i	0.160	. . . . .	0.200	0.225	0.075
Radius at Root of External Tooth, R_e	Normal dedendum, min. rad.	0.075	. . . . .	0.200	0.225	0.100
	Full dedendum, min. rad.	0.250	0.350	0.250	0.250	. . . . .
Minimum No. Teeth with Addendum Adjustment		. . . . .	. . . . .	18	10	10
Minimum No. Teeth without Addendum Adjustment		6	6	26	16	16
Nominal Contacting Depth, h	Side-bearing fit	0.900	0.800	1.500	1.400	0.590
	Major-dia fit	0.900	. . . . .	1.500	1.400	0.590
Effective Space Thickness of Internal, S		1.571	1.771	1.600	1.600	1.625

* Based on the author's interpretations of the ASA standards covering these forms.   Standard B 5:15–1950 is under process of revision.

TABLE II—TYPICAL FIXED-SPLINE CALCULATION

1. No. of Teeth, $N = 16$
2. Diametral Pitch, $P_d = 10$
3. Working Depth, h = Table $I/P_d = 0.059$
4. Pitch Dia, $D = N/P = 1.600$
5. Pressure Angle, $\phi = 14° 30'$
6. Fit = major dia

	EXTERNAL	INTERNAL
7. Addendum	a_e = Table $I/P_d = 0.043$	a_i = Table $I/P_d = 0.025$
8. Dedendum	b_e = Table $I/P_d = 0.039$	b_i = Table $I/P_d = 0.043$
9. Major Dia	$D_o = D + 2a_e = 1.6870–1.6875$	$D_{ri} = D + 2b_i = 1.6860–1.6867$
10. Minor Dia	$D_{re} = D - 2b_e = 1.526–1.516$	$D_i = D - 2a_i = 1.550–1.552$
11. Thickness, effective	$t_v = 0.1620–0.1605$	s_v = Table $I/P_d = 0.1625–0.1640$
12. Thickness, actual	$t = 0.160–0.159$	$s = 0.164–0.165$
12. Face Width	$F = 1⅝$	$F = 1½$
14. Form Dia	$D_{fe} = D - (2b_i + \Delta h) = 1.550$	$D_{fi} = D_i + (2h + \Delta h) = 1.670$
15. Tip Chamfer, slope	β_e = (Table I) $= 45°$	. . . .
16. height	H_t = Table $I/P_d = 0.009–0.015$	. . . .
17. Root fillet, radius	R_e = Table $I/P_d = 0.010–0.015$	R_i = Table $I/P_d = 0.007$
18. height	$H_e = 0.009$ max	H_i = Table $I/P_d = 0.009–0.003$

$\triangle h$ is allowance for 'extra' involute.   It may vary from 0 to 0.050/P_d.

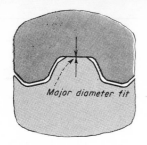

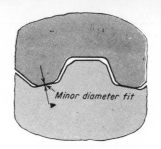

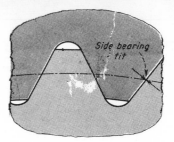

Three types of spline fits

the accompanying curve. While these values could be exceeded, the effective width should be used in calculating design stresses. Also, for fixed splines without helix modification, the effective width should not exceed:

$$F_e = 5,000 \ D^{3.5}/T$$

where F_e is the maximum effective face width, D the pitch dia, in., and T is torque, lb-in.

SPLINE TOOTH FORMS

The five involute tooth forms shown on opposite page are the most common in use today.

Form 1 is a standard spline tooth sponsored by ASME, SAE, AGMA and NMTBA, and covered by ASA B5: 15-1950 (currently under revision by the SAE Parts and Fittings Committee.) The form is a stub tooth having about 50% the working depth of a standard involute gear tooth. It is relatively easy to manufacture, has a 30° pressure angle, and can be made with as few as six teeth.

Form 2 has a 45° pressure angle with about 50% depth. It is used for fine-pitch involute "serrations." Coarsest pitch shown in the ASA standard is 10 pitch, but it can be produced with as few as six teeth. It is often employed with relatively large numbers of teeth to get many index positions in attaching parts, because it gives maximum shaft tooth dia and adds to the shaft strength of externally toothed parts.

Form 3 has a 20° pressure angle, a 75% depth, and is similar to stub-gear teeth. This type, but with 50% depth, is used in automotive applications. It has 50% more bearing area than Form 1 or 2, and enough depth to shave easily while the first two forms are not suitable for shaving.

Form 4 has a 25° pressure angle and 70% depth. This form has enough involute to shave well, and 40% more tooth-bearing area than either of the first two forms. It has a stronger centering action than Form 3, but not as strong as Form 1 or 2. Larger-dia marine couplings use this shape.

Form 5 has only 30% depth, and 14½° pressure angle. This tooth cannot be shaved and does not have much centering tendency. It is a poor choice for a

flexible coupling, but has some advantages as a fixed coupling where there is no relative motion between the tooth surfaces. The shallow depth permits a larger shaft than for the other forms, and for fixed splines the shaft is often the limiting factor in torque capacity. The low pressure angle develops less bursting stress in the internally tooth member; wall thickness can be reduced. The form also has the largest land at the top of the external tooth. It is a good choice for a fixed spline—particularly with a major-dia fit.

SPLINE DESIGN DATA

Three types of spline fits are: major dia, side bearing, and minor dia. In the major-dia fit, outside dia of the external tooth is closely fitted to root dia of the internal tooth, so addendum of the external equals dedendum of the internal. It is best to leave a small fillet at the root of the internal tooth and to chamfer the external tooth to clear this fillet. With cold-formed (rolled) external splines, the mating internal teeth can be cut with a broach to eliminate the usual internal fillet.

The side-bearing fit has the tooth thickness of the external almost equal the space width of the internal. Addendum of the external is short enough so it will not hit the root fillet of the internal. No chamfer or radius is used on the tip of the external.

The minor-dia fit spline is a fit at the root of the external. Here the dedendum of the external equals the addendum of the internal.

DESIGN TOLERANCES

Of the three, the major-dia fit is easiest to control. The outside dia of the external can be precision-ground to a tolerance of 0.0005 in. or less. On internal parts smaller than 2 in. dia, the root dia can be broached to a tolerance of 0.001 in. or less. This makes it possible to hold a tolerance range as low as 0.0015 in. For a press fit, the interference might be selected to range between 0.0005 and 0.002 in.

If the fit is made on tooth thickness, the teeth can be held to 0.0015 in. or less. However, further tooth tolerance (total index error, profile and alignment) have a composite effect which, even on small splines, can be as much as 0.005 in.

If a close fit is needed for centering purposes, the major-dia fit gives better results than the side-bearing fit. However, the latter controls backlash and is desirable in slow-speed applications where close centering is not required.

The minor-dia fit is seldom used. Its only advantage is that root dia of the external spline is formed by the same operations as the sides of the teeth. This tends

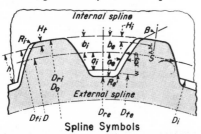

Spline Symbols

to guarantee that the root dia will be concentric with tooth sides.

Spline-tooth proportions for 1 DP are shown in Table I; Table II illustrates the typical computations for a 16-tooth spline with a press fit on the major dia.

Where a major-dia fit or a centering-ring fit is used, there is no need to be too close on clearance. In this case, clearance allowance similar to those used in gears are used, Table III. These values are appropriate for flexible coupling designs—provided an extreme amount of misalignment is not contemplated.

HOW MANY TEETH

Cost and manufacturing considerations determine the number of teeth because, contrary to gear design, doubling the number of spline teeth has no appreciable effect on tooth stress. The teeth become half as big when doubled, but there are twice as many to carry the load—the two effects tend to cancel each other.

Even-number teeth should be used wherever possible. An odd number puts a tooth opposite a space, which makes it impossible to directly measure root, bore and outside dia. Also, odd numbers of spline teeth are a nuisance to manufacture, and contribute nothing to load-carrying capacity. Recommended minimum numbers of spline teeth are shown in Table IV; employ a larger number to obtain a larger root dia on the external, to make an easier tool design, or to improve lubrication. Generally, however, the cost goes up.

INDUSTRY PRACTICES

In each industry, interchangeability of parts, customer acceptance, and industry standards establish almost mandatory practices for spline design.

In the aircraft field, an engine may have a dozen flexibly splined accessory drives, handling torque from 10 to 6,000 in.-lb. Fixed splines mount the propeller and couple the reduction gear and the engine. Generally, splines to drive accessories are specified by AND standards (see Table V). Pitch is listed as a fraction; a 20/40 pitch has a 20 diametral pitch with a stubbed-tooth height approximately equal to a 40-pitch standard full-depth gear tooth. This method does not give the exact addendum and dedendum, and is more of a hazard than a convenience.

Some designs in the table are stubbed to one-half full height while others to only two-thirds. The two-thirds height carries more load and resists wear better, but the one-half height agrees with the ASA standard.

Many designers are standardizing on Form 1 teeth; almost all avoid the keyed-on gear-tooth couplings used in commercial machinery.

Automotive splines in general have changed over from straight-sided to involute splines; transmission splines for gear shifting have the external member finished by shaving, and the internal member finished by hard broaching or lapping. For shaved splines, Form 3 with a 50% depth is favored.

Rolled splines need a high pressure angle, shallow depth and a radius root fillet. Forms 1 and 2 are preferred with a full dedendum on the external.

Automotive splines for shift gears are designed for a clearance C from about 0.0015 to 0.0045 in. Close-fit non-shifting splines with a side fit for centering are designed for 0 to 0.0030 in. clearance. Interference for press fit is about −0.0000 to −0.0020 in. For finer pitches, 10 to 24 P, clearance is reduced by 0.001 in.

In industrial machines, where cost is usually more important than weight and space, standard motors and speed reducers are used. A complete drive can be assembled on a common base with two standard spline couplings. Teeth are not as hard or as accurate as used in aircraft; Form 3 is the most popular. Catalog spline couplings are standardized by manufacturers and produced in quantity. They are not balanced for high speeds unless made to order.

TABLE III—SUGGESTED SPLINE CLEARANCES

Diametral Pitch	Clearance C, in.
1	0.024–0.036
2	0.020–0.030
3	0.016–0.020
4	0.012–0.026
6	0.010–0.018
10	0.008–0.014
14	0.006–0.012
20	0.004–0.008

TABLE IV—RECOMMENDED MINIMUM NUMBERS OF SPLINE TEETH

Pitch Dia	Broaching			Shaping				Shaving or Grinding	
Angle	30°	20° or 14½°	14½°	30°	25°	20°	14½°	25°	20°
Depth	50%	50%	30%	50%	70%	75%	30%	70%	75%
0.5	6	10	10	12	18	..	..	..	..
0.8	8	12	10	14	20	22	16	..	..
1.0	8	12	10	16	20	24	16	..	..
2.0	8	12	12	20	20	24	24	..	..
4.0	10	16	16	24	24	32	24	48	48
8.0	20	20	24	32	32	40	32	56	56
12.0	30	30	36	36	36	48	36	60	60

TABLE V—AIRCRAFT ACCESSORY SPLINE STANDARDS

Military Spec.	Type	Pitch Dia	No. Teeth	Pitch	Min Face Width*	Torque, lb-in.* Max	Continuous
AND 20001	INT.	0.600	12	20/40	0.937	2,200	500
AND 10260	EXT.	0.600	12	20/40	0.812		
AND 20002	INT.	0.800	16	20/30	1.562	2,700	600
AND 10262	EXT.	0.800	16	20/30	0.600		
AND 20002	INT.	1.200	24	20/30	1.438	11,000	2,500
AND 10262	EXT.	1.200	24	20/30	1.000	6,000	
AND 20007	INT.	1.625	26	16/32	1.063	18,500	4,200
AND 10266	EXT.	1.625	26	16/32	1.180		

* Typical values.

Splines may fail in one of these five ways:

(A) shaft of externally toothed member breaks underneath spline teeth

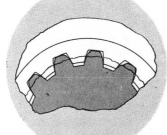

(B) teeth of spline shear off on pitch line

(C) teeth break at roots in a cantilever type failure similar to that of gear teeth

(D) contacting surface of spline teeth wear away by fretting corrosion

(E) shell of the internally toothed member ruptures

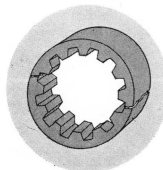

WHEN SPLINES NEED STRESS CONTROL

DARLE W. DUDLEY

Manager, Advance Gear Engineering
General Electric Co., Lynn, Mass.

Correct diagnosis of spline-failure problems is a vital first step toward correct design. This article supplies charts and formulas for computing four important types of stresses that must not be exceeded.

SPLINE SYMBOLS

A	height of crown
B	misalignment of spline, in./in.
D	pitch dia, in.
D_h	bore dia of shaft, in.
D_{oi}	outside dia of internally toothed part, in.
D_{ri}	major dia of internally toothed part, in.
D_{re}	root dia, also minor dia of external toothed member, in.
F	full face width, in.
F_e	effective face width, in. (may be less than F; see Editor's Note)
h	radial height of the tooth in contact, in.
K_a	application factor
K_m	load distribution factor
L_f	life factor limited by fatigue
L_w	life factor limited by wear
n	rpm
N	number of spline teeth
S_c	compressive stress, psi
S'_c	allowable compressive stress, psi
S_s	shear stress, psi
S'_s	allowable shear stress, psi
S_i	total stress, psi
t_c	chordal thickness at pitch line, in. (approximately equal to $D/2N$)
t_w	wall thickness, in.
T	torque, in.-lb = 63,000 hp/n
ϕ	pressure angle, deg

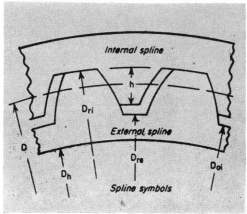

Spline symbols

Spline design follows a simple procedure. Calculate the various stresses, then adjust size of the spline so that all stresses come within design limits. These four types of stresses are the major ones:

1. Shear stresses in spline shaft
2. Shear stresses in spline teeth
3. Compressive stresses in spline teeth
4. Bursting stresses in internal spline parts

SHAFT STRESSES

The externally toothed spline is cut on either solid or hollow shaft. Shear stresses induced in a solid shaft when transmitting a torque are:

$$S_s = \frac{16T}{\pi D_{re}^3} \qquad (1)$$

For a hollow shaft:

$$S_s = \frac{16TD_{re}}{\pi(D_{re}^4 - D_h^4)} \qquad (2)$$

Whether or not the shaft can withstand these induced stresses depends on the shaft material. Table I gives the maximum allowable shear stress S_s' for various types of steel. Computed stresses S_s from Eq (1) and (2) must not exceed the allowable stress S_s' in Table I, modified by application factor K_a and fatigue life factor L_f:

$$S_s' \geq S_s \frac{K_a}{L_f} \qquad (3)$$

Application factors are listed in Table II. Note that relatively high service factors are used where both the driving and driven apparatus are relatively rough-running. Under normal conditions a generator represents a smooth drive; but a short circuit through a generator may develop a severe overload torque—and this possibility requires a high value of K_a in the shear-stress formula.

Fatigue-life factor L_f is based on the number of torque cycles. Generally, spline teeth get a cycle of load on and load off only when a machine is started up, and then later turned off. But some machinery—for example, power shovels, bulldozers, planers, shapers and airplane rudder controls—subject a spline drive to more frequent cycles of loading. Life factors for different torque-cycle values are shown in Table III. Unity life factor is taken as 10,000 cycles. If the torque cycles are known to be frequent, but difficult to determine accurately, a life factor for 10 million cycles should be used.

The allowable stresses given in tables represent the author's "middle of the road" recommendation for design standards. At present there are no generally recognized standards for intensity of spline loading. Perhaps these published values may become the basis for effective standards.

SHEAR STRESSES IN TEETH

Assuming now that the spline shaft can transmit the desired torque, the next potential failure spot is in the teeth. There are two ways they can fail: They may shear; or they may wear excessively because of high compressive stresses. Both possibilities must be examined.

Teeth are assumed to shear at the pitch line. The

TABLE I—ALLOWABLE SHEAR STRESSES FOR SPLINES

Material	Hardness: Brinell	Hardness: Rockwell C	Max Allowable Shear Stress, S_s', psi
Steel	160–200	.....	20,000
Steel	230–260	.....	30,000
Steel	302–351	33–38	40,000
Surface-hardened steel	....	48–53	40,000
Case-hardened steel	....	58–63	50,000
Thru-hardened steel (aircraft quality)	....	42–46	45,000

TABLE II—SPLINE APPLICATION FACTORS, K_a

Power Source	Type of Load			
	Uniform (generator, fan)	Light Shock (oscillating pumps, etc.)	Intermittent Shock (actuating pumps, etc.)	Heavy Shock (punches, shears, etc.)
Uniform (turbine, motor)	1.0	1.2	1.5	1.8
Light shock (hydraulic motor)	1.2	1.3	1.8	2.1
Medium shock (internal combustion engine)	2.0	2.2	2.4	2.8

TABLE III—FATIGUE-LIFE FACTOR FOR SPLINES

No. of Torque, Cycles	Life Factor, L_f: Unidirectional	Life Factor, L_f: Fully-reversed
1,000	1.8	1.8
10,000	1.0	1.0
100,000	0.5	0.4
1,000,000	0.4	0.3
10,000,000	0.3	0.2

TABLE IV—LOAD DISTRIBUTION FACTOR FOR SPLINES

Misalignment	Factor K_m			
	½-in. Face Width	1-in. Face Width	2-in. Face Width	4-in. Face Width
0.001 in./in.	1	1	1	1½
0.002 in./in.	1	1	1½	2
0.004 in./in.	1	1½	2	2½
0.008 in./in.	1½	2	2½	3

induced shear stresses are:

$$S_s = \frac{4TK_m}{DNF_e t_o} \qquad (4)$$

In a spline, contrasted with a gear, tooth failure cannot stop the drive until all teeth are broken on both members. The constant 4 in the above equation assumes that, because of spacing errors, only half the teeth carry the load. With poor manufacturing accuracies, it is best to increase the factor to 6. Values for load distribution factor K_m, given in Table IV, are based on the amount of misalignment; K_m is 1.0 for a fixed spline (misalignment in a fixed spline is zero).

After calculating tooth shear stress, Eq (3) can again be used—to relate the calculated stresses with the allowable stresses.

COMPRESSIVE STRESSES

Compressive stresses act on the sides of the spline teeth and can be calculated from:

$$S_c = \frac{2TK_m}{DNF_e h} \qquad (5)$$

Factor h is radial height in inches of the contacting tooth. If the teeth do not have a substantial tip radius or chamfer, then h is the sum of the addendum of the external tooth and the addendum of the internal tooth. If with chamfer or radius, subtract its average radial height from the sum of the addendums.

The constant 2 in Eq (5) assumes all teeth to be working, which becomes true after some initial wear.

Computed compressive stress should be compared with the allowable compressive stresses in Table V, by using these equations:

$$\text{Flexible splines,} \quad S'_c \geqq \frac{S_c K_a}{L_w} \qquad (6)$$

$$\text{Fixed splines,} \quad S'_c \geqq \frac{S_c K_a}{9L_f} \qquad (7)$$

The allowable compressive stress values are very much lower than those used for gear teeth because splines do not distribute the load uniformly between teeth. With typical misalignment, straight splines (a term used to differentiate from crowned splines) have only one end of the tooth loaded at a time. The full surface area of the spline is rather inefficiently used. Exceeding the allowable compressive-stress values results in fretting corrosion. The life factor L_w for Eq (6) is a wear-life factor, Table VI; the life factor L_f for Eq (7) is a fatigue-life factor, Table III. Wear life factors are based on revolutions of the spline—not cycles of torque on and off. Each time a flexible spline makes a revolution, there is a back-and-forth rubbing of the teeth which causes wear. In Eq (7), factor 9 indicates that a fixed spline has about nine times more ability to carry compressive stress than a flexible spline.

BURSTING STRESSES

Internally toothed spline parts tend to burst because of three different kinds of tensile stresses: (1) bursting stress caused by the radial force component at the pitch line, (2) bursting stress caused by centrifugal force, (3) tensile stress caused by the tangential force com-

TABLE V—ALLOWABLE COMPRESSIVE STRESS FOR SPLINES

Material	Hardness Brinell	Hardness Rockwell C	Max Allowable Compressive stress, S'_c, psi Straight	Max Allowable Compressive stress, S'_c, psi Crowned
Steel	160–200	—	1,500	6,000
Steel	230–260	—	2,000	8,000
Steel	302–351	33–38	3,000	12,000
Surface-hardened steel	—	48–53	4,000	16,000
Case-hardened steel	—	58–63	5,000	20,000

TABLE VI—WEAR LIFE OF FLEXIBLE SPLINES

No. of Revolutions	Life Factor, L_w
10,000	4.0
100,000	2.8
1 million	2.0
10 million	1.4
100 million	1.0
1 billion	0.7
10 billion	0.5

TABLE VII—ALLOWABLE TENSILE STRESS FOR SPLINES

Material	Hardness Brinell	Hardness Rockwell C	Max Allowable Stress S'_t, psi
Steel	160–200	—	22,000
Steel	230–260	—	32,000
Steel	302–351	33–38	45,000
Surface-hardened steel	—	48–53	45,000
Case-hardened steel	—	58–63	55,000
Thru-hardened steel	—	42–46	50,000

TABLE VIII—COMPARISON OF SPLINE MANUFACTURING METHODS

Method	Used on External	Used on Internal	General Comments
Broaching	No	Yes	Low cost for quantity production; high accuracy
Shaping	Yes	Yes	Low cost for small lot production
Hobbing	Yes	No	Low cost for small lot production
Shaving	Yes	Yes	For high accuracy, high speed couplings
Grinding	Yes	Yes	For highest capacity couplings
Casting	Yes	Yes	For small-dia nonmetal splines
Cold-forming	Yes	No	High-production, low-cost process for automotive parts

ponent at the pitch line. The bursting stress caused by radial forces is:

$$S_1 = \frac{T \tan \phi}{\pi D t_w F} \qquad (8)$$

The wall thickness of the internally toothed member is obtained by subtracting the major dia from the OD of the coupling sleeve, and dividing the result by two.

Bursting stress caused by centrifugal force may be estimated by the formula for a simple cylinder which assumes that the face width is less than one-third the pitch dia.

$$S_2 = 0.828(10^{-6})(n^2)(2D_{oi}^2 + 0.424D_{ri}^2)$$

The tensile stress due to beam loading of the teeth is

$$S_3 = \frac{4T}{D^2 F_e Y} \qquad (10)$$

The Y-factor is the same as in the Lewis equation for gear design. Internal teeth of splines have a rather high Y-factor compared with gear teeth; usually equal to 1.5 or more. In most spline designs, it is not worth the trouble to make layouts to get an exact value for the Y-factor. Factor 4 in Eq (10) compensates for the load being carried on half the teeth.

All of the above stresses are tensile stresses at the root dia of the internally toothed part. The total stress tending to burst the rim is

$$S_t = K_a K_m (S_1 + S_3) + S_2 \qquad (11)$$

This total can be compared with allowable tensile stress values in Table VII using the relation,

$$S'_t \geqq S_t / L_f \qquad (12)$$

CROWNED TEETH

Crowned teeth on splines can handle a high amount of misalignment. Radii in the illustration can be obtained from these equations:

$$r_1 = 0.90(D/2)(\tan \phi) \qquad (13)$$
$$r_2 = r_1 / \tan \phi \qquad (14)$$

If the spline is fully crowned, the contact will stay fairly close to the tooth center even when the spline rocks through an appreciable angle. Fully crowned splines are successful with even as much as 3° misalignment.

If only slight misalignment is expected, the curvature of the teeth may be reduced, providing that dimension A is made greater than the quantity BF/2. This means that a spline of 1 in. face width misaligned 0.002 in. per in. should have a height of crown equal to at least 0.001 in.

When the crown is specified by the height of crown method, the approximate radius, r_2, may be calculated by

$$r_2 = F^2 / 8A \qquad (15)$$

When a flexible coupling has sufficient crown, the width of the band of contact remains constant; Eq (5) is no longer valid for calculating the surface compressive stress. Instead a Hertz-stress formula must be used.

$$S_c = 2290 \sqrt{\frac{2T}{DNhr_2}} \qquad (16)$$

The computed value of S_c obtained above should be compared with the allowable values of S'_c for crowned splines, using Table V and Eq (6). Crowned splines

never exceed the capacity of precisely aligned straight splines; the crown is only valuable in helping a spline carry a high amount of misalignment.

HOW DO SPLINES FAIL?

Splines transmit more torque for their size than any other type of coupler or joint. Frequently they are operating with a relatively small-dia shaft. Thus, the most common type of fixed spline failure is that of the shaft shearing.

Next most common is failure of tooth shear. Generally, the "fixed" spline (which mounts a gear, pulley wheel, turbine wheel, etc., in contrast with the "flexible" spline used as a coupling between shafts) has no relative motion between internal and external teeth to cause wear.

Fixed splines frequently carry high compressive (or "bearing") stresses and yet show no wear. When there is wear, it usually results from abnormal vibrations of the apparatus, or severe misalignment.

Fixed-spline teeth seldom fail in bending fatigue as in gear teeth because spline teeth are usually shorter and stubbier. Also a gear tooth is subjected to fatigue loading in which it receives a full on-and-off load each revolution—unlike fixed-spline teeth which do not have much change in loading per revolution.

Flexible splines are usually employed as couplings and do not carry as much torque per dia. There is little danger of the teeth shearing off. But there is some relative motion between teeth which makes them vulnerable to wear.

In larger spline couplings, the internally toothed member may burst because of applied torque. In addition, high speeds create severe centrifugal forces which add to bursting stresses. Even in small couplings, a thin wall on the internal member, a high tooth pressure angle, or a high torque loading for the size of the coupling can burst the internal member.

Crowned-tooth splines . . .

permit high misalignments—are often needed in couplings. Design procedure is same as straight splines except when determining radii define amount of crown.

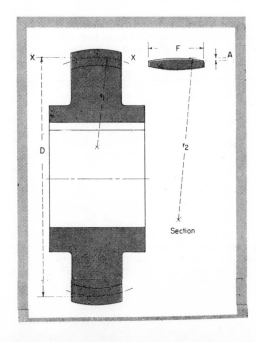

Section

Racks for Enclosed Spindles and Sleeves

FRED ROGERS

Racks and pinions are frequently used for imparting a straight forward and reverse motion to machine spindles and sleeves, many of which are enclosed in the machine housing. The accompanying examples show methods of cutting rack teeth on such spindles and sleeves; ways for aligning, assembling, and fastening component parts; and some arrangements for taking end thrust on the rack.

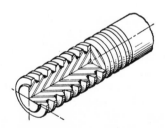

Fig. 1—Teeth cut directly across surface of spindle or bar. Engaging pinion meshes completely only at areas where the teeth of the rack are cut full depth. Used where the load requirements are relatively light.

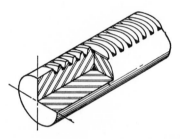

Fig. 2—Bar or spindle that is both rotated and moved axially. Rack teeth are cut circularly around bar with no lead. Design is limited to light axial thrust loads since area of rack tooth contact with pinion is small.

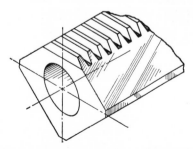

Fig. 3—Rack cut on square sleeve or ram. Cutting the teeth across one corner, rather than across one of the flat sides, requires flattening the corner to pinion width. Rack is covered in housing by a cap.

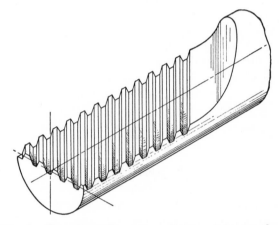

Fig. 4—To reduce the center distance between the pinion and spindle, the spindle is flatted as required. Cutting the rack at an angle and using a helical pinion increases the number of teeth in mesh thus permitting heavier loading. Flat is cut with small mill.

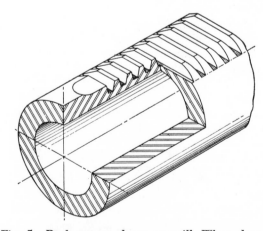

Fig. 5—Rack cut on sleeve or quill. When sleeve is flatted to obtain wide tooth bearing surfaces, sufficient wall thickness should remain after cutting teeth to prevent sleeve distorting under load.

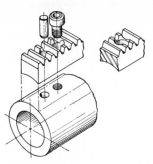

Fig. 6—Rack attached to thin wall sleeve that does not permit a flat or slot on its periphery. Rack end thrust is taken by close fitting dowel pin. Several fillister or socket head cap screws in drilled and counterbored holes in the rack fasten the rack to sleeve.

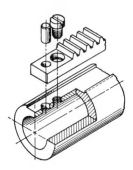

Fig. 7—Sleeve milled to receive rack. When teeth are cut as straight spurs, there is no side thrust, therefore, the sides of the milled groove do not need to be high. Screw and dowel are flush fits in rack end.

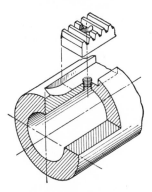

Fig. 8—Heavy wall thickness permits cutting a groove deep enough to keep bottoms of rack teeth flush with slot sides. Then headed screws in closely reamed body size holes make the use of dowel pins unnecessary.

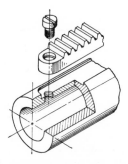

Fig. 9—Rack snugly fitted into an end milled slot also eliminates the need for dowel pins. Side of the slot can be shallow but enough metal should be left at end of slot to support end thrust against the rack.

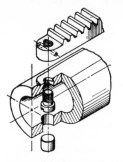

Fig. 10—To increase length of thread engagement, the screw holds the rack from the underside. To prevent vibration loosening the screw, a pin lock can be added to keep the screw from turning in rack.

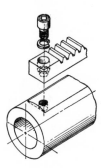

Fig. 11—In this design, the rack must be aligned and held on center. Body size hole is reamed in rack so that the upper body of the screw acts as a dowel.

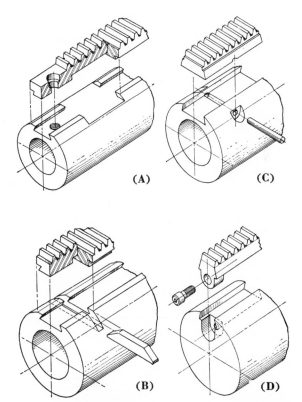

(A)   (C)

(B)   (D)

Fig. 12—Dovetail designs.—(A) Dovetails are needed only at the ends of the rack. Center portion is cut away to facilitate assembly. Retaining screw serves as a dowel.—(B) Milled key replaces retaining screws to carry thrust.—(C) Tapered pin takes end thrust of rack.—(D) Boss at one end of rack fits into a milled opening at end of spindle. This arrangement avoids drilling and counterboring through the teeth.

Sheet Metal Gears, Sprockets,

When a specified motion must be transmitted at intervals rather than continuously, and the loads are light, these mechanisms are ideal because of their low cost and adaptability to mass production. Although not generally considered precision parts, ratchets and gears can be stamped to tolerances of ±0.007 in. and if necessary, shaved to closer dimensions. Sketches indicate some variations used on toys, household appliances and automobile components.

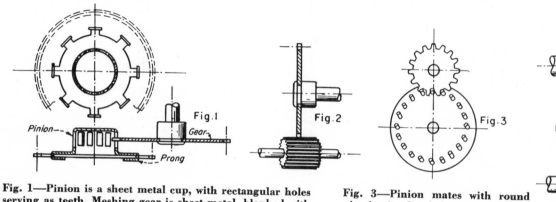

Fig. 1—Pinion is a sheet metal cup, with rectangular holes serving as teeth. Meshing gear is sheet metal, blanked with specially formed teeth. Pinion can be attached to another sheet metal wheel by prongs, as shown, to form a gear train.

Fig. 2—Sheet metal wheel gear meshes with a wide face pinion, which is either extruded or machined. Wheel is blanked with teeth of conventional form.

Fig. 3—Pinion mates with round pins in circular disk made of metal, plastic or wood. Pins can be attached by staking or threaded fasteners.

Fig. 4—Two blanked gears, conically formed after blanking, make bevel gears meshing on parallel axes. Both have specially formed teeth.

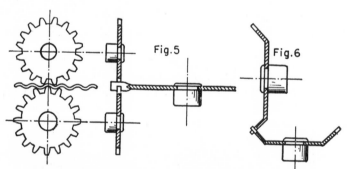

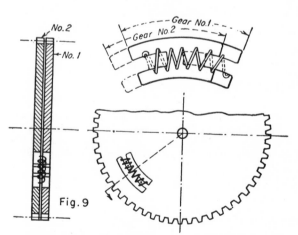

Fig. 5—Wheel with waves on its outer rim to replace teeth, meshes with either one or two (shown) sheet metal pinions, having specially formed teeth, and mounted on intersecting axes.

Fig. 6—Two bevel type gears, with specially formed teeth, mounted for 90 deg intersecting axes. Can be attached economically by staking to hubs.

Fig. 7—Blanked and formed bevel type gear meshes with solid machined or extruded pinion. Conventional form of teeth can be used on both gear and pinion.

Fig. 8—Blanked, cup-shaped wheel meshes with solid pinion for 90 deg intersecting axes.

Fig. 9—Backlash can be eliminated from stamped gears by stacking two identical gears and displacing them by one tooth. Spring then bears one projection on each gear taking up lost motion.

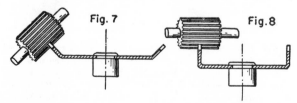

Worms, and Ratchets

HAIM MURRO
Bayonne, N. J.

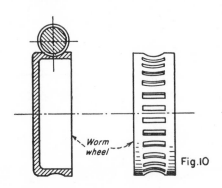

Fig. 10—Sheet metal cup which has indentations that take place of worm wheel teeth, meshes with a standard coarse thread screw.

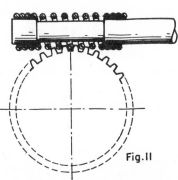

Fig. 11—Blanked wheel, with specially formed teeth, meshes with a helical spring mounted on a shaft, which serves as the worm.

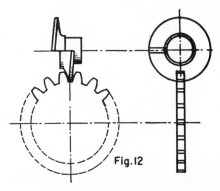

Fig. 12—Worm wheel is sheet metal blanked, with specially formed teeth. Worm is made of sheet metal disk, split and helically formed.

Fig. 13—Blanked ratchets with one sided teeth stacked to fit a wide, sheet metal finger when single thickness is not adequate. Ratchet gears can be spot welded.

Fig. 14—To avoid stacking, single ratchet is used with a U-shaped finger also made of sheet metal.

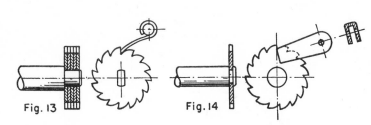

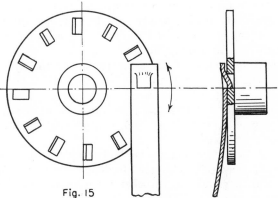

Fig. 15—Wheel is punched disk with square punched holes to serve as teeth. Pawl is spring steel.

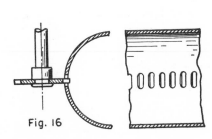

Fig. 16—Sheet metal blanked pinion, with specially formed teeth, meshes with windows blanked in a sheet metal cylinder, to form a pinion and rack assembly.

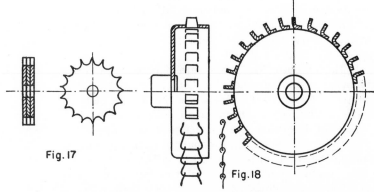

Fig. 17—Sprocket, like Fig. 3, can be fabricated from separate stampings. Fig. 18—For a wire chain as shown, sprocket is made by bending out punched teeth on a drawn cup.

DESIGN OF PRECISION SPROCKETS

ARTHUR HILL

Daco Instrument Co, Brooklyn, NY

From American Machinist

The sprocket will accommodate standard 70 mm perforated film, manufactured in accordance with Military Standard MS33525 (see Fig. 1). The sprocket has 56 teeth, a roll diameter of 3.3273 in. that will provide a mean diameter to the film (0.006 in. thick) of 3.3333 in.

In the design of this particular sprocket, the teeth are shaped to meet two different conditions: the film can leave the sprocket on a 1-in. dia roller, or the film can be pulled off in a tangential path. These two extremes call respectively for a tooth contour not exceeding an epicycloid or an involute using the roll diameter of the sprocket as a base circle, see Figs. 2a and 2b.

A tooth radius is chosen coincident with both curves at their base and lying just inside at the tip of the tooth. A radius of 0.280 in. with its center 0.010 in. below the roll diameter is found to be satisfactory. It is now necessary to compute data to design and make a templet that can be used on the Panto-Crush grinder. The templet is used for dressing a wheel to grind the space between the sprocket teeth as shown in Figs. 3 and 4. Note: the computations and process described here are not restricted to any specific kind of gear grinder. In fact, this information can also be used for any kind of a gear hobbing machine by simply making a hob to the same proportion and with the same profile as the grinding wheel shown in Fig. 7. It will be noticed that the tooth fillet radius R must originate below the roll diameter, and that the circumference of the wheel may be straight in section, i.e., parallel to the axis of the wheel.

A better production method was devised as a result of a study conducted to determine if it is practical to grind the sprocket teeth on the Reishauer gear grinder. This method requires dressing the grinding wheel to fabricate a tooth form that would satisfy the conditions as stated above. The tooth now has an involute form instead of only a radius.

The first step in this preferred production method is to establish the correct pressure angle for the involute tooth form. As indicated above, the tooth profile must be generated in such a manner that there will not be any interference with the epicycloid or involute take-off.

If the involute has a base circle below the roll diameter of the sprocket the tooth form will clear the perforations in the film while the film is being loaded or unloaded tangent to the roll diameter.

To help determine the correct pressure angle, it is necessary to establish how many degrees of rotation on the sprocket are needed to satisfy an epicycloid profile tooth. This information can be established as follows (See Fig. 5) and these compuations:

R_o—Outside radii of sprocket teeth, 1.7146 in.

r —Mean radii of the film rolled on 1-in. dia roller, 0.503 in.

C —Center distance between the sprocket and roller with film in between, 2.1697 in.

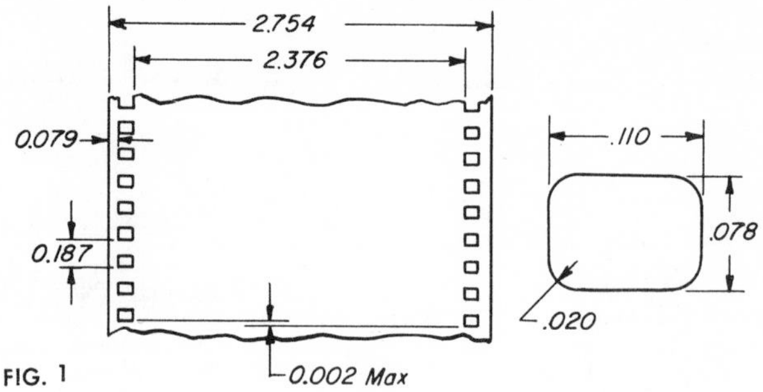

FIG. 1

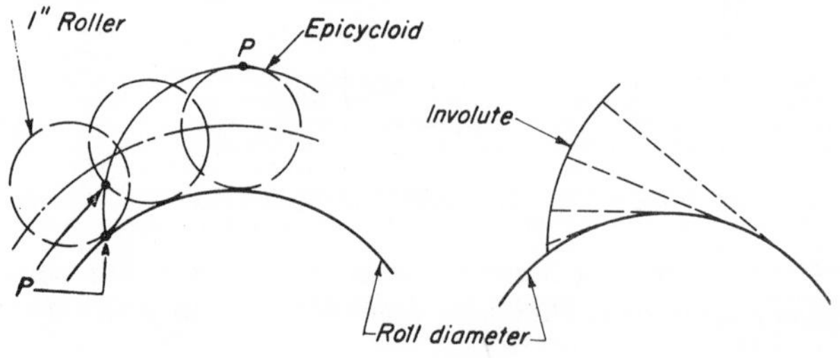

FIG. 2 (a) (b)

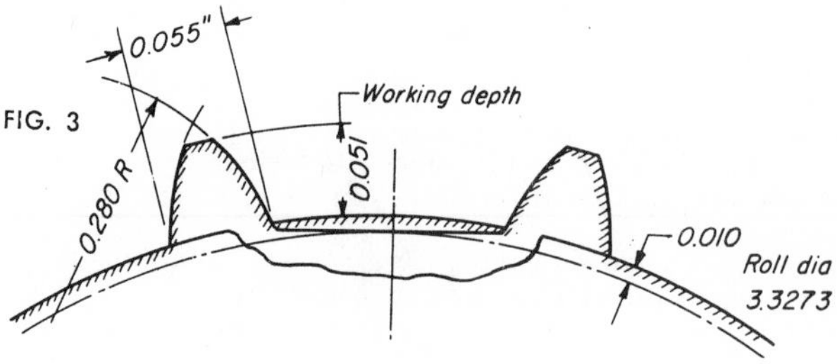

FIG. 3

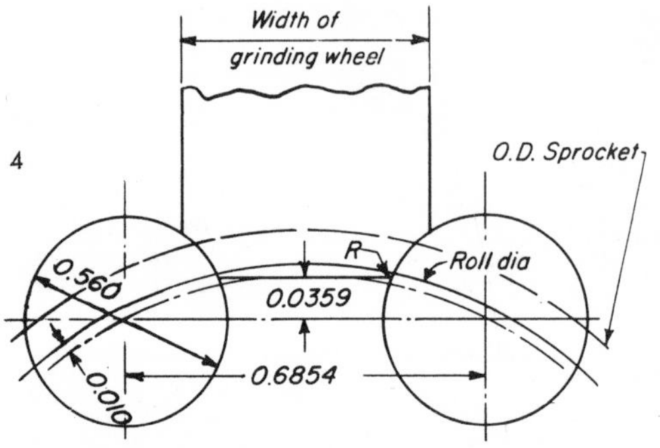

FIG. 4

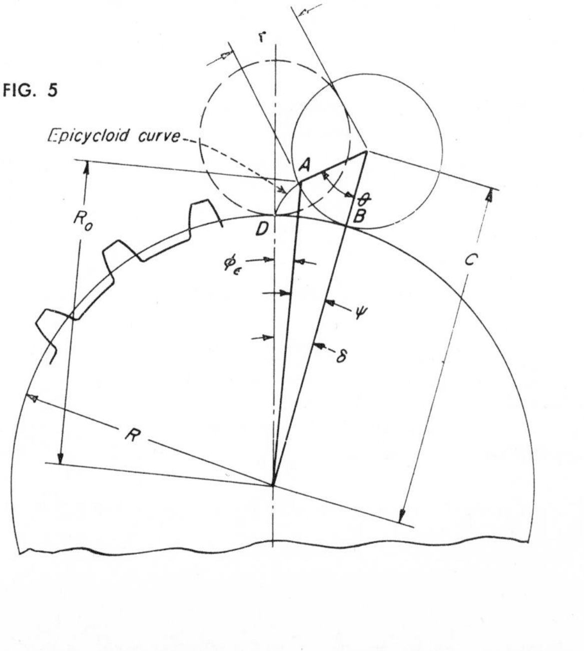

FIG. 5

R —Mean radii of film rolled on roll diameter of sprocket, 1.6667 in.

$$\cos \theta = \frac{r^2 + C^2 - R_o^2}{2\,r\,C}$$

$$\theta = \frac{0.503^2 + 2.1697^2 - 1.7146^2}{2 \times 0.503 \times 2.1697}$$

$$= 0.92567$$

$$\theta = 22.2302°$$

$$\cos \Psi = \frac{1.7146^2 + 2.1697^2 - 0.503^2}{2 \times 2.1697 \times 1.7146}$$

$$= 0.99382$$

$$\Psi = 6.37202°, \text{ or } 0.1112 \text{ radians}$$

Because the roller r rolls on the radius R and does not slip, they both roll off an equal amount of their circumference. Therefore, their arcs AB and BD are equal. Employing the theorem—radius multiplied by the included angle expressed in radians equals the length of arc in the included angle; then because the two arcs are equal to each other, their equations are also equal to each other.

$$\frac{\theta\,\pi}{180} \times r = \frac{\delta\,\pi}{180}\,R$$

$$\delta = 6.709°, \text{ or } 0.1171 \text{ radians}$$

$$\phi_E = \delta - \Psi = 0.0059 \text{ radians}$$

It is important to make certain that the pressure angle of the teeth at the outside diameter of the sprocket when generated is an involute greater than 14°47′, which would be the pressure angle of an involute tooth whose involute function is equal to ϕ_E. Because the pitch diameter of the gear in the following computations is just about equal to the outside diameter of the sprocket, a pressure angle of 15° 3½′ is selected because the base circle for this gear would then fall approximately 0.011 in. below the roll diameter. Furthermore, by providing a minimum radius of 0.004 in. on the wheel, the sprocket tooth will not be undercut.

This data is now converted into information similar to gear calculations in order to setup the Reishauer gear grinder, or any other gear-generating machine tool. From the Reishauer manual PZA 75 a gear train can be setup as follows:

$$\frac{12}{DP} = \frac{G_1}{G_2} \times \frac{G_3}{G_4} = \frac{40}{70} \times \frac{54}{42} = 16\,1/3$$

Computing the imaginary gear

No. of teeth	N	$= 56$
Diametral Pitch	P	$= 16\,1/3$
Pressure angle	ϕ_2	$= 15°\,3½′$
Pitch diameter	D_3	$= 3.4286$
Circular pitch	$C = \dfrac{\pi}{P}$	$= 0.1923$
Outside diameter	$OD = \dfrac{N+2}{P}$	$= 3.5510$

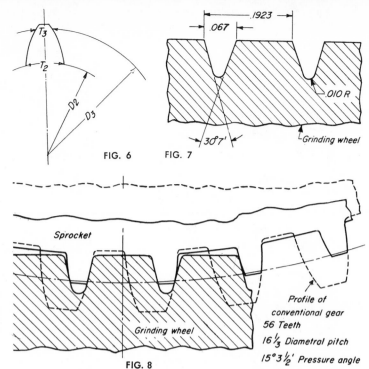

FIG. 6   FIG. 7

Sprocket

Grinding wheel

Profile of conventional gear
56 Teeth
16 1/3 Diametral pitch
15° 3 ½′ Pressure angle

FIG. 8

Referring to Fig. 3, the sprocket tooth shows a height of 0.051 in. and an undercut of 0.10 in. below roll diameter. Therefore, the wheel will penetrate 0.061 in. below the outside diameter of the sprocket. Also note that the tooth has a chordal thickness of 0.055 in. at the roll diameter. The arc tooth thickness is 0.055 in. at the point of contact with the mean thickness of the film. However, for the purpose of dimensioning the grinding wheel the arc tooth thickness must be determined at the pitch diameter of the imaginary gear.

T_2 = Arc tooth thickness of tooth at $D_2 = 0.055$
T_3 = Arc tooth thickness of tooth at D_3
ϕ_1 = Pressure angle at point where the mean diameter of the film makes contact with the tooth
D_2 = mean dia of film = 3.3333
D_3 = pitch dia = 3.4286

$$\text{Cos } \phi_2 = 15°\,3½′ = 0.9639$$

$$\text{Cos } \phi_1 = \frac{D_3 \text{ Cos } \phi_2}{D_2} = \frac{3.4286 \times 0.9639}{3.3333} = 0.99145$$

$$\phi_1 = 7°\,30′$$

$$\text{Inv } \phi_1 = 0.00075 \quad \text{Inv } \phi_2 = 0.00622$$

$$T_3 = D_3 \left[\frac{T_2}{D_2} \times \text{Inv } \phi_1 - \text{Inv } \phi_2 \right] = 0.0343$$

The root diameter of the sprocket is equal to the roll diameter minus 0.020 in. as indicated in Fig. 3, or 3.3073 in. This figure is 0.1213 in. less than the pitch dia of the imaginary gear. Therefore, to determine the dimension for the width of the groove in the grinding wheel at the point of deepest penetration, Fig. 7, multiply 0.1213 in. by the tangent of ϕ_2 and add this value to T_3.
ie: $0.26904 \times 0.1213 + 0.0343 = 0.067$ in.

From this information the grinding wheel can now be dimensioned.

How to Specify Precision Knurls

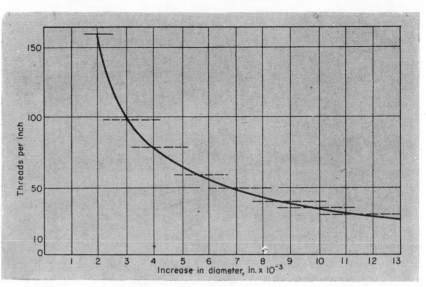

Fig. 1—Data for steel, brass and aluminum for selecting a tool pitch to give the desired increase in outside diameter or buildup.

GEORGE BARROW
Westinghouse Electric Corp.

It takes more than a single adjective or short note on a drawing to specify a straight knurl on which tolerances are critical. An accurate method of designation such as the dimensioning system proposed by the author herein is required.

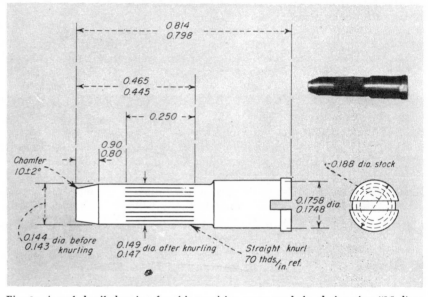

Fig. 3—Actual detail drawing for this precision parts used the designation "Medium Straight Knurl." Preferred system of specifying the information is shown.

FREQUENTLY, knurls are specified on drawings by the general classifications of fine, medium or coarse. Since knurled parts are seldom trade items and there is no standardization as such, this system of designation tends to put the responsibility for the end-product on the machine operator rather than on the designer. Furthermore, what is considered a medium knurl by one operator can be a fine, or even a coarse one, to an operator in a different shop or another machine operator in the same shop. This leads to non-uniformity of products, and does not provide control over critical dimensions.

Of course, precision is not always required. Some types of knurls, such as diagonal tooth or diamond shaped, are widely used to hold studs or inserts

in die castings and molded plastic products. Close tolerances are unnecessary in these cases since the material will always mate intimately with the surface. The same condition exists if the serrations are used for decorative or grip purposes.

But some products, particularly small machine elements that employ straight knurling to obtain a press fit assembly, require a rigid system of designation. The degree of this minimal fit must be sufficient to meet performance requirements and forms the basis for the maximum press fit that occurs with the maximum diameter of the shaft and the minimum diameter of the hole in the hub or bushing. The following specification procedure was developed primarily for applications of this type.

Specification Procedure

Establishing a designation system that is both accurate and practical from a production standpoint involves two factors. One is the relation between the diameter of stock and pitch of the teeth to prevent double knurling or splitting of serrations on repeated revolutions of the work against the tool. This requires an approximately even division of the circumference of the stock by the tooth pitch. Except in special cases where conditions warrant tool development it is unnecessary to compute the exact pitch size and supply a special tool for each diameter of stock encountered. A more practical solution is for the designer to concentrate on the necessary buildup, or increase in diameter, and allow the shop freedom in select-

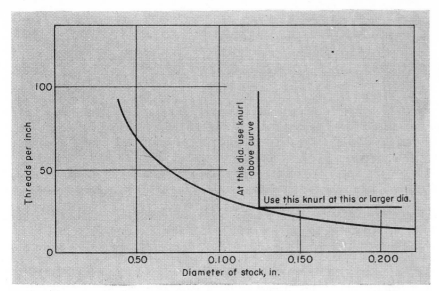

Fig. 2—Approximate range of tool pitches that can be used with given stock diameters. Combinations that fall beneath the curve are to be avoided.

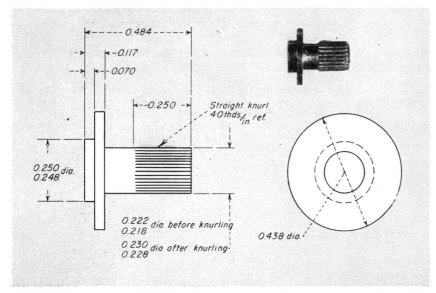

Fig. 4—This knurl was originally specified as "Standard Fine." Following the authors suggestions, the part is detailed above to eliminate such arbitrary classifications.

ing the final pitch. An approximate pitch can be suggested if modifications are allowed to prevent double-knurling. Thus, a change from 40 to 41 or from 47 to 50 threads per inch, as determined by the setup man, is usually better assurance of satisfactory results than the use of an untried calculated pitch for a given diameter.

The other factor, which was referred to previously, is the final diameter of the knurled area. For this, the classifications fine, medium and coarse, are entirely inadequate. From a practical basis, twelve is the minimum number of serrations that can be made on a given diameter of stock and yet retain good workmanship. This gives the maximum increase in diameter and can be designated the coarsest knurl. Reducing the pitch reduces the buildup

and increases the number of serrations for a given diameter stock until at some point the result can be termed medium or fine. However, the buildup in diameter over the final surface is the important factor and the resulting number of serrations is secondary.

Data on the buildup of steel, brass and aluminum stock is given in Fig 1 to guide in the selection of a knurl tool. For a given pitch, the buildup increases slightly with increase in diameter of stock. The range is indicated by the short lines through the curve and parallel to the abscissa. The left end of the line represents the minimum of twelve serrations and indicates the buildup on a diameter of stock derived by dividing a constant 3.82 by the threads per inch. The other end of the line indicates the

maximum buildup for the same knurl on a diameter of stock approximately 6.4 divided by threads per inch. Further increases in the diameter of the stock will not affect the amount of buildup.

Knowing the diameter of stock to be worked and the buildup required, in thousandths, an approximate tool pitch can be determined from the curve. The derivation of the constant 3.82 is as follows:

$$\frac{C}{N} = p$$

$$P = \frac{1}{p} = \frac{N}{C} = \frac{N}{\pi D}$$

When $N = 12$
$PD = 3.82$

N = number of serrations
D = diameter of stock before knurling
C = circumference of stock before knurling
P = $1/p$, threads per inch
p = pitch

The use of constant 3.82 permits further enlargement of data as demonstrated by Fig. 2. This curve shows the approximate minimum limit for a given diameter of stock or a given pitch of tool to produce the twelve-serration type of area. For example, with a tool having 30 threads per inch the minimum diameter of stock on which it can be used is 0.127 in., but it can be used on any larger diameter stock, in which case it will make more than twelve serrations. Likewise it is seen that, at 0.127-in. diameter, 30 is the largest pitch that can be used, but tools with a smaller pitch can be used and if they are, more than twelve serrations will be made. Therefore, the area under the curve represents combinations of threads per inch and stock diameter that should be avoided.

A suggested designation for specifying this information on engineering drawings is shown in Figs. 3 and 4. The nominal increases in the diamaters of the knurled areas for these two precision parts are 0.0045 and 0.009 in., respectively. From Fig. 1, the suggested pitches are 70 and 40 threads per inch, respectively. These sizes are shown on the drawing for reference only. Alternatively, these values could have been specified by the designer with an appropriate tolerance. The diameter of the knurled area, which in the final analysis is the important detail, is specified definitely in both the before and after conditions. Sufficient information and latitude are given to permit manufacture and inspection of the critical areas without further questions or possibility of misinterpreting the designer's needs and specifications.

Design of All-Steel Friction Drives

J. KAPLAN and D. BEAUMONT

The usual rubber rollers are absent in this novel friction drive. Instead, there are spring-loaded idlers that permit making all parts from hardened steel. Like other friction drives, this arrangement transmits torque much smoother than a gear drive can. But without rubber, which tends to creep, it provides a well-defined transmission ratio.

The floating idlers are loaded against the shafts by a spring-and-lever arrangement and furnish proper pressure between the rolling elements. The idlers also obviate the need for a high degree of accuracy when locating the shaft centerlines.

Like other friction drives, this one is equivalent to a gear drive without teeth. But here, power from the input shaft rotates an idler (or pair of idlers) which, in turn, rotates an output disk. This disk is pinned to the output shaft or made an integral part of it. Speed ratio between the shafts is inversely proportional to ratio between diameters of output disk and input shaft.

As with gears, the output can drive a secondary system for still higher speed-reduction. This is done in a film-magazine drive for Fairchild Camera and Instrument Company's high-resolution camera.

Whether to use one or two idlers between input and output shaft depends on the application. Paired idlers will obviously double the torque transmission capacity for the same maximum contact stress of a single idler. Also, a judicious location of the axes of the two idlers can eliminate the reaction due to spring loading on the bearings—of particular value when transmitted motion is from shafts mounted in bearings not designed to withstand large radial loads.

The new friction drive can produce an output torque of several in.-lb while retaining these other advantages:
- Motion is smooth and free from jitter.
- There is a complete absence of backlash.
- Transmission efficiency is high.
- The drive can be designed to give large speed reductions in a single pass.

If the drive must operate under conditions of high contact stresses between its elements, careful consideration should be given to material, finish, alignment and concentricities. The rolling parts should be made of hardened steel; surfaces which are in rolling contact should have a good finish; input, output and idler shafts should be parallel and aligned axially. Where a pair of idlers is employed, eccentricities of the rolling surfaces should be held to minimum—such variations will result in slightly different surface speeds of the disk. Since the output disk is rigid, either one, or both, of the idlers must slip and this in time will produce wear of the rolling surfaces.

DESIGN EQUATIONS

For simplification, centers of input and output shafts are laid out in line with the pivot point of the crank arm. Distance between axes of the input and output shafts equals the sum of the radii of the driving shaft and the driven disk, and the clearance between them. For the double-idler arrangement, axes of the idlers and axis of the input shaft are made coplanar, so that the resulting reaction on the input-shaft bearings is zero. Reactions on the crank-arm pivot can also be eliminated by making its axis coplanar with the idler-link pivot axes.

Design procedure is similar to that followed with gears. Generally, one should know:
(1) Maximum output torque that must be produced.
(2) Speed ratio between input and output.
(3) Over-all size of the unit. For example, should the disk diameters for a ratio of 10 be about 10 in. and 1 in., or 2.5 in. and 0.25 in.
(4) Material to be employed, and the coefficient of friction between surfaces. If the coefficient ranges between two extremes, smallest value should be used to be on the safe side.

Knowing these values, the following equations will determine the link dimensions, required spring loading, and resulting surface stresses.

Center distance d (see diagram next page) between input and output shafts is:

$$d = \sqrt{(r_o + r_d)^2 - (r_i + r_d)^2} \qquad (1)$$

The output torque that could be transmitted through one idler is:

$$T_o = \mu N r_o \qquad (2)$$

The maximum permissible output torque is determined by the smaller μN product of the two reaction points of the idler (using the smallest predicted value for μ). The minimum resultant force R, necessary to produce a given output torque capacity, is obtained when μN is equal at both reaction points. Because the coefficients of friction at both contact points can be considered the same, the reactions must also be equal to produce the optimum conditions described in the previous statement. Direction of the external force which is to produce equal reactions, N, is along the angle bisector R of angle θ. Inasmuch as the idler link transmits the external load, direction of the idler-link axis must coincide with direction of the external load.

The small force diagram in the sketch on next page shows that load on the bearings of this idler is

$$\text{Cos}\,\frac{\theta}{2} = \frac{1}{2}\,R/N$$

or

$$R = 2N \cos \frac{\theta}{2} \qquad (3)$$

By taking moments around point L, the spring force acting on the crank arm, which is necessary to produce force R is:

$$F = \frac{Rc \sin \alpha}{a} \qquad (4)$$

The linkage could be designed so that line JL is not perpendicular to line KL; normally, however, it is designed so that angle $\alpha = \theta/2$ (line JL is perpendicular to line KL). Combining Eq (2), (3) and (4) gives

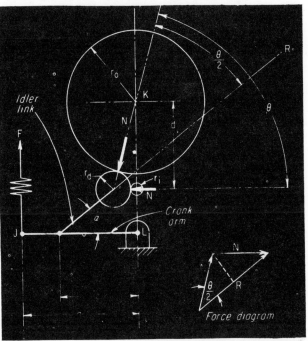

Force diagram

Symbols

a = moment arm of spring about crank arm pivot, in.

α = angle between crank arm and link, deg

b = face width of disks, in.

c = crank arm, in.

E = modulus of elasticity, psi

F = spring force, lb

μ = coefficient of friction

N = normal force, lb

R = resultant force on bearing of idler, lb

r_o = output disk radius, in.

r_i = input shaft radius, in.

r_d = idler roller radius, in.

S = contact stress, psi

T_o = output torque, lb-in.

$$F = \frac{cT_o}{\mu a r_o} \sin \theta \qquad (5)$$

Based on Hertz equations, maximum stress in the contact surfaces of the shaft and disk is:

$$S = \sqrt{\frac{0.35N(1/r_1 + 1/r_2)}{b(1/E_1 + 1/E_2)}} \qquad (6)$$

For materials of the same modulus of elasticity, Eq (6) reduces to:

$$S = \sqrt{\frac{0.175NE(1/r_1 + 1/r_2)}{b}} \qquad (7)$$

Maximum stress occurs at the contact points between idler and input shaft.

PROBLEM—TO DESIGN A FILM DRIVE

Given data are:

• Minimum output torque (2-idler system) = 4.5 in.-lb

• Coefficient of friction = 0.1

• $E = 30 \times 10^6$

• $r_o = 1.5$, $r_i = 0.125$, $r_d = 0.375$ (this gives a transmission ratio of 1.5/0.125 or 12:1)

• Thickness of disks = 0.25 in.

Computations:

From Eq (2), $\quad N = \dfrac{4.5}{2(0.1)(1.5)} = 15$

From Eq (7), $\quad S = 58,000$ psi

This stress is satisfactory. To find θ:

$$\cos \theta = \frac{r_d + r_i}{r_o + r_d} = \frac{0.375 + 0.125}{1.5 + 0.375} = 0.266$$

$$\theta = 74.5°$$

From Eq (5), keeping in mind that for one idler, $T_o = 4.5/2$, the spring should be loaded to a force of

$$F = 14.5 \text{ lb}$$

Reaction load on idler link, from Eq (3):

$$R = 24 \text{ lb}$$

This value is employed when selecting bearings for the idler link.

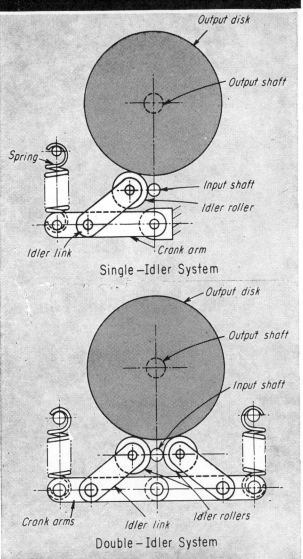

Single—Idler System

Double—Idler System

Clusters of rollers boost ratios in friction drives

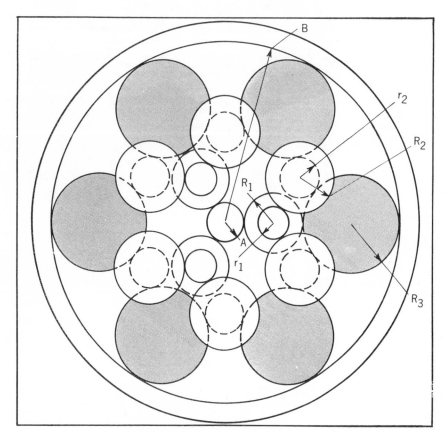

1. **Three rows** of planet rollers in a compact drive circulate around a centered sun roller to provide high speed-reducing or speed-increasing ratios with minimum noise and vibration. An undersized outer ring maintains contact pressure. The same principle as that shown in this drawing is applied in the mechanisms pictured on these pages and on page 331. The cluster arrangement of rollers was worked out by A. L. Nasvytis of TRW, Inc., to solve earlier problems of such drives.

2. **This 120:1, 3-hp drive** boosts speeds to 480,000 rpm.

3. **Multi-roller principle** is applied in a bearing.

Multi-roller planetary systems run more smoothly
and quietly than gears, and at higher speeds. Clusters work
well, too, in bearings, brakes, and geared drives

Planetary friction drives have been taken a step further in design usefulness by a cluster arrangement of rollers, developed by A. L. Nasvytis of TRW, Inc., Cleveland. Multiple rows of stepped rollers around a sun roller produce very high ratios of speed increase or speed reduction in a remarkably compact package sketched in Fig. 1 at left.

For example, the drive shown in Fig. 2 has the ultra-high speed-increasing ratio of 120:1. The same concept is being successfully applied to the development of three other mechanical components:

• A new type of roller bearing for high-speed or high-temperature applications (Fig. 3).

• A new design of shoe brake, with multi-roller clusters acting as toggles to distribute braking forces evenly (Fig. 7).

• A hybrid type of planetary gear drive with combination "roller-gear" elements in place of the usual gears (Fig. 4).

Overcoming problems. The idea of a planetary friction drive is at-

tractive. Such a drive can operate with much less noise and vibration than similar geared units, and it is cheaper to make. But it tends to slip under starting torque if pressure between rollers is insufficient.

In the few planetary friction drives that have been built, rollers are simply substituted for gears in a ring that is slightly undersized, to provide the necessary contact pressures. In one drive for a vertical grinder (*PE*—Oct. 12 '59, p 79), Pratt & Whitney improved the contact pressure by using an undersized outer ring that is expanded by heat during the assembly of the rollers.

Nasvytis, however, gets around the need for preheating by the use of a fixture he developed to spread the outer ring by cam action, with a second fixture to keep the rollers in position for easy assembly.

Another drawback of these drives also stemmed from the slippage problem. Sun rollers could not be made as small as equivalent sun gears, else they would slip excessively. Thus, the speed-change ratio

of a planetary drive of a given size was limited.

Nasvytis overcame this problem by replacing the single row of planetary gears by either two rows (Fig. 5) or three rows (Figs. 1, 2, and 6) of planetary rollers. The stepping boosts the speed-change ratio, and the multiplicity of contacts causes the rollers to remain ideally parallel with each other during operation— a feature not contained in previous planetary friction drives.

Three-row types. Nasvytis found two practical ways of arranging the three-row types: the N-N-N drive in which all three rows have the same number of planets (five in the drive shown in Fig. 6), and the N-2N-2N type in which the second and third rows contain twice as many planets as the first row (Figs. 1 and 2). The N-N-N type has more contact points and better stability, but the N-2N-2N type offers higher speed-change ratios.

A cage or "spider" usually encompasses one row of planets (in Fig. 2, for example, the row with the protruding shafts). Power can be applied or taken off either the planet cage, the sun shaft, or the outer ring, depending on the drive requirements (as with geared planetaries).

The speed ratio, N, of a three-row arrangement in which the cage is stationary can be computed from:

$$N = \frac{B}{A}\left[\frac{R_1\,R_2\,R_3}{r_1\,r_2\,r_3}\right]$$

where R and r are roller radii and A and B are sun and ring radii, as shown in Fig. 1.

High-power drives. Theoretically there are no limits to the number of rows in a drive, but Nasvytis finds it hard to see a need for more than four rows.

TRW has built a 3-6-6 drive with a 5-in.-dia. ring and a ¼-in.-dia. sun roller that produces a 120:1 speed-increasing ratio and transmits

4. **Power capacity** is increased by substituting gear teeth for part of rollers.

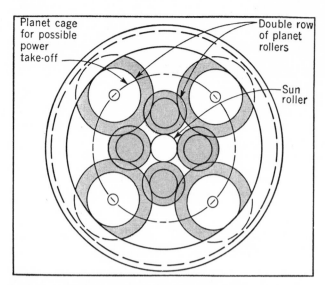

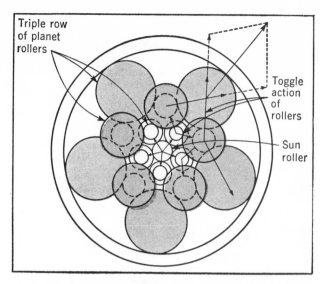

5. Two-row drives are lighter than three-row types, but they overload the sun roller and have lower efficiencies.

6. Three-row drive in this version has 5-5-5 planet design with better stability than the 3-6-6 type shown in Fig. 1.

3 hp at 480,000 rpm. This drive has run 48 hr. without lubrication, meeting a requirement of this application. Another 3-6-6 drive has transmitted 5 hp at 150,000 rpm with 86% efficiency. Many other high-speed drives have been built by TRW, and a 500,000-rpm gas-turbine drive is under consideration.

High-speed bearings. By stabilizing the multi-roller planet cluster with a special ring (Fig. 3), Nasvytis has been able to apply his design principle to roller bearings. He cites two benefits:

• Instead of orbiting at about 40% of the input rpm, the rollers orbit only 5% to 10% as fast. Stress from centrifugal forces is reduced 100 to 200 times, making it possible to build large roller bearings capable of very high speeds.

• Such bearings don't need the usual cages to separate the rollers, so they can be operated without lubrication even at high temperatures.

Power brakes, too. The multi-roller clusters can be employed as toggle devices, amplifying the radial forces from the center outwards, to actuate brake segments that bear on a rotary drum (Fig. 7). With this type of brake, there are no bending forces. No pins, shafts, or other rotary elements are needed.

Best of both. Geared drives can transmit higher torques than friction elements, so Nasvytis was intrigued by the possibility of combining the best features of both toothed and friction drives. The result was his

"roller-gear" element (Fig. 8), in which much of the rolling surface is replaced by gear teeth. The remaining rolling surface now functions as a bearing to keep the roller-gear in its correct position in the cluster.

This type of drive is significantly lighter than friction drives. The two prototypes built by TRW, including the 1100-hp, 38:1 drive in Fig. 4, have shown unusually high efficiencies of 98.5% to 99%.

For accelerated testing. The multi-roller concept is also a new tool for investigation of fatigue, surface durability, and friction characteristics of materials. Nasvytis found that his ultra-high-speed drives can perform roller fatigue testing 20 to 30

times as fast as existing models, with better control over preload.

It is easy to build a pure roller drive with six contacts on the sun roller, 300,000 psi or even more Hertz stress on each contact, and a 500,000-rpm rotation. The result is 50,000 cps on the sun roller at considerable loading.

Because of the high parallelity of the rollers, the multi-roller drive, Nasvytis says, also presents new possibilities for measuring accurate rolling friction coefficients.

The multi-roller drive can also achieve surface velocities of 2,000 fps. At present, the behavior of most materials at such high speeds and contact conditions is not known.

—*Nicholas P. Chironis*

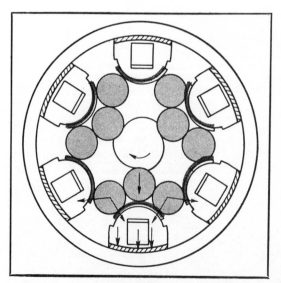

7. Brake has toggle-action rollers that amplify radial force from centered cam for smooth action.

8. Gear teeth in combination with roller surfaces make up hybrid design that increases torque-transmission ability of drives.

12
FASTENING AND SHIFTING TECHNIQUES FOR GEARS

15 ways to FASTEN GEARS TO SHAFTS

So you've designed or selected a good set of gears for your unit— now how do you fasten them to their shafts? Here's a roundup of methods—some old, some new— with a comparison table to help make the choice.

L M RICH, *project engineer, Librascope Inc, Glendale, Calif*

1 PINNING

Pinning of gears to shafts is still considered one of the most positive methods. Various types can be used: dowel, taper, grooved, roll pin or spiral pin. These pins cross through shaft (A) or are parallel (B). Latter method requires shoulder and retaining ring to prevent end play, but allows quick removal. Pin can be designed to shear when gear is overloaded.

Main drawbacks to pinning are: Pinning reduces the shaft cross-section; difficulty in reorienting the gear once it is pinned; problem of drilling the pin holes if gears are hardened.

Recommended practices are:
• For good concentricity keep a maximum clearance of 0.0002 to 0.0003 in. between bore and shaft.
• Use steel pins regardless of gear material. Hold gear in place on shaft by a setscrew during machining.
• Pin dia should never be larger than $\frac{1}{3}$ the shaft—recommended size is $0.20\,D$ to $0.25\,D$.
• Simplified formula for torque capacity T of a pinned gear is:

$$T = 0.787\,Sd^2D$$

where S is safe shear stress and d is pin mean diameter.

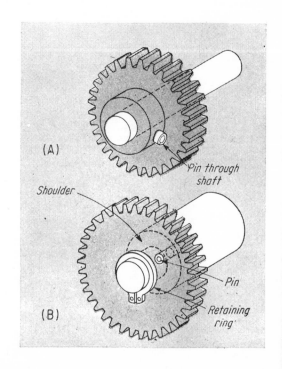

2 CLAMPS AND COLLETS

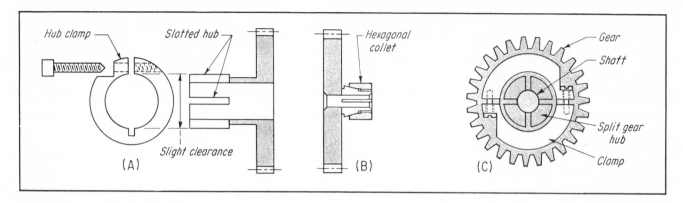

Clamping is popular with instrument-gear users because these gears can be purchased or manufactured with clamp-type hubs that are: machined integrally as part of the gear (A), or pressed into the gear bore. Gears are also available with a collet-hub assembly (B). Clamps can be obtained as a separate item.

Clamps of one-piece construction can break under excessive clamping pressure; hence the preference for the two-piece clamp (C). This places the stress onto the screw threads which hold the clamp together, avoiding possible fracture of the clamp itself. Hub of the gear should be slotted into three or four equal segments, with a thin wall section to reduce the size of the clamp. Hard-

ened gears can be suitably fastened with clamps, but hub of the gear should be slotted prior to hardening.

Other recommendations are: Make gear hub approximately same length as for a pinned gear; slot through to the gear face at approximately 90° spacing. While clamps can fasten a gear on a splined shaft, results are best if both shaft and bore are smooth. If both splined, clamp then keeps gear from moving laterally.

Material of clamp should be same as for the gear, especially in military equipment because of specifications on dissimilarity of metals. However, if weight is a factor, aluminum-alloy clamps are effective. Cost of the clamp and slitting the gear hub are relatively low.

3 PRESS FITS

Press-fit gears to shafts when shafts are too small for keyways and where torque transmission is relatively low. Method is inexpensive but impractical where adjustments or disassemblies are expected.

Torque capacity is:

$$T = 0.785 \, f D_1 \, LeE \left[1 - \left(\frac{D_1}{D_2} \right)^2 \right]$$

Resulting tensile stress in the gear bore is:

$$S = eE/D_1$$

where f = coefficient of friction (generally varies between 0.1 and 0.2 for small metal assemblies), D_1 is shaft dia, D_2 is OD of gear, L is gear width, e is press fit (difference in dimension between bore and shaft), and E is modulus of elasticity.

Similar metals (usually stainless steel when used in instruments) are recommended to avoid difficulties arising from changes in temperature. Press-fit pressures between steel hub and shaft are shown in chart at right (from Marks' Handbook). Curves are also applicable to hollow shafts, providing d is not over 0.25 D.

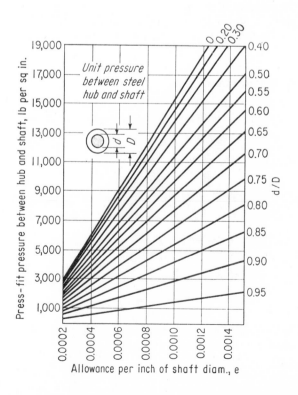

COMPARISON OF GEAR-FASTENING METHODS

Method	Torque Capacity	Ease of Replacing Gear	Reliability Under Operation	Versatility in Applications	Ability to Meet Environment Specs	Machining Requirements	Ability to Use Prehardened Parts	Relative Cost
1—Pinning	Excellent	Poor	Excellent	Excellent	Excellent	High	Poor	High
2—Clamping	Good	Excellent	Fair	Fair	Good	Moderate	Excellent	Medium
3—Press fits	Fair	Fair	Good	Fair	Good	Moderate	Excellent	Medium
4—Loctite	Good	Good	Good	Excellent	Excellent	Little	Excellent	Low
5—Setscrews	Fair	Excellent	Poor	Good	Fair	Moderate	Good	Low
6—Splining	Excellent	Excellent	Excellent	Fair	Excellent	High	Excellent	High
7—Integral shaft	Excellent	Poor	Excellent	Good	Excellent	High	Excellent	High
8—Knurling	Good	Poor	Good	Poor	Good	Moderate	Poor	Medium
9—Keying	Excellent	Excellent	Excellent	Poor	Excellent	High	Excellent	High
10—Staking	Poor	Fair	Poor	Poor	Good	Moderate	Poor	Low
11—Spring washer	Poor	Excellent	Good	Fair	Good	Moderate	Excellent	Medium
12—Tapered shaft	Excellent	Excellent	Excellent	Good	Excellent	High	Excellent	High
13—Tapered rings	Good	Excellent	Good	Excellent	Good	Moderate	Excellent	Medium
14—Tapered bushing	Excellent	Excellent	Excellent	Good	Good	Moderate	Excellent	High
15—Die-cast assembly	Good	Poor	Good	Excellent	Good	Little	Fair	Low

4 RETAINING COMPOUNDS

Several different compounds can fasten the gear onto the shaft—one in particular is "Loctite," manufactured by American Sealants Co. This material remains liquid as long as it is exposed to air, but hardens when confined between closely fitting metal parts, such as with close fits or bolts threaded into nuts. (Military spec MIL-S-40083 approves the use of retaining compounds).

Loctite sealant is supplied in several grades of shear strength. The grade, coupled with the contact area, determines the torque that can be transmitted. For example: with a gear $\frac{3}{8}$ in. long on a $\frac{5}{16}$-in.-dia shaft, the bonded area is 0.22 in.² Using Loctite A with a shear strength of 1000 psi, the retaining force is 20 in.-lb.

Loctite will wick into a space 0.0001 in. or less and fill a clearance up to 0.010 in. It requires about 6 hr to harden, 10 min. with activator or 2 min. if heat is applied. Sometimes a setscrew in the hub is needed to position the gear accurately and permanently until the sealant has been completely cured.

Gears can be easily removed from a shaft or adjusted on the shaft by forcibly breaking the bond and then reapplying the sealant after the new position is determined. It will hold any metal to any other metal. Cost is low in comparison to other methods because extra machining and tolerances can be eased.

5 Setscrews

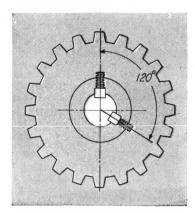

Two setscrews at 90° or 120° to each other are usually sufficient to hold a gear firmly to a shaft. More security results with a flat on the shaft, which prevents the shaft from being marred. Flats give added torque capacity and are helpful for frequent disassembly. Sealants applied on setscrews prevent loosening during vibration.

6 GEARS INTEGRAL WITH SHAFT

Fabricating a gear and shaft from the same material is sometimes economical with small gears where cost of machining shaft from OD of gear is not prohibitive. Method is also used when die-cast blanks are feasible or when space limitations are severe and there is no room for gear hubs. No limit to the amount of torque which can be resisted—usually gear teeth will shear before any other damage takes place.

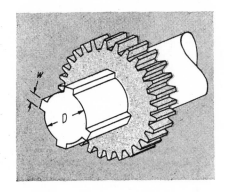

	4-spline	6-spline
D	w	w
1/2	0.120	0.125
3/4	0.181	0.188
7/8	0.211	0.219
1	0.241	0.250
1-1/4	0.301	0.313

7 SPLINED SHAFTS

Ideal where gear must slide in lateral direction during rotation. Square splines often used, but involute splines are self-centering and stronger. Non-sliding gears are pinned or held by threaded nut or retaining ring.

Torque strength is high and dependent on number of splines employed. Use these recommended dimensions for width of square tooth for 4-spline and 6-spline systems; although other spline systems are sometimes used. Stainless steel shafts and gears are recommended. Avoid dissimilar metals or aluminum. Relative cost is high.

8 KNURLING

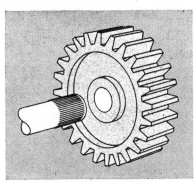

A knurled shaft can be pressed into the gear bore, to do its own broaching, thus keying itself into a close-fitting hole. This avoids need for supplementary locking devices such as lock rings and threaded nuts.

The method is applied to shafts 1/4 in. or under and does not weaken or distort parts by the machining of grooves or holes. It is inexpensive and requires no extra parts.

Knurling increases shaft dia by 0.002 to 0.005 in. It is recommended that a chip groove be cut at the trailing edge of the knurl. Tight tolerances on shaft and bore dia are not needed unless good concentricity is a requirement. The unit can be designed to slip under a specific load—hence acting as a safety device. See "Small Shaft Becomes Its Own Broach for Retaining Gears,"

9 KEYING

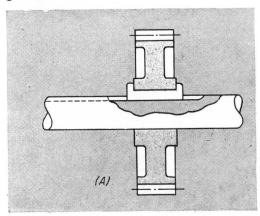

(A)

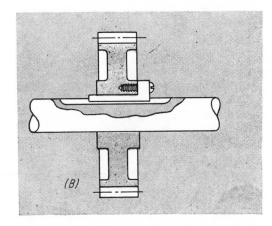

(B)

Generally employed with large gears, but occasionally considered for small gears in instruments. Feather key (A) allows axial movement but keying must be milled to end of shaft. For blind keyway (B), use setscrew against the key, but method permits locating the gear anywhere along length of shaft.

Keyed gears can withstand high torque, much more than the pinned or knurled shaft and, at times, more than the splined shafts because the key extends well into both the shaft and gear bore. Torque capacity is comparable with that of the integral gear and shaft. Maintenance is easy because the key can be removed while the gear remains in the system.

Materials for gear, shaft and key should be similar, preferably steel. Larger gears can be either cast or forged, and the key either hot- or cold-rolled steel. However, in instrument gears, stainless steel is required for most applications. Avoid aluminum gears and keys.

10 STAKING

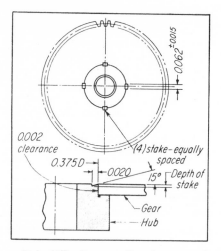

No. of stakes	Depth	Clearance	Torque in.-lb.
4	0.015	0.0020	27
4	0.015	0.0025	20
4	0.020	0.0020	28
4	0.020	0.0020	30
8	0.020	0	52

It is difficult to predict the strength of a staked joint—but it is a quick and economical method when the gear is positioned at the end of the shaft.

Results from five tests we made on gears staked on 0.375-in. hubs are shown here with typical notations for specifying staking on an assembly drawing. Staking was done with a 0.062-in. punch having a 15° bevel. Variables in the test were: depth of stake, number of stakes, and clearance between hub and gear. Breakaway torque ranged from 20 to 52 in.-lb.

Replacing a gear is not simple with this method because the shaft is mutilated by the staking. But production costs are low.

11 SPRING WASHER

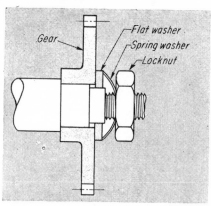

Assembly consists of locknut, spring washer, flat washer and gear. The locknut is adjusted to apply a predetermined retaining force to the gear. This permits the gear to slip when overloaded—hence avoiding gear breakage or protecting the drive motor from overheating.

Construction is simple and costs less than if a slip clutch is employed. Popular in breadboard models.

12 TAPERED SHAFT

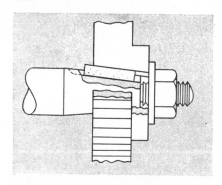

Tapered shaft and matching tapeı in gear bore need key to provide high torque resistance, and threaded nut to tighten gear onto taper. Expensive but suitable for larger gear applications where rigidity, concentricity and easy disassembly are important. A larger dia shaft is needed than with other methods. Space can be problem because of protruding threaded end. Keep nut tight.

13 TAPERED RINGS

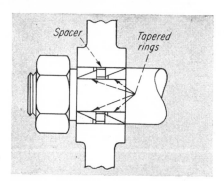

These interlock and expand when tightened to lock gear on shaft. A purchased item, the rings are quick and easy to use, and do not need close tolerance on bore or shaft. No special machining is required and torque capacity is fairly high. If lock washer is employed, the gear can be adjusted to slip at predetermined torque.

14 TAPERED BUSHINGS

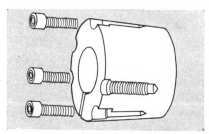

This, too, is a purchased item—but generally restricted to shaft diameters $\frac{1}{2}$ in. and over. Adapters available for untapered bores of gears. Unthreaded half-holes in bushing align with threaded half-holes in gear bore. Screw pulls bushing into bore, also prevents rotational slippage of gear under load.

15 DIE-CAST HUB

Die-casting machines are available, which automatically assemble and position gear on shaft, then die-cast a metal hub on both sides of gear for retention. Method can replace staked assembly. Gears are fed by hopper, shafts by magazine. Method maintains good tolerances on gear wobble, concentricity and location. For high-production applications. Costs are low once dies are made.

SMALL SHAFT
Becomes Its Own Broach for Retaining Gears

This neat method developed by the author also keys links to shafts in such applications as computing devices.

WILLARD J OPOCENSKY, *Staff Engineer, Librascope Inc, Glendale, Calif*

In this fastening technique, a knurled shaft does its own broaching and keys itself into a close-fitting hole. This avoids need for press fits or locking devices when retaining gears and links on small shafts.

The method, being applied to assemblies where shaft size is ¼ in. or under, has these definite advantages:

• It does not take up space nor does it weaken or distort parts by the machining of grooves or holes.

• It is inexpensive, simple and requires no extra parts.

Here is how the method works when self-broaching an aluminum or brass gear onto a stainless steel shaft.

The shaft is machined to a tolerance of plus zero minus 0.0002 in. Bore of gear is made to a tolerance of plus 0.0004 in. to minus zero. These tight tolerances may not be needed in other applications as they are employed here mainly to get high concentricity.

The shaft is now given a full 80-pitch knurl. Shaft diameter, gear thickness and knurl width are equal. Knurling increases shaft dia by 0.003 to 0.005 in.

A chip groove is then cut at the leading edge of the knurl. This cut, made cleanly at right angles to the axis of the shaft and to a depth of a few thousandths, sharpens the teeth of the knurl so that it acts as a broach when pressed into the hole—the groove provides space for the displaced chips.

The method has also been tested with steel shafts up to ¼ in. dia in aluminum and brass links for electromagnetic computing devices; and it was found that the shaft can be twisted until it snaps off at the chip-groove with no evidence of loosening of the joint. Thickness of the link has, in all cases, been equal to shaft diameter. It has been estimated that a 20- to 30-lb pull is required to remove a shaft or gears. There will be variation in pull depending on the materials. The holding power can also be increased by increasing the width of the knurled area or heat-treating the shaft before assembly to permit using harder gear link materials. However, the pull-out force should not be too high or it will increase possibility of distortion when disassembly is required.

Application in a Differential

The knurl-broach method is used in a differential to retain gears that must withstand a temperature range of −65 to 160 F.

Several factors complicate such an assembly problem, and rule out other methods. The side gears are specified by customers, are usually added after the differential has

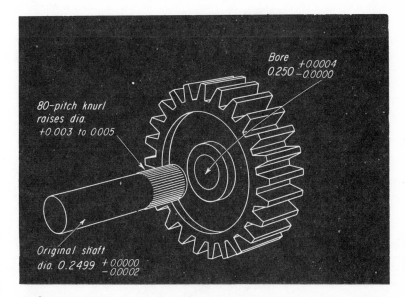

Tolerances on stainless steel shaft and aluminum gear before assembly.

(figure labels:) Bore 0.250 +0.0004 −0.0000 — 80-pitch knurl raises dia. +0.003 to 0.005 — Original shaft dia. 0.2499 +0.0000 −0.0002

been assembled, and are often too thin to permit using pins or setscrews. Also, cutting through the gear to insert a pin, key or screw would cause loss of concentricity when thermal expansion took place; and the pressure exerted by a screw would bring further distortion. The precision needed in many applications of the differential assembly—in computers, flow totalizers, fire-control mechanism and other high-accuracy instrumentation—requires that no distortion be present.

Again, an 80-pitch knurl is used to increase the shaft hub dia 0.003 to 0.005 in. Although the differential is designed for max torque of 12 in.-oz in the gear train, the bond between the hub and gear will withstand over 800 in.-oz of torque.

If one of the ring gears has to be replaced, it is only necessary to clean out the chip-groove before installing a new one. This can be done at least twice with no measurable loss of bond strength.

Because the knurl-broach method gives strength far in excess of the application requirements, it has not received much mathematical analysis. Considering that a full-depth knurl is used in all cases, and that the self-broaching feature insures exact mating of the shaft and the hole, it should be possible for the theoretically-inclined to apply conventional gear formulas to determine the resistance to torque. ■

SINGLE-PLATE GEAR TRAINS:

11 ways to

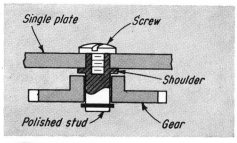

1 Screw-held

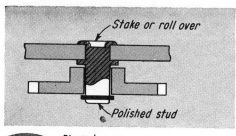

2 Riveted

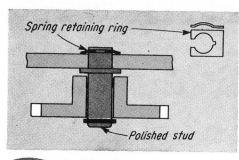

3 Retaining ring

Polished studs . . .
may be held in several ways. The gear-hub length is sufficient to provide, with the polished stud-surface, adequate bearing support

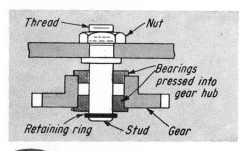

4

Two ball bearings . . .
pressed into gear hub give more precision than plain journal-bearing type of stud mount in previous examples.

Precision gear trains mounted on single plates are easier to assemble and inspect than the more conventional double-plate mounts. Practical plate-thicknesses and commercially available bearings, however, are too thin to provide adequate bearing support for the gear shaft.

Gear mounts illustrated here solve the problem. Some typical design restrictions are:

1. Precision
2. Space available on gear plate
3. Gearing load (torsional or axial)

But one or another of the following arrangements will usually satisfy any combination of requirements.

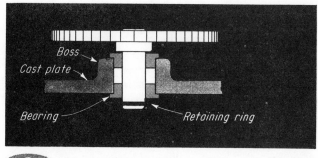

5

Spacer plate . . .
attached before boring hole for the bearing allows mounting two ball bearings in line on virtually a single plate.

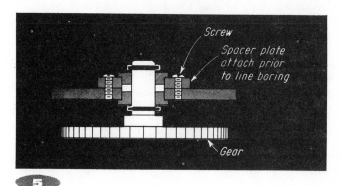

6

Boss in cast plate . . .
performs same function as spacer plate, but eliminates much assembly. Also, bearing misalignment is virtually impossible.

MOUNT SHAFTS

FRANK WILLIAM WOOD JR, *Design Engineer*
Vitro Laboratories, Silver Spring, Md

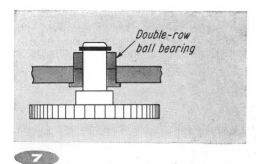

Double-row . . .
ball-bearing unit in single plate has enough bearing length but is more expensive than two single-row bearings.

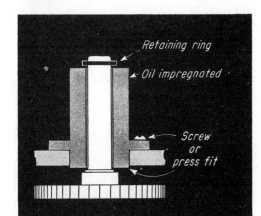

Oil-impregnated . . .
bearing of sintered material provides self-lubricating pedestal that supports shaft and functions as bearing.

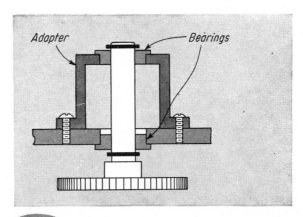

Adapter assembly . . .
must be located accurately unless mounting holes for the bearings are made after adapter has been fastened to the mounting plate.

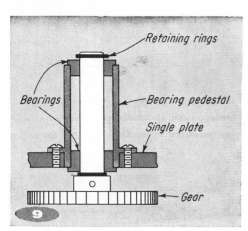

Bearing pedestal . . .
of flanged tube obviates bearing-alignment difficulty, can be assembled separately as a unit.

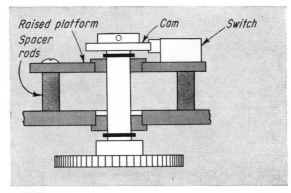

Raised platform . . .
accommodates switch or other component as well as bearing. Gear shaft is also accessible during operation.

Gear-Shift Arrangements

1 Keyed to shaft

2 Not keyed to shaft (rotatable on shaft)

3 Sliding gear keyed to shaft

4 Clutch (shown with clutch keyed to shaft, and gear not keyed to shaft)

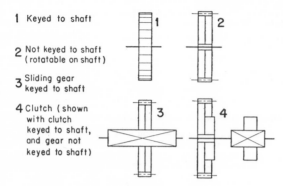

Fig. 1. Schematic symbols used in the illustrations to represent gears and clutches.

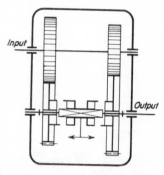

Fig. 2. Double-clutch drive. Two pairs of gears permanently in mesh. Pair I or II transmits motion to output shaft depending on position of coupling; other pair idles. Coupling shown in neutral position with both gear pairs idle. Herring-bone gears recommended for quiet running.

Fig. 3. Sliding-change drive. Gears meshed by lateral sliding. Up to three gears can be mounted on sliding sleeve. Only one pair in mesh in any operating position. Drive simpler, cheaper and more extensively used than drive of Fig. 2. Chamfering side of teeth facilitates engagement.

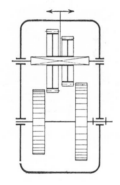

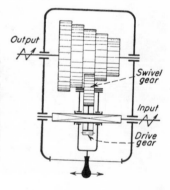

Fig. 4. Swivel-gear drive. Output gears are fastened to shaft. Handle is pushed down, then shifted laterally to obtain transmission through any output gear. Not suitable for transmission of large torques because swivel gear tends to vibrate. Over-all ratio should not exceed 1:3.

Fig. 5. Slide-key drive. Spring-loaded slide key rides inside hollow output shaft. Slide key snaps out of shaft when in position to lock a specific change gear to output shaft. No central position is shown.

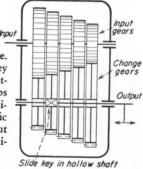

Slide key in hollow shaft

Fig. 6. Combination coupling and slide gears. Three ratios: direct mesh for ratios I and II; third ratio transmitted through gears II and III which couple together.

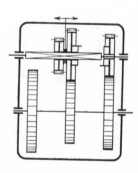

Fig. 7. Double-shift drive. One shift must always be in a neutral position which may require both levers to be shifted when making a change. However only two shafts are used to achieve four ratios.

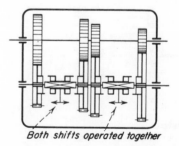

Both shifts operated together

SIGMUND RAPPAPORT

Project Supervisor, Ford Instrument Company
Adjunct Professor of Kinematics
Polytechnic Institute of Brooklyn

13 ways of arranging gears and clutches to obtain changes in speed ratios.

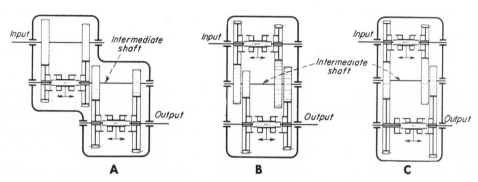

Fig. 8. (A) Triple shaft drive gives four ratios. Output of first drive serves as input for second. Presence of intermediate shaft obviates necessity for always insuring that one shift is in neutral position. Wrong shift lever position can not cause damage. (B) Space-saving modification. Coupling is on shaft *A* instead of intermediate shaft. (C) Still more space saved if one gear replaces a pair on intermediate shaft. Ratios can be calculated to allow this.

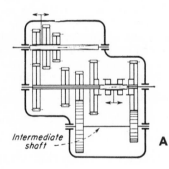

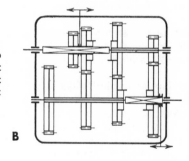

Fig. 9. Six ratios available with two couplings and (A) ten gears, (B) eight gears. Up to six gears in permanent mesh. It is not necessary to insure that one shift is in neutral.

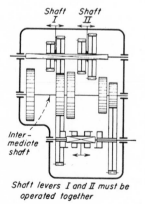

Shaft levers *I* and *II* must be operated together

Fig. 10. Eight-ratio drive uses two slide gears and a coupling. This arrangement reduces number of parts and meshes. Position of shifts I and II are interdependent. One shift must be in neutral if other is in mesh.

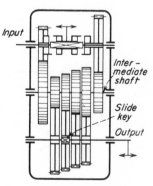

Fig. 11. Eight ratios; coupled gear drive and slide-key drive in series. Comparatively low strength of slide key limits drive to small torque.

Data based on material and sketches in AWF und VDMA Getrieblaetter, published by Ausschuss fuer Getriebe beim Ausschuss fuer Wirtschaftiche Fertigung, Leipzig, Germany.

Gears and friction disks make a
Fast-reversing drive

A slight shift of linkage causes friction disks to reverse the output shaft even under full load.

J BOEHM, A L HERRMANN, J F BLANCHE, *George C Marshall Space Flight Center, Huntsville, Ala*

HERE is a mechanical device capable of reversing the direction of a high-speed rotating shaft within a fraction of a second. It employs friction disks in combination with gears which eliminate the need for a clutch system to absorb the shock caused by an abrupt change in direction. Reversal can be accomplished while the shaft is under full load.

Key characteristics of the new device are:

• Capable of handling high power at high speeds. It's been tested at 2.5 hp and 10,000 rpm—but the principle can be employed for either higher or lower power requirements.

• Designed for fast start and reversal times. Because of its low inertia it can reverse a 300 ft-lb torque in 0.009 sec.

• Slippage in the friction disks is greatly reduced. A linkage arrangement automatically increases the friction force between disks when load torque is increased.

• Linkage movement is kept to a minimum. A shift in linkage of only a few thousands of an inch will cause reversal in direction. This helps reduce reversal time.

The gear and friction disk arrangement permits easy adjustment, low-cost manufacturing and long-time storage ability. The device was specifically designed for control and guidance systems of missiles. The controls must respond promptly to error signals from computers—even when high torques are required for operation of the control surfaces. This new device should also find wide application in industrial servos.

Basic configuration

It is best to insert the reversing transmission at the high-speed level of the driver. This keeps transmitted torques to a minimum, and the elements accomplishing the reversal can be extremely lightweight which improves the response time.

In the basic arrangement of the device (below), the input shaft continuously drives two hardened, wear-resistant steel disks in opposite directions. Direction of rotation of the driven shaft is reversed by shifting the driven disk to the right or left.

In the actual layout a motor drives twin pairs of opposite-turning disks, (diagram on the opposite page). The disks (numbers 1 through 4) are grooved wheels set just enough apart so that when the output wheel 5 contacts the counterclockwise-turning pair (1 and 3) there is a few thousandths of an inch clearance between wheel 5 and the clockwise-turning pair (2 and 4). Solenoids 17 and 18 turn shaft 15 through a small arc, and the eccentrically mounted pivot controls the linear position of the linkage and disk 5. When solenoid 17 is energized, disk 5 is forced against wheels 2 and 4. When solenoid 18 is energized,

disk 5 is shifted over against wheels 1 and 3. When neither solenoid is energized the output disk assumes a neutral position engaging neither pair of wheels and there is no power transmission.

The output torque produces a lateral force on gear 12 and shaft 13. This makes a driven disk bear more heavily against its driving wheels (2 and 4 in this position) as the output torque increases. This effect increases as the diameter of gear 12 is reduced. A value of $\theta = 60°$ produces an optimum force holding the driving and driven disks in contact. A smaller angle provides a higher holding force but at the expense of greater distance between the two pairs of driving disks, greater solenoid travel, and therefore an increase in the reversal time.

Solenoid characteristics

The arrangement of the friction

1·· Basic arrangement of gears and disks

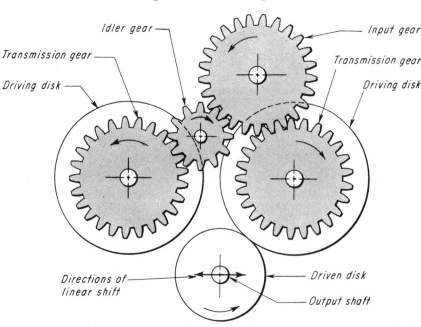

Idler gear · Input gear · Transmission gear · Transmission gear · Driving disk · Driving disk · Directions of linear shift · Driven disk · Output shaft

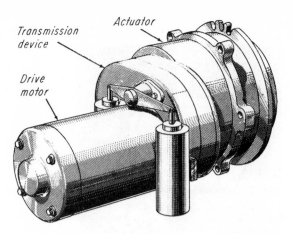

Transmission device

Actuator

Drive motor

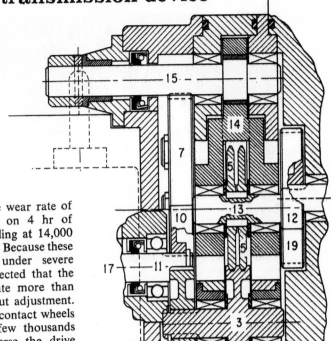

disks also reduces the amount of force that must be applied by the solenoid for fast start, stop and reversing characteristics. However for optimum solenoid performance other modifications must be made.

Standard commercial solenoids were modified as follows: The plungers were drilled out and slotted to reduce inertia and eddy currents. A solenoid with a long coil was selected to further reduce self-induction. Use of a transistor and capacitor allows an increase of the starting current without overheating the solenoid.

Materials and design details

The most satisfactory combination found so far is titanium carbide against hardened tool steel. The driven disk was fabricated of a machinable titanium carbide hardened to Rc-72 and the driving wheels were tool steel hardened to Rc-62. The wear rate of this combination based on 4 hr of testing under shock loading at 14,000 per hr was 0.0003 in./hr. Because these results were obtained under severe shock loading, it is expected that the device will easily operate more than 1.5 million cycles without adjustment.

Exact position of the contact wheels is essential if only a few thousands displacement is to reverse the drive reaction. We mounted two of the wheel bearings in eccentric bushings which may be adjusted to obtain the correct center distances.

To obtain better contact conditions between disk and wheels, V-shape grooves are cut into the rim of the wheels and the disk is tapered to fit the groove. However we have also used wheels with several grooves, as can be noted in the mechanical layout, previous page. This increases power transmitted and reduces wear rate.

2·· Double-pair arrangement

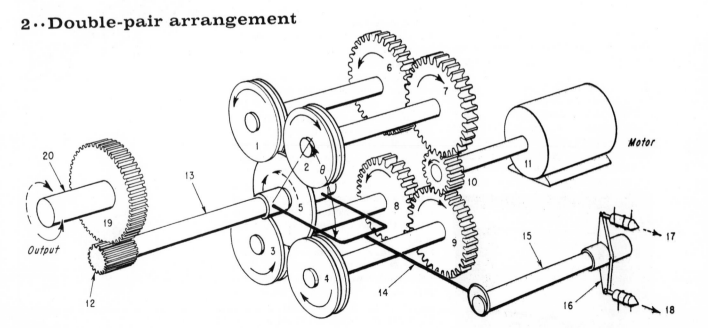

Motor

Output

LOCOMOTIVE GEAR-SHIFT ARRANGEMENT

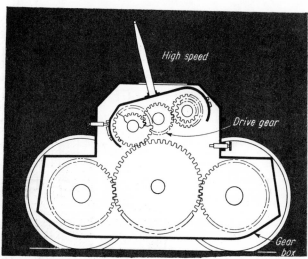

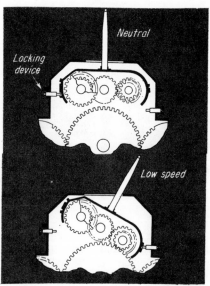

Gear changes . . .

are made by a lever on each bogie which is accessible from the corridor joining the cabs at each end of the locomotive. Three gear positions may be selected: High, Low and Neutral. A mechanical locking device insures that the gears are properly meshed and stay in mesh. Also, an elec-

trical interlock keeps the motor supply circuit open until the, locking device is fully engaged. Neutral position permits the locomotive to be towed or dead-headed without stripping gears or damaging electrical equipment. Gear shifts are made by the regular train crew—this requires stopping the train for about 3 minutes.

MULTISPEED GEAR ARRANGEMENT

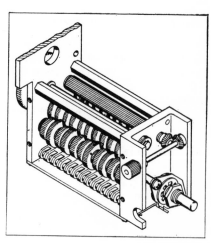

Variable-speed strip recorders make the tester's life simpler all round: First off, you can choose a chart speed that will produce easily interpreted data. Then you save on expensive paper since you get a longer recording period. These various advantages are to be had with a new strip chart recorder made by F. L. Mosley Co, Pasadena, Calif — model 7100A — which provides 12 chart speeds at the twist of a knob. Speeds range from 1 in./hr to 2 in./sec plus a 10-1 reducer that provides a 3-in./day speed for long-term recording.

Drive mechanism. With 22 nylon gears in continuous mesh, a series of cam-controlled idler spur gears are used to engage selected speed reduction gear sets. The synchronous motor is connected either to the optional 10

to 1 speed reducer or through transfer gears to the motor drive gear.

Any of the 12 output gears may be coupled to the output spline and chart drive by the manually selected idlers. Mounted on spring-loaded arms, the idlers are controlled by a 12-position detented cam. By rotating the cam the appropriate idler is pivoted into simultaneous mesh with the output of the selected speed reducer and the output drive spline.

Zener-reference solid-state circuits are used in the model 7100A to eliminate drift and hold accuracy to 0.2% of full scale on all ranges. Maximum sensitivity is 5 mv full scale or 1 mv optional and full-scale balance time of less than 0.2 sec is achieved. Standard input impedance is one megohm at null on both channels, providing a constant-impedance load, regardless of range setting. Potentiometric operation with zero current drain from the source, at null, is available by removing an internal strap for either or both channels. The 10 in. chart, dual channel recorder weighs 24 lb and is less than ⅔ cu ft. Cost, says Mosley, is higher than other recorders, but the flexibility of operation should save interpretation time and allow one recorder to be used for many operations.

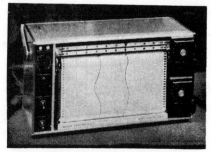

Dial the speed you want

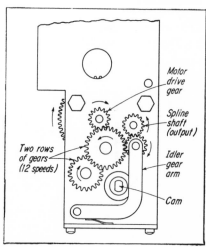

13

GEAR MECHANISMS
AND SPECIAL DRIVES

Geared Machinery Mechanisms

PREBEN W. JENSEN, *Mechanism Consultant, Associate Professor, University of Bridgeport, Bridgeport, Conn*

1. Double-link reverser

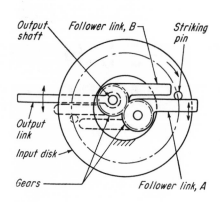

Automatically reverses the output drive every 180-deg rotation of the input. Input disk has press-fit pin which strikes link *A* to drive it clockwise. Link *A* in turn drives link *B* counterclockwise by means of their respective gear segments (or gears pinned to the links). The output shaft and the output link (which may be the working member) are connected to link *B*.

After approximately 180 deg of rotation, the pin slides past link *A* to strike link *B* coming to meet it—and thus reverses the direction of link *B* (and of the output). Then after another 180-deg rotation the pin slips past link *B* to strike link *A* and start the cycle over again.

2. Toggle-link reverser

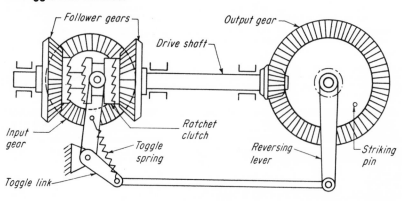

This mechanism also employs a striking pin—but in this case the pin is on the output member. The input bevel gear drives two follower bevels which are free to rotate on their common shaft. The ratchet clutch, however, is spline-connected to the shaft—although free to slide linearly. As shown, it is the right follower gear that is locked to the drive shaft. Hence the output gear rotates clockwise, until the pin strikes the reversing level to shift the toggle to the left. Once past its center, the toggle spring snaps the ratchet to the left to engage the left follower gear. This instantly reverses the output which now rotates counterclockwise until the pin again strikes the reversing level. Thus the mechanism reverses itself every 360-deg rotation of the input.

3. Modified-Watt's reverser

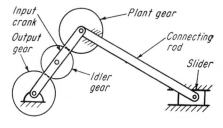

This is actually a modification of the well-known Watt crank mechanism. The input crank causes the planet gear to revolve around the output gear. But because the planet gear is fixed to the connecting rod it causes the output gear to continually reverse itself. If the radii of the two gears are equal then each full rotation of the input link will cause the output gear to oscillate through same angle as rod.

4. Gear-belt reverser

Reversing transmission uses gear-belt combination. Designed for applications requiring high-speed reversing, positioning and sequencing of loads, this transmission produced by Airborne Accessories Corp employs gears to drive output (lower shaft in photo above) in one direction, and a silent belt to reverse. Shifting is done through two opposing magnetic clutches. Speed can be reversed in less than 0.2 sec.

Motor operates constantly at rated speed in one direction. When clutch at right is energized, the drive operates through gear train—belt runs free. Switching to clutch at left causes drive to operate through belt drive and cuts out the first clutch—gears now run free. Rated power is 5 hp at 1750 rpm. Gear ratio is 1:1.

5. Epicyclic dwell mechanism

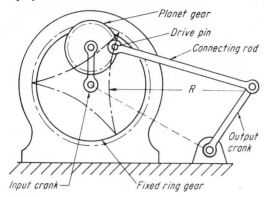

Here the output crank pulsates back and forth with a long dwell at the extreme right position. Input is to the planet gear by means of the rotating crank. The pin on the gear traces the epicyclic three-lobe curve shown. The right portion of the curve is almost a circular arc of radius *R*. If the connecting rod is made equal to *R*, the output crank comes virtually to a standstill during a third of the total rotation of the input. The crank then reverses, comes to a stop at left position, reverses, and repeats dwell.

6. Cam-worm dwell mechanism

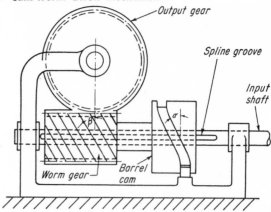

Without the barrel cam, the input shaft would drive the output gear by means of the worm gear at constant speed. The worm and the barrel cam, however, are permitted to slide linearly on the input shaft. Rotation of the input shaft now causes the worm gear to be cammed back and forth, thus adding or subtracting motion to the output. If barrel cam angle α is equal to the worm angle β, the output comes to a stop during the limits of rotation illustrated, then speeds up to make up for lost time.

7. Cam-helical dwell mechanism

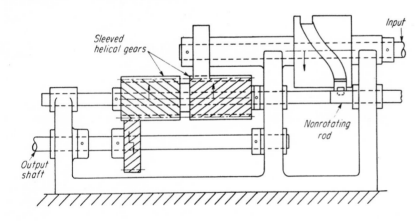

When one helical gear is shifted linearly (but prevented from rotating) it will impart rotary motion to the mating gear because of the helix angle. This principle is used in the mechanism illustrated. Rotation of the input shaft causes the intermediate shaft to shift to the left, which in turn adds or subtracts from the rotation of the output shaft.

8. Three-gear drive

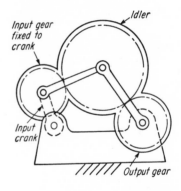

This is actually a four-bar linkage combined with three gears. As the input crank rotates it takes with it the input gear which drives the output gear by means of the idler. Various output motions are possible. Depending on proportions of the gears, the output gear can pulsate, or come to a short dwell—or even reverse briefly.

9. Modified motion geneva

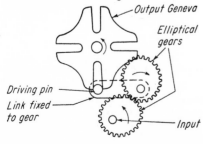

With a normal geneva drive the input link rotates at constant velocity, which restricts flexibility in design. That is, for given dimensions and number of stations the dwell period is determined by the speed of the input shaft. Use of elliptical gears produces a varying crank rotation which permits either extending or reducing the dwell period.

10. Linear reciprocator

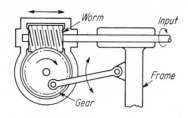

The objective here is to convert a rotary motion into a reciprocating motion that is *in line* with the input shaft. Rotation of the shaft drives the worm gear which is attached to the machine frame by means of a rod. Thus input rotation causes the worm gear to draw itself (and the worm) to the right—thus providing a back and forth motion. Employed in connection with a color-transfer cylinder in printing machines.

11. Cross-bar reciprocator

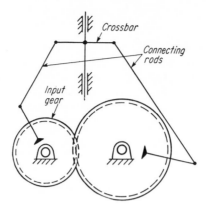

Although complex-looking, this device has been successful in high speed machines for transforming rotary motion into a high-impact linear motion. Both gears contain cranks connected to the cross bar by means of connecting rods.

12. Shaft synchronizer

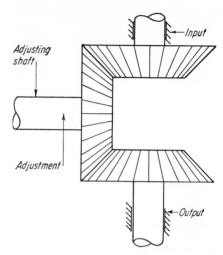

Actual position of the adjusting shaft is normally kept constant. The input then drives the output by means of the bevel gears. Rotating the adjusting shaft in a plane at right angle to the input-output line changes the relative radial position of the input and output shafts, used for introducing a torque into the system while running, synchronizing the input and output shafts, or changing the timing of a cam on the output shaft.

13. Gear-rack drive

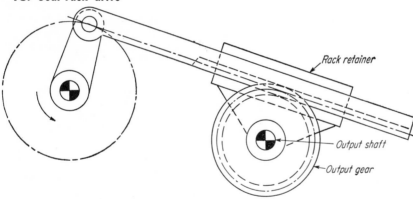

This mechanism is frequently employed to convert the motion of an input crank into a much larger rotation of the output (say, 30 to 360 deg). The crank drives the slider and gear rack, which in turn rotates the output gear.

14. Chain drive

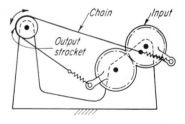

Springs and chains are attached to geared cranks to operate a sprocket output. Depending on the gear ratio, the output will produce a specified oscillation, say two revolutions of output in each direction for each 360 deg of input.

15. Gear-sector drive

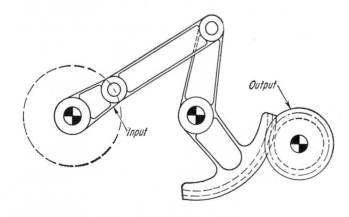

This is actually a four-bar linkage combined with a set of gears. A four-bar linkage usually can get no more than about 120-deg maximum oscillation. The gear segments multiply the oscillation in inverse proportion to the radii of the gears. For the proportions shown, the oscillation is boosted 2½ times.

16. Rack and gear sector

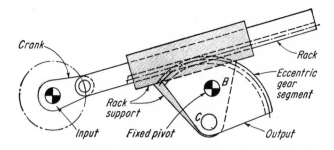

Rotary motion of the input is translated into oscillating motion of the output. The rack support and gear sector are pinned at C, but the gear itself oscillates around B.

17. Oscillating crank and planetary drive

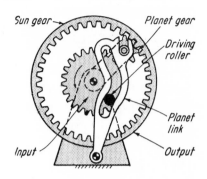

Here the planet is driven with a stop-and-go motion. The driving roller is shown entering the circular-arc slot on the planet link, hence the link and the planet remain stationary while the roller travels this portion of the slot. Result: a rotating output motion with a progressive oscillation.

18. Reciprocating-table drive

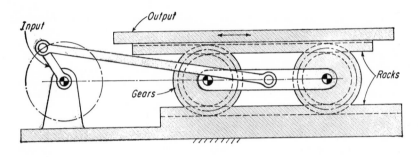

Two gears rolling on stationary bottom rack, drive the movable top rack which is attached to a printing table.

When the input crank rotates, the table will move to a distance of four times the crank length.

19. Gear-slider crank

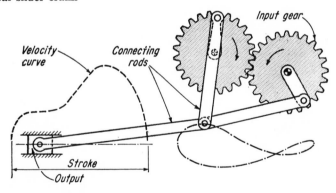

Employed in metal-drawing presses where the piston must move with a low constant velocity. The input drives both gears which in turn drive the connecting rods to produce the velocity curve shown.

20. Gear oscillating crank

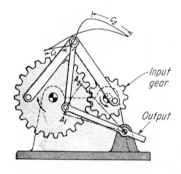

Similar arrangement to the one previously shown, but the curve described by the pin connection has two portions, C_1 and C_2, which are very close to circular arc with centers at A_1 and A_2. Hence the driven link will have a dwell at both extreme positions.

21. Rectangular-motion drive

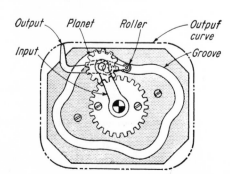

For producing closed curves consisting of several sections of straight lines. Rectangular-shaped curve is shown, but the device is capable of producing many sided curves. The output member is eccentrically mounted on a planet gear and simultaneously guided by the roller which runs in a stationary cam groove.

22. Single tooth indexer

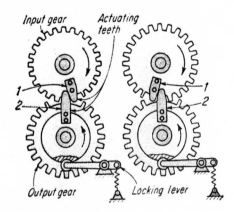

Input gear Actuating teeth

1
2

1
2

Output gear Locking lever

Key factor in this device is the use of an input gear which is smaller than the output gear—hence it can complete its circuit faster than the output when both are in mesh. In the left diagram, the actuating tooth of the input, tooth 1, strikes that of the output, tooth 2, to roll both gears into mesh. After one circuit of the input (right diagram), tooth 1 is now ahead of tooth 2, the gears go out of mesh, and the output comes to a stop (kept in position by the bottom locking detent) for almost 360 deg of the input.

Shifting the controlled rod linearly twists the propellers around on the common axis by means of the rack and gear arrangement. Note the use of a double rack, one above and on either side of the other to obtain an *opposing* twisting motion required for propellers.

23. Adjustable-pitch device

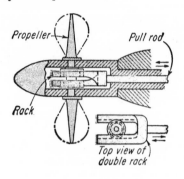

Propeller Pull rod

Rack

Top view of double rack

24. Spherical-gear coupling

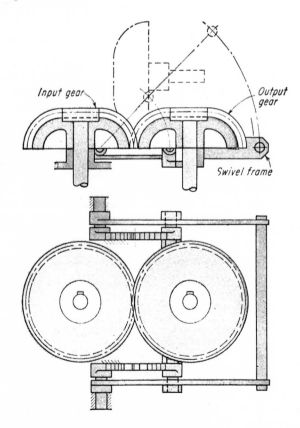

Input gear Output gear

Swivel frame

The angle between input and output shafts can be varied at will (up to 180 deg) during the transmission of uniform rotation. Useful also for applications such as drilling rigs in which the long drill on the output shaft must be moved out of the way when not in use.

25. Elliptical-gear planetary

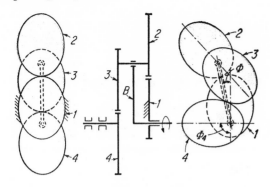

26. Cammed-gear speed variator

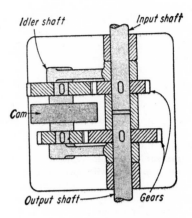

By employing elliptical gears instead of the usual circular gears, a planetary drive is obtained which can provide extra-large variations in the angular speed output. The difference between input speed ω_B, and the output speed ω_4, at any particular position of the input crank ρ, is given by the equation:

$$\frac{\omega_4}{\omega_B} = \frac{1}{\left(\dfrac{1-e}{1+e}\right)^2 \cos^2\dfrac{\phi}{2} + \left(\dfrac{1+e}{1-e}\right)^2 \sin^2\dfrac{\phi}{2}} - 1$$

where a = major axis of elliptics
b = minor axis of elliptics
$e = (a^2 - b^2)^{\frac{1}{2}}/a$

This is a normal parallel-gear speed reducer, but with cam actuation to provide a desired variation in the output speed. If the center of the idler shaft were stationary, the output motion would be uniform, but the cam attached to the idler shaft gives the shaft an oscillating motion which varies the final output motion.

27. Planetary gear geneva

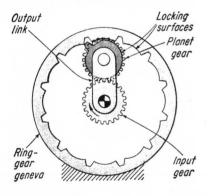

28. Triple-harmonic drive

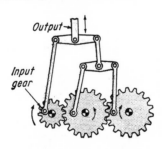

29. Three-gear stroke multiplier

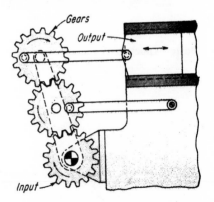

The output link remains stationary while the input gear drives the planet gear with single tooth on the locking disk, which is part of the planet gear, and which meshes with the ring-gear geneva to index the output link one position.

The input drives three gears with connecting rods. A wide variety of reciprocating motions of the output can be obtained by selective proportioning of the linkages, including one to several dwells per cycle.

Rotation of the input gear causes the connecting link. attached to the machine frame, to oscillate. This produces a large-stroke reciprocating motion in the output slider.

Many novel designs in this roundup of
Cycloid gear mechanisms

Cycloidal motion is becoming popular for
mechanisms in feeders and automatic machines. Here
are arrangements, formulas, and layout methods

PREBEN W. JENSEN, Mechanism Consultant

THE appeal of cycloidal mechanisms is that they can easily be tailored to provide one of these three common motion requirements:

- **Intermittent motion**—with either short or long dwells
- **Rotary motion with progressive oscillation**—where the output undergoes a cycloidal motion during which the forward motion is greater than the return motion
- **Rotary-to-linear motion with a dwell period**

All the cycloidal mechanisms covered in this article are geared; this results in compact positive devices capable of operating at relatively high speeds with little backlash or "slop." The mechanisms can also be classified into three groups:

Hypocycloid—where the points tracing the cycloidal curves are located on an external gear rolling inside an internal ring gear. This ring gear is usually stationary and fixed to the frame.

Epicycloid—where the tracing points are on an external gear which rolls in another external (stationary) gear

Pericycloid—where the tracing points are located on an internal gear which rolls on a stationary external gear.

HYPOCYCLOID MECHANISMS

1. Basic hypocycloid curves

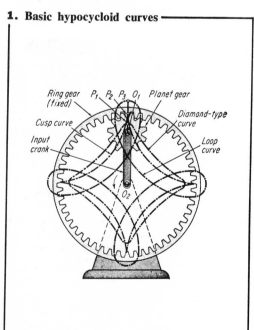

2. Double-dwell mechanism

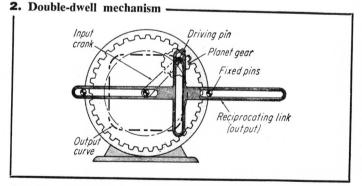

Coupling the output pin to a slotted member produces a prolonged dwell in each of the extreme positions. This is another application of the diamond-type hypocycloidal curve.

Input drives a planet in mesh with a stationary ring gear. Point P_1 on the planet gear describes a diamond-shape curve, point P_2 on the pitch line of the planet describes the familiar cusp curve, and point P_3, which is on an extension rod fixed to the planet gear, describes a loop-type curve. In one application, an end miller located at P_1 was employed in production for machining a diamond-shape profile.

3. Long-dwell geneva drive

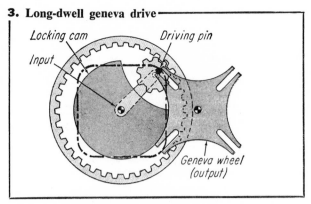

Locking cam · Driving pin
Input
Geneva wheel (output)

As with standard four-station genevas, each rotation of the input indexes the slotted geneva 90 deg. By employing a pin fastened to the planet gear to obtain a rectangular-shape cycloidal curve, a smoother indexing motion is obtained because the driving pin moves on a noncircular path.

4. Internal-geneva drive

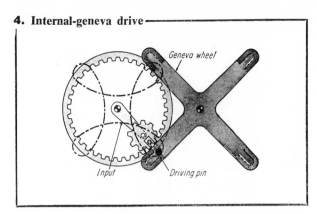

Geneva wheel
Input · Driving pin

Loop-type curve permits driving pin to enter slot in a direction that is radially outward from the center, and then loop over to rapidly index the cross member. As with the previous geneva, the output rotates 90 deg, then goes into a long dwell period during each 270-deg rotation of the input.

5. Cycloidal parallelogram

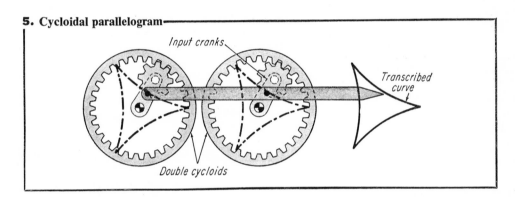

Input cranks
Transcribed curve
Double cycloids

Two identical hypocycloid mechanisms guide the point of the bar along the triangularly shaped path. They are useful also in cases where there is limited space in the area where the curve must be described. Such double-cycloid mechanisms can be designed to produce other types of curves.

6. Short-dwell rotary

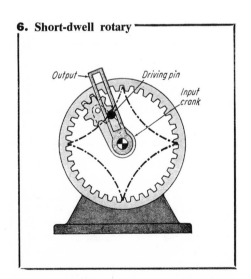

Output · Driving pin
Input crank

Here the pitch circle of the planet gear is exactly one-quarter that of the ring gear. A pin on the planet will cause the slotted output member to have four instantaneous dwells for each revolution of the input shaft.

7. Cycloidal rocker

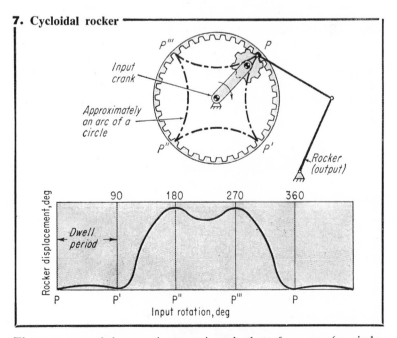

P''' · P
Input crank
Approximately an arc of a circle
P'' · P'
Rocker (output)

Rocker displacement, deg
90   180   270   360
Dwell period
P P' P'' P''' P
Input rotation, deg

The curvature of the cusp is approximately that of an arc of a circle. Hence the rocker comes to a long dwell at the right extreme position while point P moves to P'. There is then a quick return from P' to P'', with a momentary dwell at the end of this phase. The rocker then undergoes a slight oscillation from point P'' to P''', as shown in the displacement diagram.

8. Cycloidal reciprocator

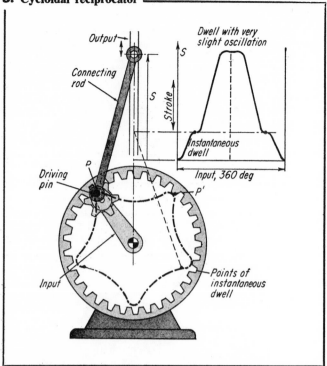

Portion of curve, P-P' produces the long dwell (as in previous mechanism), but the five-lobe cycloidal curve avoids a marked oscillation at the end of the stroke. There are also two points of instantaneous dwell where the curve is perpendicular to the connecting rod.

By making the pitch diameter of the planet equal to half that of the ring gear, every point on the planet gear (such as points P_2 and P_3) will describe elliptical curves which get flatter as the points are selected closer to the pitch circle. Point P_1, at the center of the planet, describes a circle; point P_4 at the pitch circle describes a straight line. When a cutting tool is placed at P_3, it will cut almost-flat sections from round stock, as when machining a bolt. The other two sides of the bolt can be cut by rotating the bolt, or the cutting device, 90 deg. (Reference: H. Zeile, *Unrund- und Mehrkantdrehen*, VDI-Berichte, Nr. 77,1965.)

9. Adjustable harmonic drive

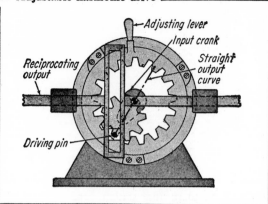

By making the planet-gear half that of the internal gear, a straight-line output curve is produced by the driving pin which is fastened to the planet gear. The pin engages the slotted member to cause the output to reciprocate back and forth with harmonic (sinusoidal) motion. The position of the fixed ring gear can be changed by adjusting the lever, which in turn rotates the straight-line output-curve. When the curve is horizontal, the stroke is at a maximum; when the curve is vertical, the stroke is zero.

10. Elliptical-motion drive

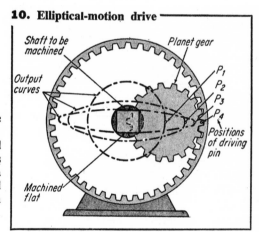

EPICYCLOID MECHANISMS

11. Epicycloid reciprocator

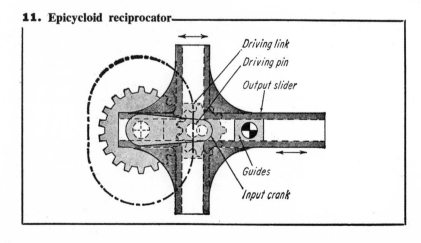

Here the sun gear is fixed and the planet gear driven around it by means of the input link. There is no internal ring gear as with the hypocycloid mechanisms. Driving pin P on the planet describes the curve shown which contains two almost-flat portions. By having the pin ride in the slotted yoke, a short dwell is produced at both the extreme positions of the output member. The horizontal slots in the yoke ride the end-guides, as shown.

12. Progressive oscillating drive

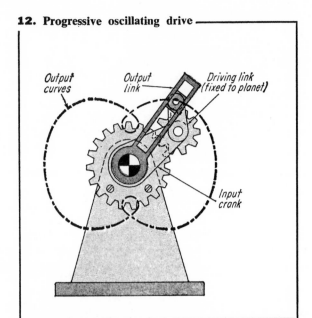

By fixing a crank to the planet gear, a point P can be made to describe the double loop curve illustrated. The slotted output crank oscillates briefly at the vertical portions.

13. Parallel-guidance mechanisms

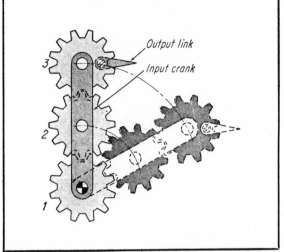

The input crank contains two planet gears. The center sun-gear is fixed as in the previous epicycloid mechanisms. By making the three gears equal in diameter and having gear 2 serve as an idler, any member fixed to gear 3 will remain parallel to its previous positions throughout the rotation of the input ring crank.

MOTION EQUATIONS

14. Equations for epicycloid drives

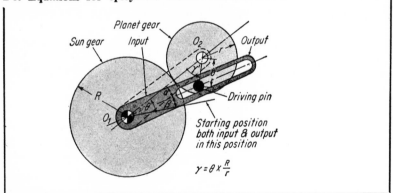

$$\gamma = \theta \times \frac{R}{r}$$

The equations for angular displacement, velocity and acceleration for basic epicyclic drive are given below. (Reference: Schmidt, E. H., "Cycloidal Cranks," *Transactions of the 5th Conference on Mechanisms*, 1958, pp 164-180):

Angular displacement

$$\tan \beta = \frac{(R + r) \sin \theta - b \sin (\theta + \gamma)}{(R + r) \cos \theta - b \cos (\theta + \gamma)} \tag{1}$$

Angular velocity

$$V = \omega \frac{1 + \dfrac{b^2}{r(R + r)} - \left(\dfrac{2r + R}{r}\right)\left(\dfrac{b}{R + r}\right)\left(\cos \dfrac{R}{r} \theta\right)}{1 + \left(\dfrac{b}{R + r}\right)^2 - \left(\dfrac{2b}{R + r}\right)\left(\cos \dfrac{R}{r} \theta\right)} \tag{2}$$

Angular acceleration

$$A = \omega^2 \frac{\left(1 - \dfrac{b^2}{(R + r)^2}\right)\left(\dfrac{R^2}{r^2}\right)\left(\dfrac{b}{R + r}\right)\left(\sin \dfrac{R}{r} \theta\right)}{\left[1 + \dfrac{b^2}{(R + r)^2} - \left(\dfrac{2b}{R + r}\right)\left(\cos \dfrac{R}{r} \theta\right)\right]^2} \tag{3}$$

Symbols

A = angular acceleration of output, deg/sec²

b = radius of driving pin from center of planet gear

r = pitch radius of planet gear

R = pitch radius of fixed sun gear

V = angular velocity of output, deg/sec

β = angular displacement of output, deg

$\gamma = \theta R / r$

θ = input displacement, deg

ω = angular velocity of input, deg/sec

357

15. Equations for hypocycloid drives

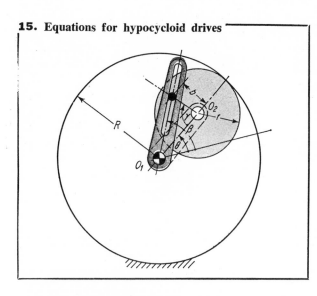

$$\tan \beta = \frac{\sin \theta - \left(\dfrac{b}{R-r}\right)\left(\sin \dfrac{R-r}{r}\theta\right)}{\cos \theta + \left(\dfrac{b}{R-r}\right)\left(\cos \dfrac{R-r}{r}\theta\right)} \quad (4)$$

$$V = \omega \frac{1 - \left(\dfrac{R-r}{r}\right)\left(\dfrac{b^2}{(R-r)^2}\right) + \left(\dfrac{2r-R}{r}\right)\left(\dfrac{b}{R-r}\right)\left(\cos \dfrac{R}{r}\theta\right)}{1 + \dfrac{b^2}{(R-r)^2} + \left(\dfrac{2b}{R-r}\right)\left(\cos \dfrac{R}{r}\theta\right)} \quad (5)$$

$$A = \omega^2 \frac{\left(1 - \dfrac{b^2}{(R-r)^2}\right)\left(\dfrac{b}{R-r}\right)\left(\dfrac{R^2}{r^2}\right)\left(\sin \dfrac{R}{r}\theta\right)}{\left[1 + \dfrac{b^2}{(R-r)^2} + \left(\dfrac{2b}{R-r}\right)\left(\cos \dfrac{R}{r}\theta\right)\right]^2} \quad (6)$$

DESCRIBING APPROXIMATE STRAIGHT LINES

16. Gear rolling on a gear—flatten curves

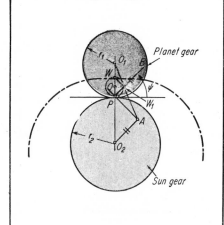

It is frequently desirable to find points on the planet gear that will describe approximately straight lines for portions of the output curve. Such points will yield dwell mechanisms, as shown in Fig 2 and 11. Construction is as follows (shown at left):

1. Draw an arbitrary line PB.
2. Draw its parallel O_2A.
3. Draw its perpendicular PA at P. Locate point A.
4. Draw O_1A. Locate W_1.
5. Draw perpendicular to PW_1 at W_1 to locate W.
6. Draw a circle with PW as the diameter.

All points on this circle describe curves with portions that are approximately straight. This circle is also called the inflection circle because all points describe curves which have a point of inflection at the position illustrated. (Shown is the curve passing through point W.)

17. Gear rolling on a rack—vee curves

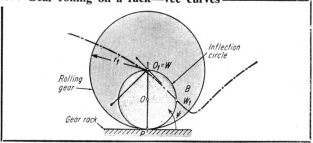

This is a special case. Draw a circle with a diameter half that of the gear (diameter O_1P). This is the inflection circle. Any point, such as point W_1, will describe a curve that is almost straight in the vicinity selected. Tangents to the curves will always pass through the center of the gear, O_1 (as shown).

18. Gear rolling inside a gear—zig-zag

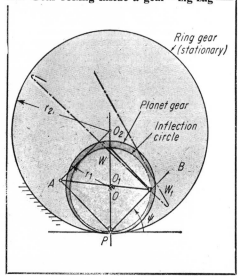

To find the inflection circle for a gear rolling inside a gear:

1. Draw arbitrary line PB from the contact point P.
2. Draw its parallel O_2A, and its perpendicular, PA. Locate A.
3. Draw line AO_1 through the center of the rolling gear. Locate W_1.
4. Draw a perpendicular through W_1. Obtain W. Line WP is the diameter of the inflection circle. Point W_1, which is an arbitrary point on the circle, will trace a curve of repeated almost-straight lines, as shown.

19. Center of curvature—gear rolling on gear

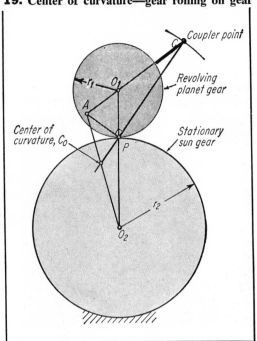

Coupler point

Revolving planet gear

r_1 O_1

A

Center of curvature, C_0

Stationary sun gear

P

r_2

O_2

20. Center of curvature—gear rolling on a rack

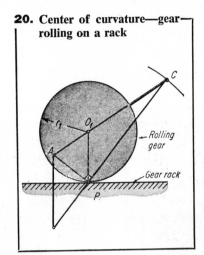

C

r_1 O_1

Rolling gear

A

Gear rack

P

Construction is similar to that of the previous case.

1. Draw an extension of line CP.

2. Draw a perpendicular at P to locate A.

3. Draw a perpendicular from A to the straight suface to locate C_o.

By locating the centers of curvature at various points, one can then determine the proper length of the rocking or reciprocating arm to provide long dwells (as required for the mechanisms in Fig 7 and 8), or proper entry conditions (as for the drive pin in the mechanism in Fig 3).

In the case of a gear with an extended point, point C, rolling on another gear, the graphical method for locating the center of curvature is given by these steps:

1. Draw a line through points C and P.

2. Draw a line through points C and O_1.

3. Draw a perpendicular to CP at P. This locates point A.

4. Draw line AO_2, to locate C_o, the center of curvature.

21. Center of curvature—gear rolling inside a gear

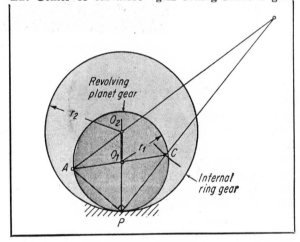

Revolving planet gear

r_2

O_2

O_1 r_1 C

A

Internal ring gear

P

1. Draw extensions of CP and CO_1.

2. Draw a perpendicular of PC at P to locate A.

3. Draw AO_2 to locate C_o.

22. Analytical solutions

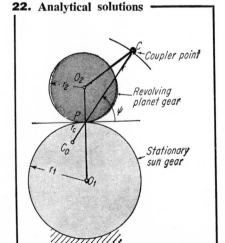

C Coupler point

O_2

r_2

Revolving planet gear

P

ψ

r_c

C_o

Stationary sun gear

r_1

O_1

The centure of curvature of a gear rolling on a external gear can be computed directly from the Euler-Savary equation:

$$\left(\frac{1}{r} - \frac{1}{r_c}\right)\sin\psi = \text{constant} \quad (7)$$

where angle ψ and r locate the position of C.

By applying this equation twice, specifically to point O_1 and O_2 which have their own centers of rotation, the following equation is obtained:

$$\left(\frac{1}{r_2} + \frac{1}{r_1}\right)\sin 90° =$$

$$\left(\frac{1}{r} + \frac{1}{r_c}\right)\sin\psi$$

or

$$\frac{1}{r_2} + \frac{1}{r_1} = \left(\frac{1}{r} + \frac{1}{r_c}\right)\sin\psi$$

This is the final design equation. All factors except r_c are known; hence solving for r_c leads to the location of C_o.

For a gear rolling inside an internal gear, the Euler-Savary equation is

$$\left(\frac{1}{r} + \frac{1}{r_c}\right)\sin\psi = \text{constant}$$

which leads to

$$\frac{1}{r_2} - \frac{1}{r_1} = \left(\frac{1}{r} - \frac{1}{r_c}\right)\sin\psi$$

CAM-CONTROLLED PLANETARY GEAR SYSTEM

JOSEPH KAPLAN, Machine Components, Inc, Farmingdale, NY

By incorporating a grooved cam you get a novel mechanism that's able to produce a wide variety of output motions.

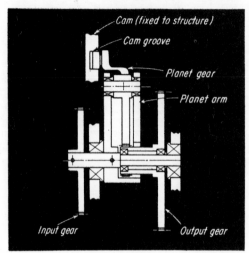 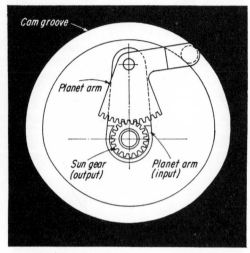

Construction details of cam-planetary mechanism employed in film drive.

Do you want more variety in the kinds of output motion given by a planetary gear system? You can have it by controlling the planet with a grooved cam. The method gives the mechanism these additional features:
- Intermittent motion, with long dwells and minimum acceleration and deceleration.
- Cyclic variations in velocity.
- Two levels, or more, of constant speed during each cycle of the input.

The design is not simple because of need to synchronize the output of the planetary system with the cam contour. However, such mechanisms are now at work in film drives and should prove useful in many automatic machines. Here are equations, tables and a step-by-step sequence that will make the procedure easier.

How the Mechanism Works

The planet gear need not be cut in full—a gear sector will do because the planet is never permitted to make a full revolution. The sun gear is integral with the output gear. The planet arm is fixed to the input shaft, which is coaxial with the output shaft. Attached to the planet is a follower roller which rides in a cam groove. The cam is fixed to the frame.

The planet arm (input) rotates at constant velocity and makes one revolution with each cycle. Sun gear (output) also makes one revolution during each cycle. Its motion is modified, however, by the oscillatory motion of the planet gear relative to the planet arm. It is this motion that is controlled by the cam (a constant-radius cam would not affect the output, and the drive would give only a constant one-to-one ratio).

Comparison with Other Devices

A main feature of this cam-planetary mechanism is its ability to produce a wide range of non-homogeneous functions. These functions can be defined by no less than two mathematical expressions, each valid for a discrete portion of the range. This feature is not shared by the more widely known intermittent mechanisms: the external and internal Genevas, the three-gear drive, and the cardioid drive.

Either three-gear or cardiod can provide a dwell period —but only for a comparatively short period of the cycle. With the cam-planetary, one can obtain over 180° of dwell during a 360° cycle by employing a 4-to-1 gear ratio between planet and sun.

And what about a cam doing the job by itself? This has the disadvantage of producing reciprocating motion. In other words, the output will always reverse during the cycle —a condition unacceptable in many applications.

Design Procedure

Basic equation for an epicyclic gear train is:

$$d\theta_S = d\theta_A - n\,d\theta_{P\text{-}A}$$

where:
$d\theta_S$ = rotation of sun gear (output), deg
$d\theta_A$ = rotation of planet arm (input), deg
$d\theta_{P\text{-}A}$ = rotation of planet gear with respect to arm, deg
n = ratio of planet to sun gear.

The required output of the system is usually specified in the form of kinematic curves. Design procedure then is to:
- Select the proper planet-sun gear ratio
- Develop the equations of the planet motion (which also functions as a cam follower)
- Compute the proper cam contour

INTERMITTENT SPUR GEARS

E. N. SWANSON, Brown & Sharpe Mfg. Co.

INTERMITTENT SPUR GEARS have often been used for the transmission of intermittent rotary motion. Simplicity of design, relative ease of manufacture, and a wide range of applications, are important advantages which these gears have when compared to other types of mechanisms that provide non-uniform motion. There is almost no limit to the possible combinations of motion which can be obtained, just as the possible ratios in conventional gearing are practically numberless.

Obviously, the shock of the follower being picked up by the first driver tooth limits the practical applications of intermittent gears to those cases where speeds are low or kinetic energy values are reasonable. However, within these limits, a variety of motions can be developed by varying the number of lands on the driver and follower. For example the driver will make three revolutions for every one of the follower, if the follower has three lands and the driver one. Each of the three toothed segments of the follower in this case would have the same number of teeth as spaces in the toothed segment of the driver.

The new form has advantages in that the follower is completely and precisely indexed by gear tooth action as shown. Formerly, an index was completed by the first portion of the driver land at a point which could vary over a range of several degrees of driver rotation. Another important feature of this form is that, just before the driver picks up the follower for an index, the follower cannot "rock in" and cause a bind or an interference.

In laying out the follower land for this form, the number of circular pitches which the land will include must be such that the radius of the land creates sufficient bearing surface on the two end teeth. It is possible to select a number of teeth for the land, resulting in little or no bearing surface.

Also, it is good practice to lay out the gears in the vicinity of the lands to as large a scale as possible on heavy paper. If these layouts are then cut out and mounted at scale center distance, their operation can be observed and any interference found and corrected before the gears are made.

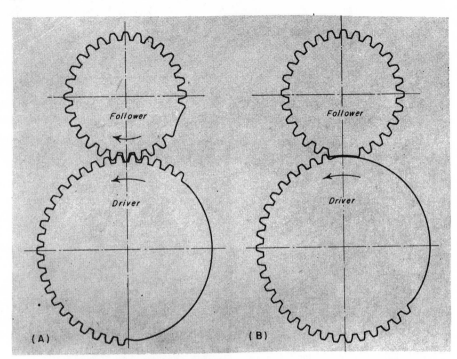

1—Operating cycle of a pair of intermittent spur gears. In (A) action is identical to that of conventional spur gears. The follower begins its dwell at (B), while the driver continues to rotate. The follower remains in this locked position until the first tooth of the toothed segment of the driver contacts the flank of the follower land.

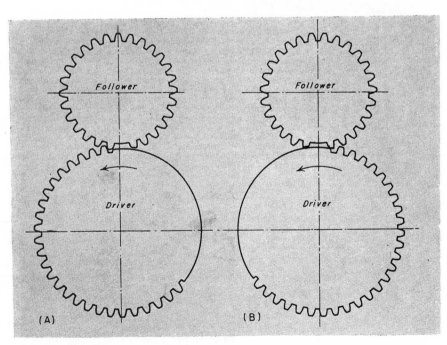

2—New concept of designing intermittent spur gears does away with most interference and binding problems. The follower land consists of two conventional teeth whose tops are formed by the radius of the land. The driver land does not go below the pitch line but remains at this line until it meets the tooth space preceding the first tooth.

12 SPECIAL GEARING

A mixed bag of unusual gear arrangements to answer various tricky design problems. Federico Strasser of Santiago, Chile, is credited for much of this material. A further selection will appear in a coming issue.

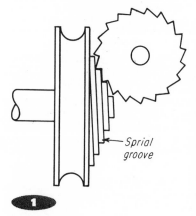

1

Worm gear . . .
has spiral groove. At least two or three pinion teeth are always in mesh.

2

Two-tooth pinion . . .
consists of two diametrically opposed, specially shaped teeth. Gear is locked through part of the pinion revolution: by friction type (above), or by cardioid-shape pinion tooth.

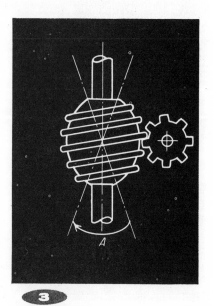

3

Globoid gear . . .
allows shaft to be swung through angle A without varying rotation speed.

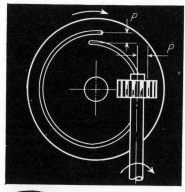

4

Faceplate worm . . .
has spiral ridge on flat face, meshing with worm wheel that turns forward one tooth per faceplate revolution.

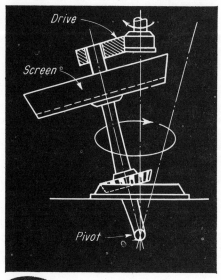

5

Conical-rotary gear . . .
causes shaft to rotate as it is swung about its pivot. Arrangement is used in reaping machines and other applications where screening action is required.

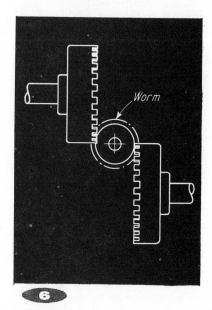

6

Worm and crown-gear . . .
gives slow, simultaneous feed to two shafts, which rotate in opposite directions. An application for this device is in chaffing machines.

DEVICES

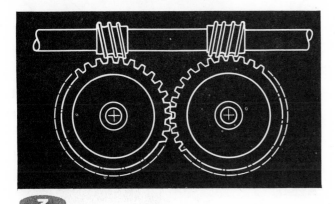

7

Double worms . . .
with opposite-hand threads neutralize end-thrust. Meshing the two gear-wheels gives greater stability to the setup.

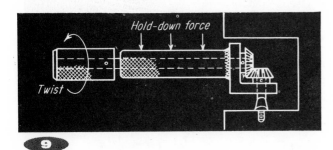

9

Right-angle screwdriver . . .
uses two small bevel-gears to transmit torque through 90°.

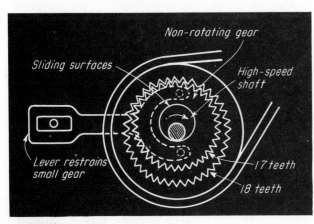

11

"Wobble" gear . . .
provides large speed reduction, depending on numbers of teeth. E.g. for arrangement shown, speed ratio will be 18 to 1— for one shaft revolution the large gear will rotate one tooth space. A 16-to-19 tooth ratio would give a 19-to-3 speed ratio

12

Bevel-gear differential . . .
in analog computors solves the equation: $z = c(x \pm y)$, where c is scale factor, x and y are inputs and z is output. Motion of x and y in same direction results in addition; opposite direction gives subtraction.

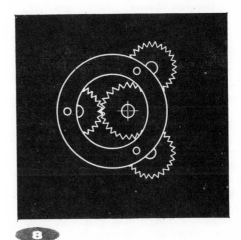

8

Planetary gears . . .
have ring pinned to them at eccentric points. As the planets rotate, the ring-center rotates about a circle with a radius equal to eccentricity of the planet ring-mounting pin.

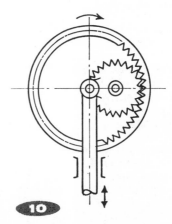

10

For straight-line motion . . .
epicyclic gear has pinion with pitch diameter equal to pitch radius of sun gear. Pivot-point on pinion, at pitch line, generates straight-line motion as sun gear rotates.

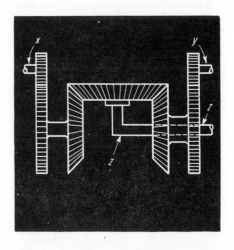

SPECIAL-MOTION GEAR MECHANISMS

SIGMUND RAPPAPORT
Ford Instrument Company

Straight Line Motion

FIG. 1—No linkages or guides are used in this modified hypocyclic drive which is relatively small in relation to the length of its stroke. The sun gear of pitch diameter D is stationary. The drive shaft, which turns the T-shaped arm, is concentric with this gear. The idler and planet gears, the latter having a pitch diameter of $D/2$, rotate freely on pivots in the arm extensions. Pitch diameter of the idler is of no geometrical significance, although this gear does have an important mechanical function. It reverses the rotation of the planet gear, thus producing true hypocyclic motion with ordinary spur gears only. Such an arrangement occupies only about half as much space as does an equivalent mechanism containing an internal gear. Center distance R is the sum of $D/2$, $D/4$ and an arbitrary distance d, determined by a particular application. Points A and B on the driven link, which is fixed to the planet, describe straight-line paths through a stroke of $4R$. All points between A and B trace ellipses, while the line AB envelopes an astroid.

Parallel Motion

FIG. 2—A slight modification of the mechanism in Fig. 1 will produce another type of useful motion. If the planet gear has the same diameter as that of the sum gear, the arm will remain parallel to itself throughout the complete cycle. All points on the arm will thereby describe circles of radius R. Here again, the position and diameter of the idler gear are of no geometrical importance. This mechanism can be used, for example, to cross-perforate a uniformly moving paper web. The value for R is chosen such that $2\pi R$, or the circumference of the circle described by the needle carrier, equals the desired distance between successive lines of perforations. If the center distance R is made adjustable, the spacing of perforated lines can be varied as desired.

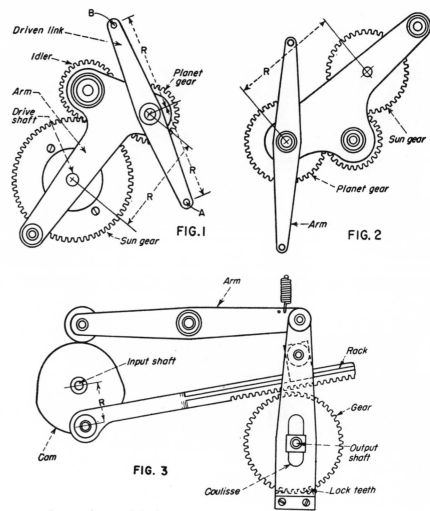

FIG. 1

FIG. 2

Intermittent Motion

FIG. 3—An operating cycle of 180 deg motion and 180 deg dwell is produced by this mechanism. The input shaft drives the rack which is engaged with the output shaft gear during half the cycle. When the rack engages, the lock teeth at the lower end of the coulisse are disengaged and, conversely, when the rack is disengaged, the coulisse teeth are engaged, thereby locking the output shaft positively. The change-over points occur at the dead-center positions so that the motion of the gear is continuously and positively governed. By varying R and the diameter of the gear, the number of revolutions made by the output shaft during the operating half of the cycle can be varied to suit requirements.

Rotational Motion

FIG. 4—The absence of backlash makes this old but little used mechanism a precision, low-cost replacement for gear or chain drives otherwise used to rotate parallel shafts. Any number of shafts greater than two can be driven from any one of the shafts, provided two conditions are fulfilled: (1) All cranks must have the same length r; and (2) the two polygons formed by the shafts A and frame pivot centers B must be identical. The main disadvantage of this mechanism is its dynamic unbalance, which limits the speed of rotation.

FIG. 3

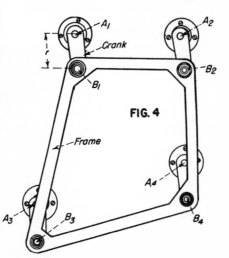

FIG. 4

Fast Cam-Follower Motion

FIG. 5–Fast cam action every n cycles when n is a relatively large number, can be obtained with this manifold cam and gear mechanism. A single notched cam geared $1/n$ to a shaft turning once a cycle moves relatively slowly under the follower. The double notched-cam arrangement shown is designed to operate the lever once in 100 cycles, imparting to it a rapid movement. One of the two identical cams and the 150-tooth gear are keyed to the bushing which turns freely around the cam shaft. The latter carries the second cam and the 80-tooth gear. The 30- and 100-tooth gears are integral, while the 20-tooth gear is attached to the one-cycle drive shaft. One of the cams turns in the ratio of 20/80 or 1/4; the other in the ratio 20/100 times 30/150 or 1/25. The notches therefore coincide once every 100 cycles (4 x 25). Lever movement is the equivalent of a cam turning in a ratio of 1/4 in relation to the drive shaft. To obtain fast cam action, n must be broken down into prime factors. For example, if 100 were factored into 5 and 20, the notches would coincide after every 20 cycles.

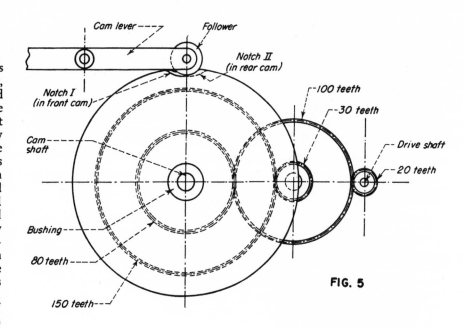

FIG. 5

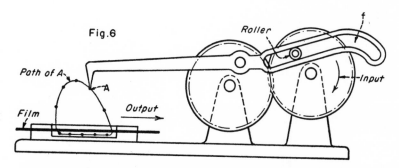

Fig.6

Fig. 6—This mechanism is intended to accomplish the following: (1) Film hook, while moving the film strip, must describe very nearly a straight line; (2) Engagement and disengagement of the hook with the perforation of the film must take place in a direction approximately normal to the film; (3) Engagement and disengagement should be shock free. Slight changes in the shape of the guiding slot f enable the designer to vary the shape of the output curve as well as the velocity diagram appreciably.

Fig. 7—"Multilated tooth" intermittent drive. Driver b is a circular disk of width w with a cutout d on its circumference and carries a pin c close to the cutout. The driven gear, a of width $2w$ has standard spur gear teeth, always an even number, which are alternately of full and of half width (mutilated). During the dwell period two full width teeth are in contact with the circumference of the driving disk, thus locking it; the multilated tooth between them is behind the driver. At the end of the dwell period pin c comes in contact with the mutilated tooth and turns the driven gear for one circular pitch. Then, the full width tooth engages the cutout d and the driven gear moves one more pitch, whereupon the dwell period starts again and the cycle is repeated. Used only for light loads primarily because of high accelerations encountered.

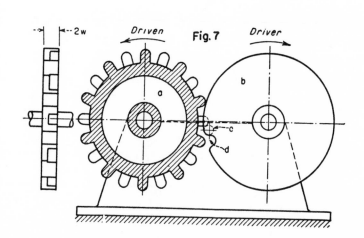

One-way Output from Gear Reducers

When input reverses, these 5 slow-down mechanisms continue supplying a non-reversing rotation.

LOUIS SLEGEL
Head, Dept of Mechanical Engineering
Oregon State College
Corvallis, Ore

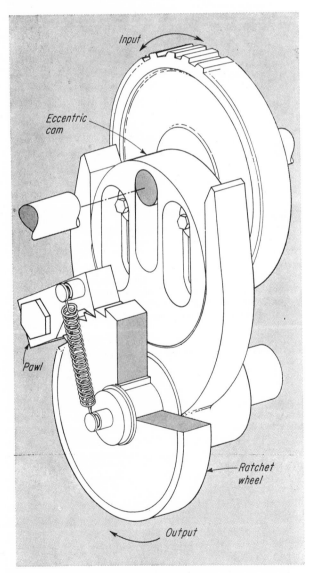

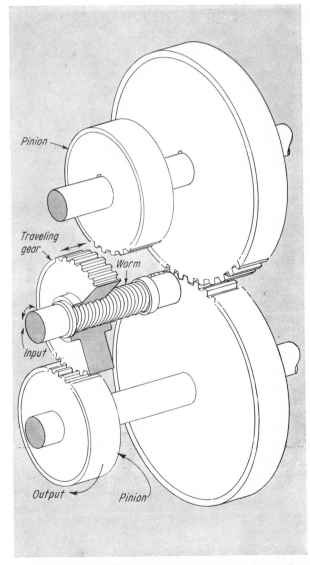

1 ECCENTRIC CAM adjusts over a range of high-reduction ratios, but unbalance limits it to low speeds. When direction of input changes, there is no lag in output rotation. Output shaft moves in steps because of ratchet drive through pawl which is attached to U-follower.

2 TRAVELING GEAR moves along worm and transfers drive to other pinion when input rotation changes direction. To ease engagement, gear teeth are tapered at ends. Output rotation is smooth, but there is a lag after direction changes as gear shifts. Gear cannot be wider than axial offset between pinions, or there will be interference.

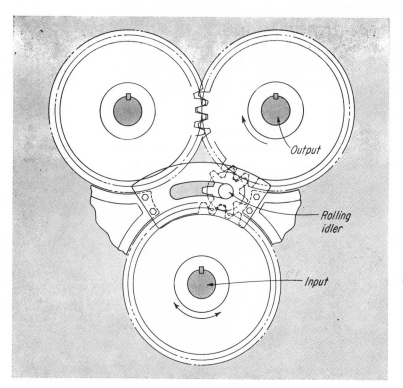

3 ROLLING IDLER also gives smooth output and slight lag after input direction changes. Small drag on idler is necessary, so that it will transfer into engagement with other gear and not sit spinning in between.

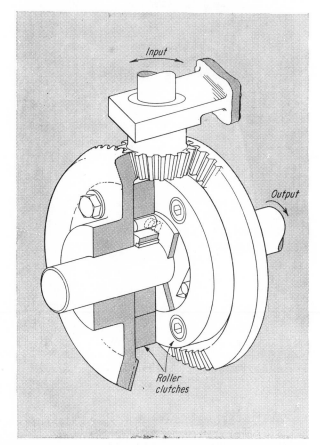

4 TWO BEVEL GEARS drive through roller clutches. One clutch catches in one direction; the other catches in the opposite direction. There is negligible interruption of smooth output rotation when input direction changes.

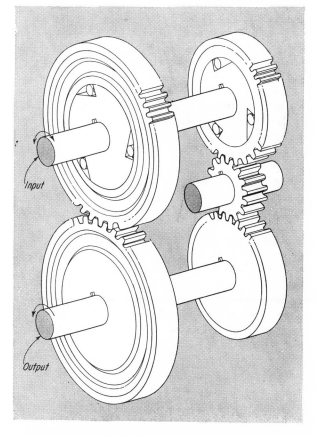

5 ROLLER CLUTCHES are on input gears in this drive, again giving smooth output speed and little output lag as input direction changes.

for self-locking at high efficiency . . .

the TWINWORM GEAR

NICHOLAS CHIRONIS

Developed by an Israeli engineer, this innovation in gearing combines two worm-screws to give self-locking characteristics, or to operate as a fast-acting brake when power is shut off. Model shows basic components.

The term "self-locking" as applied to gear systems denotes a drive which gives the input gear freedom to rotate the output gear in either direction—but the output gear locks with the input when an outside torque attempts to rotate the output in either direction. This characteristic is often sought by designers who want to be sure that loads on the output side of the system cannot affect position of the gears. Worm gears are one of the few gear systems that can be made self-locking, but at the expense of efficiency—they seldom exceed 40%, when self-locking.

An Israeli engineer displayed a simple, dual-worm gear system that not only provided self-locking with over 90% efficiency, but exhibited a new phenomenon which the inventor calls "deceleration-locking."

A point in favor of the inventor—B. Popper, an engineer with the Scientific Department of the Israel Ministry of Defense in Tel Aviv—is that his "Twinworm" drive has been employed in Israel-designed counters and computers for several years and with marked success.

The Twinworm drive is quite simply constructed. Two threaded rods, or "worm" screws, are meshed together. Each worm is wound in a different direction and has a different pitch angle. For proper mesh, the worm axes are not parallel, but slightly skewed. (If both worms had the same pitch angle, a normal, reversible drive would result—similar to helical gears.) But by selecting proper, and different, pitch angles, the drive will exhibit either self-locking, or a combination of self-locking and deceleration-locking characteristics, as desired. Deceleration-locking is a completely new property best described in this way.

When the input gear decelerates (for example, when the power source is shut off, or when an outside force is applied to the output gear in a direction which tends to help the output gear) the entire transmission immediately locks up and comes to an abrupt stop moderated only by any elastic "stretch" in the system.

Almost any type of thread will work with the new drive—standard, 60° screw threads, Acme threads, or any arbitrary shallow-profile thread. Hence, the worms can be produced on standard machine-shop equipment such as lathes and automatic screw machines, although a gear-milling machine may be best for the more precise applications.

JOBS FOR THE NEW DRIVE

Applications for Twinworm can be divided into two groups:

(1) Those employing self-locking characteristics to prevent the load from affecting the system.

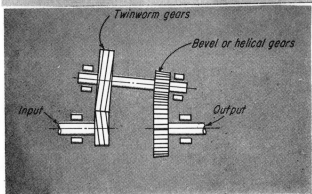

ANGLE BETWEEN SHAFTS IN NEW DRIVE is easily compensated by pairing with bevel or helical gears.

(2) Those employing deceleration-locking characteristics to brake the system to an abrupt stop if the input decelerates.

Self-locking Applications

Mechanical counters. This is the application that led to development of Twinworm gears. Popper was

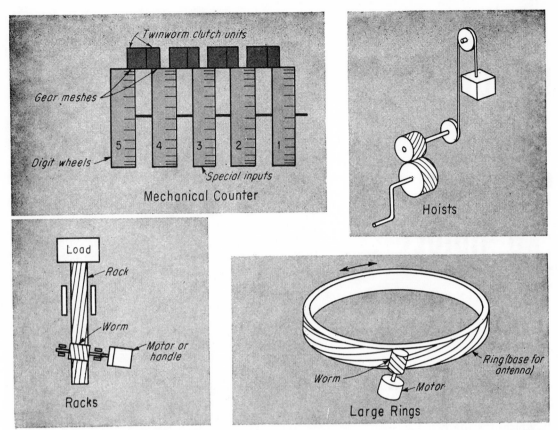

Mechanical Counter

Hoists

Racks

Large Rings

MECHANICAL COUNTER was first application. Sketches show three other Twinworm possibilities.

given the problem of developing a gear system that would permit inputs to be made directly to any one of five digit-wheels. The inputs were to affect higher-digit wheels (see sketch of the mechanical counter above), but not the lower digits. This was accomplished by coupling Twinworm drives to slip clutches. An impulse to digit-wheel 3 would cause it to also rotate wheel 4, but the Twinworm between wheels 2 and 3 would lock up, causing it to slip against its clutch. Standard worm gears were originally employed but their low efficiency was compounded because of their series-arrangement. If each worm has 40% efficiency, the over-all efficiency, when impulse is to wheel *1*, is 0.4 x 0.4 x 0.4 x 0.4 = 0.0256 or only 2½%.

Hoists and lifts—Popper believes the drive could be employed advantageously wherever loads must be raised, such as in hoists, elevators, lift trucks, mechanisms for adjusting car windows, and so forth. The drive not only prevents the load from rotating the gears but can be so designed that the same input torque is required both for raising and lowering the load (even without a counterweight). In the illustrations above, for example, one can crank the load to any position with the same force—and remove his hand without fear that the load will fall. Also, the power unit (if a motor is employed) will be the smallest possible.

The principle of the drive could also be applied to large external or internal rings; for example, for rotation of antennas. Wind loads on the antenna cannot shift the ring. Also, the principle is applicable to the bevel-gear system.

Self-locking occurs as soon as tan ϕ_1 is equal to or smaller than μ, or when

$$\tan \phi_1 = \frac{\mu}{S_1}$$

Here, S_1 represents a "safety factor" (selected by the designer) which must be somewhat greater than one to make sure that self-locking is maintained even if μ should fall below an assumed value. Neither ϕ_2 nor the angle $(\phi_2 - \phi_1)$ affects self-locking.

Deceleration-locking occurs as soon as tan ϕ_2 is also equal to or smaller than μ; or, if a second safety factor S_2 is employed (where $S_2 > 1$), when

$$\tan \phi_2 = \frac{\mu}{S_2}$$

For the equations to hold true, ϕ_2 must always be made greater than ϕ_1. Also, μ refers to the idealized case where the worm threads are square. If the threads are inclined (as with Acme-threads or V-threads) then a modified value of μ must be employed, where

$$\mu_{modified} = \frac{\mu_{true}}{\cos \theta}$$

Here, θ is the pressure angle of the thread profile—approximately 30° for V-threads and 14½° for Acme threads.

Relationship between input and output forces during rotation is:

$$\frac{P_1}{P_2} = \frac{\sin \phi_1 + \mu \cos \phi_1}{\sin \phi_2 + \mu \cos \phi_2}$$

Efficiency for the drive is:

$$\eta = \frac{1 + \mu/\tan \phi_2}{1 + \mu/\tan \phi_1} = \frac{1 + S_2}{1 + S_1}$$

NEW EQUATIONS
locate
DWELL POSITION
of 3-GEAR DRIVES

These drives can produce three different motion patterns—depending on how far the output gear is located from the input gear. Author develops equations which—for the first time—pinpoint this all-important dimension.

DR J HIRSCHHORN,

Senior lecturer in Mechanical Engineering
University of New South Wales, Australia

There has been a marked surge of interest recently in the three-gear drive as an intermittent mechanism. Although it is better known for its cyclic characteristics, proper design can give it a smooth momentary pause that is ideal for a quick automatic-transfer or fabrication operation. In contrast with other intermittent-motion

BASIC COMPONENTS OF 3-GEAR DRIVE

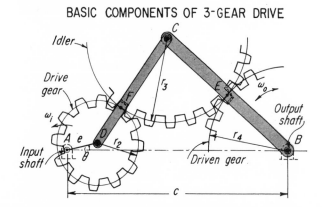

mechanisms, the three-gear drive can pick up a load without slightest shock because both velocity and acceleration of the output gear are zero when it is in the dwell position.

Success in designing such a drive depends on finding correct center-to-center distance between input and output shafts in order to get the desired pause. Initial investigation of the kinematic properties of the three-gear drive has already been performed. But the equations developed here provide, for the first time, a direct solution of this critical center distance.

THREE PATTERNS OF MOTION

The drive is built around a gear mounted eccentrically on the input shaft, plus an idler gear and an output gear. Two links keep the idler in mesh with the output and input gears, but the gears are free to turn with respect to the links.

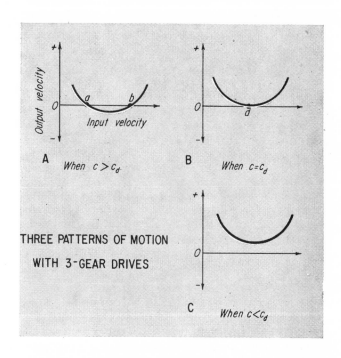

THREE PATTERNS OF MOTION WITH 3-GEAR DRIVES

A When $c > c_d$.
B When $c = c_d$
C When $c < c_d$

The above comparison of motion patterns shows that when the distance c between input and output shafts is considered adjustable, three patterns of motion are available:

● If c is made larger than a critical value, c_d, the output gear stops for an instant, reverses for a finite period, stops again, and then resumes its original sense of rotation, while the driving gear is completing one revolution at constant speed.

● If c is made equal to c_d, the output gear dwells for an instant, and then continues to rotate in its original sense. Although, theoretically, it pauses only an instant, in and actual mechanism this may last about 45° of the cycle—and is the motion pattern which is usually required.

● If c is made smaller than c_d, the output gear slows down and then accelerates, but does not actually stop.

The output gear comes to rest—positions a, b in (A) and d in (B)—when the instantaneous center of its rotation relative to the driver coincides with the input shaft center. This is the case when points A, F and E are in line, as shown in the dwell-position diagram below. This diagram shows one of the two dwell conditions that occur when c is larger than c_d. As c is made smaller the

IN DWELL POSITION

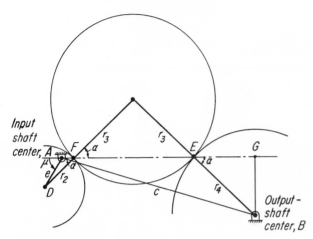

two dwell positions occur closer together; therefore, the critical value c_d can be defined as being the smallest center distance which, for a given mechanism, allows points A, F and E to be in line.

THE CENTER-DISTANCE EQUATION

The center distance c can be expressed in terms of trigonometric quantities:

$$c^2 = (AF + FE + EG)^2 + (GB)^2 \tag{1}$$

with

$$AF = r_2 \cos \gamma - e \cos \mu, \tag{2}$$

$$FE = 2r_3 \cos \gamma, \tag{3}$$

$$EG = r_4 \cos \gamma, \tag{4}$$

$$GB = r_4 \sin \gamma, \tag{5}$$

Eq (1) becomes:

$$c^2 = [(r_2 + 2r_3 + r_4) \cos \gamma - e \cos \mu]^2 + r_4{}^2 \sin^2 \gamma \tag{6}$$

but,

$$e \sin \mu = r_2 \sin \gamma \tag{7}$$

hence,

$$e \cos \mu = \sqrt{r_2{}^2 \cos^2 \gamma - (r_2{}^2 - e^2)} \tag{8}$$

Substituting Eq (8) into Eq (6) gives:

$$c^2 = [(r_2 + 2r_3 + r_4) \cos \gamma -$$
$$\sqrt{r_2{}^2 \cos^2 \gamma - (r_2{}^2 - e^2)}]^2 + r_4{}^2 \sin^2 \gamma \tag{9}$$

To obtain $c_d = c_{min}$, Eq (9) is differentiated with respect to γ, and equated to zero. After some simple but tedious algebraic operations, the following quadratic equation can be obtained involving the angle for the dwell position γ_d:

$$K \cos^4 \gamma_d - L \cos^2 \gamma_d - M = 0 \tag{10}$$

where $K = [(r_4{}^2 - r_2{}^2)^2 - 2(r_4{}^2 + r_2{}^2)(r_2 + 2r_3 + r_4)^2 +$
$$(r_2 + 2r_3 + r_4)^4]r_2{}^2 \tag{11}$$

$L = [(r_4{}^2 - r_2{}^2)^2 - 2(r_4{}^2 + r_2{}^2)(r_2 + 2r_3 + r_4)^2 +$
$$(r_2 + 2r_3 + r_4)^4](r_2{}^2 - e^2) \tag{12}$$

$$M = (r_2 + 2r_3 + r_4)^2(r_2{}^2 - e^2)^2 \tag{13}$$

Knowing γ_d, all other quantities, in particular c_d, can be determined.

SAMPLE PROBLEM

Determine the dwell center distance for the three-gear drive with dimensions $r_2 = 1$ in. $e = 0.8$ in., $r_3 = 2$ in., $r_4 = 2$ in.

Substituting these values into Eq (10) gives:

$$1920(\cos^4 \gamma_d) - 691(\cos^2 \gamma_d) - 6.35 = 0$$

which gives

$$\cos^2 \gamma_d = \frac{1417}{3840} \qquad \text{and} \qquad \gamma_d = 52°36'$$

from Eq (7)	$\mu_d = 83°14'$
from Eq (2)	$AF = 0.51$ in.
from Eq (3)	$FE = 2.43$ in.
from Eq (4)	$EG = 1.21$ in.
from Eq (5)	$GB = 1.59$ in.
from Eq (1)	$c_d = 4.45$ in.

CONSTRUCTION DETAILS FOR SAMPLE PROBLEM

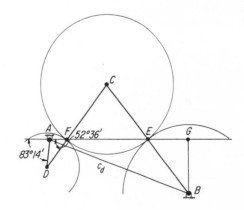

Therefore the distance between input and output shafts should be 4.45 in. The sketch above shows such a mechanism in its dwell position.

Gear Arrangements for Amplifying Motion

FEDERICO STRASSER

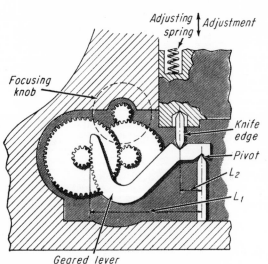

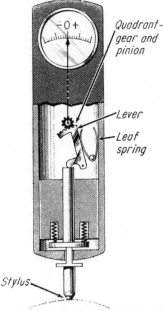

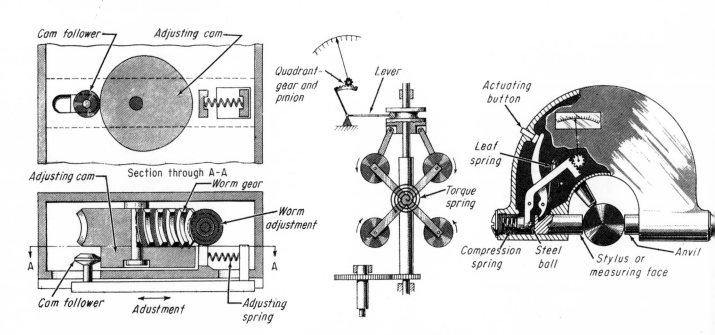

LEVER AND GEAR train amplify the microscope control-knob movement. Knife edges provide frictionless pivots for lever.

DIAL INDICATOR starts with rack and pinion amplified by gear train. The return-spring takes out backlash.

COMBINATION LEVER AND GEARED quadrant are used here to give the comparator maximum sensitivity combined with ruggedness.

MICROSCOPIC ADJUST-MENT is achieved here by employing a large eccentric-cam coupled to a worm-gear drive. Smooth, fine adjustment result.

QUADRANT-GEAR AND PINION coupled to an L-lever provide ample movement of indicator needle for small changes in governor speed.

ZEISS COMPARATOR is provided with a special lever to move the stylus clear of the work. A steel ball greatly reduces friction.

INDEX

AGMA classification system for gears, 176–182
 (*See also* Gear tolerances)
Aluminum gears, 66, 181, 255, 289–293
Angular errors in gears, 228–233
Anodizing, 291–293

Backlash, as affected by mounting tolerances, 234–235
 devices for control, 235–242, 322
 effect on angular errors, 228–233
 lost-motion calculations, 236
 relationship to center distance, 18–19
 (*See also* Gear tolerances)
Bearing eccentricity, effect of, 17
Bearing loads on geared shafts, 265–266
Bevel gears, Coniflex, 45, 49, 58
 design formulas, 56–68
 dimensions, 65
 force analysis, 92
 Humpage's, 103
 minimum-weight design, 25–32
 mounting details, 68
 planetary, 103
 power capacity, 56–57
 Revacycle, 46–49
 sheet metal, 322–323
 smallest-size design, 2–12
 spiral, 7–8, 44–55
 straight, 44–68
 strength factors, 67
 tooth proportions, 62–64
 Zerol, comparison data, 49
  contact stress factors, 7
  derating factors, 8
  description, 44–46
  design formulas, 56–68
  power capacity, 56–57
Beveloid gears, 48–54
Brake employing multi-roller clusters, 330, 332
Brass gears, 181
Bronze gears (*see* Gear materials and finishes, bronzes)

Cadmium plating, 303
Case-carburized gears (*see* Gear materials and finishes)
Cast irons for gears, 10, 59, 66, 268–284
Cast metal, 284–285
Center distance, basic formula, 25
 effect on contact ratio and backlash, 15–24, 229
Cermets, 308
Clamps and collets, 334
Compressive stress between gear teeth, 4
Computers for gear design, 39–42
Cone drive, 47, 49, 52, 69–77
Contact ratios, 15–24, 61, 138, 146–147, 224–225
Contact stress factors, 6–7

Coupling gear mechanism, 352
Critical speeds of geared shafts, 263–264
Crossed-helical gears, 47, 49, 52, 78–82
Cushioned gear drives, 259–262
Cycloid gear mechanisms, 354–359
Cylinders, moment of inertia of, 251

Daimler preselective drive, 101
Deformation factor, 33–34
Delrin gears, 297
Derating factors, 6–9
Differential gears, 363
 (*See also* Planetary gears)
Diophantine equations to find gear ratios, 219
Durability of gears (*see* Surface durability)
Dwell geared mechanisms, 348, 349, 370–371
Dynamic factors, 6–7
Dynamic loading on gears, 33–34

Eccentric gears, 158–165, 169–173
Efficiency of gears (*see* Gear efficiency)
Electrolizing, 291–292, 302–303
Elliptical gears, 158–168
Enlarged-tooth pinions, 144–148
Epicyclic gears (*see* Planetary gears)
Errors, angular, 228–233
Extruded stock for gears, 285

Face gears, 46, 49, 87–91
Face width, for bevel gears, 61
 versus diametral pitch, 42
 for spur and helical gears, 9–11
Fastening of gears, 333–341
Fordomatic transmission planetary, 99
Friction drives, 328–332
Friction losses in gears, 53, 116–117, 139, 246–248

Gear-blank tolerances, 24
Gear design, avoiding tooth interference, 144–147, 154–156
 basic procedures for bevel gears, 7–10, 56–68
 computers for, 39–42
 efficiency of gear train, 94–103, 244–245
 gear-center coordinates, 183–185
 gear-tooth ratios, 210–221
 helical gears, 2–12
 involute curve constants, 222–223
 life predictions, 249–258, 301
 long-and-short addendum, 136–153
 minimum-weight design, 25–32
 number of revolutions before all gears return to original alignment, 210–213
 planetary gears, 94–122
 size estimates, 2–12, 33–34

Gear design, spur gears, 2–12
Gear efficiency, of bevel gears, 53
 of gear trains, 244–245
 of hypoid gears, 53
 of planetary gears, 94–103
 of spur gears, 139
 of worm gears, 53
Gear failures, 4–5
Gear fastening, 333–341
Gear life, 249–258, 280, 289–297, 301
Gear lubrication, 68, 304–308
Gear manufacturing, 55
 tolerances, 14–15
Gear materials and finishes, aluminum, 66, 181, 255, 289–293
 anodizing, 291–293
 brass, 181
 bronzes, aluminum bronze, 254, 269, 277
  cast bronze, 277
  leaded bronze, 269, 277
  manganese bronze, 254
  nickel bronze, 269, 278
  phosphor bronze, 254, 269, 277
  SAE bronze, 277
  silicon bronze, 269, 276, 299
  tin bronze, 66, 181, 269, 276
 cadmium plating, 303
 cast irons, 10, 59, 66, 268–284
 cast metal, 284–285
 cermets, 308
 chrome plating, 291–293, 302–303
 cold-drawn, 286
 Delrin, 297
 electrolizing, 291–293, 302–303
 extruded stock, 285
 gold plating, 302–303
 hardened gears (*see* steel, *below*)
 Hastelloy, 308
 Haynes, 308
 high-temperature alloys, 308
 indium plating, 303
 laminates, 269–279
 Lexan, 297
 machinability, 11
 magnesium, 255
 molybdenum disulfide, 302–303
 nickel aluminum, 181
 nitriding, 300–303
 Nylasint, 289–293
 nylon, 289–291
 phenolic glass, 279
 plastics, 269, 278–279, 287–288
 platinum plating, 302–303
 powdered metals, 269, 279, 281–284, 298
 PTFE coatings, 302–303
 rawhide, 269–279
 Rene, 308
 Rulon, 289–293
 sheet metal, 322–323
 shot peening, 300
 silver plating, 302–303
 Sinite, 289–293

Gear materials and finishes, sintered metals
(see powdered metals, above)
stainless steels, 18
steel, including heat-treated steel, 5, 10,
59, 66, 180, 253–254, 268–284,
289–293, 300–303
Stellite, 308
sulfurized treatment, 302–303
tool alloy, 308
tungsten-carbide treatment, 303
Gear mechanisms, 348–372
coupling, 352
cycloid, 354–359
dwell, 348, 349, 359, 370–371
geneva, 350, 351, 357
indexing, 349, 352, 353, 361, 364–365
motion amplifiers, 353, 365, 372
oscillating, 350–351
reciprocating, 350, 351, 354–356, 363
reversing, 342–346, 348
shaft synchronizer, 350
straight-line motion, 364–365
stroke multiplier, 353, 365, 372
three-gear drive, 349, 370–371
Gear noise, 109–262
cushioned gears, 259–262
critical speeds, 263
Gear quality numbers (see Gear tolerances)
Gear ratios, diophantine equations to find,
219
matrix arithmetic to find, 220–221
Gear-shift arrangements, 342–346
Gear specifications (see Gear tolerances)
Gear teeth, compressive stress between, 4
recommended number of, 10
sizes of, 226
Gear testing, 201–206, 252
Gear tolerances, 15–19, 176–182, 186–
200
Gear-tooth geometry, 21–24, 58, 140, 146–
147, 198–200, 222–226
Gear-tooth interference, avoidance of,
144–147, 152–156
Gear-tooth profiles, comparison of tooth
size and profiles, 226
effect, on angular errors, 228
on efficiency, 245
involute-curve ratio constants, 222–223
profile diagrams, 198–200
Gear-tooth ratios, 210–221
Geometry factors, 63, 67
Globoid gear, 362
Gold plating, 302, 303

Hardened gears (see Gear materials and
finishes, steel)
Hardness, versus K-factors, 9
versus machinability, 11
in reducing gear size, 2–12
versus stress, 10
(See also Gear materials and finishes)
Harmonic drive, 103, 120–122
Hastelloy, 308
Haynes, 308
Helical gears, for boosting reduction ratios
in planetary gears, 82
changing gear ratio, 81
comparisons to other gear systems, 121
compensating for center-distance errors,
80
crossed-, 47, 49, 52, 78–82
derating factors, 8
design for smallest pair, 2–12
helix-angle variations, 35–38, 79–82

Helical gears, K-factors, 4
replacing module spur gears, 208
special angle to simplify calculations,
35–38
Helicon gears, 48–54
Helix angle of helical gears, effect on
other parameters, 35
Herringbone gears, 121
(See also Helical gears)
Hertzian stress, 4
High-temperature gear alloys, 308
Hindley worm gears, 69–77
Humpage's bevel gears, 103
Hydramatic planetary drive, 100
Hydraulic torque applier, 206
Hypoid gears, 47, 49, 53

Indexing geared mechanisms, 349, 352,
353, 361, 364, 365
Indium plating, 303
Involute-curve ratio constants for plotting
gear-tooth profiles, 222–223
(See also Gear-tooth profiles)

K-factors, for gear-teeth durability, 4
replacing module spur gears, 208
special angle to simplify calculations,
35–38
table, 9
Keying of gears, 336–337
Knurls, 326–327, 336–337, 339

Laminates for gears, 269, 279
Lewis Y factor, 21, 152
Lexan, 297
Life predictions of gears (see Gear life)
Line-of-action, 23, 138
Load-distribution factor, 6
Load testing of gears, 201–206, 252
Logarithmic-spiral gears, 158–165
Long-and-short addendum gears, 136–153
Lost motion in gear trains, 236
Lubrication, 68, 304–308

Machinability of steel or iron gears, 11
Magnesium for gears, 255
Material factors for gear mesh, 59
Materials for gears (see Gear materials and
finishes)
Matrix arithmetic to find gear ratios, 220–
221
Measurement over wires, 36–37, 189–200
Mechanisms, geared (see Gear mecha-
nisms)
Minimum-weight design, 25–32
Minute-cover drive, 98
Modified-tooth gear systems, 123–154
Molybdenum disulfide, 302–303
Moment of inertia of cylinders, 251
Motion amplifiers, 353, 365, 372
Multi-roller bearings, 330–332

Nickel aluminum, 181
Nitriding of gears, 300–303
Noise (see Gear noise)
Noncircular gears, 158–173
Novikov gears, 124–135
Nylasint, 289–293

One-way output from gear reducers, 366–
367

Overload factors, 6–8, 59
Over-pin measurements, 36–37, 189–200

Phenolic glass for gears, 279
Pinning of gears, 334, 336
Pitting, 3–4
Planetary gears, 94–122
bearing problems, 109
cam-controlled, 360
comparisons, to helical gears, 110, 121
to other systems, 121
efficiency formulas, 94–103
friction drives, 330–332
helical-spur, 82
lightest-weight design, 25–32, 110–113
method for finding speed ratios, 104–109
noise in, 109
non-standard spur, 149–151
planetary gear systems, 82, 94–103
planetary torque applier, 204, 205
speed-ratio formulas, 94–103, 114
spur-helical, 82, 115–117
synchronic index, 210–213
torque analysis, 94–103
volume requirements, 110–113
Planocentric drive, 102
Planoid gears, 48–54
Plastic gears (see Gear materials and finishes)
Platinum plating, 302–303
Powdered metals, 269–279, 281–284, 298
Power capacity, of bevel gears, 56–68
of spur and helical gears, 2–12
(See also other types of gears)
Press fit of gears to shafts, 335–336
Pressure angle, 149–151, 154–156
Profile diagrams, 198–200
(See also Involute-curve ratio constants)
PTFE coatings, 302–303

Racks, 320–321
Ratchets, 322–323
Rawhide for gears, 269–279
Recess-action gears, 136–143
Reciprocating mechanisms, 350, 351, 354–
356, 363
Rene gear material, 308
Retaining rings, 338, 340–341
Revacycle bevel gears, 46–49
Reversing mechanisms, 342–346, 348,
366–367
Right angle systems, 43–92
(See also Bevel gears; Face gears; Heli-
cal gears, crossed-; Worm gears)
Rolling fixture for gear checking, 18–19
Rotary-to-linear gear system, 120
Rulon, 289–293

Sample gears, calculating design data
from, 207
Scale effect on accuracy, 16
Scoring, 3–4
Self-locking gears, 54
worms, 54, 368–369
Service factors for motors and engines, 12
Shaft-misalignment effects on angular
errors, 231
Sheet-metal gears, 322–323
Shot peening, 300
Silver plating, 302–303
Sine-function gears, 158–165
Sinite, 289–293

Sintered-metal gears, 269–279, 281–284, 298
Size factor, 6
Speed reducers, design, of bevel-gear types, 56–68
    of helical-gear types, 2–12
    of planetary gear types, 94–122
    of spur-gear types, 2–12
    of worm-gear types, 69–78
Spiral bevel gears, 7–8, 44–55
Spiroid gears, 48–54
Spline design and types, 310–319, 336, 337
Spring washer for fastening gears, 338
Sprockets, 322–325
Spur-gear derating factors, 8
Spur gears, intermittent, 361
    meshing, with bevel gears, 83–86
    with face gears, 87–91
    with worm gears, 70–71
    non-standard, 149–153
    paired, with bevel gears, 103
    with helical gears for high-reduction ratios, 80
    (See also Gear design)
Staking of gears, 338
Steel gears (see Gear materials and finishes, steel)

Stellite, 308
Strength factors, for bevel gears, 58–67
    for spur gears, 33–34
Sulfurized treatment, 302–303
Surface condition factor, 6
Surface durability, for bevel-gear design, 59
    influencing gear size, 2–12
Surface treatments for gears, 300–303
Synchronic index of gear trains, 210–213

Tapered-gear design, 83–86
Testing of gears, 201–206, 252
Three-gear drive, 349, 370–371
Tool alloy, 308
Tooth breakage, 4
Tooth proportions, 13–23, 140, 146–147, 198–200, 226
Tooth-thickness tolerances, 21–22
    effects on angular errors, 229
Tooth-to-tooth and tooth composite errors, 15, 19, 176–182
Torque-speed relationship, 7
Total-weight equations for gear trains, 28
Twinworm gear system, 368–369
Two-tooth gear systems, 148–362

Undercutting of teeth, avoidance of, 144–147, 152–156

Wear of gear teeth (see Gear life)
Weight estimates for gears, 25–32
Wildhaber-Novikov gears, 124–135
Wildhaber worm gears, 69–77
Wobble effects on angular errors, 231
Wobble-gear drive, 102, 363
Worm gears, comparison data, 47, 49, 70–71, 121
    cone drive, 47, 49, 52, 69–77
    cylindrical, 47, 49, 51, 69–77
    design procedure, 76–77
    double, 69–77
    double-enveloping, 48–49, 53, 69–77
    efficiency equation, 53
    faceplate, 87–91
    Hindley, 69–77
    rubbing-speed formula, 53
    self-locking characteristics, 54, 368–369
    sheet-metal, 323
    types, 69–77
    Wildhaber, 69–77
    worm-and-crown, 362

Zerol bevel gears (see Bevel gears, Zerol)
Zinc plating, 302–303

WITHDRAWN

JUN 1 0 2024

DAVID O. McKAY LIBRARY
BYU-IDAHO